Handbook of
ENVIRONMENTAL AND SUSTAINABLE FINANCE

Handbook of
ENVIRONMENTAL AND SUSTAINABLE FINANCE

Edited by

VIKASH RAMIAH AND GREG N. GREGORIOU

Amsterdam • Boston • Heidelberg • London • New York • Oxford
Paris • San Diego • San Francisco • Singapore • Sydney • Tokyo
Academic Press is an imprint of Elsevier

Academic Press is an imprint of Elsevier
125 London Wall, London EC2Y 5AS, UK
525 B Street, Suite 1800, San Diego, CA 92101-4495, USA
225 Wyman Street, Waltham, MA 02451, USA
The Boulevard, Langford Lane, Kidlington, Oxford OX5 1GB, UK

Notices
Knowledge and best practice in this field are constantly changing. As new research and experience broaden our understanding, changes in research methods, professional practices, or medical treatment may become necessary.

Practitioners and researchers may always rely on their own experience and knowledge in evaluating and using any information, methods, compounds, or experiments described herein. In using such information or methods they should be mindful of their own safety and the safety of others, including parties for whom they have a professional responsibility.

To the fullest extent of the law, neither the Publisher contributors, or editors, assume any liability for any injury and/or damage to persons or property as a matter of products liability, negligence or otherwise, or from any use or operation of any methods, products, instructions, or ideas contained in the material herein. Each author(s) is responsible for his/her chapter. Neither the editors nor the publisher can guarantee the accuracy of each chapter in this book and each chapter is the sole responsibility of its author(s).

ISBN: 978-0-12-803615-0

British Library Cataloguing-in-Publication Data
A catalogue record for this book is available from the British Library

Library of Congress Cataloging-in-Publication Data
A catalog record for this book is available from the Library of Congress

For information on all Academic Press publications
visit our website at http://store.elsevier.com

Working together
to grow libraries in
developing countries

www.elsevier.com • www.bookaid.org

Publisher: Nikki Levy
Acquisition Editor: Scott Bentley
Editorial Project Manager: Susan Ikeda
Production Project Manager: Jason Mitchell
Designer: Mark Rogers

Typeset by TNQ Books and Journals
www.tnq.co.in

Printed and bound in the United States of America

Vikash dedicates this book to Emma Campbell (partner in crime), late Professor Tony Naughton (life mentor), Professor Imad Moosa (guru) and Professor Petko Kalev (the true believer in 1502) and Greg dedicates this book to his mother Evangeline Gregoriou and to his deceased father, Nicholas Gregoriou.

CONTENTS

LIST OF CONTRIBUTORS

John Anderson
University of New England, Armidale, NSW, Australia

Sukanta Bakshi
School of Accounting, RMIT University, Melbourne, VIC, Australia

Henning Bjornlund
School of Commerce, University of South Australia, Adelaide, SA, Australia

Davide Bonzanini
School of Management, Politecnico di Milano, Milan, Italy

Steven A. Brieger
Center for Leadership and Values in Society, University of St. Gallen, St. Gallen, Switzerland; Institute of Corporate Development, Leuphana University of Lüneburg, Lüneburg, Germany

Albie Brooks
Department of Accounting, University of Melbourne, Melbourne, VIC, Australia

Massimiliano Caporin
Department of Economics and Management "Marco Fanno", University of Padova, Padova, Italy

Robert Christopherson
Department of Economics & Finance, School of Business & Economics, The State University of New York at Plattsburgh, Plattsburgh, NY, USA

Fulvio Fontini
Department of Economics and Management "Marco Fanno", University of Padova, Padova, Italy

Michael Gangemi
School of Economics, Finance & Marketing, RMIT University, Melbourne, VIC, Australia

Philip Gharghori
Department of Banking and Finance, Monash Business School, Monash University, Berwick, VIC, Australia

Giancarlo Giudici
School of Management, Politecnico di Milano, Milan, Italy

Michael Graham
Stockholm Business School, Stockholm University, Stockholm, Sweden

Greg N. Gregoriou
State University of New York, Plattsburgh, NY, USA

Anil Gupta
SJVN Limited, Shimla, India

Kartick Gupta
Centre for Applied Financial Studies (CAFS), School of Commerce, UniSA Business School, University of South Australia, Adelaide, SA, Australia

Anton Hasselgren
Stockholm Business School, Stockholm University, Stockholm, Sweden

Tehmina Khan
RMIT University, Melbourne, VIC, Australia

Chandrasekhar Krishnamurti
Faculty of Business, Education, Law and Arts, University of Southern Queensland, Toowoomba, QLD, Australia

Abraham Lioui
Department of Finance, EDHEC Business School, Nice, France

Min Liu
School of Economics, Finance & Marketing, RMIT University, Melbourne, VIC, Australia

Ronald McIver
Centre for Applied Financial Studies (CAFS), School of Commerce, UniSA Business School, University of South Australia, Adelaide, SA, Australia

Imad A. Moosa
School of Economics, Finance and Marketing, RMIT University, Melbourne, VIC, Australia

Camillo von Müller
Center for Leadership and Values in Society, University of St. Gallen, St. Gallen, Switzerland

Judy Oliver
Department of Accounting, Economics and Finance, Swinburne University of Technology Melbourne, VIC, Australia

Elizabeth Ooi
University of Western Australia Business School, University of Western Australia, Perth, WA, Australia

Andrea Patrucco
School of Management, Politecnico di Milano, Milan, Italy

Jarkko Peltomäki
Stockholm Business School, Stockholm University, Stockholm, Sweden

Dhimitri Qirjo
Department of Economics & Finance, School of Business & Economics, The State University of New York at Plattsburgh, Plattsburgh, NY, USA

Vikash Ramiah
School of Commerce, University of South Australia, Adelaide, SA, Australia

Colin Read
State University of New York College at Plattsburgh, Plattsburgh, NY, USA

Peter Rossini
School of Commerce, University of South Australia, Adelaide, SA, Australia

Harald Sander
Technische Hochschule Köln, Cologne, Germany and Maastricht School of Management, Limburg, Netherlands

Raja Vinesh Sannassee
Department of Finance & Accounting, Faculty of Law & Management, University of Mauritius, Reduit, Mauritius

Boopen Seetanah
Department of Finance & Accounting, Faculty of Law & Management, University of Mauritius, Reduit, Mauritius

Claire Settre
Global Food Studies, University of Adelaide, Adelaide, SA, Australia

Belinda Spagnoletti
Nossal Institute for Global Health, Melbourne School of Population and Global Health. University of Melbourne, Melbourne, VIC, Australia

Gillian Vesty
School of Accounting, RMIT University, Melbourne, VIC, Australia

Sarah Ann Wheeler
School of Commerce, University of South Australia, Adelaide, SA, Australia; Global Food Studies, University of Adelaide, Adelaide, SA, Australia

EDITORS' BIOGRAPHY

Vikash Ramiah is an Associate Professor of Finance at RMIT University. He has a Diploma of Management, BSc (Hons) Economics, Master of Finance program, and Doctor of Philosophy from RMIT University. He has received numerous awards for outstanding performance in teaching, research, and supervision. He has been teaching economics and finance courses at RMIT, University of Melbourne, La Trobe University, and Australian Catholic University, since 1999. He has published in academic journals (e.g., Journal of Banking and Finance, Journal of Behavioral Finance, European Journal of Finance, Applied Economics, Pacific Basin Finance Journal, and Journal of International Financial Market, Institution and Money), industry reports, one book, book chapters, and conference papers. He supervises numerous PhD students and regularly attracts research funding. He is an expert reviewer for several finance journals and for the Mauritius Research Council. He serves on the editorial board of finance journals. He was an elected board member of the RMIT University Business Board, Program Director of Open Universities Australia, and acting board member at the Australian Centre for Financial Studies. He was as a junior auditor at H&A Consultant, manager at Intergate PTY Limited, quantitative analyst at ANZ, Investment Banking Division, provided consultancy services to the Australian Stock Exchange and worked in collaboration with the Finance and Treasury Association of Australia and the Australian Centre for Financial Studies. Other services to the community include being a research fellow for Institute of Global Business and Society, Cologne University of Applied Sciences and Tianjin Academy of Environmental Sciences, Academic Advisers and Collaborators at Behavioural Finance Australia, External Examiner at the University of Mauritius and reviewer of the University of Technology Mauritius. He is the founder of Researchers Sans Frontiere Network, Environmental Finance Cluster at RMIT, and his research areas are financial markets, behavioral finance, and environmental finance.

Greg N. Gregoriou is a native of Montreal and he obtained his joint PhD in finance at the University of Quebec at Montreal, which merges the resources of Montreal's four major universities McGill, Concordia, UQAM, and HEC. He is a Professor of Finance at State University of New York (Plattsburgh) and has taught a variety of finance courses such as Alternative Investments, International Finance, Money and Capital Markets, Portfolio Management, and Corporate Finance. He has also lectured at the University of Vermont, Universidad de Navarra, and at the University of Quebec at Montreal. He has published 50 books, 65 refereed publications in peer-reviewed journals, and 24 book chapters since his arrival at SUNY Plattsburgh in August 2003. His books have

been published by McGraw-Hill, John Wiley & Sons, Elsevier-Butterworth/
Heinemann, Taylor and Francis/CRC Press, Palgrave-MacMillan, and Risk Books.
Four of his books have been translated into Chinese and Russian. His academic articles
have appeared in well-known peer-reviewed journals such as the Review of Asset
Pricing Studies, Journal of Portfolio Management, Journal of Futures Markets, European
Journal of Operational Research, Annals of Operations Research, Computers and
Operations Research, etc. He is the derivatives editor and editorial board member for
the Journal of Asset Management as well as editorial board member for the Journal of
Wealth Management, the Journal of Risk Management in Financial Institutions, Market
Integrity, IEB International Journal of Finance, and the Brazilian Business Review. His
interests focus on hedge funds, funds of funds, commodity trading advisors, managed
futures, venture capital, and private equity. He has also been quoted several times in
the New York Times, Barron's, the Financial Times of London, Le Temps (Geneva),
Les Echos (Paris), and L'Observateur de Monaco. He has done consulting work for
numerous clients and investment firms in Montreal. He is a part-time lecturer in finance
at McGill University, an advisory member of the Markets and Services Research Centre
at Edith Cowan University in Joondalup (Australia), a senior advisor to the Ferrell Asset
Management Group in Singapore, and a research associate with the University of
Quebec at Montreal's CDP Capital Chair in Portfolio Management. He is also a Fellow
at the Research Center for Productivity and Management at the Hefei University of
Technology in Hefei, Anhui, China.

CONTRIBUTORS' BIOGRAPHY

John Anderson holds a BFinAdmin from the University of New England, a Master of Business by Research from SCU and a PhD in finance from RMIT University and is currently senior lecturer in finance at the University of New England. He has previously been an honorary fellow of both the Cass Business School, City University, London, and University of Birmingham. Prior to joining academia, he was a floor trader and local member of the Sydney Futures Exchange, worked as a treasury dealer in interest rate products and derivatives and also served as Futures and Derivatives Specialist with the Australian Securities and Investments Commission. He has taught in the Banking and Finance fields at the British University in Dubai and various universities in Australia including Queensland University of Technology, RMIT University, and University of New England including adjunct appointments to Griffith University and the Macquarie Graduate School of Management. He has consulted extensively to numerous major Australian and International Financial Institutions in addition to holding international board positions in the funds management sector. His research covers a variety of financial markets and computational finance fields appearing in international journals, conferences, and books including citations by other authors in professional financial books. He has also coauthored the second edition of the banking textbook "Financial Institutions Management" through leading international publishers McGraw-Hill. He also appears in media providing commentary on a range of financial issues in both print and radio.

Sukanta Bakshi holds a BSc and is currently undertaking a Masters by Research in Accounting at RMIT University. Over the last 14 years, Sukanta has worked as a cost accountant in both manufacturing and the public sector. Sukanta's research interests include full cost accounting and sustainable waste management practices.

Henning Bjornlund is an adjunct professor at the Centre for Regulation and Market Analysis and a Canada Research Chair in Water Policy and Management at University of Lethbridge, Alberta, Canada. He has researched water management and policy issues in Australia since 1993 and in Canada since 2005. Henning is a director of the International Water Resources Association and serves on the board of the International Water Resource Economics Consortium. He also serves on the editorial board of four international journals: Agricultural Water Management, Water International, Sustainable Development and Planning, and Water, the Journal on the Ecology and Management of Water Resources.

Davide Bonzanini graduated with a Master of Science in Management Engineering, with major in Finance, at Politecnico di Milano (Italy). He is research fellow at the School of Management at Politecnico di Milano.

Steven A. Brieger (M.A.) is a PhD student in economics (Dr rer. pol.) at Leuphana University of Lüneburg, Germany, and research fellow of the Center for Leadership and Values in Society at the University of St. Gallen, Switzerland. His research focuses on value formation, public values and public opinion, the history of economic and management thoughts, and business ethics. Steven, who is also teaching at the Leuphana University, is author of a book about rating agencies and developed the World Economic Opinion Survey to research on cultural and regional differences in economic thinking.

Albie Brooks is a senior teaching fellow in management accounting at the University of Melbourne. He has wide-ranging teaching experiences and has both authored and presented materials for the profession, as well as coauthored two university-level textbooks. Albie's research and scholarship interests include contemporary management accounting issues, the development of innovative educational materials, accounting education, and corporate governance issues.

Massimiliano Caporin is associate professor of econometrics at the University of Padova, Italy, where he teaches econometrics and finance. His research interests include the development of quantitative models for risk management and asset allocation, the study of the dynamic behaviors of financial time series, and, in general, the use of quantitative tools for financial analyses. His researches appeared in the Journal of Financial Econometrics, Econometric Reviews, the Journal of Banking and Finance, the Journal of Empirical Finance, Energy Economics, the Journal of Economic Surveys, Quantitative Finance, and Computational Statistics and Data Analysis, among others.

Robert Christopherson is interim associate dean and professor of economics and finance at SUNY Plattsburgh. He is the coauthor of a book on Globalization and has published widely on topics related to pensions, corporate governance, tax-exempt organizations, state and local government, and PILOT arrangements for nonprofit agencies. Dr Christopherson received his PhD from Wayne State University in 1990 and has numerous articles in referred journals, book chapters, and other scholarly articles.

Fulvio Fontini is professor of economics at the University of Padua, Italy, and cochair of the Electricity Security of Supply Task Force of the Council of European Energy Regulators, Brussels, Belgium. He has been visiting professor at the University of Paris 1 Pantheon-Sorbonne (France) and visiting researcher at Curtin University (Australia) and Saarbruecken University (Germany). His research interests are in energy economics, environmental economics, and application of ambiguity theory.

Michael Gangemi is a teaching and research academic at RMIT University in Melbourne, Australia, who has a diverse research history and has published articles in the areas of modeling of industry and country betas, input-out economic impact assessment, drivers of student academic performance, determinants of research and development spending, player selection biases in professional football leagues, and the financial and economic impacts on employer organizations of teleworking programs. Michael also coordinates and teaches the Business Finance course at RMIT University.

Philip Gharghori is a senior lecturer in finance at Monash University in Melbourne, Australia. Prior to his full time appointment in 2006, he was a PhD student and tutor at Monash. He is the author of the finance chapters of Principles of Accounting and Finance, a text used in undergraduate business and commerce courses. Phil's main research area is asset pricing. In particular, Phil's research has focused on the performance of multifactor models, stock market anomalies, the consumption CAPM, and the intertemporal CAPM. His research also encompasses socially responsible investments, funds management, default risk analysis, and the market reaction to stock splits. He has published in a wide range of journals and research books, which include the Journal of Banking and Finance and the Journal of Business Ethics.

Giancarlo Giudici is associate professor of corporate finance at Politecnico di Milano—School of Management. His research interests are in the field of equity capital raising (initial public offerings, equity issues, crowdfunding) and renewable energy investments. He has published several books and articles in refereed journals and has coordinated several research projects on innovation, energy management, entrepreneurship, and finance funded by public and private entities. He is member of the European Finance Association.

Michael Graham is an associate professor of finance at the Stockholm Business School, Stockholm University, Sweden. His primary areas of research include financial markets and corporate finance. Dr Graham is published in a diverse range of international journals including the *European Journal of Finance, Quantitative Finance, Journal of Multinational Financial Management, Journal of Economics and Business, Corporate Governance: An International Review and Research in International Business and Finance*.

Anil Gupta is a mechanical engineering graduate from National Institute of Technology, Kurukshetra, India, of the year 1983. He also has an MBA from IGNOU, New Delhi, India, with specialization in Finance. In 2012, he was awarded his PhD in Management Studies (Carbon Finance) by Himachal Pradesh University, Shimla, India. He has researched and authored various articles and papers in the area of Carbon Finance, Project Appraisal Systems, Modified IRR, Working Capital Finance etc. Presently, he is head of Electrical Contracts Department of SJVN Limited, Shimla, a Public Sector Power Corporation of India engaged in implementation and operation of a large number of hydropower projects in India and neighboring countries. In the area of Carbon Finance and CDM of Kyoto Protocol, he has been instrumental in registration of 412 MW Rampur Hydroelectrical Project of his organization as CDM Project for annual Certified Emission Reductions of 1.40 million tons of carbon dioxide equivalent with UNFCCC on September 29, 2011 (largest public sector project of India registered with CDM Executive Board).

Kartick Gupta (MA (Finance and Investment) (Nottingham), PhD (Finance) (Waikato), CFA®, Certified FRM) joined the University of South Australia as a senior lecturer in finance starting July 2013. Previous to this, he was a member of the Department of Finance, Auckland University of Technology, New Zealand from 2009 to 2013. Kartick has won a number research and teaching awards, including the best paper award at the Ninth International Conference on Corporate Governance and the second best paper award at the 2011 Asian FMA, sponsored by the CFA Institute. His main research interests are in the areas of corporate finance, corporate social responsibility, corporate governance, behavioral finance, financial crises, and stock market anomalies.

Anton Hasselgren is a PhD candidate in finance at Stockholm Business School, Stockholm University, Sweden. Prior to enrolling into the PhD program, Anton graduated from Stockholm Business School with a master's degree in Business and Economics. His research interests include international finance, foreign exchange markets, and emerging markets. Anton Hasselgren also has practical experience in stock trading having worked in customer service for Avanza Bank, the largest online stock broker in Sweden.

Tehmina Khan obtained her PhD in 2007 from Victoria University's and her primary research areas are sustainability accounting, accountability, online reporting, and disclosures. She is a faculty member at RMIT University. Dr Khan has published in internationally recognized journals including the Accounting, Auditing and Accountability journal and the International Journal of Sustainability in Higher Education. Dr Khan is currently undertaking research in the area of accountability for female representations in social media. Dr Khan supports the encouragement and recognition of females' advancements in careers and fields; this chapter being a tribute to this. She is a full time academic and a busy mom of three.

Chandrasekhar Krishnamurti is the Professor and Head of Finance Discipline at the University of Southern Queensland. His research interests include asset pricing, corporate governance, market microstructure, and corporate finance, and he has won eight awards for his research work. He has published 49 research articles in outlets such as the Journal of Corporate Finance, Journal of Banking and Finance, Financial Management, Journal of Financial Research, Pacific-Basin Finance Journal, Corporate Governance: An International Review, Australian Journal of Management, Journal of the Asia Pacific Economy, and the Journal of International Financial Markets, Institutions and Money. He has consulted for business corporations and served as the principal consultant in the Financial Institutions Reform and Expansion (FIRE) project funded by USAID. He has served on the board of the Asian Finance Association and has been the Vice President (Program).

Abraham Lioui is professor of finance at EDHEC Business School. He has served as a consultant to various financial institutions on questions related to performance measurement, option pricing, and market making. His research interests in finance revolve around the valuation of financial assets, portfolio management, and risk management. His economics research looks at the relationship between monetary policy and the stock market. He published widely in, and refereed for, leading journals and received several research grants. He has recently coauthored a book on the use of derivatives for dynamic asset allocation. He is an experienced graduate and executive education instructor and is regularly invited to present at international scientific conferences.

Min Liu is a PhD candidate and teaching associate in the School of Economics, Finance, and Marketing at RMIT University in Melbourne, Australia, and her thesis focuses on contrarian profits associated with environmental regulations, while her other research interests include trading strategies. Min has had a conference paper accepted for presentation at Oxford University, and prior to commencing her PhD project Min worked at the Bank of China.

Ronald McIver is lecturer in financial economics within the UniSA's Business School's School of Commerce. His research encompasses banking, securities markets, corporate finance, and financial education. His published work includes papers on: banking system regulation, reform and rehabilitation and efficiency, particularly in transition economies; governance and institutional structures' roles in delivering efficient outcomes in transition and emerging-market economies; the operation of water markets, including for environmental applications; the impact of ownership, particularly state control, on stock return performance in emerging market and transition economies; and teaching and learning methodologies based on student-centered and activity-based learning.

Imad A. Moosa is currently a professor of finance at RMIT, Melbourne. Before taking on the present position, he was a professor of finance at Monash University and La Trobe University, and a lecturer in economics and finance at the University of Sheffield.

Prior to becoming an academic in 1991, he was a professional economist and a financial journalist for over 10 years, and he also worked as an economist at the Financial Institutions Division of the Bureau of Statistics, the International Monetary Fund (Washington DC). Professor Moosa has published 16 books and over 200 papers in scholarly journals. His most recent book is "Quantitative Easing as a Highway to Hyperinflation," which was published by World Scientific. He has served in a number of advisory positions, including his role as an economic advisor to the U.S. Treasury.

Camillo von Müller (PhD (HSG), MA (JHU), MA (HU)) is an economist at the German Federal Ministry of Finance in Berlin. He obtained a PhD in Management/Finance at the University of St. Gallen, Switzerland, and has been a visiting and teaching fellow at Harvard University's economics department having also taught at the economics and social science departments at the universities of Zurich, St. Gallen, and Leuphana University. Camillo has published widely in the field of management and finance. Prior to joining the Federal Ministry of Finance, he has worked and consulted for nonprofit, public, and private sector institutions including Finance Watch in Brussels, the Ministry of Finance and Economics of Baden-Württemberg, and Deutsche Börse.

Judy Oliver is a senior lecturer in the Swinburne Business School at Swinburne University. After 18 years in industry Judy joined academia and has taught at both undergraduate and postgraduate levels both locally and internationally. She has coauthored two university-level textbooks and her research interest is in understanding more about how management accounting tools and practices are used within organizations.

Elizabeth Ooi is a PhD candidate and teaching associate in the Department of Banking and Finance at Monash University in Melbourne, Australia. Her thesis focuses on the governance of pension funds. Her other research interests include socially responsible investing. She has published in the Journal of Business Ethics. Prior to commencing her PhD, Elizabeth worked in the financial planning industry.

Andrea Patrucco graduated with a Master of Science in Management Engineering, with major in Finance, at Politecnico di Milano (Italy). He is research fellow at the School of Management at Politecnico di Milano.

Jarkko Peltomäki is an associate professor at the Stockholm Business School, Stockholm University. His research interests focus on hedge funds, investment strategies, emerging markets, and performance measurement. His articles have appeared in the *Journal of Behavioral Finance, Managerial Finance, Journal of Wealth Management,* and *Emerging Markets Review.*

Dhimitri Qirjo is an assistant professor in the Department of Economics and Finance at SUNY, Plattsburgh, where he has been a faculty member since 2012. From June 2010 to June 2012, Dhimitri served as a postdoctoral research and teaching fellow in the Vancouver School of Economics at University of British Columbia in Vancouver, Canada. He completed his PhD in economics at Florida International University, in Miami, FL and his undergraduate studies at Aristotle University in Thessaloniki, Greece. His research interests lie mainly in the fields of international trade and economics of immigration.

Colin Read has a PhD in economics from Queen's University in Kingston, Ontario, a law degree, and a BSc in physics. He has taught environmental economics for more than 20 years and authored the recent series *Great Minds in Finance*, published by Palgrave Macmillan Press. He currently chairs the finance and economics department at the State University of New York College at Plattsburgh. He has also lectured at Harvard and MIT. In addition, he is a columnist for a newspaper as well as guest on several PBS programs.

Peter Rossini is a senior lecturer in property in the School of Commerce, UniSA.

Harald Sander holds a Jean-Monnet Chair on "Europe in the Global Economy" and is professor of economics and International Economics at Hochschule Köln where he is also Director of the Institute of Global Business and Society. He also holds a position as professor of economics at Maastricht School of Management. He specializes in international economics, international finance, and environmental economics. He is author of six books and about 50 refereed scholarly articles. Harald is also involving himself actively into public policy debates as a blogger for outlets such as the Conversation (Australia), LSE Europp, and Roubini's Economonitor.

Raja Vinesh Sannassee is an associate professor of the Department of Finance at the University of Mauritius. He holds a PhD in economics from the University of Reading, UK, and currently lectures International Business both at undergraduate and postgraduate levels. In addition, Dr Sannassee has several publications in the areas of finance, economics, and trade. He is also presently the Director of Program for the MSc in Social Protection Financing, a joint initiative by the UoM and the ILO, funded by the IDRC. In addition, Dr Sannassee is due to take over the Deanship of the Faculty of Law and Management in March of next year. Furthermore, he presently sits on various committees both at the University and at national level, and he is due to take up the Chair of the WTO Chairs Program in July of this year. Finally, Dr Sannassee has also acted as a consultant for various international organizations which include the World Bank, the UNDP, and the AfDB, among others.

Boopen Seetanah is an associate professor at the University of Mauritius and a former faculty research advisor at the Faculty of Law and Management. He is currently the Head of the Department of Finance and Accounting. His research interests are transport and tourism economics, Development, Trade, and applied financial economics. Boopen has published several articles in the above fields in leading journals and he is also a reviewer for a number of refereed international journals (including Annals of Tourism Research, Tourism Management, Tourism Economics, Journal of Transport Economics, and Policy and Empirical Economics, among others) as well as an editorial board member for a few journals. He has been consulting with both the Government of Mauritius and with international organization including UNCTAD, World Bank, UNDP, UNEP, African Development Bank, and COMESA, among others.

Claire Settre is a PhD student at The University of South Australia Business School's Centre for Regulation and Market Analysis in the School of Commerce working on the environmental benefits of water markets across a variety of countries. Claire has a Bachelor of Civil and Environmental Engineering with Honours and a Bachelor of Arts (with a major in international politics).

Belinda Spagnoletti is a research assistant at University of South Australia and University of Melbourne. She holds a Bachelor of Arts with Honours in International Studies and is currently completing her PhD in the Faculty of Medicine at the University of Melbourne.

Gillian Vesty is a senior lecturer in the School of Accounting, RMIT University. Gillian has a PhD from the University of Melbourne. She is also a member of CPA Australia. In addition to her academic role, Gillian has had 20 years experience working in the public healthcare sector. Her research interests include management accounting for sustainability as well as accounting for public health care, underpinned by a keen interest in the sociological of organizations.

Sarah Ann Wheeler is an associate professor of applied economics and an ARC future fellow with the Centre for Regulation and Market Analysis, in the School of Commerce, UniSA. She graduated with her PhD in 2007, and her research interests are organic farming, water markets, crime, and gambling. Sarah is currently an associate editor of the Australian Journal of Agricultural and Resource Economics, a guest editor for a special edition for Agricultural Water Management, and on the editorial board of Agricultural Science and Water Resources and Economics. She has published (had accepted to publish) almost 80 peer-reviewed publications, most in high-quality outlets.

ACKNOWLEDGMENTS

Working on this book would not have been possible without the help and encouragement we received from family, friends, and colleagues. We thank the contributors and researchers working in this field for stimulating discussions of some of the issues exposed in this handbook. We would like to thank (1) our research students who played a key role in the quality of our research namely Minhua Yang, Nguyen Anh Huy Pham, and Ammar Asbi at UNISA and Min Lui, Guang Ping Huang, Jassim Aladwani, and Justin Nguyen from RMIT, (2) Tianjin Academy of Environmental Sciences and our post doc candidate Dr Wentao Chang, and (3) other research collaborators who extend to many domestic and international universities including RMIT (Dr Michael Gangemi), Monash (Dr John Vaz), UNE (Dr John Anderson), VU (Professor Terrence Hallahan), Curtin University (Dr Kelly Burns), UNISA (Dr Braam Lowies, Dr Jeremy Gabe, Dr Kym Thorne, Mr Ron McIver, and Dr John Wilson), Adelaide University (Associate Professor Sarah Wheeler), TH Köln (Professor Harald Sander), Stockholm University (Associate Professor Michael Graham), Beijing Technology and Business University (Dr Xiaoming Xu), Politecnico di Milano (Associate Professor Giancarlo Giudici), and the University of Mauritius (Associate Professor Boopen Seetanah). UNISA has provided a research-conducive environment for us to continue our work and we are thankful to Professor Marie Wilson, Professor Christine Helliar, Professor Petko Kalev (CAFS), Dr Ilke Onur, Dr Marguerite Kolar, Marie Feo, Kathryn Pickering, Peter Edwards, Doreen Michalski, and many others. In addition, we would like to thank the following contributors from the Economics and Finance department in the School of Business and Economics at State University of New York (Plattsburgh), Dr Dhimitri Qirjo, Dr Robert Christopherson and Dr Colin Read.

Vikash's gratitude goes to his family (Narain, Radha, Kavita, Vishal, Harry, Raunak, Nadisha, Vrisha, Zizu, Chanel, Mylo and Bozo) who had to bear the opportunity but our utmost gratitude goes to the Elsevier team who did an excellent job in putting everything together namely Susan Ikeda, Jason Mitchell, J. Scott Bentley, and Jackie Zhou.

Environmental Regulations Post the Kyoto Protocol on Climate Change

CHAPTER 1

Climate Change and Kyoto Protocol: An Overview

Anil Gupta
SJVN Limited, Shimla, India

Contents

1.1 INTRODUCTION

Scientists worldwide accept that climate change or global warming is a manmade phenomenon due to industrial growth because of increase in greenhouse gases concentration. They advocate strong necessity for reduction in greenhouse gases emission to save the

Handbook of Environmental and Sustainable Finance
ISBN 978-0-12-803615-0

world from this menace of global warming, the main characteristics of which are increase in average global temperature; changes in cloud cover and precipitation particularly over land; melting of ice caps, glaciers and reduced snow cover; and increase in ocean temperature.

Acknowledging the fact that global warming is a complicated problem involving the entire world tangled up with different issues such as poverty, economic development, and population growth and also knowing well that tackling the issue of global warming shall not be an easy task but at the same time ignoring the issue will be disastrous, 189 countries joined hands to frame an international treaty known as United Nations Framework Convention on Climate Change (UNFCCC).

UNFCCC is an international legal framework adopted in June 1992 at the "Rio Earth Summit" to address issue of climate change with an objective of stabilization of greenhouse gases concentration in the atmosphere at a level that would prevent dangerous interference with climate system.

Even after the adoption of the UNFCCC in 1992, the greenhouse gas emissions levels continued to rise around the world. It became increasingly evident that only a firm and binding commitment by developed countries to reduce emissions could send a signal strong enough to convince business communities and individuals to act on climate change. Accordingly, member countries of the UNFCCC began negotiations on a protocol or an international agreement linked to the existing treaty, which led to adoption of Kyoto Protocol in 1997.

Kyoto Protocol assigned the legally binding numerical targets for reduction of greenhouse gas emissions to industrialized countries of the world for achievement during the commitment period 2008–2012. Three mechanisms have been established by this protocol for meeting the emission reduction targets with option to use any one. Clean Development Mechanism (CDM) is one of the most important mechanisms provided in the Protocol for emission reductions wherein industrialized countries need to involve the developing countries for achieving their targets by implementing greenhouse gases mitigation projects in such countries.

The Protocol created a new globally traded commodity, that is, carbon credits expressed in tons of carbon dioxide equivalent having a market value similar to other commodities like wheat or oil, which can be traded across borders on the market. The revenue realized by sale of carbon credits is generally termed as carbon finance.

Recognizing well that if the world fails to renew the Kyoto Protocol, beyond period 2008–2012, a huge part of the global emission trading regime will get disrupted, the efforts were on worldwide for its extension by carbon market proponents like Parties to UNFCCC, NGOs, and climate change activists. The historic decision to launch a second commitment period of eight years "2013–2020" commencing from January 01, 2013 was taken at UN Climate Change Conference in Doha, Qatar during December 2012 by the countries which are Parties to the UNFCCC and Kyoto

Protocol, just couple of days ahead of the date by which the Kyoto Protocol was valid as per original agreement.

1.2 GLOBAL WARMING

The frequency and intensity of extreme weather events such as tropical cyclones including hurricanes and typhoons, floods, droughts are rising and expected to rise further because of global warming. It has wide-ranging effects on the environment including water resources, agriculture and food security, human health, terrestrial ecosystem, bio diversity, and coastal zones. The current global warming trend is expected to disturb life on the earth in a big way. Numerous plant and animal species which are already weakened by pollution and loss of habitat are not expected to survive the next hundred years. Human beings, while not threatened in this way, but are likely to face mounting difficulties.

Gupta (2011) highlights that the average temperature of the earth's surface has risen by 0.85 °C since the late 1800s. It is expected to increase by another 1.8 °C—4 °C by the year 2100—a rapid and profound change. Even if the minimum predicted increase takes place, it will be larger than any century long trend in the last 10,000 years. The principal reason for the mounting temperature is industrialization during the last 150 years leading to the burning of greater quantities of fossil fuels and the cutting of forests, etc. These activities have increased the amount of greenhouse gases in the atmosphere.

According to UNEP (2003), the average sea level rose by 10—20 cm during the twentieth century, and an additional increase of 9—88 cm is expected by the year 2100. The higher temperature causes ocean volume to expand and at the same time, it leads to melting of glaciers and ice caps which results into addition of more water in the oceans. It is likely that in times to come, the sea could overflow the heavily populated coastline of the countries like Bangladesh and cause the disappearance of some of the nations entirely such as the island state of the Maldives.

IPCC (2001) assessed that out of total sea level rise since 1993, the contribution of thermal expansion of the oceans is about 57%. Decrease in glaciers and losses from polar ice sheets have contributed about 28% and 15%, respectively.

The concept of global warming can be better appreciated through understanding of greenhouse effect.

Greenhouse Effect: Life on earth is made possible by energy from the sun that arrives mainly in the form of visible light. About 30% of the sunlight scatters back into space by the outer atmosphere and the rest 70% reaches the surface of the earth, which is a slow moving type of energy called infrared radiation. These infrared radiations are carried slowly aloft by air currents, and greenhouse gases delay its eventual escape into space. The greenhouse gases make up only about 1% of the atmosphere and act like a blanket around the earth or like the glass roof of a greenhouse. The greenhouse gases occur naturally and are critical for life on the earth, these decline some of the sun's warmth from reflecting back into space. In other words, these gases trap heat and keep the planet warmer by around 30 °C than it would be otherwise—a cold and barren place. Human activities are making the blanket thicker, which increases the global temperature. The natural levels of greenhouse gases are being increased by emissions of carbon dioxide from the burning of coal, oil, and natural gas; by adding methane and nitrous oxide produced by farming activities and changes in land use; and by emission of several long-lived industrial gases that do not occur naturally.

Carbon dioxide is responsible for around 70% of the total greenhouse gas emissions, methane accounts for about 20%, and nitrous oxide for about 9%. Half of the carbon dioxide emissions are from industry, half of the methane emissions are from agriculture, and 60% of nitrous oxide emissions are from agriculture (World Bank, 2011).

To realize how dangerous the global warming could be for humankind, it is important to deliberate upon the current assessment of climate change by the scientist world over and the predictions of its future effects to outline the measures required to slow-down the rate of global warming and to help the world manage with the climate shift.

1.2.1 Current Evidence of Climate Change

The observations of IPCC (2007) in respect to current evidences of climate change, some of which are given below, have propelled climate change into popular consciousness.

Agriculture yield: Maize, wheat, and other major crops have experienced significant yield reductions at the global level of 40 megatons per year between 1981 and 2002 due to warmer climate.

Snow cover: The Arctic's sea ice extent has shrunk in every successive decade since 1979, with 1.07 million km^2 of ice loss every decade. Mountain glaciers and snow cover, on an average, have declined in both hemispheres.

Rain and drought: Since the Industrial Revolution, there have been significant changes in precipitation patterns globally. It now rains much more in eastern parts of North and South America; northern and central Asia; and northern Europe but less in the Sahel, Mediterranean, southern Africa, and parts of southern Asia. Globally, the area affected by drought has increased since the 1970s.

A hotter world and extreme weather: Over the past 50 years, cold days, cold nights, and frosts have become less frequent over most land areas, and hot days and hot nights, more frequent. Warm air being the fuel for cyclones and hurricanes, an increase in intense tropical cyclone activities in the North Atlantic has been observed since about 1970.

The seasons: Spring events come earlier, and plants and animals are moving upward and poleward because of recent warming trends. Climate-induced changes have been observed in at least 420 physical processes and biological species.

1.2.2 Future Effects

Scientific understanding and computer models have improved the level of many projections which can now be made with greater certainty, and the future predictions show that shifts in climate for the twenty-first century are likely to be significant and disruptive. Following collective projections by IPCC (2007), paint a clear and serious picture:

Agricultural yields are expected to drop in most tropical and subtropical regions and in temperate regions too, if the temperature increase is more than a few degrees.

Diseases, especially those carried by vectors like mosquitoes, could spread to new areas in the world. Many mosquito species, such as those, which carry malaria and dengue, survive and breed more efficiently in hotter temperatures.

Millions of people are expected to expose to increasing water stress as ice packs that feed melt water into river are shrinking progressively over the decades. During summer, the extra water gets pumped into the rivers causing damages and unprecedented floods.

More intense weather-related disasters combine with rising sea levels and other climate-related stresses make the lives of those living on coastlines; particularly the world's poor, more miserable. Extinctions of large number of plant and animal species are expected because of the current warming trends.

1.2.3 Measures Required

Though highly dependent on teamwork and political will, adoption of various measures are required to tackle the climate change. Burning oil and coal more efficiently, switching to renewable form of energy such as solar and wind power, encouraging hydropower development, and developing new technologies for industry and transport can attack the problem at the source. Expanding forests is another way to cut down emissions as trees remove carbon dioxide, the dominant greenhouse gas from the atmosphere. Changing life styles and rules also matters. The cultures and habits of millions of people, whether they waste energy or use it efficiently, also influence the climate change. Similarly, the government policies and regulations do have a major impact on climate change.

Above deliberations on global warming advocate for a concerted international effort for reduction in greenhouse gases concentration worldwide to save the world from adverse effects of global warming.

1.3 UNITED NATIONS FRAMEWORK CONVENTION ON CLIMATE CHANGE

UNFCCC, commonly referred as "Convention" is an international legal framework adopted in June 1992 to address the issue of climate change. The objective of formation of UNFCCC was to begin considering what can be done to reduce global warming.

The countries, who are Parties to the UNFCCC, committed themselves to establish human-induced greenhouse gases emission at level that would prevent dangerous man-made interference with the climate system.

The Parties to UNFCCC acknowledged the followings:

1. The change in the earth's climate and its adverse effects are common concerns of humankind.
2. The global nature of climate change calls for the widest possible cooperation by all countries, their participation in an effective manner and their appropriate international response, in accordance with their common but differentiated responsibilities and their social economic conditions.

The Convention sets an overall framework for intergovernmental efforts to tackle the challenges posed by climate change. It recognizes that the climate system is a shared resource whose stability can be affected by industrial and other emissions of carbon dioxide and other greenhouse gases. Under the Convention, the participating governments decided to gather and share information on greenhouse gases emissions, national policies and practices, and to launch national strategies for addressing greenhouse gas emissions including the provision of financial and technological support to developing countries. The UNFCCC entered into force on March 21, 1994 and enjoys near universal membership of 196 Parties (195 Nations/States and one regional economic integration organization-European Union).

1.3.1 Objectives of UNFCCC

The ultimate objectives of UNFCCC described in United Nations (1992) are (1) stabilization of greenhouse gases concentration in the atmosphere at a level that would prevent dangerous interference with the climate system and (2) to achieve such a level within a time frame sufficient to allow ecosystems to adapt naturally to climate change. The aim is to ensure that food production is not threatened, and economic development proceeds in a sustainable manner.

1.3.2 UNFCCC—Secretariat

A secretariat staffed by international civil servants supports the UNFCCC and its supporting bodies. The secretariat is at Bonn, Germany, which is known as the Climate Change Secretariat. The secretariat is institutionally linked to the United Nations without being integrated in any program and is administered under United Nations Rules and Regulations. It employs around 200 staff including short-term staff and consultants from all over the world. Its Head, the Executive Secretary, is appointed by the Secretary General of the United Nations. The secretariat makes practical arrangements for meetings, compiles and distributes statistics and information, and assists member countries in meeting their commitments under the Convention.

1.3.3 Bodies of UNFCCC

Conference of the Parties (COP) is the prime authority of the UNFCCC. It is an association of all member countries (or Parties) and usually meets annually for a period of 2 weeks. Several thousand government delegates, observer organizations, and journalists attend these sessions. The COP evaluates the status of climate change and the effectiveness of the treaty. It examines the activities of member countries particularly by reviewing national communications and emissions inventories.

Subsidiary Body for Scientific Technological Advice (SBSTA) counsels the COP on matters of climate, the environment, technology, and methods. It meets twice a year. Subsidiary Body for Implementation (SBI) helps review how the Convention is being applied by analyzing the national communications submitted by member countries. It also deals with financial and administrative matters and meets twice a year.

Several expert groups also exist under the Convention. A Consultative Group of Experts helps developing countries in preparation of national reports on climate change issues, a Least Developed Country Expert Group advises least developed countries on establishing programs for adapting to climate change and an Expert Group on Technology Transfer seeks to spur the sharing of technology with less advanced nations.

The Global Environment Facility (GEF) and the Intergovernmental Panel on Climate Change (IPCC), the independent bodies that are not part of UNFCCC, can be termed as partner agencies to the Convention. GEF, which exists since 1991 to fund projects having global environmental benefits in developing countries, has been delegated the job of channelizing grants and loans to poor countries to help them address climate change because of its established expertise. The IPCC provides services to the Convention through publishing comprehensive reviews of the status of climate change along with special reports and technical papers. IPCC reviews worldwide research, issues regular assessment reports, compiles special reports and technical papers. IPCC reports are frequently used as the basis for decisions made under the Convention.

1.4 KYOTO PROTOCOL

After 2 1/2 years of intense negotiations, the Kyoto Protocol, commonly referred as Protocol, was adopted at the third COP to the UNFCCC in Kyoto, Japan on December 11, 1997. This Protocol has divided the countries of the world into two groups, Annex-I countries comprising of industrialized countries who have historically contributed the most to climate change, and non–Annex-I countries, primarily the developing countries (UNFCCC, 1998).

The Kyoto Protocol binds developed countries to reduce their greenhouse gas emission below levels specified for each of them. The major distinction between the Convention adopted in 1992 and this Protocol is that while the Convention encouraged developed countries to stabilize greenhouse gas emissions, the Protocol binds them to do so. The Protocol provides for legally binding emission reduction commitments for the Annex-I countries. Each industrialized country has been assigned individual emission reduction targets to reduce the total greenhouse gas emission by at least 5% below the 1990 level during the commitment period 2008–2012. The Protocol placed a heavier burden on the developed nations because of two main reasons: (1) these countries can more easily pay the cost of cutting emissions and (2) developed countries have historically contributed more to the problem of global warming by emitting larger amounts of greenhouse gases per person compared to developing countries.

The details of Annex-1 countries along with their emission reduction targets are given in Table 1.1.

The detailed rules for implementation of Kyoto Protocol were adopted at COP to the UNFCCC in Marrakesh in 2001 and are called the "Marrakesh Accords." The Kyoto Protocol took a considerable period in becoming politically acceptable by the countries targeted for committed emission reductions and it came into force only on February 16, 2005 after most of the world's countries eventually agreed to it.

The Kyoto Protocol recognizes six main greenhouse gases, each with different impact on the global climate. The common currency of the Kyoto Protocol target is reduction of one metric ton of carbon dioxide equivalent. Each of the other greenhouse gases can be expressed in this form on a weight-for-weight basis by multiplying with its global warming potential (GWP). The six main greenhouse gases along with their GWP are given in Table 1.2.

1.4.1 Mechanisms of Kyoto Protocol

The effect on the global environment is the same irrespective of whether the greenhouse gases emission reductions are achieved in Annex-I country or non-Annex-I country. Therefore, the countries are given an option to meet their targets through one of the following three mechanisms provided under Kyoto Protocol:

- Joint Implementation (JI)
- International Emissions Trading (IET)
- Clean Development Mechanism (CDM)

JI allows industrialized countries to meet part of their required targets of emission reductions by paying for project that reduce emissions in other industrialized countries. The mechanism thus allows generation of carbon credits from projects within the Annex-I countries. Practically this means, projects built in the countries of Eastern Europe and former Soviet Union—the economies in transitions, paid for by Western European and North American countries. The sponsoring government will receive carbon credits

Table 1.1 Emission Reduction Targets for Annex-I Countries

Annex-I Country (or Party)	Quantified Emission Limitation or Reduction Commitment (% of Base Year or Period)
Austria[a]	−8
Belgium[a]	−8
Denmark[a]	−8
Finland[a]	−8
France[a]	−8
Germany[a]	−8
Greece[a]	−8
Ireland[a]	−8
Italy[a]	−8
Luxembourg[a]	−8
The Netherlands[a]	−8
Portugal[a]	−8
Spain[a]	−8
Sweden[a]	−8
UK of Great Britain and Northern Ireland	−8
Bulgaria[b]	−8
Czech Republic[b]	−8
Estonia[b]	−8
Latvia[b]	−8
Liechtenstein	−8
Lithuania[b]	−8
Monaco	−8
Romania[b]	−8
Slovakia[b]	−8
Slovenia[b]	−8
Switzerland	−8
USA[c]	−7
Canada	−6
Hungary[b]	−6
Japan	−6
Poland[b]	−6
Croatia[b]	−5
New Zealand	0
Russian Federation[b]	0
Ukraine[b]	0
Norway	+1
Australia	+8
Iceland	+10

[a]The 15 countries of European Union (EU) members in 1990, who are allowed to redistribute their targets among themselves with overall target for EU 15 to remain same.
[b]Represents the Economies in Transition (EIT), which are having a base line other than year 1990. All other countries have year 1990 as base line.
[c]The country that has yet not ratified the Kyoto Protocol.
Source: UNFCCC (1998).

Table 1.2 Greenhouse Gases Recognized by Kyoto Protocol with Respective Global Warming Potential

Name of the Greenhouse Gas	Global Warming Potential
Carbon dioxide (CO_2)	1
Methane (CH_4)	21
Nitrous oxide (N_2O)	310
Hydrofluorocarbons (HFCs)	140–11,700
Perfluorocarbons (PFCs)	6500–9200
Sulfur hexafluoride (SF_6)	23,900

Sources: Six Greenhouse Gases, UNFCCC (1998). GWP IPCC, 1995. Second Assessment Report on Climate Change 1995. Available at: http://www. cambioclimatico.gob.do/eng/en/BasicInformationCDM/BackgroundInformation (assessed 24.12.14.).

that can be adjusted to their emission reduction targets, and the recipient nations will gain foreign investment but not the credits toward meeting their own emission caps.

IET allows trading directly between Annex-I countries by permitting these countries to transfer parts of their allowed emissions to each other. Individual industrialized countries have mandatory emission targets but it is understood that some may do better than expected or may reduce emission more than the targets, while others may not meet the targets. This mechanism allows Annex-I countries that have emission units to spare, to sell this excess capacity to Annex-I countries that are falling short of their targets.

CDM is a mechanism whereby Annex-I countries may buy the emission reductions arising out of the implementation of projects located in non-Annex-I countries to meet their emission reduction targets. Out of three mechanisms of Kyoto Protocol, CDM is the only mechanism where developing countries can participate and join in mitigation of the climate change. Through CDM, the developed countries can implement greenhouse gases mitigation projects in developing countries at reduced costs. For example, the mitigation of one ton of carbon dioxide equivalent in developed countries cost around US$50, while in developing countries it cost around US$15 only, as per general estimates. It is against this background that CDM projects assumed significance.

The concept of CDM projects provides a win–win situation to both developed and developing countries. It allows industrialized countries to invest in "clean" projects in developing countries and gain carbon credits, which they can use to either offset its own emission reduction targets during a given period or sell them to another country. On the other hand, for developing countries the CDM projects provides an opportunity to achieve sustained development. Additional revenue through sale of carbon credits help developing countries to improve the viability of the projects. Since these investments are viewed in a positive light, they also add to the reputation of project developers and investors for taking up clean and green projects.

1.5 CDM: PROJECT ELIGIBILITY, PROJECT CYCLE, AND EXECUTIVE BOARD

1.5.1 CDM Project Eligibility

For a project to be considered for registration as CDM project, it should fulfill the following eligibility criteria:

- The project contributes to the sustainable development of the host country,
- The project results in real, measurable, and long-term benefits in terms of climate change mitigation, and
- The reduction in the emissions must be additional to any that would have occurred without the project, meaning thereby that the project should not be business as usual but needs to be an additional project.

Proving "additionality" is most critical and challenging eligibility criteria. A project needs to provide or demonstrate to CDM authorities that the project is not a business as usual but is additional in the sense that it has been taken up as a substitute of some essential project which would have resulted into carbon dioxide emissions. It is important to understand that no project results in reduction of already existing emissions in the atmosphere. CDM projects basically replace a project which would have resulted into more emissions. In fact, environmental integrity is essential for overall climate change mitigation to ensure that the project is really beneficial for the environment and actually results in the reduction of emissions. Under CDM, the environmental integrity is preserved through the concept of additionality. The CDM Executive Board issues the tools for the determination and assessment of additionality and these are termed as additionality tools. These tools require project entities to explain how and why the project is additional and therefore the same is not business as usual or a base line scenario. Essentially, a project has to demonstrate that either (1) it is less profitable than the most attractive alternative through an investment analysis and/or (2) it must overcome prohibitive barriers through barrier analysis. In addition, it must also be demonstrated that the project is not a "common practice" in the country where it is being commissioned.

The broad categories of the project activities, which are eligible for registration as CDM projects, are given as follows:

1. Energy Industry (renewable/non-renewable sources)
2. Energy Distribution
3. Energy Demand
4. Manufacturing Industries
5. Construction
6. Transport
7. Mining/Mineral Projects
8. Fugitive Emissions (FE) from Fuels
9. FE from Production and Consumption of Halocarbons and SF_6

10. Solvent Use
11. Waste Handling and Disposal
12. Afforestation and Reforestation
13. Agriculture

1.5.2 CDM Project Cycle

A project has to undergo rigorous process documentation and approval as specified under the CDM modalities and procedures for getting itself registered as a CDM project. A typical CDM project cycle involves following steps from inception to issuance of carbon credits, the Certified Emission Reductions (CERs):

1. Project Identification
2. Preparation of Project Design Document (PDD)
3. Validation of PDD by Designated Operational Entity (DOE)
4. Host Country Approval
5. Registration by the CDM Executive Board
6. Verification and Monitoring of Emission Reductions during Implementation Stage
7. Certification and Issuance of CERs

Project identification involves identification of a green project by the host and the investor; host refers to the project entity and investor the Party from Annex-I country. Preparation of Project Idea Note giving description of the project and initial estimation of the emission reductions from the project is also part of this stage. PDD provides all the technical documentation of the project including its location, financial projection ascertaining technical and economic viability of the project, and precise estimation of the emission reductions from the project. PDD describes in detail through various additionality tools that the project is eligible and additional. Validation of PDD by DOE is the process of independent evaluation of a project activity, on the basis of the PDD, by a third Party UN accredited auditor known as DOE, against the various requirements of CDM particularly the eligibility as well as additionality. Host Country approval refers to clearance of the project from the government of the country in which project is being implemented through its Designated National Authority. Registration is the formal acceptance by the CDM Executive Board of a validated project as a CDM project activity. This step is a prerequisite for the verification, certification, and issuance of the CERs from that project activity. Verification is the periodic independent review and ex-post determination by the UN accredited DOE, other than the one involved in validation of PDD, of the CERs as monitored and reported by the project entity during the given verification period. Certification is the written assurance by the DOE that during a specified period, a project activity has achieved generation of CERs as verified. The CDM Executive Board thereafter reviews and approves this certification and subsequently issues CERs.

1.5.3 CDM Executive Board

The CDM Executive Board (CDM EB) supervises Kyoto Protocol's CDM under the authority and guidance of the COP to the Kyoto Protocol. The CDM EB is the ultimate point of contact for CDM project participants for the registration of projects and issuance of CERs. The role of the CDM EB is to develop procedures for the CDM; approve new methodologies; accredit DOEs; register projects; issue CERs; and maintain a public database of CDM project activities. The Executive Board comprises of 10 members and 10 alternate members from the Parties to the Kyoto Protocol. Five members and alternates are from five UN Regional groups, two from Annex-I countries, two from non-Annex-I countries, and one from small island developing states. The Executive Board may establish committees, panels, or working groups to assist it in the performance of its functions.

1.6 CARBON FINANCE AND TYPES OF EMISSION REDUCTION CREDITS

The Kyoto Protocol provides flexibility to the Annex-I countries to frame the domestic policies and measures, they wish to implement, to meet their respective emission reduction targets. In addition, as discussed earlier also, the Protocol allows these countries to meet some proportion of their quantified emission reductions through three market-based Kyoto Mechanisms by transfer or trading of various types of emission reduction credits provided under three mechanisms.

The Protocol has thus created a new globally traded commodity, that is, the carbon credits expressed in tons of carbon dioxide equivalent, which can be traded across borders, and the revenue realized by sale of carbon credits is generally termed as carbon finance. The common currency of the Kyoto Protocol target is Emission Reduction (ER) equivalent to reduction of one metric ton of carbon dioxide. In respect of CDM, it is CER and for JI and IET mechanisms of Kyoto Protocol, the respective currencies are Emission Reduction Unit (ERU) and Assigned Amount Unit (AAU). Carbon finance is the generic name for the revenue stream generated by the projects through sale or trading of any of the above stated Emission Reduction Credits i.e., CERs or ERUs or AAUs.

In addition to above three emission reduction credits, the Verified Emission Reductions (VERs) are also traded in the voluntary market. VER is a unit of greenhouse gas emission reductions that has been verified by an independent auditor but has not undergone the procedures for verification, certification and issuance of CERs in the case of the CDM, and ERUs in the case of JI under the Kyoto Protocol. Buyers of VERs assume all carbon-specific policy and regulatory risks (the risk that the VERs are not ultimately registered as CERs or ERUs). Buyers therefore tend to pay a discounted price for VERs, considering the inherent regulatory risks.

1.7 CLIMATE FINANCE

Climate finance refers to local, national, or transnational financing, which may be drawn from public, private, and alternative sources of financing. Climate finance is critical to addressing climate change because large-scale investments are required to reduce emissions significantly, notably in sector that emit large quantities of greenhouse gases. Climate finance is equally important for adaptation, for which significant financial resources will be similarly required to allow countries to adapt to the adverse effects and reduce the impacts of climate change. Climate finance should not be confused with carbon finance as the two are totally different; climate finance refers to the funds required for addressing the climate change whereas carbon finance is the revenue realized by projects through sale of carbon credits earned.

In accordance with the principle of common but differentiated responsibility and respective capabilities set out in the Convention, developed countries (Annex-I countries) are to provide financial resources to assist developing countries (non-Annex-I countries) in implementing the objectives of the UNFCCC. It is important for all governments and stakeholders to understand and assess the financial needs the developing countries have so that such countries can undertake activities to address climate change. Governments and all other stakeholders also need to understand the sources of this financing, in other words, how these financial resources will be mobilized.

Equally significant is the way in which these resources are transferred to and accessed by developing countries. Developing countries need to know that financial resources are predictable, sustainable, and that the channels used allow them to utilize the resources directly without difficulty. For developed countries, it is important that developing countries are able to demonstrate their ability to effectively receive and utilize the resources. In addition, there needs to be full transparency in the way the resources are used for mitigation and adaption activities. The effective measurement, reporting, and verification of climate finance are keys to building trust between Parties to the Convention, and for external actors.

At international level, the Adaptation Fund (AF), the Green Climate Fund (GCF) and the GEF are major instruments of climate finance at present.

The AF was established in the year 2001, through a decision taken by COP to UNFCCC, to finance concrete adaptation projects and programs in developing countries that are particularly vulnerable to the adverse effects of climate change. The AF is financed through financial contributions from developed countries and with a share of proceeds from the CDM project activities. The share of proceeds amounts to 2% of CERs issued for a CDM project activity. This fund has achieved its goal of raising a fund of US$100 million by end of the year 2013 with target for the resource mobilization of US$80 million per calendar year in 2014 and 2015 to support the approved projects and programs.

The GCF was established by the COP in its 16th session during November 29 to December 10, 2010 as an operative entity to the financial mechanism of the Convention. The GCF supports projects, programs, policies, and other activities in developing countries, which are Party to Kyoto Protocol. The World Bank is the interim trustee of the GCF. This fund has mobilized US$10.2 billion to date by contributing Parties, enabling it to start its activities in supporting developing countries and making it the largest dedicated climate fund.

The GEF, a partnership for international cooperation where 183 countries work together with international institutions, civil society organizations, and the private sector, to address global environmental issues, also serves as financial mechanism for UNFCCC and its Kyoto Protocol.

1.8 NATIONAL LEVEL EMISSION TRADING SYSTEMS

Beside Parties to the IET program of the Kyoto Protocol, which provides for trading across nations, some of the countries have developed and adopted their own national emission trading systems as one of the strategies for mitigating climate change through promotion of investment in clean energy projects. Emission trading or "cap and trade" system is a market-based approach used to control pollution by providing economic incentives for achieving reductions in the emission of pollutants. A central body, usually a governmental body, sets a limit or cap on the amount of a pollutant that may be emitted. The cap is allocated and/or sold by the central authority to firms in the form of emission permits which represent right to emit a specific volume of the specified pollutant. Permits can also be sold in the secondary market.

The European Union Emissions Trading Scheme (EUETS) is largest multinational well-established greenhouse cap and trade scheme of the world, which trades in European Union Allowances (EUAs). Swiss ETS and New Zealand ETS (NZ ETS) are other prominent national emission trading systems.

Emerging economies like China, India and South Korea have also come forward to implement market based domestic emission trading systems aimed at reducing greenhouse gas emissions. These countries are seeking to promote cleaner technologies and behavior change while also promoting economic development and growth.

National emission trading schemes of various countries are briefed below.

1.8.1 European Union: EUETS

EUETS launched in the year 2005 is the first and biggest scheme of the world operative in European Union where "EUAs" are traded. EUAs are rights to emit greenhouse gases granted to private entities by national governments through auctions, regulations, or specific decree. Under the EUETS, the governments of EU member states agree on national emission caps, which are approved by EU Commission. These countries then allocate allowances to their industrial operators mainly large emitters of carbon dioxide. The operators may assign or trade their allowances within the Emission Trading Scheme (ETS).

Emissions trading programs such as the EUETS complement the country to country trading provided for in the Kyoto Protocol by permitting private party trading of emissions permits. Under such programs which are generally coordinated with the national emissions targets provided within the framework of the Kyoto Protocol, a national or international authority allocates emissions permits to individual companies based on established criteria, with a purpose to achieve national and/or regional Kyoto targets at the lowest overall economic cost.

EUETS operates in 28 European Union countries plus Iceland, Liechtenstein, and Norway. The scheme limits greenhouse gas emissions from (1) more than 11,000 heavy energy-using installations in power generation and manufacturing industry and (2) operators of flights to and from EU, Iceland, Liechtenstein, and Norway.

After successfully completing its two trading periods, the scheme is in the third trading period 2013–2020 wherein from the year 2013 itself, the cap on emission from power stations and other fixed installations is being reduced by 1.74% every year. This means that in the year 2020, greenhouse emission from these sectors will be 21% lower than in the year 2005.

The price of the allowance is determined by supply and demand. As many as 7.9 billion allowances have been traded in the year 2012 with a total value of Euro 56 billion.

1.8.2 Switzerland: Swiss ETS

Switzerland Emission Trading Scheme (Swiss ETS) was introduced on January 01, 2008 as an alternative option for complying with the national carbon dioxide levy on heating, industrial process, and transport fuels. For the period 2008–2012, firms covered by the levy had two choices: (1) pay the carbon dioxide levy or (2) voluntarily set a verified absolute emissions target and participate in the Swiss ETS, which exempted them from paying the carbon dioxide levy. In essence, the carbon dioxide levy functioned as a hard price ceiling for covered entities, and the option for ETS participation allowed firms to potentially pay a lower rate for emissions reductions than this ceiling price. For the year 2008 and 2009, the carbon dioxide levy was Swiss Franc (CHF) 12 per ton of carbon dioxide equivalent, which was increased to CHF 36 per ton of carbon dioxide equivalent from the year 2010. The Swiss ETS became mandatory for large emitters from February 2013 while medium-sized companies can voluntarily participate in it.

The Federal Office for the Environment (FOEN) of Switzerland predetermines the quantity of available emission allowances based on a cap that is lowered every year. Emission allowances are then issued every year to the companies participating in the ETS

(ETS companies). The quantity of emission allowances required for carbon dioxide efficient operations is issued free to ETS companies. This quantity is calculated based on benchmarks. Emission allowances that are not issued free are auctioned off by the FOEN via the national Emission Trading Registry. Auction is carried out several times a year.

To offset their emissions, ETS companies are also allowed to use a limited number of certificates from projects carried out abroad, provided they are issued according to the rules of the Kyoto Protocol and meet specific quality criteria.

For commitment period 2008–2012, the penalty for companies that failed to achieve their Swiss ETS targets was retroactive payment of the carbon levy (plus interest) for each ton of carbon dioxide emitted since the company's exemption. The companies that fail to surrender enough emission allowances and/or emission reduction certificates to meet the required targets during period 2013–2020 need to pay CHF 125 per ton of carbon dioxide equivalent as penalty by the following year.

1.8.3 New Zealand: NZ ETS

The NZ ETS launched in January 2008 is a partial-coverage all-free allocation scheme, which was covering only the forestry sector at its launch. The NZ ETS was expanded in July 2010 to include the stationery energy, industrial processes, and liquid fossil fuel sectors. It is highly linked to international carbon markets as it allows the import of most of the Kyoto Protocol emission units. Domestic unit, the New Zealand Unit (NZU), is issued by free allocation to emitters that vary by sector. Participants in the NZ ETS are required to surrender one emission unit (either an international Kyoto Protocol unit or an NZU) for every two tons of carbon dioxide equivalent emissions reported. For excess emission, NZUs can be bought from the government at a fixed price of New Zealand dollar (NZ$) 25.

1.8.4 Australia: Australian ETS

Australian Emission Trading Scheme (Australian ETS) was set to begin its operation from the year 2015 with provision to issue permits initially at a fixed price of Australian dollar 10 per ton of carbon dioxide equivalent and thereafter at floating rates. The amount of permits, the cap, was to be guided by the overall national commitment. The planned ETS, as well as the existing carbon tax, was repealed by government of Australia on July 16, 2014, considering it to be extra burden on households who are expected to benefit by $550 a year on an average from the repeal, with gas prices to fall by 7% and electricity by 9%. However, Australia is Party to Kyoto Protocol and even submitted its second commitment period's quantified emission limitation or reduction obligation (QELRO) of 99.5%, consistent with the unconditional target to reduce emissions by 5% below 2000 levels by the year 2020.

1.8.5 Kazakhstan: ETS

ETS of the republic of Kazakhstan was introduced in the year 2013. The scheme covers plants in the manufacturing, energy, mining, metallurgy, chemicals, agriculture, and transport industries, which emit more than 20,000 tons of carbon dioxide per annum.

1.8.6 China: Pilot Carbon Trading System

In June 2011, China announced that it would seek to implement carbon trading systems in pilot regions covering seven cities and provinces with different prices in each region. Thereafter based on the experience of pilot system a unified national system is proposed to be established over the year 2016–2020. China has made progress in its carbon trading programs but still has a long way to commence national trading expected in the year 2016.

1.8.7 South Korea: Korean ETS

South Korea is planning to establish Korean Emission Trading System (Korean ETS) in the year 2015, covering about 470 companies from all sectors that together produce about 60% of the country's emissions. The South Korean government has set an emissions reduction target of 30% below projected "business as usual" levels by the year 2020.

1.8.8 India: PAT Scheme

The launch of the Perform, Achieve, and Trade (PAT) scheme is an important initiative taken by the Indian Government during July 2012 to meet its pledge of reduction of 20–25% in emission intensity, carbon dioxide emissions per unit of gross domestic product, from 2005 level by the year 2020. It is a mandatory trading scheme aimed to reduce energy consumption in 478 large industries in eight sectors namely thermal, power, aluminum, cement, fertilizer, iron and steel, pulp and paper, and textile which are responsible for 54% of India's industrial energy consumption. The central government has notified the energy reduction target for all 478 industries that are to be achieved within a period of 3 years by the year 2015. If a company is unable to meet the energy reduction targets specified, it can either pay the prescribed penalty, which is dependent upon the units by which it has failed to meet its target. Alternatively, energy-saving certificates can be purchased that are to be issued under the PAT scheme to the industries having saved more energy than their targeted requirement for additional units of energy saved. These certificates may then be traded on the two power exchanges of India. The assessment of the energy consumption of the companies is to be carried out by third party auditors, who will submit their verification report to Energy Efficiency Services Limited, a company promoted by Government of India for implementing this scheme.

1.9 KYOTO PROTOCOL: JOURNEY AND CURRENT DEVELOPMENTS

As on date, the Kyoto Protocol stands ratified/accepted/approved by 192 countries including all Annex-1 countries except USA and Canada. The United States, although a signatory to the Kyoto Protocol, has never accepted the Protocol. In March 2001, the US Administration announced that it would not implement the Kyoto Protocol that requires nations to reduce their greenhouse gas emissions, claiming that ratifying the treaty would create economic setbacks in the US. According to them, the protocol has not put enough pressure to limit emissions from developing nations. In February 2002, the US announced its alternative to the Kyoto Protocol, by bringing forth a plan to reduce the intensity of greenhouse gases by 18% over 10 years. The intensity of greenhouse gases specifically is the ratio of greenhouse gas emissions and economic output, meaning that under this plan, emissions would still continue to grow, but at a slower pace. It was stated that this plan would prevent the release of 500 million metric tons of greenhouse gases, and this target would be achieved by providing tax credits to businesses that use renewable energy sources.

Canada ratified the Protocol on February 16, 2005 but renounced it effective from December 15, 2012 and ceased to be a member from that date thus becoming the first and only nation to withdraw from the Protocol (UNFCCC, 2007). Canada signed the Kyoto Protocol but could not meet the targets and the Canadian Government decided to withdraw from the Protocol to save the country from paying Canadian $14 billion in penalties for not achieving the assigned emission reduction targets. Calling Kyoto Protocol radical and irresponsible, Canada claimed that penalties would cost it the loss of thousands of jobs and transfer of billions of dollars from Canadian tax payers to other countries.

Australia was the last Annex-I country to ratify the Kyoto Protocol on December 03, 2007, accepting QELRC provided in the Protocol, limiting Australia's emission growth over the first commitment period to 108% of 1990 level.

The Kyoto Protocol market-based mechanisms for emission reduction, specifically the CDM has exceeded the expectations not only in terms of projects registered but also in terms of the awareness and capacity building it has generated. The Kyoto Protocol was valid up to December 31, 2012 with no reference about its future thereafter. The absence of clear policy and regulatory signals beyond 2012 has negatively impacted the carbon market during the year 2011 and 2012.

Activities under the Kyoto Mechanisms have largely focused, at least initially, on the simplest projects with lowest abatement costs and largest volume potential to bring through the CDM system. Renewable energy projects like Hydro, Wind, and Biomass are the most popular types of projects in the CDM. As on October 31, 2014, the energy industries (renewable/nonrenewable sources) sector has attracted the largest number of CDM projects with a share of 73.06% of total registered 7772 projects.

Waste management and manufacturing industries sectors are the other most popular sectors with a share of 12.24% and 4.03%, respectively.

Asia and the Pacific Region accounts for 84.02% share of the total registered CDM projects, followed by Latin America and Caribbean having a share of 12.84%. China is the biggest beneficiary of Kyoto Protocol with a share of CDM-registered projects at 49.67% followed by India with a share of 20.23%.

In spite of all-out efforts by the various market proponents, there were doubts about extension of Kyoto Protocol. In the absence of carbon finance revenue, the basic viability of most of the already registered CDM projects was bound to get adversely affected resulting in wastage of investment of billions of dollars in these projects. Ultimately, the efforts succeeded and a decision was taken to extend Kyoto Protocol for a second commitment period of 8 years "2013—2020" commencing from January 1, 2013 at UN Climate Change Conference in Doha, Qatar, during December 2012 by the countries to the UNFCCC and Kyoto Protocol.

Doha Amendment to the Kyoto Protocol adopted on December 08, 2012 at Doha, Qatar, UNFCCC (2012), includes the following:

1. New commitments for Annex-I countries to the Kyoto Protocol to reduce greenhouse gases emissions by at least 18% below 1990 levels in a second commitment period of 8 years from January 01, 2013 to December 31, 2020;
2. A revised list of greenhouse gases to be reported on by the Parties in the second commitment period; and
3. Amendments to several articles of the first commitment period of Kyoto Protocol which needs to be updated for the second commitment period.

UNFCCC (2013) decided that each Party included in Annex-I will revise its QELRO for the second commitment period within 2 years. The countries may provisionally apply the amendment pending its entry into force. All the Parties have been requested to submit the instruments of acceptance with a view to expedite its entry into force.

1.10 CONCLUSION

The Kyoto Protocol has developed overtime into a massive success, much bigger than expected at the time it was conceived. The protocol has catalyzed the design and implementation of over 7500 CDM projects in just over a decade. These registered CDM projects have potential of annual emission reductions of around 983 million tons of carbon dioxide equivalent (UNFCCC, 2014). With respect to actual emission reduction, 1.511 billion CERs have been generated and issued from more than 2500 already commissioned projects by the end of November 2014, thus helping the world in reduction of 1.5 gigatons of carbon dioxide equivalent (UNFCCC, 2015).

The Kyoto Protocol through its project-based mechanisms is successful in enabling the initial flow of capital into abatement projects around the world, and in creating a new mechanism that has led to actual generation of emission reduction credits. In addition to its contribution in meeting greenhouse gases commitments cost effectively, the Protocol has generated other noteworthy benefits like raising the climate change awareness worldwide, which led to building capacity in developing countries to use carbon finance to support greenhouse gas reductions.

CDM, the largest, most widely recognized crediting mechanism of the world has leveraged private investment 10 times the public funds invested and led to investment of around US$138 billion in mitigation of climate change and sustainable development (UNFCCC, 2015). CDM has proved that it can incentivize emission reductions and contribute significantly to development, technology transfer, and enhanced well-being. It has since evolved into a reliable, sophisticated, and remarkable flexible tool in the international response to climate change. The challenge facing the Kyoto Protocol is now related to scaling up the system through early enforcement of second commitment period with revised firmed up modalities and procedures.

REFERENCES

Gupta, A., 2011. A Study of Global Carbon Finance Market with Special Reference to India. Himachal Pradesh University, Shimla.

IPCC, 2001. Climate Change 2007: Synthesis Report. Intergovernmental Panel on Climate Change, Geneva, Switzerland.

IPCC, 2007. Climate Change 2007: Assessment Report (AR4). Intergovernmental Panel on Climate Change, Geneva, Switzerland.

United Nations, 1992. Full text of the convention, New York.

UNEP, 2003. How Will Global Warming Affect My World? United Nations Environment Program, Nairobi, Kenya.

UNFCCC, 1998. Text of Kyoto Protocol to UNFCCC. United Nations Framework Convention on Climate Change, Bonn, Germany.

UNFCCC, 2007. Status of Ratification of Kyoto Protocol. Available at: http://unfccc.int/kyoto_protocol/status_of_ratification/items/2613.php (assessed 26.12.14.).

UNFCCC, 2012. Doha Amendment to the Kyoto Protocol. Bonn, Germany.

UNFCCC, 2013. Report of the Conference of the Parties Serving as the Meetings of the Parties to the Kyoto Protocol on its Eighth Session. Decision I/CMP.8, Bonn, Germany.

UNFCCC, 2014. CDM: CDM Insights—Intelligence about the CDM at the End of Each Month. Available at: http://cdm.unfccc.inf/statistics/public/CDMinsights/Index.html#iss (assessed 13.01.15.).

UNFCCC, 2015. CDM Fact Sheet: Leveraging Private Finance, Delivering Verified Results. Bonn, Germany.

World Bank, 2011. Climate Change E-learning Course for Parliamentarians and Parliamentary Staff. Unit 1. Available at: http://parlimentarrystrengthening.org/climatechangemodule/index.html (assessed 16.01.15.).

CHAPTER 2

Environmental Policies Post the Kyoto Protocol on Climate Change: Evidence from the US and Japan

Vikash Ramiah[1], Michael Gangemi[2], Min Liu[2]
[1]School of Commerce, University of South Australia, Adelaide, SA, Australia; [2]School of Economics, Finance & Marketing, RMIT University, Melbourne, VIC, Australia

Contents

Handbook of Environmental and Sustainable Finance
ISBN 978-0-12-803615-0

2.1 INTRODUCTION

The mission of the Environmental Protection Agency (EPA; also referred to as "the Agency," http://www.epa.gov/) in the United States is to protect human health and the environment. As of 2013, the priorities of EPA are to make a significant "Visible Difference" in communities across the country, address the issues of climate change, improve air quality, take action on toxics and chemical safety, and protect water. The Agency works alongside other agencies and ensures that the US President's policies on climate change are enforced, and a recent example of such activity is the application of the Significant New Alternatives Policy (EPA 1) program. With regards to chemical safety, the objective of the EPA is to implement the existing Toxic Substances Control Act (EPA 2) to the maximum extent while simultaneously modernizing the law. The EPA is embracing "Next Generation" tools and processes whereby e-government delivers transparent, readily available, and understandable data for the agency which is then able to conduct business electronically in an integrated way. The Agency believes that these new techniques streamline decision-making processes, in turn, increasing efficiency and environmental benefits, and reducing costs. To achieve its goals, the EPA develops and enforces regulations after Congress has written and passed environmental laws. Furthermore, the EPA provides funding to state environmental programs and nonprofit institutions to undertake scientific studies and projects that improve the environment. On the official Web site of EPA, we find various topics for which there are regulatory frameworks, including in the areas of acid rain, asbestos, agriculture, electric utilities, climate change, oil and gas extraction, toxic release inventory, construction, transportation, oil spill and hazardous substance releases, risk management planning, and water and in this chapter, we explore the regulations that govern these areas.

In Section 2.2 we begin discussion of the EPA's environmental actions, looking at its acid rain program, followed in Section 2.3 we discuss its laws relating to asbestos, while in Section 2.4 we examine its laws related to agriculture. In Section 2.5 we investigate the EPA's regulations related to electric utilities, followed in Section 2.6 by discussion of its actions on climate change, and in Section 2.7 its laws and actions regulating the oil and gas extraction sectors. In Section 2.8 we turn our attention to the EPA's toxic release inventory, followed in Section 2.9 with discussion of the EPAs regulations concerning the construction sector, while in Section 2.10 the focus is on the transportation sector. In Section 2.11 we take a look at the EPAs efforts and laws concerning oil spills and hazardous substance releases, followed in Section 2.12 with discussion of its risk management plan rule, and in Section 2.13 we conclude our examination of the EPA with a look at its water regulations. Sections 2.14−2.20 document the environmental regulations in Japan.

We also look at environmental regulation and protection in Japan where the MoE (Ministry of Environment) plays a central role in the Japanese government's environmental protection and conservation policy. The official Web site of the MoE (www.env.go.jp) states that its main aim is to create a sustainable society by ensuring that each individual in Japanese society participates in action designed to solve environmental problems. The work of the MoE can be divided into three basic types: (1) Work for which the Ministry is fully responsible; (2) Work for which the Ministry shares responsibility with another ministry; and (3) Work where the Ministry provides advice from the perspective of environmental conservation. And, again, according to its official Web site, in order to perform its work efficiently, the MoE cooperates with environmental offices in different regions of Japan, as well as with other government organizations, such as the National Institute for Minamata Disease, a research organization, the National Institute for Environmental Studies, and the Environmental Restoration and Conservation Agency, both of which are independent administrative institutions, and the Japan Environmental Safety Corporation, a special company wholly owned by the government.

In order to achieve its objectives, the MoE targets seven distinct areas of the environment, enacting laws, regulations, and procedures in each area to ensure protection and conservation of the environment. In Section 2.14 we take a look at the first of the MoE's seven targeted areas with examination of its laws related to environmental policy, followed in Section 2.15 by discussion of the second key area of the global environment, then in Section 2.16 the focus is on the key area of waste and recycling. The fourth of the MoE's key areas of air and transportation is discussed in Section 2.17, and in Section 2.18 the focus is on key area five of water, soil, and the ground environment, followed by discussion of key area six of health and chemicals in Section 2.19. In Section 2.20 we examine the seventh key area of nature and parks and finally Section 2.21 concludes this chapter.

2.2 ACID RAIN

The Acid Rain Program was created in Title IV of the 1990 Clean Air Act (CAA) (EPA 3) Amendments with a view to reduce emissions of sulfur dioxide (SO_2) and nitrogen oxides (NOx). It is estimated that power plants (coal and heavy oil) produce over 60% of the US's annual SO_2 emissions, and transportation (cars, buses, and trucks) contributes around 50% of the NOx emissions. SO_2 and NOx are the main pollutants causing acid precipitation resulting in acid rain, acid snow, acid fog or mist, acid gas, and acid dust which, in turn, cause human health problems and other environmental-related problems. Unlike the two control zones policy in China, the US uses a market-based cap and trade approach—for example, introducing a cap on the total amount of SO_2 that can be emitted by electric power plants. Emissions trading programs are not new in the US and have been in existence since the 1970s. The addition of the Acid Rain Program (EPA 4) to the CAA gave birth to the cap and trade system (the Clean Air Market) for power plants which has proven successful in that there has been a 41% reduction in emissions from power plants. The Acid Rain Program has an incentive in terms of bonus allowances for power plants that install clean coal technology that reduces the release of SO_2 or for using renewable energy sources. One of the reasons why the cap and trade system is so effective in the United States is because it directly targets the main polluters, who are usually power plants within an economy.

2.3 ASBESTOS

When humans are exposed to asbestos it increases the risk of lung cancer and many other serious diseases, and regulations have emerged to control the use of asbestos in buildings and at cleanup sites. Examples of laws related to asbestos are the Asbestos Hazard Emergency Response Act (AHERA) (EPA 5), the Asbestos Information Act (AIA) (EPA 6), the Asbestos School Hazard Abatement Reauthorization Act (ASHARA) (EPA 7), CAA (EPA 3), and the Safe Drinking Water Act (EPA 8). AHERA mandates EPA to circulate regulations requiring local educational agencies to inspect school buildings for asbestos-containing building materials, prepare asbestos management plans, and perform asbestos response actions to prevent asbestos hazards. AIA is designed to provide transparency in terms of identifying companies engaged in the manufacturing of any kind of asbestos and forcing these companies to report their production to the EPA. ASHARA is a law that provides funding for asbestos abatement loan and grant programs for schools. Based on these laws, the EPA drafted a series of regulations relating to asbestos in the form of the Containing Materials in Schools Rule (EPA 9), the Worker Protection Rule (EPA 10), the Asbestos Ban and Phaseout Rule (EPA 9), and the asbestos-related National Emission Standard for Hazardous Air Pollutants (NESHAP) (EPA 11) regulations.

2.4 AGRICULTURE

The cultivation of crops, farming, and harvesting of fish fall within the agriculture sector is a vital sector producing food for human consumption. Various regulations (air, general, pesticides, water, and compliance) have emerged to ensure that mass production of food does not harm human health and the environment. In November 1997, the Agency proposed the NESHAP (EPA 11) for pesticide active ingredient (PAI) production under the authority of Section 112 of the CAA (EPA 3), where the objective is to reduce emissions of hazardous air pollutants (HAPs) from the production of PAIs. Direct exposure to HAPs usually occurs through inhalation, soil ingestion, the food chain, and dermal contact leading to a variety of adverse health effects (including death). The EPA and the states have registered or licensed more than 1055 pesticides for use in the United States under the Federal Insecticide and Rodenticide Act (FIFRA) (EPA 12), and the main regulatory activities revolve around the process of involving the public to participate in the review of registered pesticides, evaluation of new pesticides, provision for special needs and emergency situations, and enforcing pesticide requirements. The Pesticide Registration Improvement Act (EPA 13) of 2003 establishes pesticide registration service fees, and the Federal Food, Drug, and Cosmetic Act (FFDCA) (EPA 14) sets out maximum levels/tolerances of pesticide usage in food and animal feed. The FFDCA monitors and controls for pesticide residues in foods, such as in fruits, vegetables, seafood, meat, milk, poultry, and eggs and has strict rules to protect infants and children. When pesticide residues are above legal limits, the government seizes the commodity. The Food Quality Protection Act of 1996 (FQPA) (EPA 15) amended both FIFRA and FFDCA to alter the way the EPA regulates pesticides. As per FQPA, the assessment of pesticide must consider (1) aggregate exposures—that is, the total exposure to pesticides through inhalation, oral, or optic contact; (2) cumulative effects—that is, interactions between pesticides; (3) the protection of infants and children through the incorporation of a 10-fold safety factor for infants, and (4) the establishment of a tolerance reassessment program. The Agency has worked with the Tolerance Reassessment Advisory Committee to identify nine science policy issues that are important to the implementation of the FQPA and they are the FQPA 10-Fold Safety Factor, Dietary Exposure and Risk Assessment, Cumulative Risk Assessment for Pesticides with a Common Mechanism of Toxicity, Aggregating Exposure and Risk Assessment, Threshold of Regulation, Drinking Water Exposure, Residential Exposure, Cholinesterase Inhibition End Point and Use and Usage Information. FQPA dictates that an additional 10-fold margin of safety is added to the chemical residue for infants and children as potential pre- and postnatal toxicity must be taken into account, with pre- and postnatal toxicity defined as adverse effects on developing organisms that may result from exposure prior to conception, during prenatal development, or postnatally to the time of sexual maturation. In other words, adverse effects may be detected at any point in the human life span and may cause death of organisms, sudden infant death syndrome, structural abnormalities, birth defects, growth retardation, functional deficiencies, and cancer. In 2009, the EPA decided to withdraw the pesticide science policy document "Guidance for Submission of Probabilistic Human Health Exposure Assessments to the Office of Pesticide Programs" as the policy was superseded by the "Guidance Cumulative Risk Assessment for Pesticides with a Common Mechanism of Toxicity" (EPA 16) and by the "Guidance for Performing Aggregate Exposure and Risk Assessment" (EPA 17). In simple terms, these policies are set to analyze how chemicals interact with humans and focus on pesticide residues in food and drinks.

2.5 ELECTRIC UTILITIES

Electric power generation, transmission, and distribution fall within the utilities sector, which uses fossil fuels (coal, petroleum, and gas), and laws and regulations covering areas like air, waste, water, and compliance aim to have "Cleaner Power Plants." At the end of 2011, the Agency set up national standards to reduce toxic air pollution originating from coal and oil power plants as in the US power plants emit 50% of released mercury levels, 75% of acid gases, and many other toxic metals. The Mercury and Air Toxics Standards (MATS) (EPA 18) provides a regulatory framework for power plants in that it limits emissions levels and sets standards for all HAPs emitted from these plants and is part of the NESHAP. New power plants must adhere to these standards, while existing plants were given up to 4 years to comply with MATS.

In the CAA there is a "good neighbor" provision which requires the EPA to investigate interstate air pollution so that National Ambient Air Quality Standards (NAAQS) (EPA 19) are met. Table 2.1 shows the standards for six pollutants that the EPA considers harmful with these pollutants being carbon monoxide, lead, nitrogen dioxide, ozone, particle pollution, and sulfur dioxide. In this area, primary standards provide health protection, and secondary standards provide welfare protection. For instance, the review of NAAQS for carbon monoxide (in August 2011) shows that carbon monoxide can cause serious damage to health as it is part of the primary standards. In an 8-h average carbon monoxide levels must not exceed 9 parts per million (ppm), and must not exceed 35 ppm over a 1-h interval (see Table 2.1). Lead has both primary and secondary standards, and the rolling 3-month average cannot exceed more than 0.15 μg per cubic meter of air ($\mu g/m^3$), while nitrogen dioxide cannot exceed 100 parts per

Table 2.1 National Ambient Air Quality Standards in the United States

Pollutant	Standards	Interval (Average)	Level	Regulatory Year
Carbon monoxide	Primary	8 h	9 ppm	2011
		1 h	35 ppm	
Lead	Primary and secondary	Rolling 3 month	0.15 μg/m^3	2008
Nitrogen dioxide	Primary	1 h	100 ppb	1996 and 2010
	Primary and secondary	Annual	53 ppb	
Ozone	Primary and secondary	8 h	0.075 ppm	2008
Particle pollution	Primary	Annual	12 μg/m^3	2012
	Secondary	Annual	15 μg/m^3	
	Primary and secondary	24 h	35 μg/m^3	
	Primary and secondary	24 h	150 μg/m^3	
Sulfur dioxide	Primary	1 h	75 ppb	1973 and 2010
	Secondary	3 h	0.5 ppm	

billion (ppb) in 1-h interval, and Table 2.1 shows the standards for the remaining pollutants.

When power plants burn fossil fuels they generate waste, such as ash and slag, referred to as fossil fuel combustion (FFC) wastes. The EPA has classified FFC as a "special waste," exempted from federal hazardous waste regulations under Subtitle C of the Resource Conservation and Recovery Act (RCRA) (EPA 20). FFC wastes are subcategorized into "Large-volume coal combustion wastes (CCWs)" and "All remaining FFC wastes," and although these two categories are not regarded as hazardous, the EPA mandates that CCWs are disposed of in landfills and surface or underground mines, with the minefields regulated under the RCRA or the Surface Mining Control and Reclamation Act (EPA 21). Power plants use water as a coolant and Section 316(b) of the Clean Water Act (CWA) warrants power plants to construct their cooling systems using the best technology available in order to minimize environmental damage.

2.6 CLIMATE CHANGE

The EPA has adopted a "common-sense" approach to develop standards for GHG emissions where the regulations fall under the CAA. The EPA commenced its effort to reduce GHG emissions by targeting the transportation industry as this sector accounts for more than a quarter of U.S. GHG emissions. The aim is to reduce oil consumption by increasing the fuel efficiency of cars and light trucks for new models (2012–2016), and the target for manufacturers has been set at 54.5 miles per gallon by 2025. In 2008, congress directed the EPA to establish a mandatory reporting system for GHG emission, and in 2009 the EPA promulgated the Greenhouse Gas Reporting Rule (GGRR) (EPA 22), and over 8000 (over 80%) GHG emitters are covered by the GGRR, and the EPA has created an online Greenhouse Gas Data Publication Tool (EPA 23) where users have access to data by facility, industry, and location. Table 2.2 provides a highlight of the greenhouse gas reporting program for stationary sources and it is apparent that power plants emit the most carbon dioxide. There were 1582 power plants reporting their carbon emissions in 2010, with the number increasing to 1594 in 2011. Interestingly, over that 1-year period carbon emissions were reduced[1] from 2326 million metric tons to 2221 million metric tons for power plants. In 2012, the EPA proposed the Standards of Performance for Greenhouse Gas Emissions for New Stationary Sources: Electric Utility Generating Units (EPA 24) with uniform national limits on the amount of carbon pollution that new plants can emit, with the limit being 1000 pounds of CO_2 emission per megawatt hour, although this does not apply to

---

[1] Refineries, other industrial, and pulp and paper are the other main sectors which recorded a reduction in carbon emissions.

Table 2.2 Highlights of the Greenhouse Gas Reporting Program

Industry Sector	Number of Reporters		Emissions (Million Metric Tons CO$_2$e)	
	2010	2011	2010	2011
Power plants	1582	1594	2326	2221
Refineries	146	145	183	182
Chemicals	549	458	172	180
Adipic acid production	2	3	5.9	12
Ammonia manufacturing	22	22	25.2	25
Hydrogen production	101	103	32.5	34
Nitric acid production	37	36	12.7	12
Petrochemical production	63	64	50.1	53
Phosphoric acid production	10	13	1.9	2
Silicon carbide production	1	1	0.1	0.1
Soda ash manufacturing	4	4	5.1	5.1
Titanium dioxide production	7	7	2.5	2.4
Other chemicals	321	213	29.1	34.4
Other Industrial	1782	1377	156	126
Oil and natural gas	973	N/A	78.9	N/A
Food processing	277	299	31.4	30
Ethanol production	158	162	17.4	18
Manufacturing	169	280	8.1	17
Other industrial	205	636	20	61
Metals	270	297	100	115
Aluminum production	9	10	4.8	6.7
Ferroalloy production	10	10	2.3	2.3
Iron and steel production	123	128	82.8	91
Lead production	12	13	0.9	1
Zinc production	6	6	0.8	0.9
Other metals	109	130	8.5	12.5
Minerals	354	362	95	98
Cement production	98	96	54.1	56
Glass production	108	110	8.1	8.4
Lime manufacturing	71	73	29.1	31
Other minerals	77	83	3.4	3.6
Pulp and Paper	228	230	47	44
Pulp and paper manufacturers	110	110	38.9	31
Other paper producers	118	120	13.7	13
Universities	108	109	9.5	9.4
Military	38	43	3	2.7

existing power plants. The EPA employed a greenhouse gas Tailoring Rule (EPA 25) in May 2012 and provided permits to the largest greenhouse gas emitters, such as power plants, refineries, and cement production companies, with the tailored solution referred to as the "common-sense approach."

2.7 OIL AND GAS EXTRACTION SECTOR

The oil and gas extraction sector extracts coal and ores, petroleum and gases; and there are various laws and regulations that prevent this sector from polluting air and water, as well as guidelines for this sector to follow regarding waste management and other compliance issues. The greenhouse gas reporting program, NESHAP, reciprocating internal combustion engines, and stationary combustion turbines have specific guidelines for this sector to abide by regarding air pollution, and NESHAP has a subsection dictating rules and implementation information for oil and natural gas production. The main HAPs emitted by this sector include benzene, toluene, ethyl benzene, mixed xylenes (collectively referred to as BTEX), and n-hexane. Benzene is a known human carcinogen which has a tendency to cause leukemia. Additionally, short-term inhalation of high benzene levels has the potential to cause drowsiness, dizziness, headaches, and even death, while lower concentrations may irritate the skin, eyes, and respiratory system. Long-term inhalation of benzene can cause blood disorders and may affect the female reproductive system. Oil and natural gas production have been included on the EPA's list of categories of major sources of HAP emissions established under Section 112(c) (1) of the Act, and firms operating in this sector have the obligation to reduce HAP emissions to the EPA's standard and meet air emission control equipment requirements. Similar to power plant waste, waste generated during oil and gas extraction processes is categorized as "special wastes" and thus exempt from federal hazardous waste regulations.

The guidelines and standards for oil and gas extraction were written in 1979 and then amended by the EPA in 1993, 1996, and 2001, and these regulations cover wastewater discharge from field exploration, drilling, production, and well treatment, and the guidelines and standards dictate levels of chemicals/pollutants that are acceptable in water discharge and are incorporated into the National Pollutant Discharge Elimination System (NPDES) (EPA 26). As water pollution affects drinking water, fishing, and swimming, the CWA (EPA 27) and the NPDES Permit Program (EPA 28) control water pollution by regulating point sources that discharge pollutants into waters.

2.8 TOXIC RELEASE INVENTORY

Following the toxic methyl isocyanate gas leak in Bhopal, India, in 1984, a similar incident occurred in 1985 in West Virginia with around 135 residents treated for eye, throat, and lung irritation; thus in 1986 Congress passed the Emergency Planning and Community Right-to-Know Act (EPCRA) (EPA 29) where the objective is to have an emergency plan for and to provide the public with information about the release of toxic chemicals in communities. Section 313 of EPCRA established the Toxics Release Inventory (TRI) which tracks the management of certain toxic chemicals that have the potential to harm human health and the environment and is a mandatory program where certain firms must report their annual release of chemicals into the environment and/or managed through recycling, energy recovery, and treatment. The 2012 TRI toxic chemical list (EPA 30) contains 682 chemicals and the EPA updates the list annually through an EPA-initiated review and the chemical petitions process. Not all organizations are required to report to TRI, with reporting firms typically including larger facilities involved in manufacturing, metal mining, electric power generation, chemical manufacturing, and hazardous waste treatment, and in 2011 there were over 80,000 facilities reporting their data. The TRI data are used by other programs, such as the risk management plan (RMP) (EPA 31), RCRA (EPA 32), and NPDES (EPA 33), to reduce environmental damage and play a critical role as the EPA tries to identify and hold polluters accountable for their action. Analysis of the data shows that total toxic air releases in 2011 decreased by 8% from 2010, mostly due to decreases in HAP (coming from hydrochloric acid and mercury) emissions as a result of the installation of control technologies at coal-fired power plants and a shift to other fuel sources, as well as a steady decline in the amount of TRI chemicals released into the air since 1998, and the TRI program and rigorous efforts by industry, regulators, and public interests account for the successful reduction of airborne emissions.

2.9 CONSTRUCTION SECTOR

The construction sector comprises of establishments engaged in the construction of buildings, highways, and utility systems, and firms involved in subdividing new land for sale, and this industry is governed by asbestos laws and regulations, such as asbestos for building owners and managers, and for school buildings, and ozone layer protection-regulatory programs regarding air pollution. Construction companies must consider Federal Environmental Requirements for Construction (EPA 34) which have environmental regulations for businesses engaged in clearing, grading, or excavation activities; building roads, golf courses, playing fields, homes, or buildings; demolition activities; discharging dredged or filling material to a waterway or wetland; and involved in tunnel or pipeline projects. Penalties for violating these regulations range from fines of up to $27,500 per day, and criminal penalties of up to $250,000 and 15 years imprisonment.

In the event that a construction project disturbs one or more acres of land, the owner of the construction business must obtain CWA permit coverage for discharge of storm water runoff from the construction site, with the permits issued through the NPDES program. If the project involves discharging dredged or fill material (replacing an aquatic area with dry land or changes to the bottom elevation of a water body), then a permit under Section 404 of the CWA must be obtained. Section 404 permits are issued by either the U.S. Army Corps of Engineers or a state with an approved Section 404 permitting program, and permit decisions are made in accordance with environmental standards developed by the EPA. Spent cleaners (organic solvents), paints, used oil, paint thinners, wastes that contain ignitable and corrosive materials, and wastes that contain certain toxic pollutants are classified as hazardous wastes and businesses must follow the RCRA when managing, treating, and disposing of these hazardous materials. Producers of large amounts of hazardous wastes are subject to more regulatory requirements than those of small amounts, and the RCRA defines what is a "large" or "small" amount of hazardous wastes, and the regulatory framework also extends to the storage and transportation of these types of hazardous waste. Construction businesses must also be mindful of endangered species as the Endangered Species Act (EPA 35) requires that federally listed species and habitat are not to be adversely affected during any construction activity.

2.10 TRANSPORTATION SECTOR

The transportation sector is comprised of industries providing transportation of passengers and cargo through aircraft, railway, and by sea. In 2003, the Agency issued a final rule to reduce emissions of HAPs from engine test cells/stands where an engine test cell/stand is defined as any apparatus used for testing uninstalled stationary or uninstalled mobile engines. Engines emit air toxins from combustion of gaseous and liquid fuels and the primary air toxins present are toluene, benzene, mixed xylenes, and 1,3-butadiene which can cause cancer, respiratory irritation, and damage to the nervous system. The final rule covers four subcategories of engines, namely internal combustion engines of 25 horsepower or more, internal combustion engines of less than 25 horsepower, combustion turbine engines, and rocket engines.

In 1994, the EPA issued a rule to reduce air toxin emissions from gasoline distribution facilities across the US and the rule is now known as the "Rule and Implementation Information for Gasoline Distribution MACT and GACT" (EPA 36) (see 40 CFR 63, Subpart R). Gasoline bulk terminals and pipeline breakout stations transfer and store various petroleum products in the process of distributing these products to refineries and service stations and around 10 air toxins (including benzene and toluene) are generally released during these processes and the purpose of the regulation is to reduce the emissions of these toxins. Under the CAA, the EPA is required to regulate emissions of listed HAPs/toxins and it is for this reason that the EPA developed standards for gasoline distribution facilities. The control strategy is preventive in nature and emissions of air toxins are prevented by improving seals on storage tanks, detecting leaks, and repairing leaks from equipment used to transfer gasoline, as well as involving the use of vapor processors to recover vapors displaced during cargo tank loading operations. The standard requires cargo tank loading rack emissions at both new and existing facilities be collected and processed to limit emissions to no more than 10 mg total organic compound per liter (mg TOC/liter), and according to the standard, floating-roof gasoline storage tanks must be equipped with specified types of primary and secondary rim seals. Furthermore, fixed-roof storage tanks must be equipped with internal floating roofs with specific types of primary seals and installation of gaskets on floating roof fittings is also required. A bulk gasoline terminal loading rack(s) with gasoline (of more than 250,000 gallons per day) must (1) equip its loading rack(s) with a vapor collection system designed to collect TOC vapors displaced from cargo tanks during product loading; (2) reduce emissions of TOC to less than or equal to 80 mg/L of gasoline loaded into gasoline cargo tanks at the loading rack; (3) design and operate the vapor collection system to prevent any TOC vapors collected at one loading rack or lane from passing through another loading rack or lane into the atmosphere; and (4) limit the loading of gasoline into gasoline cargo tanks that are vapor tight using specified procedures.

The EPA proposed the NESHAP: Organic Liquids Distribution (OLD) (Non-Gasoline) (EPA 37) in April 2002 which is to be carried out at storage terminals, refineries, crude oil pipeline stations, and various manufacturing facilities. The EPA estimates that approximately 70,200 Mg per year (Mg/yr) of HAP are emitted from facilities in this source category and this new regulation will reduce HAP emissions by around 28%. These standards fall under Section 112(d) of the CAA and require all OLD operations at plant sites to meet HAP emission standards reflecting the application of the maximum achievable control technology (MACT)—the MACT floor is the minimum control level allowed for NESHAP.

The EPA adopted emission standards for aircraft gas turbine engines with rated thrusts greater than 26.7 kN (predominantly on commercial passenger and freight aircraft) and these standards are similar to International Civil Aviation Organization (ICAO) (http://www.icao.int) regulations. There are two new tiers (Tier 6 standards referred to as CAEP/6 (ICAO 1) and Tier 8 standards or CAEP/8 (ICAO 2)) of more stringent emission standards for NOx, with Tier 6 standards more applicable to newly manufactured aircraft. Engine models that were certificated prior to the effective date of the rule may continue production without meeting Tier 6 standards, and as of December 31, 2012 these engines must comply with the new Tier 6 standards, while engine models that were originally certificated beginning on or after January 1, 2014 must comply with the CAEP/8. The goal of this regulation is to reduce NOx emissions from taxiing, take off, landing, idling, and flight for certain gas turbofan engines; and implementation of Tier 6 standards is expected to reduce emissions by 12% below current Tier 4 standards and Tier 8 standards are expected to reduce emissions by 15% below Tier 6 standards. According to cost/benefit analysis, this regulation will cost around 10 h $365 per manufacturer as aircraft turbofan engines are already designed and built to ICAO standards.

Marine diesel engines are used in different kinds of vessels and the new marine diesel engines must abide by strict emission requirements in order to decrease emissions of NOx and particulate matter (PM). In 2004, the EPA instigated initiatives that decreased allowable levels of sulfur in marine diesel fuel by 99% as part of the Nonroad Diesel Tier 4 Rule (EPA 38), and these initiatives have paid off with significant reduction in PM after coming into effect in 2007. In 2008, the EPA introduced a three-part program that targeted small fishing boats, towboats, tugboats and Great Lake freighters, and marine auxiliary engines ranging from small generator sets to large generator sets on ocean-going vessels; and the program has the potential to cut PM emissions by around 90% and NOx emissions by around 80% when fully implemented. Emission standards for existing commercial marine diesel engines have also been set and apply to engines larger than 600 kW and Tier 3 emission standards apply to engines that built in 2009. The rule also establishes Tier 4 standards for newly (from 2014) built commercial marine diesel engines above 600 kW, based on application of high-efficiency catalytic after-treatment technology. Table 2.3 provides a description of Tier 1, Tier 2, and Tier 3 standards, with the bottom two rows of Table 2.3 referring to diesel engines above 560 kW; and if the engine model was year 2000, Tier 1 standards apply, and it shows that emission targets for NOx and PM are 9.2 g/kW-h and 0.54 g/kW-h, respectively. If the model was year 2006, Tier 2 standards apply, meaning no emission of NOx and a PM level of 0.20 g/kW-h.

Table 2.3 Emission Standards for Diesel Engines (g/kW-h)

Rated Power (kw)	Tier	Model Year[1]	NOx	HC	NMHC + NOx	CO	PM
kW < 8	Tier 1	2000	—	—	10.5	8.0	1.0
	Tier 2	2005	—	—	7.5	8.0	0.80
8 ≤ kW ≤ 19	Tier 1	2000	—	—	9.5	6.6	0.80
	Tier 2	2005	—	—	7.5	6.6	0.80
19 ≤ kW ≤ 37	Tier 1	1999	—	—	9.5	5.5	0.80
	Tier 2	2004	—	—	7.5	5.5	0.60
37 ≤ kW ≤ 75	Tier 1	1998	9.2	—	—	—	—
	Tier 2	2004	—	—	7.5	5.0	0.40
	Tier 3	2008	—	—	4.7	5.0	
75 ≤ kW ≤ 130	Tier 1	1997	9.2	—	—	—	—
	Tier 2	2003	—	—	6.6	5.0	0.30
	Tier 3	2007	—	—	4.0	5.0	
130 ≤ kW ≤ 225	Tier 1	1996	9.2	1.3	—	11.4	0.54
	Tier 2	2003	—	—	6.6	3.5	0.20
	Tier 3	2006	—	—	4.0	3.5	
225 ≤ kW ≤ 450	Tier 1	1996	9.2	1.3	—	11.4	0.54
	Tier 2	2001	—	—	6.4	3.5	0.20
	Tier 3	2006	—	—	4.0	3.5	
450 ≤ kW ≤ 560	Tier 1	1996	9.2	1.3	—	11.4	0.54
	Tier 2	2002	—	—	6.4	3.5	0.20
	Tier 3	2006	—	—	4.0	3.5	
kW > 560	Tier 1	2000	9.2	1.3	—	11.4	0.54
	Tier 2	2006	—	—	6.4	3.5	0.20

NOx, Nitrogen Oxides; HC, Hydrocarbons; NMHC, Nonmethane hydrocarbon; CO, Carbon monoxide; and PM, Particulate matter.
[1]The model years listed indicate the model years for which the specified tier of standards takes effect.

Diesel locomotives (line-haul, switch, and passenger rail) are also regulated by the EPA, and in 2008 the EPA finalized a three-part program intended to reduce emission of PM and NOx. However, these standards apply only to engines manufactured after 2015 which must use high-efficiency catalytic after-treatment technology, which is a state of the art technology significantly reducing undesirable emissions and providing effective pollution abatement while being cost-effective. The EPA has set standards for existing locomotives but these standards will apply when they are remanufactured. There are also regulations aiming to reduce idling for new and remanufactured locomotives. Table 2.4 shows the standards that apply to locomotives that are propelled by engines with a horsepower (hp) of 750^2 kilowatts (kW) or more, with five tiers (Tier 0—Tier 4). Tier 0 for line-haul applies to locomotives that built from 1973 to 1992 and the emission standard for hydrocarbons is 1 g/hp-h, for NOx is 9.5 g/bhp-h, for PM is 0.22 g/bhp-h, and for carbon monoxide (CO) is 5.0 g/bhp-h. Opacity is the degree to which visibility of a background (i.e., blue sky) is reduced by particulates (smoke) with measurements made as a percent in increments of five, and at opacity of 5% smoke blocks enough of the background to just be visible. Tier 0 for line-haul has opacity of 30, 40, and 50.

2.11 OIL SPILLS AND HAZARDOUS SUBSTANCE RELEASES

Another important role that the EPA plays is to respond to emergencies such as oil spills that occur in the United States. The EPA is the lead federal response agency for oil spills occurring inland, and the US Coast Guard is the lead response agency for spills in coastal waters and deep-water ports. The Discharge of Oil regulation (EPA 39) is commonly known as the "sheen rule" and it falls under the legal authority of the CWA and provides the framework for determining whether an oil spill to inland and coastal waters and/or their adjoining shorelines should be reported to the federal government. According to this regulation, persons in charge of a facility or vessel are responsible for notifying the authorities (National Response Center) in the event of a spill, and according to the Facility Response Plan (FRP) rule (EPA 40), facilities that have the potential to cause "substantial harm" to the environment by discharging oil into or on navigable waters are required to prepare and submit FRPs showing that the facility is prepared to respond to a worst-case oil discharge. The Oil Pollution Prevention regulation contains two methods to identify facilities that can cause substantial harm, namely through a self-selection process or by a determination of the EPA Regional Administrator. The National Oil and Hazardous Substances Pollution Contingency Plan (NCP) (EPA 41) requires the EPA to have a list of dispersants, chemicals, oil spill mitigating devices and substances that can be used to remove or control oil discharges; and the NCP Product

---

[2] Equivalent to 1006 hp.

Table 2.4 Emission Standards for Locomotives

Duty-Cycle	Tier	Year	HC	NOx	PM	CO	Smoke (%)	Minimum Useful Life (hours/years/miles)
Line-haul	Tier 0	1973–1992	1	9.5	0.22	5	30/40/50	(7.5 × hp)/10/750,000
	Tier 1	1993–2004	0.55	7.4	0.22	2.2	25/40/50	(7.5 × hp)/10/750,000
	Tier 2	2005–2011	0.3	5.5	0.1	1.5	20/40/50	(7.5 × hp)/10/—
	Tier 3	2012–2014	0.3	5.5	0.1	1.5	20/40/50	(7.5 × hp)/10/—
	Tier 4	2015+	0.14	1.3	0.03	1.5	—	(7.5 × hp)/10/—
Switch	Tier 0	1973–2001	2.1	11.8	0.26	8	30/40/50	(7.5 × hp)/10/750,000
	Tier 1	2002–2004	1.2	11	0.26	2.5	25/40/50	(7.5 × hp)/10/—
	Tier 2	2005–2010	0.6	8.1	0.131	2.4	20/40/50	(7.5 × hp)/10/—
	Tier 3	2011–2014	0.6	5	0.1	2.4	20/40/50	(7.5 × hp)/10/—
	Tier 4	2015+	0.14	1.3	0.03	2.4	—	(7.5 × hp)/10/—

Schedule (EPA 42) in August 2013 contained 19 dispersants, 52 surface washing agents, 2 surface collecting agents, 26 bioremediation agents, 19 biological additives, 18 microbiological cultures, 1 enzyme additive, 7 nutrient additives, 14 miscellaneous oil spill control agents, and nonsolidifiers.

One of the priorities of the EPA's Emergency Management program is to ensure the safety of the public and the environment in the event of a hazardous substance release or oil spill. As a result, the person responsible for notifying the authorities must follow a set of reporting requirements that differ for oil spills and hazardous substance releases. Oil discharge reporting requirements are: (1) name, organization, and telephone number; (2) name and address of the party responsible for the incident; (3) date and time of the incident; (4) location of the incident; (5) source and cause of the discharge; (6) types of material(s) discharged; (7) quantity of materials discharged; (8) danger or threat posed by the discharge; (9) number and types of injuries (if any); (10) weather conditions at the incident location; and (11) other information to help emergency personnel respond to the incident.

The federal government has established the Superfund Reportable Quantities (RQ) (EPA 43) when it comes to the release of hazardous substances and this falls under the Comprehensive Environmental Response, Compensation, and Liability Act (CERCLA) (EPA 44) and it is also known as the Superfund Law. The EPA has adopted five RQ levels of 1, 10, 100, 1000, and 5000 pounds; and any hazardous substance releases that are above the RQ levels must be reported to federal authorities, with levels for each substance reported in the "Reportable Quantities (RQs) for CERCLA Section 102(a) Hazardous Substances."

2.12 RISK MANAGEMENT PLAN RULE

The Chemical Accident Prevention Provisions (EPA 45) is a subsection of the CAA which requires facilities to develop a Risk Management Program and to prepare an RMP, with the plan explaining how the organization will produce, handle, process, distribute, or store certain chemicals, with the RMP to be submitted to the EPA. The Agency provides guidelines as to what is required in preparing the RMP and the guidelines are stipulated in the "Risk Management Program Policy and Guidance" (EPA 46). Facilities that are stationary sources, i.e., buildings, structures, facilities, or installations emitting air pollutants, use regulated substances and have regulated substances above a threshold quantity in a process are subject to this rule, with facilities then assigned program levels (1–3). Program 1 applies to processes that do not affect the public in a worst-case release and have no records of accidents with specific offsite consequences within the past 5 years. When processes are not eligible for Program 1, are not subject to Occupational Safety and Health Administration (OSHA) Process Safety Management (PSM) (USDL 1) standards, and are not listed in the North American Industrial

Table 2.5 Industries with High Accidental Release

NAICS Codes	Industry
32211	Pulp mills
32411	Petroleum refineries
32511	Petrochemical manufacturing
325181	Alkalies and chlorine manufacturing
325188	All other basic inorganic chemical manufacturing
325192	Cyclic crude and intermediate manufacturing
325199	All other basic organic chemical manufacturing
325211	Plastics material and resin manufacturing
325311	Nitrogenous fertilizer manufacturing
32532	Pesticide and other agricultural chemical manufacturing

Classification System (NAICS) (US Census Bureau 1) codes, then the facility falls under Program 2 which imposes streamlined prevention program requirements, and additional hazard assessment/management and emergency response requirements. Processes that are (1) not eligible for Program 1, (2) subject to OSHA's PSM standard, and (3) classified in 1 of 10 specified NAICS codes are placed in Program 3. The OSHA PSM standard is a set of procedures in 13 management areas designed to protect worker health and safety in case of accidental releases and represent 10 industries with high accidental release, and the NAICS and their corresponding industries are shown in Table 2.5.

In developing the risk management program and the RMP, the risk manager must (1) conduct and document a worst-case release analysis, (2) coordinate emergency response activities with local responders, and (3) sign the Program 1 certification as part of the RMP submission. For all Program 2 and 3 processes, the risk manager (1) must conduct and document at least one worst-case release analysis to cover all toxins and one to cover all flammables; (2) may be asked to conduct additional worst-case release analyses if worst-case releases from different parts of the facility would affect different public receptors; (3) must conduct one alternative release scenario analysis for each toxin and one for all flammables; (4) must coordinate emergency response activities with local responders and, if the facility uses its own employees to respond to releases, an emergency response program must be developed and implemented. There are other guidelines to follow in these two programs.

2.13 WATER

The EPA enforces clean water and safe drinking water laws in all regions of the United States and these regulations cover a range of water related matters, such as animal feeding operations (AFOs), biosolids, hydraulic fracturing, impaired waters, mercury, mountain-top mining, oceans and coastal waters, surface water (lakes, rivers, and streams), storm water, wastewater, watersheds, and wetlands.

EPA regulations protect water quality from emissions and other pollution from AFOs and concentrated animal feeding operations (CAFOs). AFOs are defined as "operations where animals have been, are, or will be stabled or confined and fed or maintained for a total of 45 days or more in any 12-month period and where vegetation is not sustained in the confinement area during the normal growing season." An AFO becomes a CAFO if it meets the regulatory definition of a large or medium CAFO, 40 CFR (EPA 47) parts 122.23(b)(4) or (6), or has been designated as a CAFO, 40 CFR part 122.23(c), by the NPDES permitting authority or by the EPA. Table 2.6 is an extract from 40 CFR parts 122.23(b)(4) and shows the number of animals that a CAFO must have to be defined as a large CAFO. For instance, a CAFO must have 55,000 turkeys to be categorized as large. CAFOs produce a significant amount of manure and wastewater which emit pollutants such as nitrogen, phosphorus, organic matter, heavy metals, hormones, ammonia, and so on. Mismanagement of these wastes results in excess nutrients in water which in turn contributes to low levels of dissolved oxygen (example is that an excess of nitrogen and phosphorus will kill fish) and contribute to toxic algal blooms. Contamination from runoff or lagoon leakage degrades water resources and damages human health. The NPDES controls this by a permit program authorizing and regulating the discharge of pollutants from point sources to waters of the US. Permits issued to a CAFO include a requirement to implement a nutrient management plan. As per the nutrient management plan CAFO has to:

1. ensure adequate storage of manure, litter, and process wastewater;
2. ensure proper management of mortalities (dead animals);

Table 2.6 Defining Large CAFOs

Number of Animals	Type of Animal
700	Mature dairy cows, whether milked or dry
1000	Veal calves
1000	Cattle, other than mature dairy cows or veal calves
2500	Swine, each weighing 55 pounds or more
10,000	Swine, each weighing less than 55 pounds
500	Horses
10,000	Sheep or lambs
55,000	Turkeys
30,000	Laying hens or broilers, if the AFO uses a liquid-manure handling system
125,000	Chickens (other than laying hens), if the AFO uses other than a liquid-manure handling system
82,000	Laying hens, if the AFO uses other than a liquid-manure handling system
30,000	Ducks, if the AFO uses other than a liquid-manure handling system
5000	Ducks, if the AFO uses a liquid-manure handling system

Source: 40 CFR parts 122.23(b)(4).

3. ensure that clean water is diverted from the production area;
4. prevent direct contact of confined animals with waters;
5. ensure that chemicals are not disposed of in any of the waste;
6. identify and establish protocols for appropriate testing of manure, litter, process wastewater, and soil;
7. identify specific records that will be maintained to document the implementation and management of minimum elements.

2.14 JAPANESE ENVIRONMENTAL POLICIES (MoE ENVIRONMENTAL POLICY)

2.14.1 Basic Environment Plan

Japan's Ministry of the Environment's (MoE) Basic Environment Law (MoE 1) sets the general principles and directions for standard environmental policies in Japan. The principal goal of this law is to build an economically sustainable environment in order to ensure a healthy system for future generations, as well as allowing Japan to make a positive contribution to global environmental harmony.

The Basic Environment Law includes three long-term plans, and the first Basic Environment Plan was adopted on the 16th of December 1994, the second Basic Environment Plan was approved on the 22nd of December 2000, and the third Basic Environment Plan was approved on the 7th of April 2006. In each of the plans the environmental responsibilities of the national and local governments, citizens, businesses, and private institutions are specified, and each plan includes the framework for long-run environmental policies, specified to the mid-twenty-first century, that are necessary to respond to current environment problems.

The first Basic Environment Plan sets out four long-term objectives, the first of which is a Sound Material Cycle, which aims to build an environmental friendly, sound recycling system and to improve the environment by reducing the mass production, consumption, and disposal of goods so as to minimize the environmental burden. The second long-term objective is referred to as Harmonious Coexistence, involving maintenance of the ecosystem, restoration of the mass environment, and the building of a sustainable environment for present and future generations and a harmonious coexistence between nature and human beings. Thirdly, there is the objective defined as Participation, which aims to ensure achievement of environmental objectives by all parties in Japan, including national and local governments, citizens, businesses, and private organizations, with each party given responsibility for reducing the burden on the environment. Finally, there is the long-term objective defined as International Activities, the aim of which is to target international cooperation in order to protect and promote the global environment beyond Japan's borders.

The second Basic Environment Plan focuses mainly on the effectiveness of environmental protection activities, with a strong emphasis on implementation of environmental regulations, as well as setting out strategies in 11 areas, including on global warming and land pollution. And the third Basic Environment Plan raised the target on achievement of a virtuous cycle between the environment and the economy by increasing strategies and more clearly defining the roles of the various parties, as well as setting long-run environmental targets to 2050.

2.14.2 Environment and Economy

In May 2004, the Japanese government's Central Environment Council made recommendations to the MoE regarding the establishment of a "Virtuous circle for the environment and economy of Japan, with the aim of building a healthy, rich, and beautiful environmentally friendly country" (MoE 2) in the belief that this will boost economic growth. In Japan steps have been taken to establish this virtuous circle, including (1) development and introduction of life-enhancing environmental techniques; (2) increase in efforts to reduce, reuse, and recycle waste; and (3) searching for and developing substitute natural energy sources that are necessary due to the planet's finite resources. The year 2025 has been set as the target to achieve the virtuous circle between Japan's environment and economy, with the perfect in that year described as achieving creation of employment opportunities from ecofriendly technologies, with an ecofriendly market worth over 100 trillion yen and creation of 2 million jobs in the sector, in conjunction with a more highly efficient use of resources, which Japan has already started to achieve. For example, Japan has improved its resource productivity, measured as GDP divided by natural resources, to 390,000 yen per ton in 2010, an improvement of 40% compared to the year 2000 levels (MoE 3).

Japan has released other guidelines to regulate the link between the environment and the economy, for example, in the accounting and finance fields. For instance, in order to support and encourage businesses to introduce environmental accounting systems, the Environment Accounting Guidelines 2005 (MoE 4) were published by the MoE, with the aim of improving user abilities regarding environmental accounting information.

2.14.3 Environment and Taxation

In order to encourage development of renewable energy sources and energy-saving activities, in October 2012 the Japanese government introduced the Tax for Climate Change Mitigation (MoE 5). About 90% of GHG emissions in Japan are carbon dioxide-based and are generated from energy use, so-called energy-related CO_2, through the using of fossil fuels, such as oil, natural gas, and coal. Hence, for realization of a low-carbon society, Japan has set a target for the cut down of CO_2 emissions of 80% by 2050.

With the tax for climate change mitigation, Japanese industry is required to measure CO_2 emissions in use of fossil fuels and must pay an additional amount of JPY289 per ton of CO_2 emissions based on the current tax rate. And to ensure further reductions in GHG emissions, the extra tax rate levied on CO_2 emissions is to be increased over three stages, with the tax first levied in October 2012, and to be increased every 2 years until April 2016. Also, as part of the tax households are now required to pay an additional JPY100 every month for average energy consumption, and revenue from the tax was estimated to be JPY39.1 billion for financial year (FY) 2012, the first year of its introduction, and is estimated to rise to JPY262.3 billion for each FY post-2016.

Other measures have also been taken to promote a low-carbon society, such as encouraging drivers to stop idling their cars for 5 min each day, which could lead to a 39 kg reduction in CO_2 emissions per year per driver, as well as encouraging the purchase of energy-saving home appliances, such as more efficient LED lightening, and also encouraging the purchase of hybrid motor vehicles.

2.14.4 Environment Assessment

Many countries have established an environmental impact assessment (EIA) system following enactment of the National Environmental Policy Act in the US in 1969. In Japan, the EIA system was first introduced in 1972, was fully enacted in June 1997, and implemented in 1999 (MoE 6).

Constructing roads and airports to improve transportation services, building dams to supply water, and establishing power plants to generate electricity are all necessary for citizens to have a comfortable life. However, these actions, obviously, generate negative environmental impacts. Consequently, the EIA system is necessary and important for the selection of environmentally responsible projects, and the EIA provides for the environmental impact of projects to be surveyed, forecasted, and evaluated in the process of designing the project, with the results open to public opinion from local governments and citizens. In this way the projects are developed by all interested parties, which is a more effective way to build a sustainable environment.

2.14.5 Environment and Education for Sustainable Development

To promote the importance of environmental conservation activities and encourage sustainable development by all parties in Japan, including corporations, citizens, and private bodies, the Japanese government has enacted the Law for Enhancing Motivation on Environment Conservation and Promoting of Environmental Education (MoE 7). The basic goals of this policy are, firstly, that the concept of a sustainable society be realized through voluntary actions and environmental education, and, secondly, to provide direction for the measures necessary to support the required voluntary actions. As such, the Japanese government has undertaken measures to support environmental education, for instance, the human resources ability certification, which provides information and support systems for the general public, private organization, and businesses to encourage the positive disclosure of environment-related information. The Government of Japan has also instigated and promoted an environmental education "Anytime, Anywhere, and Anyone" program to educate people to feel the environment, think for it, and act for it.

2.14.6 Environment and Research and Technology

The MoE and the National Institute for Environmental Studies (NIES) have jointly set up a Web site of Environmental Technology Information (NIES 1) and have begun to provide to the public information about environmental technology. Also, the Environmental Information Centre is operated by the MoE and NIES, and aims to disclosure more information about and promote technologies beneficial to the environment so as to improve public awareness, and includes information on environmental technologies as posted by developers and vendors of the technologies, updating of the latest environmental news, including announcement dates, provides easily understandable explanations and instructions of environmental technologies to attract public attention, and provides links to Web sites containing environmental technology information for those that are curious, and is an efficient way for individuals to retrieve environmental technology information they are searching for.

2.15 GLOBAL ENVIRONMENT

Japan's MoE has passed a number of laws and bills designed to protect the global environment, in addition to the Basic Environment Law and other laws and regulations discussed above. One of these laws is the Law Concerning the Promotion of Measures to Cope with Global Warming (MoE 8), which aims to stabilize greenhouse gas concentrations in the Earth's atmosphere at levels that will prevent dangerous anthropogenic interference with the planet's climate system. The MoE has stated that it is crucial for all nations to tackle the problem of global warming voluntarily and actively. In this regard the MoE aims to promote measures to cope with global warming by defining the responsibilities and measures to be taken by the national and local governments, businesses, and citizens. By establishing a basic policy on measures to cope with global warming, the MoE contributes to ensuring healthy lives for current and future generations and to the welfare of all human beings.

The MoE also aims to tackle global warming through enactment of a bill entitled the Bill on the Basic Act on Global Warming Countermeasures (MoE 9), which establishes a number of principles for global warming countermeasures. These countermeasures include (1) creating a society that can reduce greenhouse gas emissions while realizing sustained economic growth in order to ensure the prosperous lives of the people and competitiveness of industry, consistent with the MoE's establishment of a virtuous circle for the environment and economy of Japan discussed above; (2) engaging in active promotion of the countermeasures through international cooperation, and by bringing out knowledge, technology, and experience; (3) developing industries contributing to the mitigation of and adaptation to climate change, and expanding opportunities for job creation in order to ensure stable employment; (4) ensuring a stable energy supply, coordinated with energy-related matters; and (5) gaining a better understanding of the effects and impacts of global warming countermeasures on economic activities and people's daily lives.

2.16 WASTE AND RECYCLING

The MoE has made significant efforts in terms of the reduction, handling, and recycling of waste. Central to Japan's efforts in this regard is its 3Rs program, first begun in 1954 with the passing of the Public Cleansing law, that is designed to establish in Japan a sound material-cycle society. The 3Rs program is centered around Reduce, Reuse, and Recycle, meaning to reduce waste, and to reuse and recycle resources and products. In this regard, reducing means to choose to use things with care in order to reduce the amount of waste generated, while reusing involves repeated use of items or parts of items that still can be used, and recycling means the use of waste itself as a resource.

In order to become a sound material-cycle society and to achieve its aims regarding the 3Rs, the MoE has passed a number of laws. These laws include the Law for the Promotion of Sorted Collection and Recycling of Containers and Packaging (MoE 10), which is designed to promote the 3Rs for container and packaging waste by ensuring collaboration among all stakeholders, including national and local governments, business operators, and citizens, by promoting awareness among consumers and introducing measures to promote waste reduction among business owners, including retailers. Additionally, other related laws that have been passed include the Law for the Promotion of Effective Utilization of Resources (MoE 11). This law is designed to enhance measures for recycling goods and resources by implementing collection and recycling of used products by business entities, reducing waste generation by promoting resource savings and ensuring longer life products, and implementing measures for the reusing of parts recovered from collected products and reducing industrial waste by accelerating reduction of byproducts.

Also, there is the End-of-Life Vehicle (ELV) Recycling Law (MoE 12), as there are approximately four million ELVs in Japan every year, and the Construction Material Recycling Law (MoE 13), which is necessary as construction waste accounts for 20% of the total waste put into Japan's landfill sites and for over 70% of illegally dumped waste in the country.

Another such law is the Law for the Recycling of Specified Kinds of Home Appliances (Home Appliance Recycling Law) (MoE 14). This law is necessary as almost half of the postconsumer use home appliances discharged by households are discarded at landfill sites without any treatment. And since postconsumer use home appliances contain useful resources such as iron, aluminum, and glass, and that the remaining capacity at landfill sites is getting smaller and smaller, reduction of waste has become an urgent issue, and reduction and recycling of these types of wastes are all the more urgent.

Additionally, there is the Law for Promotion of Recycling and Related Activities for the Treatment of Cyclical Food Resources (Food Waste Recycling Law) (MoE 15). The need for this law has arisen because a large amount of food waste, namely dead stock and leftovers, is generated throughout the process of manufacturing, distribution, and consumption of food; and the volume of food waste is said to be approximately 30% of the total discharge of municipal solid waste. The Food Waste Recycling Law stipulates the responsibilities of all stakeholders in terms of recycling of food resources and recycling activities that should be carried out by food-related business entities and that business entities and consumers are required to be positively involved in reduction of food waste.

2.17 AIR AND TRANSPORTATION

A number of laws have been enacted by the MoE aimed at improving air quality, the most important of which is the Air Pollution Control Act (MoE 16). The aim of this act is to protect the health of citizens and the living environment from air pollution by controlling emissions of soot and smoke, volatile organic compounds, vehicle emissions, and particulates associated with business activities. Further, the Offensive Odor Control Law (MoE 17) has been enacted to regulate offensive odors emitted from businesses by designating regulated areas, establishing regulation standards, allowing for inspections and measurements of odors, and making recommendations and orders for improvements. The law also specifies legal penalties and promotes preventative measures against offensive odors in daily life by specifying the responsibilities of citizens, national and local governments, and business proprietors.

The Noise Regulation Law (MoE 18) is another law related to air quality that aims to regulate noise generated by factories and other types of work, such as construction work, and by setting maximum permissible levels of motor vehicle noise, as well as specifying penalties for breaches of the law.

Other similar laws include the Vibration Regulation Law (MoE 19) which is designed to preserve the living environment and people's health by regulating vibrations generated by the operation of factories and other work and construction sites, and by road traffic. And, finally, there is the law Concerning Special Measures for Total Emission Reduction of Nitrogen Oxides from Automobiles in Specified Areas (MoE 20). The aim of this law is to clarify the responsibilities of national and local governments, enterprises, and citizens toward preventing the problem of air pollution generated by nitrogen oxides emitted from automobiles. The law also specifies areas most adversely affected by such pollution fundamental policies and plans for reducing the total volume of vehicle-emitted nitrogen oxide, establish nitrogen oxide emission standards for specific automobiles that are registered in those areas, and restricting the amount of nitrogen oxide emissions resulting from the use of automobiles for business activities.

2.18 WATER, SOIL, AND THE GROUND ENVIRONMENT

The MoE has three main pieces of legislation related to water, soil, and the ground environment, the first of which is the Water Pollution Control Law (MoE 21). The purpose of this law is to prevent pollution of public water areas by regulating effluent discharge by factories or establishments in these areas, thereby protecting human health and preserving the living environment.

The second major piece of legislation in this area is the Law Concerning Special Measures for the Conservation of Lake Water Quality (MoE 22). This law deals with fundamental issues concerning water quality for designated lakes, including conserving and contributing to water quality through regulations and other measures, as well as implementation of projects to conserve water quality. Additionally, the law specifies regulations to reduce pollutant loads in designated lakes, with pollutant load restrictions on newly built or expanded factories and other business establishments, effluent regulations, and controls on the structure and operation of designated facilities such as livestock pens and the like.

The Soil Contamination Countermeasures Act (MoE 23) is the third main piece of legislation in this area. The purpose of this law is to implement countermeasures against soil and ground contamination by formulating means to grasp the situation of soil contamination by designated hazardous substances and prevent harm to human health resulting from such contamination. The designated hazardous substances that the law is aimed at include lead, arsenic, trichloroethylene, and similar substances that have harmful effects on human health when present in soil.

2.19 HEALTH AND CHEMICALS

To protect the health of humans and the living environment against the release of dangerous chemicals, the MoE has enacted a number of laws, the first being the Law Concerning Reporting, etc., of Releases to the Environment of Specific Chemical Substances and Promoting Improvements in Their Management (MoE 24). This law is designed to promote the voluntary improvement of businesses' management of specific chemical substances and to prevent any impediment of environmental protection by requiring businesses handling such substances to report the release to the environment of these substances and to provide technical information on the properties and handling of such substances. The law also states that due attention should be paid to trends in international cooperation on the management of chemical substances for environmental protection, as well as to scientific knowledge relating to chemical substances, and to the conditions relating to the manufacture, use, and other handling of chemical substances. The types of substances targeted by the law are chemical substances that may be hazardous to human health and/or may impair the life and growth of flora and fauna, including those which may form substances by naturally occurring chemical transformations or may deplete the ozone layer and increase the amount of solar ultraviolet radiation reaching the Earth.

The MoE has also passed the Agricultural Chemicals Regulation Law (MoE 25). This law is designed to improve the quality of agricultural chemicals and to ensure their safe and proper use by introducing an agricultural chemical registration system. The agricultural chemicals covered by the law include fungicides, insecticides, and other substances used to control fungi, nematodes, mites, insects, and rodents or other plants and animals that may damage crops, and substances and agents that promote or suppress the physiological functions of crops, etc.

Additionally, the Act on the Evaluation of Chemical Substances and Regulation of Their Manufacture, etc., (MoE 26) has been passed and is designed to establish a system for evaluating the properties of new chemical substances before their manufacture or import in order to prevent environmental pollution by chemical substances that pose a risk of impairing human health or of interfering with the population and/or growth of flora and fauna. The MoE has also enacted the Law Concerning Special Measures Against Dioxins (MoE 27), which is designed to consider the effects on human life and health caused by dioxins, and which specifies the basis of policies on dioxins and establishes regulations and measures relating to soil contamination to prevent and remove environmental pollution caused by dioxins.

2.20 NATURE AND PARKS

The MoE has enacted a number of pieces of legislation to protect Japan's natural and park environments, the first of which is the Natural Park Act (MoE 28), which aims to protect places of natural scenic beauty and conserve and sustain biological diversity, and the health, recreation, and culture of the Japanese people. The Act sets out the responsibilities of the state in regards to national parks, quasinational parks, and prefectural natural parks, including marine parks, the protection and utilization of these parks, ecosystem maintenance and recovery work, scenic landscape preservation, park management organization, and penal provisions for breaches of the regulations specified in the Act.

The Law for the Promotion of Nature Restoration (MoE 29) was enacted to protect and conserve nature and parks; and it establishes the basic principles of nature restoration, defines the responsibilities of effecters, and stipulates necessary matters for the implementation of nature restoration, including establishment of a basic policy for nature restoration. The intention of the law is to secure biodiversity via the creation, restoration, conservation, and maintenance of the conditions of rivers, marshes, tidal flats, seaweed and sea grass beds, community-based woods, rural landscapes, forests, and other natural environments in order to recover the ecosystem and natural environments that have been damaged or destroyed in the past.

The Invasive Alien Species Act (MoE 30) aims to regulate actions such as raising, planting, storing, carrying, and importing invasive alien species (IAS) and to mitigate IAS that are already existing in Japan to prevent damage to biodiversity, human safety, and agriculture, where "invasive alien species" means individuals (including eggs and seeds) and their organs that exist outside their original habitats as a result of introduction into Japan (i.e., nonindigenous species) from overseas and cause adverse effects on ecosystems, such as java mongooses, snapping turtles, and raccoons.

2.21 CONCLUSION

Prior to the Kyoto Protocol on Climate Change, environmental regulations tended to emerge following a particular environmental disaster. However, post the Kyoto Protocol countries have been implementing a series of environmental regulations, and in this chapter we document the environmental regulations that appeared in the United States and Japan. Although there are similarities in the type of environmental policies, we find that there are country-specific policies as environmental problems/priorities differ across countries. The implication is that policy makers must identify the environmental problems in their countries before implementing new environmental policies that will address these issues.

REFERENCES

Environment Protection Agency (EPA)
EPA 1, Significant New Alternatives Policy (SNAP), http://www.epa.gov/ozone/snap/.
EPA 2, Toxic Substances Control Act, http://www2.epa.gov/laws-regulations/summary-toxic-substances-control-act.
EPA 3, 1990 Clean Air Act (CAA), http://epa.gov/oar/caa/caaa_overview.html.
EPA 4, Acid Rain Program, http://www.epa.gov/AIRMARKETS/programs/arp/index.html.
EPA 5, Asbestos Hazard Emergency Response Act (AHERA), http://www2.epa.gov/asbestos/asbestos-laws-and-regulations.
EPA 6, Asbestos Information Act (AIA), http://www2.epa.gov/asbestos/asbestos-laws-and-regulations.
EPA 7, Asbestos School Hazard Abatement Reauthorization Act (ASHARA), http://www2.epa.gov/asbestos/asbestos-school-hazard-abatement-reauthorization-act-1990-one-hundred-first-congress-united.
EPA 8, Safe Drinking Water Act (SDWA), http://www.epa.gov/safewater/sdwa/sdwa.html.
EPA 9, Containing Materials in Schools Rule, & EPA 9, Asbestos Ban and Phase-out Rule, http://www2.epa.gov/asbestos/asbestos-laws-and-regulations.
EPA 10, Worker Protection Rule, http://www2.epa.gov/asbestos/protecting-workers-asbestos.
EPA 11, NESHAP, http://www2.epa.gov/asbestos/asbestos-neshap.
EPA 12, Federal Insecticide and Rodenticide Act (FIFRA), http://www.epa.gov/agriculture/lfra.html.

EPA 13, Pesticide Registration Improvement Act (PRIA), http://www.gpo.gov/fdsys/pkg/STATUTE-118/pdf/STATUTE-118-Pg3.pdf.

EPA 14, Federal Food, Drug, and Cosmetic Act (FFDCA), http://www2.epa.gov/laws-regulations/summary-federal-food-drug-and-cosmetic-act.

EPA 15, Food Quality Protection Act of 1996 (FQPA), http://www.epa.gov/pesticides/regulating/laws/fqpa/.

EPA 16, Guidance Cumulative Risk Assessment for Pesticides with a Common Mechanism of Toxicity, http://www.epa.gov/oppfead1/trac/science/cumulative_guidance.pdf.

EPA 17, Guidance for Performing Aggregate Exposure and Risk Assessment, http://www.epa.gov/scipoly/sap/meetings/1999/february/guidance.pdf.

EPA 18, Mercury and Air Toxics Standards (MATS), http://www.epa.gov/mats/.

EPA 19, National Ambient Air Quality Standards (NAAQS), http://www.epa.gov/air/criteria.html.

EPA 20, Resource Conservation and Recovery Act (RCRA), http://www2.epa.gov/laws-regulations/summary-resource-conservation-and-recovery-act.

EPA 21, Surface Mining Control and Reclamation Act (SMCRA), http://www.epa.gov/osw/nonhaz/industrial/special/fossil/.

EPA 22, Greenhouse Gas Reporting Rule (GGRR), http://www.epa.gov/climate/ghgreporting/basic-info/index.html.

EPA 23, Greenhouse Gas Data Publication Tool, http://www.epa.gov/ghgreporting/ghgdata/reportingdatasets.html.

EPA 24, Standards of Performance for Greenhouse Gas Emissions for New Stationary Sources: Electric Utility Generating Units, http://www2.epa.gov/carbon-pollution-standards.

EPA 25, Greenhouse Gas Tailoring Rule, http://www.epa.gov/nsr/ghgpermitting.html.

EPA 26, National Pollutant Discharge Elimination System (NPDES), http://water.epa.gov/polwaste/npdes/.

EPA 27, Clean Water Act, http://www2.epa.gov/laws-regulations/summary-clean-water-act.

EPA 28, NPDES Permit Program, http://water.epa.gov/polwaste/npdes/.

EPA 29, Emergency Planning and Community Right-to-Know Act (EPCRA), http://www2.epa.gov/epcra/what-epcra.

EPA 30, TRI Toxic Chemical List, http://www2.epa.gov/toxics-release-inventory-tri-program/tri-listed-chemicals.

EPA 31, Risk Management Plan, http://www2.epa.gov/rmp.

EPA 32, RCRA, http://www.epa.gov/epawaste/inforesources/online/index.htm.

EPA 33, NPDES, http://water.epa.gov/polwaste/npdes/.

EPA 34, Federal Environmental Requirements for Construction, http://www.epa.gov/oecaerth/resources/publications/assistance/sectors/constructmyer/myerguide.pdf.

EPA 35, Endangered Species Act, http://www2.epa.gov/laws-regulations/summary-endangered-species-act.

EPA 36, Rule and Implementation Information for Gasoline Distribution MACT and GACT, http://www.epa.gov/ttnatw01/gasdist/gasdispg.html.

EPA 37, National Emission Standards for Hazardous Air Pollutants: Organic Liquids Distribution, (Non-gasoline), http://www.epa.gov/ttnatw01/orgliq/orgliqpg.html.

EPA 38, Non-road Diesel Tier 4 Rule, http://www.epa.gov/otaq/nonroad-diesel.htm.

EPA 39, Discharge of Oil Regulation, http://www.epa.gov/oem/content/lawsregs/sheenovr.htm.

EPA 40, Facility Response Plan Rule, http://www.epa.gov/OEM/content/frps/index.htm.

EPA 41, National Oil and Hazardous Substances Pollution Contingency Plan (NCP), http://www2.epa.gov/emergency-response/national-oil-and-hazardous-substances-pollution-contingency-plan-ncp-overview.

EPA 42, NCP Product Schedule, http://www2.epa.gov/emergency-response/alphabetical-list-ncp-product-schedule-products-available-use-during-oil-spill.

EPA 43, Superfund Reportable Quantities (RQ), http://www.epa.gov/superfund/policy/release/rq/.

EPA 44, Comprehensive Environmental Response, Compensation, and Liability Act (CERCLA), http://www.epa.gov/superfund/policy/cercla.htm.

EPA 45, Chemical Accident Prevention Provisions, http://www.epa.gov/oem/lawsregs.htm.

EPA 46, Risk Management Program Policy and Guidance, http://www2.epa.gov/rmp/guidance-facilities-risk-management-programs-rmp.

EPA 47, 40 CFR, http://www2.epa.gov/laws-regulations/regulations.

ICAO

ICAO 1, CAEP/6, http://www.icao.int/environmental-protection/pages/CAEP.aspx.

ICAO 2, CAEP/8, http://www.icao.int/environmental-protection/pages/CAEP.aspx.

Ministry of the Environment (MoE)

MoE 1, The Basic Environment Law and Basic Environment Plan, http://www.env.go.jp/en/laws/policy/basic_lp.html.

MoE 2, Vision for a Virtuous Circle for Environment and Economy in Japan toward a Healthy, Rich and Beautiful Environmentally-Advanced Country, https://www.env.go.jp/en/policy/economy/vvceej.pdf.

MoE 3, Outline of the Fundamental Plan for Establishing a Sound Material-Cycle Society, https://www.env.go.jp/en/press/2004/0408a-01.pdf.

MoE 4, Environmental Accounting Guidelines 2005, http://www.env.go.jp/en/policy/ssee/eag05.pdf.

MoE 5, Details on the Carbon Tax (Tax for Climate Change Mitigation), https://www.env.go.jp/en/policy/tax/env-tax/20121001a_dct.pdf.

MoE 6, Environmental Impact Assessment in Japan, https://www.env.go.jp/en/policy/assess/pamph.pdf.

MoE 7, Law for Enhancing Motivation on Environmental Conservation and Promoting of Environmental Education, https://www.env.go.jp/en/laws/policy/edu_tt.pdf.

MoE 8, Law Concerning the Promotion of the Measures to Cope with Global Warming, http://www.env.go.jp/en/laws/global/warming.html.

MoE 9, Overview of the Bill of the Basic Act on Global Warming Countermeasures, https://www.env.go.jp/en/earth/cc/bagwc/overview_bill.pdf.

MoE 10, Law for the Promotion of Sorted Collection and Recycling of Containers and Packaging (Container and Packaging Recycling Law), http://www.env.go.jp/en/laws/recycle/07.pdf.

MoE 11, Law for the Promotion of Effective Utilization of Resources, http://www.env.go.jp/en/laws/recycle/06.pdf.

MoE 12, Law for the Recycling of End-of-Life Vehicles (End-of-Life Vehicle Recycling Law), http://www.env.go.jp/en/laws/recycle/11.pdf.

MoE 13, Construction Material Recycling Law, http://www.env.go.jp/en/laws/recycle/09.pdf.

MoE 14, Recycling of Specified Kinds of Home Appliances (Home Appliance Recycling Law), http://www.env.go.jp/en/laws/recycle/08.pdf.

MoE 15, Law for Promotion of Recycling and Related Activities for the Treatment of Cyclical Food Resources (Food Waste Recycling Law), http://www.env.go.jp/en/laws/recycle/10.pdf.

MoE 16, Air Pollution Control Act, http://www.japaneselawtranslation.go.jp/law/detail/?id=2146.

MoE 17, The Offensive Odor Control Law, http://www.env.go.jp/en/laws/air/offensive_odor/index.html.

MoE 18, Noise Regulation Law, http://www.env.go.jp/en/laws/air/noise/index.html.

MoE 19, Vibration Regulation Law, http://www.env.go.jp/en/laws/air/vibration/index.html.

MoE 20, Concerning Special Measures for Total Emission Reduction of Nitrogen Oxides from Automobiles in Specified Areas, http://www.env.go.jp/en/laws/air/amobile.html.

MoE 21, Water Pollution Control Law, http://www.env.go.jp/en/laws/water/wlaw/index.html.

MoE 22, Law Concerning Special Measures for the Conservation of Lake Water Quality, http://www.env.go.jp/en/laws/.

MoE 23, Soil Contamination Countermeasures Act, http://www.env.go.jp/en/laws/water/sccact.pdf.

MoE 24, Law Concerning Reporting, etc. Of Releases to the Environment of Specific Chemical Substances and Promoting Improvements in Their Management, http://www.env.go.jp/en/laws/chemi/prtr/index.html.

MoE 25, Agricultural Chemicals Regulation Law, http://www.env.go.jp/en/chemi/pops/Appendix/05-Laws/agri-chem-laws.pdf.

MoE 26, Act on the Evaluation of Chemical Substances and Regulation of Their Manufacture, etc. (Chemical Substances Control Law), http://www.env.go.jp/en/laws/chemi/index.html.

MoE 27, Law Concerning Special Measures against Dioxins, http://www.env.go.jp/en/laws/chemi/dioxin.pdf.

MoE 28, Natural Park Act, http://www.env.go.jp/en/laws/nature/law_np.pdf.

MoE 29, Law for the Promotion of Nature Restoration, http://www.env.go.jp/en/laws/nature/law_pnr.pdf.

MoE 30, Invasive Alien Species Act, http://www.env.go.jp/en/nature/as/040427.pdf.

National Institute for Environmental Studies (NIES)

NIES 1, Environmental Technology Information, http://2050.nies.go.jp/LCS/eng/list.html.

US Census Bureau

US Census Bureau 1, North American Industrial Classification System (NAICS), http://www.census.gov/eos/www/naics/.

US Department of Labour

USDL 1, Occupational Safety and Health Administration (OSHA) Process Safety Management (PSM), https://www.osha.gov/Publications/osha3132.html.

CHAPTER 3

Efficiency of U.S. State EPA Emission Rate Goals for 2030: A Data Envelopment Analysis Approach

Greg N. Gregoriou[1], Vikash Ramiah[2]
[1]State University of New York, Plattsburgh, NY, USA; [2]School of Commerce, University of South Australia, Adelaide, SA, Australia

Contents

3.1 INTRODUCTION

Under the Obama administration the U.S. Environmental Protection Agency (EPA) put forward the Clean Power Plan's guidelines for each State of the union to control and diminish toxic CO_2 emissions that can harm life and the health of people emanating from U.S. fossil fuel power plants under section 111(d) of the Clean Air Act (EPA, 2014). Every State will be obliged to diminish their emissions to standards set forth by the EPA in pounds per megawatts beginning in 2020, with a final goal of 30% reduction to be reached by 2030 when compared to the 2005 base levels (Makhijani and Ramana, 2014). The decision of the EPA to provide each State with a number based on their proprietary formula which considers "…efficiency improvements at coal-fired power plants; increased generation from existing natural gas combined cycle units; generation from renewable energy sources and nuclear power; and demand side energy efficiency investments" (EPA, 2014). However, the key issue is that the EPA does not offer any suggestions or recommendations on how each State should reduce its CO_2 levels to attain the EPA's 2030 goals. According to the EPA, estimates of cutting emissions by 30% could be an inaccessible goal. Moreover, the EPA (2014) stipulates that each State must implement "four building blocks" namely by "(1) increasing the efficiency of coal plants, (2) increasing the use of high efficiency natural gas combined cycle plants, (3) producing electricity from low/zero emitting facilities, and (4) increasing the efficiency of demand-side energy." Each State can use whatever technique it decides is best to attain their goal using an amalgamation of various measures that indicates its particular situation and policy objectives. Individual states are in control of these programs and can use a wide range of tools to get to their stated EPA goals.

This chapter is proactive in nature in the sense that we examine which states in the US are more likely to minimize their goal using data envelopment analysis (DEA). DEA is a nonparametric technique used to examine the productivity or efficiency of decision-making units (DMUs, stated in this chapter) and has been widely applied in areas such as airports, prison, hospitals, mutual funds, and hedge funds. One original feature of this chapter is the application of DEA to environmental problems, whereby the goal is to shed some light on how the individual states will improve on their efficiency and productivity. This chapter is organized as follows: Section 3.2 describes the literature review in this area, Section 3.3 explains the methodology used, Section 3.4 presents the data used in this study, Section 3.5 reports the results, and Section 3.6 concludes the chapter.

3.2 LITERATURE REVIEW

Prior to the testing of the efficiency of the states, we review the literature to find out if our question has been answered and consequently this section of the book chapter focuses on environmental/sustainable[1] finance and environmental economics. Interestingly, we find that efficiency measures within the environmental/sustainable finance area are at its earliest stage, although attempts have been made to evaluate the efficiency of environmental policies. Henceforth, the first part of the literature review is dedicated toward the research conducted in terms of evaluating the efficiency of environmental regulation, and the second part looks at the well-established literature in the area of environmental economics. Given that both segments of the literature fail to answer our question, we adopt the DEA methodology from the environmental economics area to answer our question.

Ramiah et al. (2013) investigate the impact of 19 announcements of environmental regulation on the equities listed on the Australian Stock Exchange over the period 2005–2011. Although their primary goal was to assess the effects of these new regulations on the risk and returns, their findings show that these policies are not achieving their desired objectives in that the biggest polluters (electricity providers) were not negatively affected by stringent regulations and demonstrate sectors that are not heavy polluters

---

[1] We note that there is no clear definition for environmental finance and sustainable finance at present, and therefore we regard these two terms as equivalent. Researchers in Europe tend to use the term sustainable finance while researchers in Australia tend to use the term environmental finance.

(beverage—wine industries) were incorrectly negatively affected. Furthermore, they show that the political discussion around these regulations lead to an unnecessary risk such as the diamond risk. Feldman et al. (1996) on the other hand, analyze a sample of 300 U.S. firms to find out if investment in environmental management leads to reduction in risk and argue that there is a risk reduction coupled with an increase in stock price.

After the Kyoto Protocol on climate change in 2005, a series of green policies emerged around the world with the intention of protecting our planet (including the US). Ramiah et al. (2015) study the effects these policies have on the U.S. capital markets and document negative abnormal returns and increase in systematic risk for the biggest polluters. However, they argue that the effects are smaller for environmentally friendly businesses. Another finding of their study is the "Obama effect," whereby they discuss how the stock markets reacted more to environmental regulation post the election of President Barrack Obama. They show that Obama was associated with four main green policies, namely the new energy for America plan, the cap-and-trade system, the clean energy funds, and offshore drilling leases. Although these findings are not a direct measure of efficiency, we gather from their results that the policies were achieving their targets on the polluters' side but less so on environmentally friendly side.

An earlier study by Kahn and Knittel (2003) examines the effects of President George H. Bush's Clean Air Act Amendment proposal of 1989 on electricity and similar to Ramiah et al. (2013) and Veith et al. (2009) they fail to observe a negative reaction to stringent environmental policies. The explanations provided in the literature for why electricity providers (biggest polluters) are insensitive to environmental policies are (1) the electricity industry has regulated prices, (2) demand for electricity is inelastic, and (3) the ability for electric utility companies to pass on the costs onto consumers. Contrary to the evidence that we just provided, Bushnell et al. (2013) show that stock prices of firms within the carbon- and electricity-intensive industries declined following the introduction of the European Union Emission Trading System.

Most studies around environmental regulation concentrate around the regulations at a national level but Ramiah et al. (forthcoming-a) assess how international policies, domestic policies (in England, Northern Ireland, Scotland, and Wales), and nuclear announcements affect the UK equity markets. According to their results, international announcements have the largest impact although the market did react to domestic and nuclear announcements. They concluded their work by stating the following: there are abnormal returns around announcements of environmental policies (giving rise to the green effect market anomaly), environmental policies create a "hexagon shape" systematic risk structure, and question the effectiveness of these policies.

Environmental regulations tend to achieve different goals in different countries. For instance, environmental regulations are designed to save water in Africa but are designed to save lives in China. Zhang et al. (2008) argue that the total number of pollution-related deaths is around 420,000 in China, and a study conducted by Tanaka (2010) shows that environmental regulation has led to significant reductions in air pollution and the infant mortality rate (25,400 fewer infants died per year). On the one hand, we can view the live savings as an indication that the policies are achieving certain efficiency level but the total number of deaths implies that a higher level of efficiency is required. Ramiah et al. (fourthcoming-b) argue that the lack of enforcement of environmental regulation, legislative shortcomings, poorly designed policy instruments, an unsupportive work environment for environmental regulators, and a progrowth political and social environment are the main reasons for the failure of environmental regulations in China.

So far, all the papers discussed provide indirect evidence of the efficiency of environmental regulations and the finance papers usually employ event study methodology and asset pricing models to provide evidence in favor or against the green policies. Nevertheless, Pham et al. (2014) have developed a model to test whether European environmental regulations are excessive in France (we consider this method as the first direct attempt to measure the efficiency of environmental regulation using financial data and methodology). Using abnormal returns from event-study methodology and asset pricing models, they compare the abnormal returns of firms with the industry average (endogenous benchmark) and the market average (exogenous benchmark) to determine whether a regulation is excessive. According to their findings, European environmental regulations are not excessive.

Zhou et al. (2008) provide the empirical evidence that DEA has been extensively applied in energy and environment policy evaluation and Emrouznejad et al. (2008) reinforce this view. For instance, Sueyoshi and Goto (2011) use DEA approach to measure the unified (operational and environmental) efficiency of energy firms where it is assumed that energy firms generate desirable outputs such as electricity and undesirable outputs in terms of toxic CO_2. Furthermore, they separate inputs into energy and nonenergy inputs to measure the efficiency of Japanese fossil fuel power generation. According to Sueyoshi and Goto, the implementation of Kyoto Protocol (2005) has not been effective on the efficiency of Japanese fossil fuel power generation. Other studies that used DEA are Dyckhoff and Allen (2001), Bevilacqua and Braglia (2002), Korhonen and Luptacik (2004), Picazo-Tadeo et al. (2005), Kumar (2006), Watanabe and Tanaka (2007), and Zhou and Ang (2008). What we gather from this rich environmental economics literature is, DEA is a well-established technique that can be applied to measure efficiency.

3.3 METHODOLOGY

DEA is a linear programming technique to estimate efficiency of DMUs via production frontiers. Using multiple inputs and outputs simultaneously as relative performance measures without the need for regression analysis in which DMUs are most and least efficient and presents the results in a single efficiency score. DEA measures the efficiency by considering a ratio that uses the outputs divided by the inputs—which is referred by Cook and Zhu (2013) to be a balanced benchmarking approach. DEA constructs a best practices frontier identifying efficient DMUs that lie on the frontier. It is important to acknowledge that we are using the DEA software from the www.deafrontier.net Web site for our experiment.

Following Charnes et al. (1978) (also referred to as the CCR model), the input-oriented simple and cross efficiency models with constant returns-to-scale (CRS) are used to rank the states that are efficient in reducing their inputs in a proportionate fashion while keeping the outputs fixed. Employing each State as a DMU and with the use of a minimization process, we identify the best practices frontier by the "efficient states" that are best at transforming the summation of outputs divided by inputs (Charnes et al., 1981; Golany and Roll, 1994). Any State that is not efficient (score of less than 1) is not on the best practices frontier, while efficient states with a score of 1 (simple efficiency) lie on the best practices frontier. The implication is that a State that is inefficient must adopt ways to reach the best practices frontier to be considered as an "efficient State." Moreover, when a State has attained a score of 75%, it implies that the State is 75% of where most efficient states are. When the score is higher than 1, it implies that the State has achieved better than efficient level (super efficiency). As suggested by Charnes et al. (1981), the minimum number of DMUs to use in an analysis must equate to three times the number of input and output variables in the experiment to generate reliable estimates. We use 49 States which are more than sufficient for the DEA analysis.

3.3.1 Envelopment Model

From a conceptual point of view, DEA is a technique that measures the relative performance of DMUs. We use the notation of an input-oriented CRS model developed by Charnes et al. (1978) and reproduce the input-oriented CRS model, text, and notation from Cook and Zhu (2013, p. 42).

$$\theta^* = \min\theta$$

$$\sum_{j=1}^{n} \lambda_j x_{ij} \leq \theta x_{io} \quad i = 1, 2, \ldots, m;$$

$$\sum_{j=1}^{n} \lambda_j y_{rj} \geq y_{ro} \quad i = 1, 2, \ldots, s;$$

$$\sum_{j=1}^{n} \lambda_j = 1$$

In our experiment, DMU_o represents one of the n DMUs under evaluation and x_{io} and y_{ro} are the ith input and rth output for DMU_o, respectively.

3.3.2 Cross Efficiency

The main idea of the cross efficiency model as first discussed in Sexton et al. (1986) is to perform peer analysis as opposed to simply using the self-appraisal approach of the envelopment model (simple DEA). Cross efficiency ranks and calculates the efficiency score of each State n times using optimal weights from the DEA frontier software. When using the cross efficiency method, a matrix is generated which is equal to the number of states that are part of the investigation. Hence, the average of each column in the matrix is computed and the peer assessment score of each State is calculated, whereby the addition of the rows and columns totals the number of states under investigation. We reproduce the input-oriented CRS cross efficiency model, text, and notation from Cook and Zhu (2013, p. 278−279):

$$Max \ E_{dd} \ = \ \frac{\sum_{r=1}^{s} u_{rd} y_{rd}}{\sum_{i=1}^{m} v_{id} x_{id}}$$

$$s.t. \ E_{dj} \ = \ \frac{\sum_{r=1}^{s} u_{rd} y_{rj}}{\sum_{i=1}^{m} v_{id} x_{ij}} \leq, \quad j = 1, 2, ..., n. \tag{3.1}$$

$$u_{rd} \geq 0, r = 1, ..., s.$$

$$v_{id} \geq 0, i = 1, ..., m.$$

where v_{id} and u_{rd} represent the ith inputs and rth output weights for DMU_d. Therefore, the cross efficiency of DMU_j, using the weights that DMU_d has selected in model 1, which is

$$E_{dj} \ = \ \frac{\sum_{r=1}^{s} u_{rd}^{*} y_{rj}}{\sum_{i=1}^{m} v_{id}^{*} x_{ij}}, d, \quad j = 1, 2, ..., n$$

where (*) denotes optimal values in Equation (3.1). For DMU_j, ($j = 1, 2,..., n$) an average of all $E_{dj}(d = 1, 2,..., n)$,

$$\overline{E}_j \ = \ \frac{1}{n} \sum_{d=1}^{n} E_{dj},$$

referred to as the cross efficiency score for DMU_j.

3.3.3 Super Efficiency

When the State which is being investigated is not part of the reference set of the simple DEA models, we obtain the super efficiency model (Zhu, 2015). Super efficiency was developed by Andersen and Petersen (1993) and enables us to break the tie of states that are efficient when using the simple DEA envelopment model. In other words, if five DMUs are efficient with a score of 1, then super efficiency will rank them from best to worst, with the highest score being the best. We reproduce the super efficiency model, text, and notation from Cook and Zhu (2013, p. 214–215):

$$\min \theta^{super}$$

subject to

$$\sum_{\substack{j=1 \\ j \neq 0}}^{n} \lambda_j x_{ij} \leq \theta^{super} x_{io} \quad i = 1, 2, \ldots, m;$$

$$\sum_{\substack{j=1 \\ j \neq 0}}^{n} \lambda_j r_{rj} \geq \gamma_{ro} \quad i = 1, 2, \ldots, s;$$

$$\lambda_j \geq 0 \quad j \neq 0.$$

3.4 DATA

Data from 49 states are obtained from the EPA Web site, and the available 2012 data set is used along with the 2030 goals of the EPA. Three inputs are used in the input-oriented CRS model and one output. The inputs are (1) emissions in million metric tons; (2) energy output terawatt hours (TWh); (3) fossil rates lbs per megawatt hour (lbs/MWh); (4) fossil, renewable, and nuclear rate (lbs/MWh) and the only output is the 2030 State goal (lbs/MWh). "The State of Vermont and the District of Columbia do not have emission rate goals, because they do not have electric generating units affected by the proposal in their jurisdictions (Ramseur, 2014, p. 1)."

3.5 RESULTS

Table 3.1 reports the findings of our analysis. When we look at the simple DEA model we find that only five states will be efficient in reducing their inputs to attain their 2030 EPA goals, while a large majority will experience difficulties in becoming efficient—implying more actions are required to achieve their objectives. The five efficient states (score equal to 1) using simple DEA are Alaska, Hawaii, Maine, North Dakota, and Rhode Island (see third column of Table 3.1).

If we use the cross efficiency model which evaluates the states on all the inputs, we find that the top five efficient states are (1) Rhode Island with a score 0.99519, (2) Hawaii with an efficiency score of 0.96146, (3) North Dakota has a score of 0.95976, (4) Montana scores 0.88576, and (5) Alaska has a score of 0.87274 (see column 4 of Table 3.1). At first

Table 3.1 Efficiency Scores

DMU No.	State (DMU)	Simple Efficiency	Cross Efficiency	Super Efficiency
1	Alabama	0.84299	0.75722	0.84299
2	Alaska	1.00000	0.87274	1.83270
3	Arizona	0.55358	0.51465	0.55358
4	Arkansas	0.63787	0.59758	0.63787
5	California	0.85988	0.73892	0.85988
6	Colorado	0.72934	0.68543	0.72934
7	Connecticut	0.80737	0.78142	0.80737
8	Delaware	0.83564	0.78746	0.83564
9	Florida	0.71182	0.61218	0.71182
10	Georgia	0.63721	0.58017	0.63721
11	Hawaii	1.00000	0.96146	1.08883
12	Idaho	0.87469	0.65119	0.87469
13	Illinois	0.75486	0.68492	0.75486
14	Indiana	0.91854	0.82874	0.91854
15	Iowa	0.93741	0.85163	0.93741
16	Kansas	0.86443	0.82631	0.86443
17	Kentucky	0.95524	0.86677	0.95524
18	Louisiana	0.69339	0.63824	0.69339
19	Maine	1.00000	0.86711	1.00080
20	Maryland	0.72684	0.70450	0.72684
21	Massachusetts	0.71121	0.67886	0.71121
22	Michigan	0.78436	0.71638	0.78436
23	Minnesota	0.66415	0.61450	0.66415
24	Mississippi	0.71212	0.66283	0.71212
25	Missouri	0.90949	0.83293	0.90949
26	Montana	0.92076	0.88576	0.92076
27	Nebraska	0.84216	0.81196	0.84216
28	Nevada	0.74535	0.70946	0.74535
29	New Hampshire	0.62049	0.59385	0.62049
30	New Jersey	0.64737	0.61908	0.64737
31	New Mexico	0.75082	0.72618	0.75082
32	New York	0.63419	0.57933	0.63419
33	North Carolina	0.68946	0.63442	0.68946
34	North Dakota	1.00000	0.95976	1.03610
35	Ohio	0.83606	0.75149	0.83606
36	Oklahoma	0.73084	0.66941	0.73084
37	Oregon	0.59069	0.53671	0.59069
38	Pennsylvania	0.78432	0.69020	0.78432
39	Rhode Island	1.00000	0.99519	1.15443
40	South Carolina	0.55034	0.51861	0.55034
41	South Dakota	0.79495	0.66500	0.79495
42	Tennessee	0.70128	0.66196	0.70128
43	Texas	0.69428	0.57251	0.69428
44	Utah	0.84115	0.80144	0.84115
45	Virginia	0.70947	0.66850	0.70947
46	Washington	0.32079	0.28116	0.32079
47	West Virginia	0.92864	0.85530	0.92864
48	Wisconsin	0.75118	0.70727	0.75118
49	Wyoming	0.92221	0.86940	0.92221

glance the results generated seem alarming due to the low number of efficient states. However, at closer look we find that few states are below the 50% mark with a handful having low scores. According to our results, the states with lower scores will have to adopt stringent environmental measures to achieve their 2030 EPA goals. The last column of Table 3.1 displays the results for the super efficiency model, whereby Alaska, Rhode Island, Hawaii, North Dakota, and Maine are super-efficient States, with the champion State being Alaska.

3.6 CONCLUSION

Environmental/sustainable finance is an emerging area, and currently it is not equipped to directly test the efficiency of environmental policies. Attempts have been made to evaluate the performance of the recent environmental policies but they are more indirect tests of performance. The first conclusion that can be drawn in this chapter is that there is a need to develop financial models and to use financial data to evaluate the performance of environmental policies.

Although DEA is a well-established technique and some empiricists still believe that it is at its infancy stages. In this chapter we successfully apply three DEA approaches to emission rates of individual states to see which ones are more likely to reach their goal as a result of their efficiency methods. Interestingly we find that very few states are likely to achieve their objectives implying that most states will have to adopt more stringent policies in order to achieve their targets in the coming years as the current ones are far from being adequate.

ACKNOWLEDGMENT

We thank Professor Joe Zhu for allowing us to use his DEA software available at www.deafrontier.net.

REFERENCES

Andersen, P., Petersen, N.C., 1993. A procedure for ranking efficient units in data envelopment analysis. Manag. Sci. 39 (10), 1261–1264.
Bevilacqua, M., Braglia, M., 2002. Environmental efficiency analysis for ENI oil refineries. J. Clean. Prod. 10, 85–92.
Bushnell, J.B., Chong, H., Mansur, E.T., 2013. Profiting from regulation: evidence from the European carbon market. Am. Econ. J. Econ. Policy 5 (4), 78–106.
Charnes, A., Cooper, W.W., Rhodes, 1978. Measuring the efficiency of decision making units. Eur. J. Oper. Res. 2 (6), 429–444.
Charnes, A., Cooper, W.W., Rhodes, E.L., 1981. Evaluating program and managerial efficiency: an application of DEA to program follow through. Manag. Sci. 27 (6), 668–697.
Cook, W.D., Zhu, J., 2013. Data Envelopment Analysis: Balanced Benchmarking. CreateSpace Independent Publishing Platform Amazon Digital Services.
Dyckhoff, H., Allen, K., 2001. Measuring ecological efficiency with data envelopment analysis (DEA). Eur. J. Oper. Res. 132, 312–325.
Emrouznejad, A., Parker, B.R., Avares, G., 2008. Evaluation of research in efficiency and productivity: a survey and analysis of the first 30 years of scholarly literature in DEA. Socio-Econ. Plan. Sci. 42, 151–157.

EPA, 2014. Factsheet: Clean Power Plan Framework: National Framework for the United States, Setting Goals to Cut Carbon Pollution. Available at: http://www2.epa.gov/carbon-pollution-standards/factsheet-clean-power-plan-framework.

Feldman, S., Soyka, P., Ameer, P., 1996. Does Improving a Firm's Environmental Management System and Environmental Performance Result in a Higher Stock Price? Washington, DC: ICF Kaiser. for Docket EPA-HQ-OAR-2013-0602. Available at: https://www.nirs.org/climate/background/ieercomments regardingepacpp12114exhibita.pdf.

Golany, B., Roll, Y., 1994. Measuring efficiency of power plants in Israel by data envelopment analysis. IEEE Trans. Eng. Manag. 41 (3), 291−301.

Kahn, S., Knittel, C.R., 2003. The Impact of the Clean Air Act Amendments of 1990 on Electric Utilities and Coal Mines: Evidence from the Stock Market. Centre for the Study of Energy Markets, University of California (Working Paper 118), Berkeley.

Korhonen, P.J., Luptacik, M., 2004. Eco-efficiency analysis of power plants: an extension of data envelopment analysis. Eur. J. Oper. Res. 154, 437−446.

Kumar, S., 2006. Environmentally sensitive productivity growth: a global analysis using Malmquist−Luenberger index. Ecol. Econ. 56, 280−293.

Makhijani, A., Ramana, M.V., 2014. Comments by the Institute for Energy and Environmental Research on the 2014 Proposed Clean Power Plan of the U.S. Environmental Protection Agency. In: Submitted to the United States Environmental Protection Agency of Data Envelopment Analysis, vol. 32. Jossey-Bass, San Francisco, 73−105.

Pham, H.N.A., Ramiah, V., Moosa, M., 2014. Are European environmental regulations excessive? In: 2014, PBFEAM Conference, Japan.

Picazo-Tadeo, A.J., Reig-Martinez, E., Hernandez-Sancho, F., 2005. Directional distance functions and environmental regulation. Resour. Energy Econ. 27, 131−142.

Ramiah, V., Martin, B., Moosa, I., 2013. How does the stock market react to the announcement of green policies? J. Bank. Finance 37, 1747−1758.

Ramiah, V., Morris, T., Moosa, I., Gangemi, M.,Puican, L. The effects of announcement of green policies on equity portfolios: evidence from the United Kingdom. Manag. Auditing J., forthcoming-a.

Ramiah, V., Pichelli, J., Moosa, M. The effects of environmental regulation on corporate performance: a Chinese perspective. Rev. Pac. Basin Financial Mark. Policies, forthcoming-b.

Ramiah, V., Pichelli, J., Moosa, M., 2015. Environmental regulation, the Obama effect and the stock market: some empirical results. Appl. Econ. 47 (7), 725−738.

Ramseur, J.L., 2014. State CO2 emission rate goals in EPA's proposed rule for existing power plants. Congressional Research Service, Washington, D.C. Available at: http://fas.org/sgp/crs/misc/R43652.pdf.

Sexton, T.R., Silkman, R.H., Hogan, A.J., 1986. Data envelopment analysis: critique and extensions. In: Silkman, R.H. (Ed.), Measuring Efficiency: An Assessment.

Sueyoshi, T., Goto, M., 2011. DEA approach for unified efficiency measurement: assessment of Japanese fossil fuel power generation. Energy Econ. 33 (2011), 292−303.

Tanaka, S., December 2010. Environmental Regulations in China and Their Effects on Air Pollution and Infant Mortality. POPOV Research Network. Available at: http://poppov.org/portals/1/documents/papers/80.tanaka.pdf.

Veith, S., Werner, J., Zimmermann, J., 2009. Capital market response to emission rights returns: evidence from the European power sector. Energy Econ. 31, 605−613.

Watanabe, M., Tanaka, K., 2007. Efficiency analysis of Chinese industry: a directional distance function approach. Energy Policy 35, 6323−6331.

Zhang, M., Song, Y., Cai, X., Zhou, J., 2008. Economic assessment of the health effects related to particulate matter pollution in 111 Chinese cities by using economic burden of disease analysis. J. Environ. Manag. 88 (4), 947−954.

Zhou, P., Ang, B.W., 2008. Linear programming models for measuring economy-wide energy efficiency performance. Energy Policy 36, 2911−2916.

Zhou, P., Ang, B.W., Poh, K.L., 2008. A survey of data envelopment analysis in energy and environmental studies. Eur. J. Oper. Res. 189, 1−18.

Zhu, Z., 2015. Quantitative Models for Performance Evaluation and Benchmarking: Data Envelopment Analysis with Spreadsheets. Springer, New York, NY.

SECTION 2

Environmental Economics

CHAPTER 4

Environmental Water Governance in the Murray-Darling Basin of Australia: The Movement from Regulation and Engineering to Economic-Based Instruments

Claire Settre[1], Sarah Ann Wheeler[1,2]
[1]Global Food Studies, University of Adelaide, Adelaide, SA, Australia; [2]School of Commerce, University of South Australia, Adelaide, SA, Australia

Contents

Handbook of Environmental and Sustainable Finance
ISBN 978-0-12-803615-0

4.1 INTRODUCTION

Many countries around the world are faced with the challenge of allocating scarce water between increasing competing demands while protecting the hydrological needs of the environment. This challenge is heightened by the nonuniform change in water supply patterns due to climate change (IPCC, 2013) and in many cases, a history of overallocation and undervaluing of water.

The Murray-Darling Basin (MDB) in Australia represents a unique case study in water policy as it has undergone a transition from command-and-control government management to a more pluralistic governance structure that incorporates the state, market, and civil sector. This process demonstrates qualities of adaptive market-based governance, meaning that institutional arrangements have evolved to satisfy the needs of the community in an environment of change (Hatfield-Dodds et al., 2007). Adaptive governance is polycentric in nature, balancing top-down and bottom-up governance in order to effectively deal with uncertainty and change through flexibility and learning (Olsson et al., 2006).

This chapter investigates the changing paradigms employed in Australian water management to illustrate the transition from being predominantly based on regulation and engineering solutions, to a movement toward economic-based instruments and environmental water governance.

4.2 WATER MANAGEMENT GOVERNANCE TOOLS

4.2.1 Demand and Supply Water Management Strategies

The actions that required meeting the challenge of increased demand for water, growing populations, and a shrinking water supply will include a range of strategies. In particular, water policy strategies can be classified as either supply strategies (building infrastructure such as new dams or piped networks to address water scarcity problems, or by designing systems to provide clean water supplies), or through demand management (changing the behavior of people in regard to water or its interrelated systems). In particular, this includes:

- *Supply strategies*: (a) improved infrastructure and network systems for water management and transport; (b) technological solutions, including improved water and land management approaches, covers for dams, etc.
- *Demand strategies*: (a) voluntary measures (e.g., change through education); (b) regulation and planning requirements (e.g., restrictions on fertilizers and chemicals applied near watercourses, and water meters); (c) economic instruments (e.g., water markets, prices, taxes, and subsidies for low-income households' water bills).

Traditionally, governments have tended to rely on supply strategies to address water policy relevant issues and they have ignored a variety of tools available to influence the use and management of water. The growing inability (and cost) to use supply strategies as a way to solve water scarcity also led to the adoption of various demand management

Table 4.1 Water Demand Management and Governance Tools

Demand Instrument	Description	Water Management Examples
Command-and-control (regulation)	This approach relies on restrictions of water use or behavior.	• Metering. • Quotas. • Restrictions on farming actively near waterways. • Gardening restrictions/bans. • Maximum water use regulations. • Timing restrictions.
Voluntary measures	Providing information to water users about the costs and benefits of their actions to encourage voluntary behavioral change.	• Education. • Signs on water savings achieved as a community. • Detailed information on water bills. • Rewards for water-saving gardens.
Economic instruments	Assumes that water consumption or production can be influenced by water charges. Subsidies change water demand for irrigation by providing greater incentives for other (less water intensive) production. Tradable rights can be created through institutional property rights being established.	• Water charges on abstraction/use and other inputs used such as electricity. • Subsidizing low-income households to adopt water-saving technologies, subsidizing dry-land production, or subsidizing more efficient water infrastructure. • Water markets—allowing water to be transferred from one user to another.

instruments. Since the 1980s, there is a growing recognition that economic instruments should be used to regulate the access to and the use of water resources. Needless to say, there are significant difficulties encountered by policy makers and water managers in using economic instruments. Some of these difficulties are highlighted in this chapter. Table 4.1 describes demand instruments in more depth.

The following section provides detail of our case study and explores the evolution of water management and governance in Australia.

4.3 THE MURRAY-DARLING BASIN

The MDB (1,061,469 km^2) is a large area covering all or part of four states and one territory of Australia. It has Australia's largest inland river system, and the country's largest

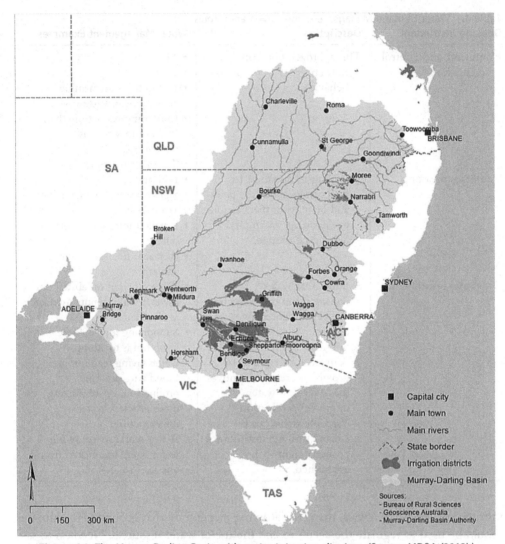

Figure 4.1 The Murray-Darling Basin with major irrigation districts. *(Source: MDBA (2012).)*

river, the River Murray. The Basin's resources support extensive agriculture which produces 30% of Australia's food. Within the MDB are a range of internationally important ecological assets, including 1 world heritage site and 16 wetlands listed on the Ramsar Convention on Wetlands of International Significance (MDBA, 2010) (Figure 4.1).

The hydrology of the MDB is characterized by high variability and aridity with long periods of prolonged drought. From 2002 to 2007, Australia was affected by the "Millennium" drought which caused significant economic and ecological damage to

Basin resources. The Millennium drought led to some of the most significant changes in water policy in Australia, which are detailed later in this chapter (Wheeler, 2014). Under future climate change scenarios, surface water availability in the Basin is likely to decrease, especially in the southern MDB, threatening ecological health and agricultural viability (CSIRO, 2008).

Historically, water management in the MDB was closely tied with national development objectives concerned with growth and productivity and was allocated through administrative or political procedures (Bell and Quiggin, 2008). Currently, there is hope that water allocation in Australian is transitioning to sustainability, and has been accompanied by a diverse array of policy mechanisms, including political and economic means of delivering environmental outcomes. In particular, an innovative policy innovation is the use of water markets to secure water for the environment (discussed further in the next section). This is achieved by government purchase of entitlement water from willing sellers which is managed to achieve ecological outcomes in the MDB. The commitment exhibited by the Australian government to the use of market mechanisms to achieve environmental outcomes is unparalleled internationally (Garrick et al., 2011). As such, the study of the changing paradigms in Australia's water policy and the resulting governance arrangements leading to this point provides key insights into the reasons and drivers of water management policy success.

4.4 PHASES IN AUSTRALIA'S WATER MANAGEMENT IN THE MDB

There are four main phases that can be traced out of Australia's MDB water management: (a) establishment and exploration phase; (b) development and supply phase; (c) scarcity phase; and (d) the application of demand strategies.

4.4.1 The Establishment and Exploration Phase

After European settlement, there followed a period of exploring and establishing water resources in the MDB. The ruling of Australia as *terra nullius* ("land belonging to no one") meant that indigenous water rights were not recognized and water management proceeded following the European traditions and objectives. At this time, riparian rights inherited from British common law formed the institutional basis for water management (Ward, 2009). The riparian doctrine permits landholders with land adjacent to a water course the right to make reasonable use of the water, provided it does not unreasonably interfere with other riparian rights holders. However, it became clear that the riparian doctrine inherited from British Common Law did not suit the largely arid and highly variable nature of Australia's water resources.

In 1886, the *Irrigation Act* was enacted and replaced the inherited riparian doctrine with state control of water, which was managed through state irrigation trusts (Crase et al., 2015). Soon after, the Australian states were federated and the Constitution of

Australia (1901) formally vested the right to water in the individual states. Each state exhibited a *command-and-control style governance* over water resources, the legacy of which still exerts influence on water management today.

Foreshadowing future conflicts, water governance was a point of debate even in the early stages of development of the Australian Constitution (Clark, 2002). Controversy about upstream water extraction (Sim and Muller, 2004) concerns about water and soil quality degradation due to overstocking and frequent drought challenged traditional European farming ideals that were inherent in early water institutions in Australia.

4.4.2 The Development and Supply Phase

Following the federation of the Australian states, water management in Australia entered the development phase characterized by state led control and a rapid and optimistic expansion of water supply infrastructure. Throughout the twentieth century, water provision was a strategic resource in the industrialization and agricultural intensification across many rapidly expanding economies, including Australia.

Throughout this time, Australia adopted a "state hydraulic model." Typical of this style of management, Australia's water policy was characterized by planning for growth, command-and-control regulation, state ownership of resources, and universal provision. The expansion of irrigated agriculture enabled by supply provision was closely entwined with national development objectives and underpinned much of the economic growth of the MDB. The close link to national policy and the state's right to water meant that state governments took a lead role in water resource management and top-down governance prevailed thought-out this phase.

This period saw the construction of large infrastructure projects across the MDB, such as Hume Dam (1934); the Goolwa Barrages (1940); Dartmouth Dam (1979); Snowy Mountain Scheme (1974) as well as 13 weirs and locks along the river (1922–1939). The development of water resources has since transformed the Basin into a series of storages, now capable of storing 22,214 GL of water at full capacity across the Basin, though all storages are seldom full (MDBA, 2014).

Often the large infrastructure projects were not subject to any cost–benefit analysis or political opposition (Ward, 2009) and went ahead without impact assessments owing to the lack of legislative obligation to consider external consequences of development (Paterson, 1987). Infrastructure development received a wide political and financial support due to their necessity for growth in the agricultural and industrial sectors.

In-line with the state hydraulic model, water was considered unequivocally as a public good (Ward, 2009). This thinking led to the generous outlay of water entitlements to promote the expansion of irrigated agriculture and establishment of towns and cities. The present difficulty in rectifying overallocation in the Basin highlights the ongoing legacy of these ideas and has led to unparalleled reforms in water resources management.

Ongoing conflict between states in the lower MDB about water use continued during the early stages of river development. In a tentative first step to coordination of resource use, the River Murray Waters Agreement passed into law simultaneously by the Commonwealth and the three states of the southern MDB in 1915. This milestone marked the first of many interstate agreements on the management of scarce water resources in the MDB (Wheeler, 2014). However, the River Murray Waters Agreement did not mention the environmental concerns that were beginning to emerge (Sim and Muller, 2004).

Similar patterns of water supply provision can be seen within this history of other OECD countries throughout the twentieth century; particularly in the US which has undergone a comparable transition from supply-led solutions to a more mature water economy (Howe, 1997). Australia and the US store the highest dam storage of water per capita around the world, followed closely by Brazil and China (Grey and Sadoff, 2006).

4.4.3 The Scarcity Phase and Factors Contributing to the Introduction of Economic Instruments

Beginning in the 1960s and gathering strength in the 1980s, a range of factors combined to provide an impetus for policy change and the formal introduction of water markets into the Australian water policy landscape. Up until this point, market mechanisms had been largely excluded from the water sector worldwide due to the fear of large externalities, such as negative impacts on return flow and downstream water users. Failure to internalize for these externalities and address additional market hurdles, such as high-transaction costs and information asymmetry, can lead to market inefficiency (Chong and Sunding, 2006). The introduction of market mechanisms was a clear sign of a change in policy objectives, signaling a transition from national development objectives toward aggregate economic efficiency objective (NWC, 2011). Selected factors contributing to the introduction of markets are discussed below.

4.4.3.1 The Rising Cost of Supply Management

A key factor in the transition to demand-based water management policies was the lack of willingness of governments to fund large-scale infrastructure projects. This was partially a result of the sharply rising cost of supply (Randall, 1981) as many of the more viable options for increasing supply had been exhausted during the development phase (NWC, 2011). Continued government involvement in supply provision was being questioned, as was the merits of the developments themselves (Davidson, 1969). The growing emphasis on demand-based policies signified that the Australian water sector was becoming a mature water economy. Characteristic of this, the cost of developing new supply infrastructure was becoming inelastic (Randall, 1981). As Australia's water economy continued to mature, the entrance of economists and environmentalists into a realm traditionally dominated by engineers meant that the involvement and remit of the water sector increased (Musgrave, 2008; Ward, 2009), leading to the development of more integrated networks. Catalyzed by government fiscal pressures (NWC, 2011), cost—benefit analyses were increasingly undertaken on large infrastructure projects and irrigation schemes for the first time.

4.4.3.2 Environmental Concern

In addition to waning political will and the cost of new generating new supplies, concerns about the environmental impact of large-scale water infrastructure were beginning to be established in public discourse (Musgrave, 2008). The issue was compounded by the recognition of overallocation of water entitlements[1] in some states of the MDB. The ecological damage and economic limits caused by the undervaluing and overallocating water in the Basin began to be questioned in the late 1960s (e.g., Davidson, 1969). In the years following, it became clear that the institutions formed throughout the establishment and the development of Australian water resources (both the constructed formal institutions and the inherited social perceptions) had left a legacy of high extractive demands far above the levels of ecological sustainability. In response, South Australia placed a moratorium on the issue of new water licenses in 1969; New South Wales placed a full embargo in 1981; and Victoria restricted the use of unregulated stream use in 1968 (Wheeler et al., 2014a). Water scarcity conditions were exasperated in 1982—1983 by severed rainfall deficiencies in the Victoria and New South Wales, known as the South East Australia drought.

In 1987, the original River Murray Waters Agreement (1915) was superseded by the MDB Agreement which established the Murray-Darling Basin Commission (MDBC). Though the MDBC had limited environmental responsibilities, the resulting Salinity and Drainage Strategy produced by the MDBC in 1989 provided one of the earliest formal environmental management schemes in the Basin and signaled changing policy direction and growing environmental concern. The 1987 agreement was in turn superseded by the MDB Agreement (1992) which displayed a pronounced shift toward the idea of balancing environmental water requirements with consumptive use (Connell and Grafton, 2008). In these early stages of environmental water management, changes in government and policy were largely spurred by exogenous variables, such as drought and salinity concerns.

More widely, environmental concerns about the rate of development were internationally institutionalized with the release of the Bruntland Report (1987) which provided a compelling definition of sustainability. In addition the 1992 Earth Summit in Rio de

---

[1] A "water entitlement" is a perpetual or ongoing right to exclusive access to a share of water from a specified consumptive pool of water. Every water entitlement is allocated a specific yearly volume each water year called a "water allocation." The amount of water allocated to each entitlement depends on the amount of water available and the reliability of the "water entitlement." Both allocations and entitlements can be traded on the water market.

Janeiro brought water scarcity and ecosystem vulnerability to the forefront of international concern with the declaration of the Dublin Principles. The four principles form the basis of integrated water resource management (IWRM) which strives to manage social and economic development without compromising the sustainability natural ecosystems (GWP, 2012). IWRM principles are prolific in many progressive water management strategies and many reforms in the MDB provide examples of IWRM.

4.4.3.3 Macroeconomic Factors

The shifting focus from supply augmentation to demand-based policies in the MDB occurred among a background of wide ranging political liberalization and changing trade agreements. Up until the late twentieth century, Australia had been largely sheltered from international prices due to high import tariffs and embargos on the importation of agricultural products (Musgrave, 2008). However, a general trend of reduced government involvement in agriculture and trade in the 1980s and 1990s meant that government became less likely to invest in agricultural supply provision. The gradual curtail of government protection over Australian agriculture cumulated in international agreement in 1994 following the Uruguay Round of Negotiations initiating reform in agricultural trade, fully exposing Australia to international commodity prices at the farm level, resulting in the need for additional flexibility to navigate global supply and demand patterns.

The trend toward market instruments over command-and-control management was institutionalized in Australia's law in 1994 through the Council of Australian Government (COAG) reforms which introduced Australia's competition policy. Despite some criticisms, water was included in Australia's competition policy and made a tradable and marketable good. As is the case in Australia, this time period signified a transition from "state to market" in most OECD countries (Haggard and Webb, 1994) resulting from the waning institutional power of the state and erosion of authority from traditional political sources (Roger and Hall, 2003).

4.5 THE DEVELOPMENT AND EXTENSION OF ECONOMIC INSTRUMENTS IN THE MDB

The addition of economic instruments to the toolkit used to address overallocation in the MDB and the resultant institutional changes are examples of market-based water governance. Governance as a broad social system is a more inclusive concept than "government," as it embraces the relationships between the state, the market, the private sector, and civil society (Roger and Hall, 2003). Far from a balanced tripartite partnership, we see that management of environmental water in the MDB has often been a contentious push-and-pull between the three. In understanding governance, it is important to conceptualize governance as a *process* rather than an *instrument* used to achieve specific objectives of a policy strategy (Castro, 2007). In doing so, we can recognize the *process* of water reform in Australia in itself as a process of governance.

In the case of the MDB, the introduction of markets meant that water allocation was no longer solely prescribed by administrative top-down regulations but distributed in a way which could flexibly respond to price signals, changing supply and demand profiles, and irrigator preferences. The more pluralistic nature of market-based governance fostered a dual focus on economic and environmental sustainability in the Basin.

Kingdon (1995) stresses the importance of timing and coordination to successfully transition to a new form of governance. The resulting policy window is a function of the combination of three factors, including (1) the recognition of a problem; (2) the availability of a solution; and (3) political will to enact change (Kingdon, 1995). When this occurs at critical points in time, this can be used to transform governance regimes. Looking at Australian water history, it could be argued that many "windows of opportunities" have arisen and overlapped to instigate change. For the purpose of this chapter, we will broadly argue that in the case of the MDB in the late twentieth century, water scarcity was among the problems; economic instruments provided part of the solution; and the combined social, economic, and environmental factors discussed in the previous section provided adequate political will. These three factors combined have allowed the transition to adaptive market-based governance and has spurred ongoing reforms concerned with the introduction and expansion of market mechanisms in Australia's environmental water management.

Additional discussion about water market function, water prices, and trade behavior within Australia's water market can be found in Chapter 5 of this book.

4.5.1 Overview of Twenty-First Century MDB Water Reform

Environmental water resource management in the Basin in the twenty-first century is a highly complex problem of sustainability (Foerster, 2011). As the complexity of issues has increased, so too has the complexity of the institutional structures and has resulted in ongoing reform to achieve effective governance of the ecological system. In the case of the MDB, the introduction of market mechanisms into the management of environmental and consumptive water began the transition to an adaptive and flexible market-based governance regime (Table 4.2).

4.5.2 Development of Water Markets in the MDB

Water markets are thought by many economists to play a vital role in the efficient allocation of water resources in the MDB (Wheeler et al., 2014b). Unofficial water trade in Australia first stemmed from water scarcity during drought. There are a number of reports of water "swapping" that occurred in the 1940s drought. South Australia stopped issuing new water licenses in the late 1960s, followed by other states later, and these restrictions

Table 4.2 Overview of Significant Water Reforms in the MDB

Year	Initiative	Description	Stage
1886	The Irrigation Act	Provides for state control of water. Water management carried out by private irrigation trusts.	Exploration and expansion
1901	Federation of the Australian States	Section 100 of the Australian Constitution allows states "reasonable use" of water resources. States exhibit command-and-control management of water.	Development and supply
1915	River Murray Waters Agreement	First transboundary water sharing agreement between Victoria, New South Wales, and South Australia.	
1985	Murray-Darling Basin Ministerial Council established	Comprises of Ministers from each Basin state responsible for decision making regarding policy water shares, funding, and integrated policy.	Maturation and scarcity
1987	The Murray-Darling Basin Agreement	Supersedes River Murray Waters Agreement and creates Murray-Darling Basin Commission (MDBC). Address limited environmental issues, mostly salinity.	
1989	Salinity and Drainage Strategy	Sets out specific salinity reduction targets and management strategies to reduce future soil and water salinization.	
1994	Council of Australian Governments (COAG) Water Reform Framework	Instructs the development of necessary institutional arrangements to facilitate water trade and the recognition of the environment as a legitimate user.	
1997	The cap arrangements put into place	Introduced to limit all future extractions to levels of 1993/94 levels in response to predictions of future unsustainable extractions.	

Continued

Table 4.2 Overview of Significant Water Reforms in the MDB—cont'd

Year	Initiative	Description	Stage
2004	National Water Initiative (NWI)	Continues and reinforces 1994 COAG reforms and prescribes state-level objectives for water sharing, water trading, and water pricing.	
2007	The National Plan for Water Security (NPWS)	Committed AUD$10 billion to return water to the environment through market mechanisms and infrastructure upgrade. Instigates legislation pertaining to the referral of state powers to the Commonwealth.	Transition to sustainability
2007	The Water Act	Establishes the Murray-Darling Basin Authority (MDBA) which replaces the MDBC. MDBA charged with the development of The Basin Plan.	
2010	Water for the Future	A restructure of NPWS, committing AUD$3.1 billion to purchasing water entitlements from willing sellers through the water market to be used in environmental watering. Referred relevant state powers to the Commonwealth.	
2012	The Basin Plan	The plan developed by the MDBA is passed through federal parliament into law and aims to return extraction in the Basin to sustainable levels.	

did lead to informal MDB markets for temporary water in the 1960s and 1970s (Connell, 2007; NWC, 2011). The first official markets for temporary water were officially trialed in the early 1980s, and trade between private diverters and district irrigators was allowed in 1995.

Water rights in Australia are defined as the right to access a share or "entitlement" of water from a consumptive pool. This water entitlement can be traded (permanent trade). Each entitlement yields a seasonal volumetric allocation which can also be traded, which

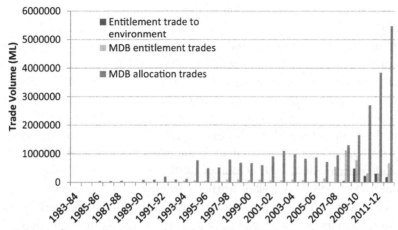

Figure 4.2 Volume of water allocation and entitlement trade in the southern MDB. *(Source: adapted from data in NWC (2011, 2013).)*

is known as water allocation trade (temporary trading). An allocation to an entitlement prescribes the amount that can be extracted by the right holder within a season and put to beneficial use. The seasonal allocation may vary annually depending on the reliability of the entitlement and the seasonal conditions.

Currently, water trading is most active in the southern MDB (the largest hydrologically connected area of the MDB) where trade in water allocations (temporary water) are bought and sold by irrigators as a risk management tool (Zuo et al., 2014). Water markets are now a common place farm management tool, with 86% of irrigators having engaged in at least one trade since 2010/11 (Wheeler et al., 2014). Entitlement trade is less prominent and entitlement sale is often a strategic decision such as farm exit or restructures (Wheeler et al., 2012).

As can be seen in Figure 4.2, trade in the southern MDB has increased significantly over time. Allocation trade is the largest form of trade. Entitlement trade to the government has increased since the inception of the Commonwealth Environmental Water Holder in 2007/08.

The use of market mechanisms in environmental management is a landmark decision, as environmental management has been historically regulatory in nature (Weal, 1992). The next sections detail the various stages of reform contributing to the transition to adaptive market-based governance in the MDB.

4.5.3 COAG Reforms and the "Cap"

In 1994, the Council of Australian Governments (COAG) instigated what would become an ongoing water reform process lasting over two decades. Notably, the water reform framework established a system of tradable property rights for water to facilitate flexible reallocation and the recognition of the environment as a legitimate water user (COAG, 1994).

As part of the COAG reforms and the ongoing concern over the environmental implications of overallocation, a water audit of the MDB was undertaken. The audit found that water extractions had significantly increased since 1988 and were likely to continue to increase unsustainably if extraction was not curbed. It asserted that modifications to existing water arrangements must be amended to prevent major adverse impacts on the health of the MDB (MDBMC, 1995). In response to the water audit, a temporary cap was introduced to limit all future extractions to levels of 1993/94 levels and was made permanent in 1997 (Connell, 2007).

The cap on extraction, as well as further restrictions such as limits on out-of-basin trade, played a role in activating underused (dozer) or unused (sleeper) licenses. This had the effect of temporarily increasing water use in the Basin as holders of sleeper and dozer licenses entered the market to meet demands of water users impacted by the cap. Further, as the aggregate availability of water was constrained by the cap, water users began to explore additional ways of obtaining water, including increased ground-water use and farm dams to capture unregulated flows. This had the ultimate effect of reducing return flow to the river and further contributing to river water availability concerns. Far from establishing a self-regulating cap-and-trade market with the introduction of the cap, state involvement became important for the role of regulating private extractions from streams, and metering and charging use of groundwater and farm dams (Bell and Quiggin, 2008).

Despite the negative changes to return flows and the temporary increase in water use, the cap represented a strong commitment by the federal government to address water scarcity in the Basin and a milestone in cooperation between the Basin states (Wheeler et al., 2014b). In a review of the cap operation it was found that while the cap did not necessarily provide for sustainable use of Basin resources, it was an essential first step without which degradation of Basin resources would have been significantly worse (MDBC, 2000).

4.5.4 The National Water Initiative (2004)

The National Water Initiative (NWI) is a transboundary agreement signed by New South Wales, Victoria, South Australia, and the Australian Federal Government. In 2004, the NWI was signed and continued the process of water reform instigated by the COAG reforms in 1994. The NWI replaced the fragmented state-based control of water resources with a state-federal management system supported by joint agreements, federal objectives, and individual state commitments enshrined in legislation (Foerster, 2011). By signing the agreement, governments made a commitment to address eight interrelated aspects of water management to achieve a more integrated approach to

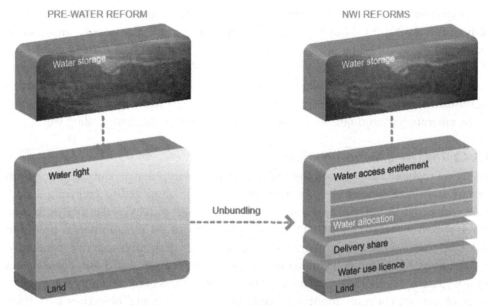

Figure 4.3 Unbundling of water rights. *(Source: NWC (2011), p. 88.)*

the planning, management, and trading of water in Australia (NWC, 2004). Among other key objectives, the NWI sought to reform the structure of water rights and address overallocation and river restoration, as discussed below.

4.5.4.1 Unbundling of Water Rights

A key objective of the NWI was the removal of barriers to trade through the restructure of water rights, such as changing the definition of a right from a volumetric entitlement to a share of a common pool resource (NWC, 2004). In addition, the NWI prescribed the removal barriers to trade such as the bundling of land and water rights. In doing so, the NWI removed the legal ties between water from land to allow each to be bought or sold separately (Crase et al., 2015). Further and more nuanced unbundling of water rights themselves was instigated and the following four were aspects of a water right were constructed: (1) a water access entitlement; (2) a water allocation account; (3) a delivery share ensuring the farmers have a right to get the water delivered to their property; and (4) a water use entitlement, giving farmers the right to use the water. The history of water licenses, commercial farming, and established user owned irrigation systems facilitated unbundled water trading with relative ease. Figure 4.3 provides an overview of the unbundling process.

The unbundling of land and water was a politically charged issue due to fears that the separation of land and water would result in the privatization of water rights and the emergence of water barons. There were also concerns about the negative impact of water markets on the environment resulting from changes to the natural timing and location of river flows (Wheeler et al., 2014b). Though concerns continue to be voiced, evidence suggests that the impact of water trading is small compared to the impacts of drought and river regulation and that key ecological assets in the southern MDB have not been significantly impacted by water trading (NWC, 2012b). In addition, tradable water rights provided policy makers with an additional policy tool to reallocated water from consumptive use to the environment through the purchase of water entitlements from willing sellers.

4.5.4.2 The Living Murray Program

The NWI required state commitment relating to the environmental concern arising from scarcity and overallocation. In particular, governments committed to preparing "water plans with provision for the environment" and to dealing "with over-allocated or stressed water systems" (NWC, 2004). However, commitment to environmental provision makes up only a single component of a comprehensive water reform agenda, concerned with a variety of objectives including urban water management, expansion of trade, and improved storage and delivery, to name a few. In the vastness of the reform agenda and its many objectives, concerns about the sidelining of environmental objectives have been validly raised (Connell, 2007).

The NWI planned to return consumptive water to the environment through dual mechanisms. Namely, an increase in water use efficiency and the use of the Australian entitlement water market to buy water entitlements from willing sellers. *The Living Murray* (TLM), initiated by the Murray-Darling Basin Ministerial Council (MDBMC) was developed in response to evidence of the continued ecological decline of the Basin environment and was a key water recovery policy mechanism included the NWI. The program identified and targeted six environmental icon sites across the Basin for environmental works and measures and the delivery of environmental water (MDBA, 2014). The first step involved a commitment of 500 GL to be delivered to the six icon sites by 2009 with a budget allocation of AU$500 million. TLM and the NWI were initially presented as a combined package as they were agreed upon in the same COAG meeting (2004). Lack of coordination and tension between the policies, particularly regarding the volume of water required to provide adequate environmental flows (Connell and Grafton, 2011). Though the NWI was effective in setting out high-level national objectives such as the watering of environmental sites, large amounts of supporting detail, and enforcement were left to the states. The degree of commitment and pace of implementation differed between states, resulting from varying state interests and benefit of reforms. Achieving a common approach on highly politicized issues of the reform agenda, such as the expansion of water markets and purchase of environmental water were particularly difficult to coordinate (Foerster, 2011). Hesitation among some states prevailed, as it is a difficult political decision for one state to unilaterally decrease water consumption if efforts are not matched by the other Basin states. As the reform continued, it became evident that individual states were unable to manage the ongoing and pressing ecological issues in the Basin and that coordinated state and federal action was required, leading to the development of the 2007 *Water Act*.

4.5.5 The Water Act and Institutional Change (2007)

The year 2007 was a watershed (excuse the pun!) and turbulent year in Australian water resource management due to the increased urgency in the need for proactive national policy to address the continued drought. Folke et al. (2005) suggest that ecological crisis or abrupt change, such as drought, can provide an opportunity to catalyze the transition to new forms of governance. In the case of governance of MDB water resources, the Millennium drought provided a catalyst for political will to embrace alternative policy tools such as market mechanisms to recover water for the environment and provide additional strategic management tools to irrigators (Wheeler, 2014).

However, the governance arrangements as prescribed by the *Water Act* are still essentially a top-down approach. The legislation established the independent Murray-Darling Basin Authority (MDBA) and endowed it with functions and powers needed to ensure that the Basin's resources are managed in a sustainable and integrated way (Water Act, 2007). The Act legislated the shift of limited state powers and responsibilities to the MDBA (Foerster, 2011). For the first time in Australian constitutional history, the Basin was managed and overseen by a federal semi-independent authority rather than the individual states. The MDBA, which replaced the MDBC in 2008, was charged with the responsibility of developing a plan (The Basin Plan) to secure the future sustainability of water resources in the MDB. For the next 3 years the MDBA worked on the Basin Plan, amid much political change and scientific debate.

In addition, the *Water Act* also strategically established the Commonwealth Environmental Water Office (CEWO) (previously the Commonwealth Environmental Water Holder) responsible for managing the water entitlements purchased and owned by the Commonwealth. While the creation of the MDBA and CEWO increased the speed of environmental water recovery and better defined objectives, the addition of a central federal governing body added an extra layer of institutional complexity to the governance landscape (Foerster, 2011), and added greater distance between the development of high-level objectives and subcatchment scale water planning. Similar policy experiences with the use of market mechanisms to recover water for the environment in the US suggest a need for a nested governance approach—one which balances local decision making with high-level policy objectives (Garrick et al., 2011).

4.5.6 Water for the Future Initiative (2010)

The Water for the Future (WFF) initiative (2010), originally called the National Plan for Water Security (NPWS) (2007) represents a milestone in Australian water market-based governance. The NPWS was released in 2007 following 5 years of ongoing drought and committed AU$10 billion over 10 years to increase production with less water and to improve environmental outcomes.

Where previously the right to water was vested in the states, relevant constitutional powers were agreed to be referred to the Commonwealth to enable it to manage the MDB in the national interest (McCormick, 2007). Additional controversy resulted from the government's emphasis on market mechanisms to address overallocation in the MDB.

Following a change of federal government in 2007, the NPWS was rebadged as a new intergovernmental agreement on water management; the WFF initiative (2010). As with the NPWS, the WFF initiative was a 10-year program involving a commitment of AU$12.9 billion from state and federal governments. The initiative built upon many of the guiding principles of the NPWS, however, differed from NPWS in the explicit reference to climate change impacts and adaptation strategies. Following the template set out by previous reform objectives, the WFF uses dual mechanisms of market instruments and infrastructure upgrades to achieve environmental watering targets. In many cases, subsidies were offered to farmers to encourage investments to modernize irrigation. To that end, AU$5.8 billion was provided to rural water use and infrastructure projects to achieve increase on-farm water use efficiency (DEWHA, 2010). An additional AU$3.1 billion was committed to purchasing water entitlements from willing sellers through the water market to be used in environmental watering for the period 2007/08 to 2016/17.

The market-based aspect of the program committed to return overallocated resources to the environment was labeled Restoring the Balance in the Murray-Darling Basin (RtB). The program aims to buy water entitlements from willing sellers, which were mostly selected using a series of competitive tenders since 2008 (Wheeler and Cheesman, 2013). The allocation water assigned to the water entitlement is used to improve the health of the Basin through watering of key environmental assets such as nationally significant wetlands, floodplains, and rivers.

The RtB program is also the key market mechanisms used to return water to the environment under the Basin Plan (see the following section). The Basin Plan aims to return water to the environment through dual mechanisms, market instruments, and infrastructure upgrades, to return water extractions to a sustainable level.

Despite the relative success of environmental buybacks, critiques of the program assert that it exploits financial difficulties faced by farmers after years of drought and therefore exploits "desperate sellers" rather than engaging "willing sellers." The counter evidence to this is that desperate sellers were able to sell their water at a high price at a time they were struggling on their farms. Wheeler and Cheesman (2013) find that the majority of irrigators who sell entitlement water to the government did so to reduce debt or increase farm income, as well as strategic planning such as farm restructure or farm exit and disposal of surplus water. Hence, the sale of water either facilitated some to leave the industry, or others to pay down debt to stay in the industry.

4.5.7 The Basin Plan (2012)

The years following the establishment of the MDBA marked a confluence of best science, policy, and participatory action aimed at developing the Basin Plan. The research leading up to the release of the Basin Plan essentially asked, among others, a question which up until that point had been neglected: what is the sustainable use of Australia's water?

In developing the Basin Plan, the MDBA endeavored to embrace the public participation principle of IWRM and inclusiveness prescribed by the NWI to achieve informed environmental water decisions. In particular, the NWI recognized, for the first time, indigenous Australians as legitimate stakeholders in water planning and prescribes greater emphasis on indigenous water issues such as access for cultural and economic purposes (NWC, 2012a). Despite engaging the indigenous community in stakeholder meeting in 2009—2010, research has shown that indigenous Australians are still largely excluded from active participation in water governance and economic development of water resources (Tan and Jackson, 2013).

The release of the draft of the Basin Plan in 2011 was met with public outcry, political debate, and prolific media attention, of which was seldom positive. Water governance literature often describes governance as a depoliticized concept (Castro, 2007). The vocal involvement from the government and civil sector during the proposed changes to governance prescribed by the Basin Plan highlights that contrary to literature, governance change is an inevitably political process in a water scarce nation.

The final Basin Plan proposed a 2750 GL reduction in surface water diversions per year (MDBA, 2012). As of September 2014, over 60% of the amount required to "bridge the gap" between current diversion and sustainable diversion limits in the Basin has apparently been recovered (DoE, 2014a). This includes water which has been recovered through infrastructure upgrades, water donations from the Queensland Government, as well as environmental water recovered by state government agencies. This represents 2195 GL of water entitlements, which translates into a long-term average annual yield of 1510 GL per year (DoE, 2014b).

The dissimilarities between the draft Basin Plan and the final plan in 2012 demonstrate that processes such as this are best characterized as a political decision-making process rather than a scientific endeavor. In November 2012, the federal parliament passed the Basin Plan into law; despite several attempts to disallow it (Loch et al., 2014).

The path to passing the Basin Plan into law in 2012 has been fraught. The heterogeneity of water users, objectives, and state histories compounded the difficulty in coordinating the water-related agencies in the six different states. The legacy of the state's constitutional right to water continues in the narrative of environmental water, as states are cautious about the sale of water to the Commonwealth (Crase et al., 2015).

Despite political, scientific, and media quarrels, the Basin Plan has achieved a globally unprecedented commitment to the return of water to environment through infrastructure and market mechanisms (Garrick et al., 2011). To achieve this, the environment was introduced as a "water user" in its own right (NWC, 2004). Environmental water used for environmental and publically beneficial outcomes was awarded statutory recognition and same degree of security as water entitlements held for consumptive use (COAG, 2004). This is a significant and indeed quite rate policy innovation in other established water markets around the world. Water purchases from willing sellers to be used to water environmental assets are managed by the CEWO, established by the *Water Act 2007*.

In addition to market mechanisms, infrastructure upgrade continues to play a vital role in water-saving measures across the Basin. Subsidies to improve on-farm irrigation efficiency have continued since 2009, most recently with an announcement of a $350 million package to fund new irrigation efficiency projects in NSW in October 2014 (DoE, 2014b).

The resulting depth of and sophistication of integrated water management, including water trading and economic instruments for the management of environmental water, testifies to the long struggle of policy makers and farmers to find a solution to living with the variability of Australia's hydrological cycle.

4.6 FUTURE POLICY DIRECTIONS

Water markets increase the productive value of scarce water supplies by facilitating the flexible allocation of water from low to high value use through trade (Wheeler et al., 2014b). Over the past few decades, water markets have emerged as a response to increasing water scarcity and have become gradually embedded in the governance arrangements of water scarce regions. More recently, markets for tradable water rights have been used as a means of reallocating water to the environment through the government purchase of water entitlements from willing sellers.

It is difficult to conclusively say if Australia's environmental water management constitutes "good governance." Roger and Hall (2003) suggest that conditions for good governance are inclusiveness, accountability, participation, transparency, predictability, and responsiveness. One case where water management clearly omits a principle of good governance is indigenous representation in water management. Although the NWI prescribes greater inclusion and participation of indigenous Australians in water management and planning, research suggests that state consultation with indigenous communities on water planning is absent and that most jurisdictions have failed to meet NWI objectives (NWC, 2012a). Further criticism has been voiced owing to the low priority awarded to indigenous rights in the NWI (Tan and Jackson, 2013). To meet the NWI objectives and the participation prescribed in Roger and Hall's (2003) definition of "good governance" future management of Australian resources requires a reemphasis on indigenous participation. However, this is complicated by the lack of targeted research, heterogeneity in objectives, and conceptual hurdles to the establishment of "cultural flows" (Tan and Jackson, 2013).

The future management of water resources in the MDB is a policy problem plagued with high levels of uncertainty and complexity. The challenge is heightened by the legacy of historic institutions and the cumulative impact of previous governance decisions on the ecological and social state of the MDB. Future governance arrangements and the institutions which support them must recognize this uncertainty and complexity, as well as the long planning horizon required to achieve sustainability (Dovers, 2003). To achieve this, future management requires continued emphasis of adaptive governance, allowing improvement to the market-based governance regime and coordination between all levels of governance. This is supported by experience from environmental water trade in the Western United States which suggests a middle path between top-down control and bottom-up management of water resources (Garrick et al., 2011).

It is likely the case that future management of water resources will include a greater diversity of tradable water products. This will address any issues surrounding a lack of willingness to sell water entitlements, and increase the choices and flexibility of the CEWO. Options contracts, allocation trade, future bonds, and water banks provide alterative water products which may serve to better engage irrigator participation in the RtB process (Wheeler et al., 2013). Experience from other countries with these institutional setups, such as water banks in Mexico and water contracts in the Colorado Basin can provide institutional insight into possible applications to Australia's future challenges in water governance.

In addition to alternative water products, environmental water recovery may in itself need to be rethought. To date, environmental water recovery has focused on the volume of water as the indicator of environmental watering success (Wheeler et al., 2014b). Such a view obscures the complexity of environmental water management. The value of environmental water is a function of the timing, location, quality, and reliability as well as the volume. Future governance arrangement must appreciate the complexity of hydrology in the MDB and respond adaptively to the changes of these qualities as we transition to a drier climate.

We suggest that greater policy recognition of the integration of the relationship between surface water and groundwater is required for the future management of environmental water resources. To date, surface water and groundwater have been largely viewed as two separate sources in the policy sector. The reality is that they are interlinked and should be dealt with as such, though there are obvious complexities in this. With surface availability predicted to decline, especially in the southern MDB, groundwater resources are likely to become increasingly exploited. There are also issues surrounding: climate change; the long-term average annual yield attached to water entitlements; irrigators selling surplus water to the Commonwealth; and the saved water attributed to irrigation infrastructure upgrades. As a consequence, the sustainability of the Basin still remains in doubt.

4.7 CONCLUSION

Australia has undergone significant reform from command-and-control regulation of resources to a market-based water governance regime. This has been accompanied by a shift in policy direction, departing from a focus on national development objectives, and focusing on aggregate economic efficiency. Significant attention has also been directed at addressing overallocation in the Basin through a variety infrastructure and market-led solutions, notably the federal government buyback programs.

Despite changing hydrological paradigms throughout Australia's history of water policy, the agricultural legacy has not been superseded by environmental concern. Policy indicates that irrigators' rights to water remain paramount. However, the recognition of scarcity and the need to restore the balance between consumptive and environmental use in the MDB is internationally unparalleled. The use of market mechanisms to achieve environmental outcomes has added new and more flexible arrangements to governance of environmental water. The ongoing reform to achieve this signals a transition from top-down government to market-based governance arrangements jointly reliant on the state, market, and civil participation to achieve positive environmental outcomes.

The study of Australia's changing paradigms in water management provides an insight into the drivers of policy change and the difficulties of reform. It highlights that although Australia has achieved an internationally unprecedented commitment to return water to the environment through market mechanisms, the path to reform has been fraught with difficulties. It demonstrates clearly that environmental change is both a catalyst and an outcome of water reform in Australia. The history of the MDB demonstrates that droughts provide an impetus for water policy reforms and an opportunity to develop new governance arrangements. Despite the relative success and formal institutionalization of the water market in Australia's environmental water governance, the process is not complete and will require continued effort, political commitment, and consumer and farmer participation to maintain the success of market-based reform in the future.

APPENDIX A: LIST OF ABBREVIATIONS

Abbreviation	Description
CEWO	Commonwealth Environmental Water Office
COAG	Council of Australian Governments
CSIRO	Commonwealth Scientific and Industrial Research Organization
DEWHA	Department of Environment, Water, Heritage and the Arts
DoE	Department of the Environment
GL	Gigalitre (1×10^9 L = 1 GL)
GWP	Global Water Partnership

Abbreviation	Description
IPCC	Intergovernmental Panel on Climate Change
IWRM	Integrated Water Resource Management
MDB	Murray-Darling Basin
MDBA	Murray-Darling Basin Authority
MDBC	Murray-Darling Basin Commission
MDBMC	Murray-Darling Basin Ministerial Council
NPWS	National Plan for Water Security
NWC	National Water Commission
NWI	National Water Initiative
OECD	Organization for Economic Co-operation and Development
RtB	Restoring the Balance
TLM	The Living Murray
WFF	Water for the Future
WTO	World Trade Organization

REFERENCES

Bell, S., Quiggin, J., 2008. Murray Darling Program Working Paper: M06–6: The Metagovernance of Markets: The Politics of Water Management in Australia. School of Economics and Political Science, University of Queensland, Brisbane.

Castro, J., 2007. Water governance in the twenty-first Century. Ambiente Soc. 10 (2), 97–188.

Chong, H., Sunding, D., 2006. Water markets and trading. Annu. Rev. Environ. Resour. 31 (2), 239–264.

Clark, S., 2002. Divided power, cooperative solutions? In: Connell, D. (Ed.), Uncharted Waters. Murray-Darling Basin Commission, Canberra, pp. 9–21.

Connell, D., 2007. Contrasting approaches to water management in the Murray-Darling Basin. Aust. J. Environ. Manage. 14 (1), 6–13.

Connell, D., Grafton, Q., 2008. Planning for water security in the Murray Darling Basin. Public Policy 3 (1), 67–86.

Connell, D., Grafton, Q., 2011. Water reform in the Murray-Darling Basin. Water Resour. Res. 47. http://dx.doi.org/10.1029/2010WR009820.

Commonwealth Scientific and Industrial Research Organization (CSIRO), 2008. Water Availability in the Murray-Darling Basin: Summary of a Report. CSIRO Murray-Darling Basin Sustainable Yields Project. CSIRO, Australia, 12 pp.

Council of Australian Government (COAG), 1994. Water Resource Policy: Council of Australian Governments, Communique 25 February 1994. http://archive.coag.gov.au/coag_meeting_outcomes/1994-02-25/docs/attachment_a.cfm (accessed 07.11.14.).

Council of Australian Government (COAG), 2004. Intergovernmental Agreement on a National Water Initiative. http://www.nwc.gov.au/__data/assets/pdf_file/0008/24749/Intergovernmental-Agreement-on-a-national-water-initiative.pdf (accessed 24.11.14.).

Crase, L., O'Keef, S., Wheeler, S., Kinoshita, Y., 2015. Water trading in Australia: understanding the role of policy and serendipity. In: Burnett, K., Howitt, R., Roumasset, J., Wada, C. (Eds.), Routledge Handbook of Water Economics and Institutions. Routledge, Oxford and New York.

Davidson, B., 1969. Australia Wet or Dry? The Physical and Economic Limits to the Expansion of Irrigation. Melbourne University Press, Melbourne.

Department of Environment (DoE), 2014a. Progress towards Meeting Environmental Needs under the Basin Plan. http://www.environment.gov.au/water/basin-plan/progress-recovery (accessed 07.11.14.).

Department of Environment (DoE), 2014b. Environmental Water Holdings. http://www.environment.gov.au/water/cewo/about/water-holdings (accessed 07.11.14.).

Department of Environment, Water, Heritage and the Arts (DEWHA), 2010. Securing Our Water Future. DEWHA, Canberra.

Dovers, S., 2003. Processes and institutions for resource and environmental management: why and how to analyse? In: Dovers, S., Rivers, S. (Eds.), Managing Australia's Environment. The Federation, Sydney, pp. 154—180.

Foerster, A., 2011. Developing purposeful and adaptive institutions for effective environmental water governance. Water Resour. Manag. 25, 4005—4018.

Folke, C., Hahn, T., Olsson, P., Norberg, J., 2005. Adaptive governance of social-ecological systems. Annu. Rev. Environ. Resour. 30, 441—473.

Garrick, D., Lane-Miller, C., McCoy, A., 2011. Institutional innovations to govern environmental water in the Western United States: lessons for Australia's Murray-Darling Basin. Econ. Pap. 32 (2), 167—184.

Global Water Partnership (GWP), 2012. IWRM Principles. http://www.gwp.org/en/The-Challenge/What-is-IWRM/IWRM-Principles/ (accessed 15.10.14.).

Grey, D., Sadoff, C., 2006. Water for growth and development. In: Thematic Documents of the IV World Water Forum. The World Bank, Mexico City.

Haggard, S., Webb, D., 1994. Voting for Reform: Democracy, Political Liberalization and Economic Adjustment. The World Bank, Washington.

Hatfield-Dodds, S., Nelson, R., Cook, D., 2007. Adaptive governance: an introduction, and implications for public policy. In: Annual Conference of the Australian Agricultural and Resource Economics Society, Queenstown, NZ, 13—16 February 2007.

Howe, C., 1997. Increasing efficiency in water markets: examples from the Western United States. In: Anderson, T., Hill, P. (Eds.), Water Marketing: The Next Generation. Rowman & Littlefield Publishers, Lanham.

Intergovernmental Panel on Climate Change (IPCC), 2013. Summary for policy makers. In: Stocker, T., Quin, D., Plattner, G., Tignor, M., Allen, S., Boschung, J., Nauels, A., Xia, Y., Bex, V., Midgley, P. (Eds.), Climate Change 2013: The Physical Science Basis: Working Group 1 Contribution to the Fifth Assessment Report of the Intergovernmental Panel on Climate Change. Cambridge University Press, New York.

Kingdon, J., 1995. Agendas, Alternatives, and Public Policy. Harper Collins, New York.

Loch, A., Wheeler, S., Adamson, D., 2014. People versus place in Australia's Murray-Darling Basin: balancing economic, social ecosystems and community outcomes. In: Squires, V., Milner, H., Daniell, K. (Eds.), River Basin Management in the Twenty-first Century: Understanding People and Place. CRC Press, Taylor & Francis Group, Boca Raton, pp. 275—303.

McCormick, B., 2007. Budget Review 2007—2008: National Plan for Water Security. Parliament of Australia. http://www.aph.gov.au/About_Parliament/Parliamentary_Departments/Parliamentary_Library/pubs/rp/BudgetReview_2007-_2008/National_Plan_for_Water_Security (accessed 15.10.14.).

Murray-Darling Basin Authority (MDBA), 2010. Guide to the Proposed Basin Plan: Technical Background, Part 1, vol. 2. MDBA, Canberra.

Murray-Darling Basin Authority (MDBA), 2012. Water Act 2007: Basin Plan. Commonwealth of Australia, Canberra.

Murray-Darling Basin Authority (MDBA), 2014. Water Storage in the Basin. http://www.mdba.gov.au/river-data/water-storage (accessed 07.11.12.).

Murray-Darling Basin Commission (MDBC), 2000. Review of the Operation of the Cap: Overview Report of the Murray-Darling Basin Commission. MDBC, Canberra.

Murray-Darling Basin Ministerial Council (MDBMC), 1995. An Audit of Water Use in the Murray-Darling Basin. MDBMC, Canberra.

Musgrave, W., 2008. Historical development of water resources in Australia: irrigation policy in the Murray—Darling Basin. In: Crase, L. (Ed.), Water Policy in Australia: The Impact of Change and Uncertainty, Resources for the Future, Washington, pp. 28—44.

National Water Commission (NWC), 2004. NWI Objectives. Available online at: http://www.nwc.gov.au/nwi/objectives (accessed 15.10.14.).

National Water Commission (NWC), 2011. Water Markets in Australia: A Short History. NWC, Canberra.

National Water Commission (NWC), 2012a. Indigenous Access to Water. Available online at: http://www. nwc.gov.au/nwi/position-statements/indigenous-access (accessed 28.10.14.).

National Water Commission (NWC), 2012b. Impacts of Water Trading in the Southern Murray-Darling Basin between 2006—07 and 2010—11. NWC, Canberra.

National Water Commission (NWC), 2013. Australian Water Markets: Trends and Drivers, 2007—08 to 2012—13. NWC, Canberra.

Olsson, P., Gunderson, L., Carpenter, S., Ryan, P., Lebel, L., Folke, C., Holling, C., 2006. Shooting the rapids: navigating transitions to adaptive governance of socio-ecological systems. Ecol. Soc. 11 (1).

Paterson, J., 1987. The privatisation issue: water utilities. In: Abelson, P. (Ed.), Privatisation: An Australian Experience. Australian Professional Publications, Sydney, pp. 181—204.

Randall, A., 1981. Property entitlements and pricing policies for a maturing water economy. Aust. J. Agric. Econ. 25 (3), 195—220.

Rogers, P., Hall, A., 2003. Effective Water Governance. Global Water Partnership Technical Committee, Background Paper no.7. Elanders Novum, Sweden.

Sim, T., Muller, K., 2004. A Fresh History of the Lower Lakes: Wellington to the Murray Mouth, 1800 to 1935. River Murray Catchment Water Management Board, Strathalbyn.

Tan, P., Jackson, S., 2013. Impossible dreaming — does Australia's water law and policy fulfil Indigenous aspirations? Environ. Plan. Law J. 30, 132—149.

Ward, J., 2009. Palisades and pathways: historic lessons from Australian water reform. In: Northern Australia Land and Water Science Review Full Report. CSIRO, Canberra.

Weal, A., 1992. The New Politics of Pollution Control. Manchester University Press, Manchester.

Wheeler, S., 2014. Insights, lessons and benefits from improved regional water security -in Australia. Water Resour. Econ. 8, 57—78. http://dx.doi.org/10.1016/j.wre.2014.05.006.

Wheeler, S., Cheesman, J., 2013. Key findings of a survey of sellers to the restoring the balance program. Econ. Pap. 32 (3), 340—352.

Wheeler, S., Zuo, A., Bjornlund, H., Lane-Miller, C., 2012. Selling the farm silver? Understanding water sales to the Australian Government. Environ. Resource Econ 52, 133—154.

Wheeler, S., Garrick, D., Loch, A., Bjornlund, H., 2013. Evaluating water market products to acquire water for the environment in Australia. L. Use Policy 30, 427—436.

Wheeler, S., Bjornlund, H., Loch, A., 2014a. Water trading in Australia: tracing its development and impact over the past three decades. In: Easter, K., Huang, Q. (Eds.), Water Markets for the Twenty First Century. Springer, New York.

Wheeler, S., Loch, A., Zuo, A., Bjornlund, H., 2014b. Reviewing the adoption and impact of water markets in the Murray-Darling Basin, Australia. J. Hydrol. 518, 28—41.

Zuo, A., Nauges, C., Wheeler, S., 2014. Farmers' exposure to risk and their temporary water trading. Eur. Rev. Agric. Econ. 42 (1), 1—4. http://dx.doi.org/10.1093/erae/jbu003.

CHAPTER 5

Damages Evaluation, Periodic Floods, and Local Sea Level Rise: The Case of Venice, Italy

Massimiliano Caporin, Fulvio Fontini
Department of Economics and Management "Marco Fanno", University of Padova, Padova, Italy

Contents

5.1 INTRODUCTION

Several coastal areas are subject to periodical floods, which depend on the interaction between the tide and the storm surge. The latter is the phenomenon of rising coastal water due to severe weather conditions and the specific (shallow) coastal shape. Some of the areas subject to repeated coastal floods are the Gulf of Mexico coastal areas, the Philippines, the Bay of Bengal, the North Sea. The frequency of the floods depends, inter alia, on the frequency of the tide and of the meteorological low-pressure systems associated with storm surge. An example is the local periodical flooding of the town of Venice (Italy), known as Acqua Alta (AA). Coastal floods can generate huge economic losses. The damages induced by the floods can be categorized into direct and indirect damages and opportunity costs. The first group contains at least the following elements: damages to real estates and disruptions due to floods; costs of interference to normal operational conditions and forced closures of economic activities; and the loss of business opportunities and lost revenues. The indirect damages refer to: the damages due to the altered ecosystem; negative externalities; the loss of future opportunities (real option values); interference with market dynamics (altered or disrupted competition, strategic reactions). Finally, opportunity costs include: expenditures needed to adapt the investments once consequences of tides are taken into account, to be compared to costs of similar investments for areas which are not subject to floods.

Interestingly enough, note that these are local or regional losses. When aggregated to national levels, a low of average applies, which implies that floods typically have little or

Handbook of Environmental and Sustainable Finance
ISBN 978-0-12-803615-0

null effects in financial markets (see Ramiah, 2013; Ramiah et al., forthcoming and references therein). Indeed, the financial literature has explored the impact of natural disasters and of the associated losses on financial markets and economic activities (Benson and Clay, 2004; Worthington and Valadkhani, 2004, 2005; Worthington, 2008; Wang and Kutan, 2013). Moreover, international organizations monitor and are interested in evaluating the economic and fiscal impact of natural disasters (Melecky and Raddatz, 2011; Haraguchi and Lall, 2013). From the financial viewpoint, natural disasters have effects on economic activities and private investments, while the protection for such kind of risks involve insurances, from both the private and public point of view (Hofman and Brukoff, 2006), and risk management and assessment procedures (G20/OECD, 2012). At the local level, as in the case of Venice, the periodical flooding represents extreme and generally unpredictable (over the long term) events, sharing some similarities with natural disasters, even though the economic and noneconomic losses are clearly lower. However, these losses have a strong impact on local economic activities, private investments, public finance (of local government offices) and translate into insurance and risk assessment and management costs. For these reasons, the evaluation of long-term impacts and damages is quite important.

It is possible to relate the value of the damages due to periodic coastal floods to the local sea level rise (LSLR) by calculating a damage function, which would depend on one or more of the elements listed above. When calculating the damage function for a specific case, data availability has to be taken into account and, as a consequence, not all of the damages and costs above mentioned can be included in the damage evaluation with the consequence of inducing a, possibly relevant, approximation or underestimation of the true economic impact of flooding. The case of the city of Venice is even more peculiar. In fact, we have repeated occurrences of coastal floods, even more than a handful per year. Repeated coastal floods are the result of several components, some of which can be predicted by harmonic analysis, like the astronomic components of the tide, and some other random components that depend on the specific geographic and climatologic characteristics of the area. The former depends on the effect on sea level rise of a well-known set of deterministic parameters, the moon's influence and the season's changes being two of most relevant. From a statistical point of view, the astronomical tide has a sinusoidal shape that can be calculated with very high precision. The latter depends on several elements, among which an important role is played by the local climatology. This can be analyzed following two approaches: a general meteorological model, taking into account a number of variables (such as wind, pressure, and others) or a reduced-form approach where a stochastic model is fit to the tide level filtered from the astronomical and deterministic component. This second methodology is the most convenient for long-run studies, where the focus is on high-frequency (i.e., daily) observations over a number of years.

We stress that such a case is of fundamental relevance in the evaluation of economic projects, like the construction of infrastructures to protect from coastal floods. Once a statistical model is designed for the sea level oscillation, it is possible to simulate the economic impact of the coastal floods defining a damage function, which allows translating floods into economic values. The stochastic model will simulate one component of the sea level rise, which should be, merely for simplicity reasons, independent from the LSLR. On the contrary, the deterministic astronomical tide component includes a trend effect, which can be related to the LSLR. Therefore, the methodology can be employed to relate the damages to the LSLR, under different scenarios. Indeed, large investment projects have often a long-time horizon, and thus their evaluation is extremely sensitive to the consequences that climate change might have on the sea level. Climate change can have a complex impact on the local meteorology and thus influence the tides and eventually the damages impacting on several components. It is difficult to take into account all those elements with a reduced-form approach, which, focusing on high-frequency data, is being used for long-term simulations. Nevertheless, it is possible to postulate that the stochastic component of the model and the approach used to simulate future tide levels, if appropriately designed, can also, at least partially, cope with that issue, and thus capture the interrelation between the storm surge and the LSLR. It is thus convenient to calculate the impact of the possible LSLR in several scenarios by considering only the possible annual increase in the tide's trend, without taking into account the possible interaction between global warming and the meteorological components of the tide. A different approach would be simulating extreme weather events, such as tropical cyclones, for instance, and evaluate the impact that climate change may have on damages induced by those meteorological events in countries around the world. See Nordhaus (2010) and Mendelsohn (2011a,b) for examples. The drawback of these approaches is that they rest on macro models, run at large geographic scales, and it is hard to downscale them to evaluate damages at specific locations.

We show here an example of the evaluation of the relationship between LSLR and the value of the investment in a large coastal project, namely the system of mobile dams aiming at protecting the town of Venice from AA. The system, called Mo.S.E. (Experimental Electromechanical Module—"Modulo Sperimentale Elettromeccanico" in Italian), is indeed a set of mobile barriers lying on the bottom of the Venice lagoon's inlets. It is currently under construction. When needed, they are lifted to separate the lagoon from the Adriatic Sea and stabilize the height of the water inside it. Its original budget, (1.6 billion euro), has been increasing manifold, the latest figure amounting to (roughly) 5.4 billion euro. To calculate its value and relate it to the LSLR, the following procedure is adopted: first, the future patterns of AA for the next 50 years are simulated, replicating the observed time trend and simulating future AA under different possible scenarios for LSLR. Then the simulations are converted into an economic value by measuring the value of the avoided damage to Venice and compared it with the reported investment costs. The approach we propose thus combines a reduced-form model based on the AA occurrences with a damage function that evaluates economic losses. By means of a statistical procedure, we show that the economic measurable costs included in the damage function we consider are lower when the Mo.S.E. is taken into account. Moreover, the contraction in the damages is larger than the investment costs for the Mo.S.E. when the LSLR takes values above the historical long-term average. The analysis is limited to 50 years since the economic evaluation is based on the actual technologies and knowledge, i.e., it does not take into account technological progress that can significantly alter the definition of the damage function. Section 5.2 describes the data and the adopted methodology, while Section 5.3 presents some results. Section 5.4 provides some concluding comments.

5.2 DATA AND METHODS

AA depends on the tide in the Venetian lagoon, which is a partially random phenomenon. It can be defined as a tide above a given threshold, conventionally set at +80 cm above the *Punta della Salute* tidal datum. The latter is the level used by the Centro Maree—the public body in charge of monitoring AA—and beyond which the town starts to be flooded. Tides are generated by the interaction of two components: the astronomical tide and the storm surge (Canestrelli et al., 2001). The former depends on the effect on sea level rise of a well-known set of deterministic parameters, the moon's influence and the season's changes being two of most relevant. The astronomical tide has a sinusoidal shape that can be calculated with very high precision. The astronomical tide, however, does not completely explain the observed sea level pattern in the lagoon. In fact, it is possible to identify (at least) three other main elements of the upper Adriatic Sea's climatology that influence the behavior of the tide: winds, the barotropic pressure, and the "seiches" (the oscillations of an almost closed basin, like the Adriatic Sea, after a perturbation). Together, these components generate the surge. Due to its nature, storm surge is a random phenomenon. Historical data show that it has had over the years an almost zero mean. The tide is thus a random variable given by the sum of a deterministic component and a stochastic component. As a result, it is possible to derive the random behavior of the tide by observing the stochastic pattern of the storm surge and fitting on it a statistical model.

Within a given day, AA, if present, is associated with a maximum daily tide level above the threshold. Clearly, the duration of AA may vary from day to day. Overall, using hourly data from the Centro Maree from 1941 to 2009, we measure the average duration of AA events and notice that AA's average duration is about 3 h, with most events lasting between 1 and 6 h.[1] Furthermore, AA took place (in almost all cases)

---

[1] Descriptive analyses of AA events duration are available upon request.

once a day (despite the fact that the astronomic tide has two peaks within the day, one of them is generally much lower than the other). Given the purpose of this chapter, we focus on the occurrence of the phenomena, and not its length, and look at its daily maxima, evaluating the Mo.S.E. system using the daily time series of Venice lagoon sea-level maxima. We stress that using the frequency rather than the length of AA to evaluate its damages is an acceptable approximation. Indeed, most of the damages occur on a time spell longer than the day or have a discrete nature (e.g., overnight stays). Moreover, even if, in principle, the length of AA could be relevant for some other economic activity (e.g., the harbor ones), the conversion of hourly data into daily maxima allows us to take into account the (economic) consequences of a hypothetical AA of average length. This procedure averages out episodes of different duration and compensates possible underestimates of daily episodes with overestimates of the night ones.

By construction, even the sequence of daily maxima is composed by the sum of two components, one being the astronomical tide and the second associated with storm surge. To evaluate the performance of the Mo.S.E. system, we need to simulate the evolution of daily maxima. To this end, we first specify a model for the daily maxima time series. Then we fit the model to data from January 1975 to December 2009, for a total of 12,784 days (sample size T). The period from 1941 to 1974 has been discarded since it includes more than 280 days with missing data. We preferred to focus on a complete set of years rather than replacing missing data with estimated values. Our aim is to provide long-range forecasts (or simulations) of the daily maxima. As a consequence, we chose to focus on a purely statistical approach, a reduced-form model, rather than an astronomical and meteorological approach, where the various components affecting the tide level are explicitly taken into account. The model we consider assumes that the daily maxima time series is given as the sum of two components, mimicking the presence of two elements affecting the tide level. If we define as m_t the sequence of daily maxima, the series is decomposed as follows:

$$m_t = a_t + \sigma_t s_t \tag{5.1}$$

where a_t is a deterministic component capturing the mean impact of the astronomical tide level on the daily maxima; σ_t is a deterministic component exploiting the effect of the astronomical tide in the dispersion of daily maxima; and finally, s_t is the stochastic component that we can associate with the storm surge. Note that in our modeling framework the storm surge has, by construction, a zero mean.

The two deterministic components, a_t and σ_t, have a similar structure and are given by the combination of a linear component and a set of harmonics. The first term, a_t is set equal to:

$$a_t = \beta_0 + \beta_1 t + \sum_{j=1}^{Q} \left(\delta_j \cos\left(\frac{2\pi t}{\omega_j}\right) + \gamma_j \sin\left(\frac{2\pi t}{\omega_j}\right) \right) \tag{5.2}$$

where t denotes a linear trend, Q is the number of harmonics considered, $\omega_j, j = 1, 2,...,$ Q, are the frequencies of the harmonics, and $\beta_0, \beta_1, \delta_j, \gamma_j, j = 1, 2,..., Q$, are parameters to be estimated. We stress that the sea rise level is associated with the value of parameter β_1. Moreover, this parameter implicitly takes into account a second component that affects the AA phenomenon, the land subsidence. We do not include explicitly measures of land subsidence in our modeling framework, but include it into the long-term trend. Therefore, even if we make explicitly reference to LSLR in the scenarios it is implicitly assumed that they include possible effects associated with land subsidence.

The harmonic frequencies are expressed in days and can be easily calibrated by looking at the periodogram of the daily maxima. Indeed, a similar approach has been used in other studies, for instance, in the analysis of wind and temperature time series (Caporin and Pres, 2012, 2013) and in the finance literature (Andersen and Bollerslev, 1997).

Note that the frequencies are mostly associated with interpretable amplitudes of daily maxima oscillations such as the yearly and half-yearly components and the monthly and half-monthly lunar components. The latter can be associated with the anomalous, sidereal, and synodic lunar month durations.

In contrast, the deterministic dynamic of σ_t is specified after a log transformation

$$\ln(\sigma_t^2) = \phi_0 + \sum_{j=1}^{P} \left(\rho_j \cos\left(\frac{2\pi t}{\omega_j}\right) + \theta_j \sin\left(\frac{2\pi t}{\omega_j}\right) \right) \tag{5.3}$$

where the number of harmonics and their frequencies can differ from those in Eqn (5.2), and the parameters to be estimated are $\phi_0, \rho_j, \theta_j, j = 1, 2,..., P$. Finally, the stochastic component, the storm surge, is modeled as an ARMA−EGARCH model:

$$A(L)s_t = B(L)\eta_t \tag{5.4}$$

$$\eta_t = h_t^{0.5} z_t \tag{5.5}$$

$$\ln(h_t) = \varphi_0 + \varphi_1 \ln(h_{t-1}) + \varphi_2 |z_{t-1}| + \varphi_3 z_{t-1} \tag{5.6}$$

where L is the lag operator, $A(L) = \sum_{j=1}^{p} a_j L^j$ and $B(L) = \sum_{j=1}^{q} b_j L^j$ are the AR and MA polynomials, h_t is the conditional variance, z_t is the standardized innovation, and the parameters to be estimated are $a_j, j = 1, 2,..., p, b_j, j = 1, 2,..., q, \varphi_0, \varphi_1, \varphi_2$, and φ_3. We prefer the EGARCH model of Nelson (1991) to other GARCH specifications due to the absence of parameter constraints ensuring variance positivity and the contemporaneous presence of volatility asymmetry.

The model parameters are estimated using a multistep approach. First, we estimated the parameters in a_t by simple linear regression methods (with robust standard errors due to the presence of heteroskedasticity in the innovations) on the following equation:

$$m_t = \beta_0 + \beta_1 t + \sum_{j=1}^{Q} \left(\delta_j \cos\left(\frac{2\pi t}{\omega_j}\right) + \gamma_j \sin\left(\frac{2\pi t}{\omega_j}\right) \right) + \varepsilon_t \tag{5.7}$$

Second, we estimate the parameters in σ_t again with least squares on the log-transformed residuals from the previous equation. In fact, $\widehat{\varepsilon}_t = m_t - \widehat{a}_t$ and $\ln(\widehat{\varepsilon}_t^2) = \ln(\sigma_t^2) + \ln(s_t^2)$, therefore:

$$\ln(\widehat{\varepsilon}_t^2) = \phi_0 + \sum_{j=1}^{P} \left(\rho_j \cos\left(\frac{2\pi t}{\omega_j}\right) + \theta_j \sin\left(\frac{2\pi t}{\omega_j}\right) \right) + \varsigma_t \tag{5.8}$$

Finally, the estimated s_t is given as $\widehat{s}_t = \widehat{\varepsilon}_t \exp(-0.5\widehat{\sigma}_t^2)$, and the ARMA–EGARCH parameters are recovered by maximum likelihood methods. The use of multistage estimation approaches is computationally convenient, but it clearly implies a loss of efficiency. Given the estimated parameters, the model can be easily used to generate long-range simulations for the evolution of the daily maxima time series. The simulation procedures we adopt are detailed in the following steps:

1. Fix the simulation length M and, if needed, calibrate the mean trend, parameter β_1;
2. Generate the innovations for the ARMA–EGARCH model, namely the sequence of values for z_{T+i}, $i = 1, 2,..., M$, in Eqn (5.6); the innovations can be generated by sampling from a given density or, alternatively, by resampling from the model residuals; we adopt the latter approach to avoid making a distributional assumption;
3. Simulate the ARMA–EGARCH sequence s_{T+i}, $i = 1, 2,..., M$;
4. Add the deterministic components to the simulated stochastic evolution, obtaining m_{T+i}, $i = 1, 2,..., M$;
5. To generate W different potential future evolutions of the daily maxima, iterate steps 2–4.

The calibration of the mean trend parameter in step 1 allows replacing the estimated LSLR value with alternative figures. Such a substitution is crucial in deriving the evolution of AA events under different LSLR scenarios. In the following, we will make use of several values for β_1, each one associated with a specific LSLR hypothesis.

We stress that we do not consider the possible interaction between global warming and the meteorological components of AA, due to lack of data to support a proper modeling of the local effects on the components of the storm surge. Therefore, we take into account the possible LSLRs under different scenarios, simulating their impact only through the (annual) increase in the tide's trend. We consider five possible scenarios in which the LSLR ranges from the lowest level, which corresponds to the historical trend, to an extremely high level. In particular, we start with a scenario in which the sea level continues to rise at the same level as the historical trend (estimated on the basis of our data sample), namely 2.5 mm/year. Clearly, such a scenario would exclude any increase in the LSLR due to global warming. There exist in the literature several possible scenarios that take the latter effect into account.

We consider three further possible scenarios in which global warming plays a progressively increasing role. In particular, we first consider a low LSLR scenario that assumes a constant yearly LSLR of 3.7 mm/year. Such a level corresponds to the common starting value of the global mean SLR for the four process-based projections (RCP2.6, RCP4.5, RCP6.0, and RCP8.5) considered by IPCC (2013) at p. 1180. In those IPCC projections, such a level increases over time; thus, keeping it fixed corresponds to the adoption of an optimistic assumption about the LSLR. Second, we construct an "average IPCC" scenario, averaging the estimated rises extracted from the four process-based IPCC (2013) scenarios mentioned above and assuming a linear rise throughout the period. The derived figure equals 5.6 mm/year. Interestingly enough, such a level coincides with the lower boundary for global SLR estimates, as reported by the National Research Council (2010). We also consider a pessimistic (high rise) scenario assuming a yearly LSLR that derives from the IPCC scenario that delivers the highest projections for the global mean SLR, namely the RCP8.5. It amounts to 7.4 mm/year.

Finally, we consider a worst-case scenario that corresponds to the highest forecast in the literature from semiempirical models (Nicholls et al., 2011), namely 2 m over the whole century. Such a figure, which also coincides with the upper bound of the forecast considered in the National Research Council report (2010), yields a (constant) yearly LSLR of 20 mm/year. Table 5.1 summarizes the five scenarios considered. We do not contemplate higher SLR for several reasons. First, even if a higher rise can be obtained assuming the occurrence of catastrophic events such as the collapse of marine-based sectors of the Antarctic ice sheet, there is no consensus in the literature on the likelihood of those events. Moreover, a sea-level rise of several meters would have such a devastating impact both globally and locally that a punctual evaluation of the cost and benefit of a specific project (such as Mo.S.E.) would not have much meaning. Finally, the very functioning of Mo.S.E. is not assured for extreme SLR. Indeed, the Mo.S.E. is designed to work for tides up to 3 m. The highest observed AA so far has been equal to 194 cm. Therefore, from a theoretical point of view, the Mo.S.E. can work for an LSLR just up to (roughly) 1 m.

We convert levels and frequencies of possible future AA episodes into economic values by measuring the damage to the municipality of Venice due to AA. We follow the methodology proposed in Fontini et al. (2010) and Vergano et al. (2010), who considered two components for the damage function. The first one depends on the

Table 5.1 Local Sea Level Rise Scenarios

Scenario	Impact on AA	LSLR
S1	Historical average	2.4 mm/year
S2	Low	3.7 mm/year
S3	Medium	5.6 mm/year
S4	High	7.4 mm/year
S5	Worst case	20 mm/year

restoration or refurbishment of real estate damaged by flooding.[2] No episode of AA has ever occurred during the months of July and August.[3]

Therefore, the period of a year is considered, starting from the 1st of July until the 30th of June of the subsequent year. During this period, it is assumed that refurbishment and restoration take place only once (during the time of year with no AA) on the basis of the highest episode observed in the time span. The height of the tide determines the surface area of the town involved by AA and thus the number of buildings affected that need work. The municipality of Venice provides data of the portion of the town that is flooded for every 10 cm rise in the tide.[4] We follow the so-called "Frassetto" altimetry, which refers to the whole surface of the historical center that is flooded for every 10 cm increase of AA, not the more recent "Insula" altimetry (see Boato et al., 2009) since the latter refers only to the public surfaces flooded, not to the whole surface. Notice, however, that in the "Frassetto" altimetry, some parts of the Venice, such as Giudecca, Murano, Burano, Torcello, Lido, and other small islands, are not included in the data. Therefore, we do not take into account these areas in our calculations.

A comparison of the two altimetries shows that, on average, 25% of the town surface that is flooded is public. For the remaining part, we assume that it is private and convert its surface area into the length of walls affected by floods using a (standard) one-to-one conversion. The cost of refurbishment (plastering) is applied to the length of walls affected by episodes of AA. A fixed height of 1 m is assumed for plastering (with a ratio of 60% external and 40% internal walls), leading to a cost of 54 euro per linear meter.

We consider constant 2013 prices and assume that 50% of buildings are of special (historical) interest and/or need specific care; for these, the costs of refurbishment are doubled. Table 5.2 reports the cumulative distribution function of the damage to real estate. Figures are in millions of euro. Note that it assumes a maximum beyond the +180 cm level since this is the level beyond which (almost) all of the town is flooded (such an extreme level has been historically observed just once, in the famous episode of the November 4, 1966).

We do not take into account the possible further costs associated with more radical restoration that could take place much less frequently than yearly plastering. For those, measurement and frequency of episodes are difficult to evaluate. Therefore, with respect to this aspect, the damage function provides an underestimate of the total costs.

[2] Venice has been constructed throughout the centuries, following a specific building technique. The types of works that are needed to maintain and possibly refurbish its buildings are described in Cellerino (1998).

[3] See http://www.comune.venezia.it/flex/cm/pages/ServeBLOB.php/L/IT/IDPagina/1754 (accessed February, 2014).

[4] See http://www.comune.venezia.it/flex/cm/pages/ServeBLOB.php/L/IT/IDPagina/2973 (accessed February, 2014).

Table 5.2 Cumulative Distribution Function of the Damage to
Real Estate from AA

AA Episodes Height	Damage (Millions Euro)
≤80	0
81—90	0.56
91—100	6.97
101—110	23.03
111—120	69.08
121—130	135.01
131—140	177.12
141—150	189.16
151—160	194.94
161—170	195.85
171—180	196.07
≥180	196.33

The second component of the damage function depends on the frequencies of the AA episodes. Two levels of AA are relevant for this component. The first one is a level of AA that is high enough to hinder everyday activities, in particular the displacement of young and elderly people and tourism activities. Only episodes of AA above +120 cm are relevant for this component.

We follow the methodology proposed in Fontini et al. (2010) that allows us to convert the number of Mo.S.E. activations above that threshold into values of lost tourism expenditures and replacement costs, for elderly people requiring day care and for babysitting (for school-age children who cannot attend school). Following Cellerino (1998), it is assumed that 10% of the population from 75 to 84 years old is confined at home and requires care during AA.

The second relevant height is the activation level of the Mo.S.E. Consorzio Venezia Nuova, the managing body that is in charge of building and operating the Mo.S.E., which has defined a tide level of +110 cm as the reference level for the tide inside the lagoon, once the Mo.S.E. has been activated. We assume that such a level will cap the height of water within the lagoon and the damage accruing from AA because of its height. On the other hand, every Mo.S.E. activation will have an impact on the Venice harbor activity. Indeed, Venice's main harbor is connected to the Adriatic Sea. Interferences with the activities of small tourist wharfs and other harbors located in Chioggia—mostly devoted to fishing—are not considered in our study because of lack of data.

When the floodgates are closed, ships will first be hosted outside the lagoon or at the wharfs and then allowed to enter or exit through a lock gate. Mo.S.E. operation will thus entail costly delays in the harbor operations. Clearly, the more frequently the floodgates are kept closed, the higher the cost will be due to the interference with harbor activity, but the lower the damage to Venice due to AA will be. Vergano et al. (2010) considered two scenarios for the interference cost of the Mo.S.E. on harbor activity. The low-cost scenario assumes that lock gates will be actively operated and ship traffic flows will remain constant over time, while the high-cost one is based on an increase in ship traffic assumption and neglects the relief of delays due to lock gate functioning.

We take the average of their figures[5] as the value of the yearly interference cost. The time series of the daily maxima shows that, on average, during the period considered in the study by Vergano et al. (2010), there has been a yearly average of 6.3 episodes of a tide maximum higher or equal to +110. This allows us to deduce an average interference cost per episode of Mo.S.E. closure that amounts to 0.168 million euro. Finally, in the evaluation exercise, we also take into account the yearly figures for operating and maintenance (O&M) costs, which amount to 13.3 million euro per year, updated from Fontini et al. (2010).

The five scenarios described above determine a different impact of LSLR on the AA. We conducted 10,000 simulations for each scenario. Figure 5.1 reports the yearly average expected number of Mo.S.E. activations for each scenario, a figure that is relevant for the second component of the damage to Venice due to AA. The first panel (Figure 5.1(A)) reports scenarios from S1 to S4, while the worst-case scenario S5 is reported in the second panel (Figure 5.1(B)). Figure 5.2 plots the expected average of AA events above 180 cm in each scenario, a figure that affects the first component of the damage function. The averages in Figure 5.1 are computed across the simulations for each year. In contrast, the averages in Figure 5.2 are derived across simulations and years.

We note that the higher the expected LSLR, the more frequent the Mo.S.E. activations will be in each year and the higher the episodes of extreme high AA will be. In particular, we can see that the frequency of Mo.S.E. activation rises exponentially for scenarios from S1 to S4; in scenario S5, it reaches a frequency so high that Mo.S.E. will (almost always) remain closed throughout the whole year in the last decade.

Obviously, such a result depends on the extremely high level of LSLR assumed in the worst-case scenario. As for the frequency of the highest tide throughout period, we can see that, even in the most conservative S1 scenario, there will be two episodes of a tide above 180 cm in 50 years. Even if such an AA is, at present, considered an exceptionally extreme event, it is not implausible given that it has been observed once in the last 50 years and because of the continuing rise of the sea level. Clearly, such a frequency increases the higher the LSLR is, reaching an overwhelming figure of 25 episodes of AA above 180 cm in 50 years. Impressive as it might be, such a figure is not implausible, taking into account that, in the S5 scenario, there will be a 1-m rise of the sea level at the end of the period, which would imply that there would be several episodes of extremely high AA per year in the last decade.

---

[5] The updated figures are 653,601 and 1,463,855 euro, respectively.

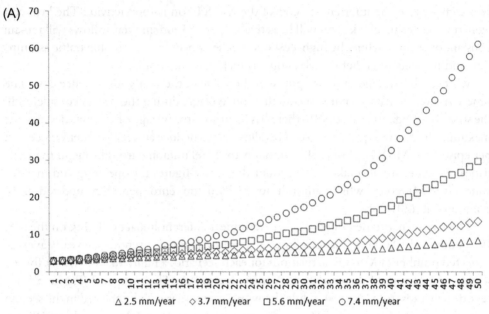

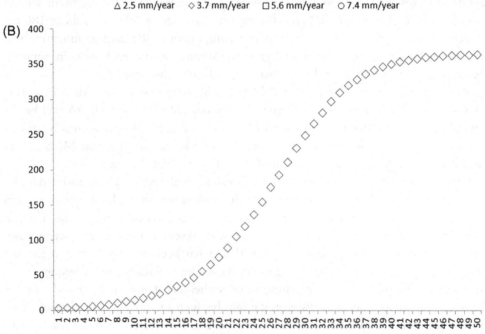

Figure 5.1 Expected number of Mo.S.E. activation on scenarios S1, S2, S3, S4 (A), and Scenario S5 (B).

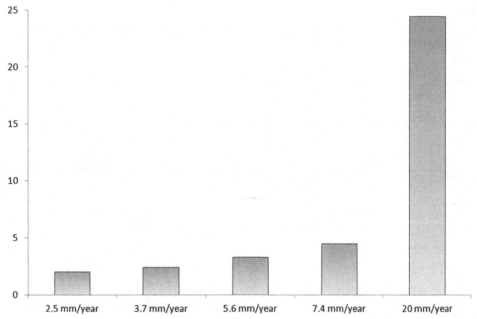

Figure 5.2 Expected number of AA events above +180 cm in scenarios S1, S2, S3, S4, and S5.

5.3 RESULTS

The difference between the damages that would occur without Mo.S.E. and those that can be experienced once the Mo.S.E. is operated provides an economic measure of Mo.S.E. benefits. For each simulation of the tide level in a 50-year horizon, we evaluate the costs associated with AA episodes both with and without the Mo.S.E. To obtain comparable monetary figures, the costs are discounted at a 2% rate, corresponding to the actual long-term target inflation rate fixed by the European Central Bank. Figure 5.3 plots the evaluation of the damages with and without Mo.S.E. for the S3 (average) scenario. The graph describes the median, the first and the third quartile and the min and the max of the simulations for the cumulated values of the first 5, 10, 25, and 50 years, with and without Mo.S.E. The dashed lines identify the Mo.S.E. total costs. All figures are in billion of euro, at constant 2013 prices.

Note that the damage increases over time as LSLR increase. Mo.S.E. functioning reduces the volatility of the estimates, since it caps the height of AA. Thus, the remaining volatility depends only on the expected numbers of its activation. We point out that, without Mo.S.E., there is an expected overall level of damages to Venice that amounts to 8.27 billion euro in 50 years. Mo.S.E. reduces it to 2.25 billion euro, thus determining

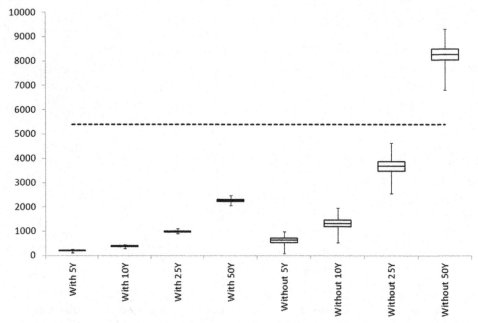

Figure 5.3 Damages to Venice in the S3 scenario with and without Mo.S.E. in 5, 10, 25, and 50 years. The dashed horizontal line represents Mo.S.E. installation costs.

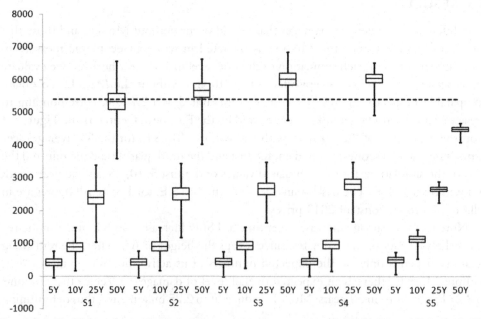

Figure 5.4 Mo.S.E. benefits to Venice (millions euro) in the scenarios S1, S2, S3, S4, and S5 in 5, 10, 25, and 50 years. The dashed horizontal line represents Mo.S.E. installation costs.

an expected benefit, i.e., avoided damages, above 6 billion euro. Such a figure is above the reported costs, which amount to 5.4 billion euro. Therefore, we can expect a net positive value for Mo.S.E. in 50 years. Clearly, the value of the benefits depends on the different LSLRs assumed. Figure 5.4 summarizes the benefits to Venice from Mo.S.E. in the S1, S2, S3, S4, and S5 scenarios. The figures (median values) are reported in Table 5.3. Descriptive statistics for the 10,000 simulations per scenario are available from the authors upon request.

Table 5.3 Damage with and without Mo.S.E. in Each Scenario in 5, 10, 25, and 50 years

Scenario S5

Years	Damage without Mo.S.E.	Damage with Mo.S.E.	Mo.S.E. Benefits
5	716	205	511
10	1576	443	1133
25	4817	2154	2664
50	16,746	12,252	4494

Scenario S4

Years	Damage without Mo.S.E.	Damage with Mo.S.E.	Mo.S.E. Benefits
5	649	197	452
10	1363	396	967
25	3848	1043	2806
50	8783	2738	6046

Scenario S3

Years	Damage without Mo.S.E.	Damage with Mo.S.E.	Mo.S.E. Benefits
5	640	196	444
10	1328	393	935
25	3674	999	2675
50	8276	2255	6020

Scenario S2

Years	Damage without Mo.S.E.	Damage with Mo.S.E.	Mo.S.E. Benefits
5	622	195	427
10	1292	388	904
25	3475	964	2512
50	7640	1976	5664

Scenario S1

Years	Damage without Mo.S.E.	Damage with Mo.S.E.	Mo.S.E. Benefits
5	613	195	419
10	1269	386	883
25	3349	946	2403
50	7197	1871	5326

All values in millions euro.

We can see that damage increases over time as expected, both without and with Mo.S.E. Its functioning reduces the damages. As a result, positive benefits accrue from the use of Mo.S.E. in each scenario. The value of the benefits increases with the LSLR for the S1, S2, S3, and S4 scenarios. The median figure ranges from 5.3 billions to 6 billion. Taking into account that the reported costs are 5.4 billions of euro, we observe a positive net benefit from Mo.S.E. under scenarios S2, S3, and S4. The benefits increase with the higher LSLRs assumed in each scenario, with a slightly negative figure for the baseline (no global warming) scenario S1.

Looking at the figures of the first quartile and the maximum, we note however, that in S1, (almost) half of the simulations report a net benefit higher than the cost. Therefore, even in S1, we can assess a nonnegative cost-benefit analysis for Mo.S.E. The figure of the extreme S5 scenario appears contradictory at a first glance, since the median estimate of benefit from Mo.S.E. is lower than the values for the other scenarios. This can be understood looking at the values of damage reported in Table 5.3. Note that the damage without Mo.S.E. in S5, even if it is higher than damage in S2 and S3, is not too high and is even comparable to the damage in S4 for the first 25 years of simulations.

However, damage both with and without Mo.S.E. explodes in the second half of the considered period for the S5 scenario compared to the damage under the other scenarios. This determines the reduced values of benefits compared to the other scenarios. The explanation depends on the frequency of Mo.S.E. activations under the S5 scenario compared with activations in the other scenarios (see Figure 5.1(A) and (B)), which determines continuous interference with harbor activity. However, we point out that a precise estimate of AA consequences in the S5 scenario cannot be considered as reliable as the estimates under the other scenarios, since S5 assumes an LSLR that is at the border of levels that are (supposed to be) compatible with Mo.S.E. operation (and, on the other hand, even the probability of an SLR higher than the one reported in the RCP8.5—our S4 scenario—cannot be reliably evaluated according to IPCC, 2013).

5.4 CONCLUSION

Several coastal areas are subject to periodic floods, depending on the interaction between the tide and the storm surge. When high frequency data are available, the sea level rise can be simulated following a statistical approach, which depurates the random behavior of sea level from the harmonics of the tide and focuses on the deviations employing a properly defined stochastic model that best fits the data. Then, such a model can be converted into economic values by defining a damage function, on the basis of the data availability of the several components of damages induced by floods: direct, indirect damages, and opportunity costs. The figures can be used to assess values of damages to real estates, for instance, or the value of investments sets up to reduce damages, such as floodgates. Such a methodology can be used also to evaluate the impact that sea level rise can have on the damages, by simulating the frequency and height of the floods on a sufficiently long-run period. The evaluation performed this way can be usefully employed to assess area-specific consequences of floods, as compared, for instance, to macroclimatic approaches. However, it has to be noted that this calculation provides an underestimate of consequences. For instance, the methodology discussed has been employed to evaluate the benefits of the system of floodgate designed to protect Venice from its periodical flooding phenomenon, Acqua Alta. The evaluation performed provides a lower boundary to the value of protecting Venice since it is based on a limited data set for direct costs (only the historical part of Venice is considered, and not all possible damages are included, such as the impact of AA on shops and warehouse inventories).

However, some interesting conclusions can be drawn from such an analysis. It is shown that the estimated benefits, which are largely higher than the original planned cost of the investment, have been greatly eroded by the increase in the budget during its construction. Nevertheless, there is still a positive net benefit, whose value depends crucially on the assumed LSLR. In particular, the benefits are higher than the costs, the higher the assumed LSLR, provided that it is not too extreme. If, on the contrary, there were a limited or null LSLR, the benefits of protecting Venice from AA would be entirely overtaken by the investment and O&M costs. Similarly, in the case of a catastrophic 1-m LSLR in 50 years, both AA and Mo.S.E. would interfere with everyday activities so frequently that the Mo.S.E. benefits would not be balanced with its costs. The analysis performed points out the importance of correct budget planning for investments of such a large scale and highlights the negative economic impact of high LSLR and consequently the importance of prophylactic methods, such as the Mo.S.E. system, to minimize it.

ACKNOWLEDGMENTS

We thank Alberto Tomasin for having made available the data set of Centro Maree and jointly with Georg Umgiesser for their help on the Acqua Alta phenomenon. Clearly, we are responsible for the contents of the article. F.F. acknowledges the research grant PRIN-MIUR 2010-11 "Climate changes in the Mediterranean area: evolutionary scenarios, mitigation policies and technological innovation."

REFERENCES

Andersen, T.G., Bollerslev, T., 1997. Intraday periodicity and volatility persistence in financial markets. J. Empir. Finance 4 (2), 115–158.

Benson, C., Clay, E.J., 2004. Understanding the Economic and Financial Impact of Natural Disasters. The World Bank, Washington.

Boato, L., Canestrelli, P., Facchin, L., Todaro, R., 2009. Venezia Altimetria. Istituzione centro previsioni e segnalazioni maree, in collaborazione con Insula spa, Comune di Venezia. Available at: http://www.comune.venezia.it/flex/cm/pages/ServeBLOB.php/L/IT/IDPagina/1754 (accessed February, 2014).

Canestrelli, P., Mandich, M., Pirazzoli, P.A., Tomasin, A., 2001. Wind, Depression and Seiches: Tidal Perturbations in Venice (1951–2000). Centro previsioni e segnalazioni maree, Comune di Venezia. Available at: http://93.62.201.235/maree/DOCUMENTI/Venti_depressioni_e_sesse.pdf (accessed February, 2014).

Caporin, M., Pres, J., 2012. Modeling and forecasting wind speed intensity for weather risk management. Comput. Stat. Data Anal. 56 (11), 3459–3476.

Caporin, M., Pres, J., 2013. Forecasting temperature indices density with time-varying long-memory models. J. Forecast. 32 (4), 339–352.

Cellerino, R., 1998. Venezia Atlantide: L'impatto Economico Delle Acque Alte. Franco Angeli, Milan.

Fontini, F., Umgiesser, G., Vergano, L., 2010. The role of ambiguity in the evaluation of the net benefits of the MO.S.E. system in the Venice lagoon. Ecol. Econ. 69 (10), 1964–1972.

G20/OECD, 2012. Disaster Risk Assessment and Risk Financing. A G20/OECD Methodological Framework.

Haraguchi, M., Lall, U., 2013. Flood Risks and Impacts Future Research Questions and Implication to Private Investment Decision-Making for Supply Chain Networks. The United Nation Office for Disaster Risk Reduction. Global Assessment Report on Disaster Risk Reduction.

Hofman, D., Brukoff, P., 2006. Insuring Public Finances against Natural Disasters - a Survey of Options and Recent Initiatives. IMF Working Paper WP/06/199.

Intergovernmental Panel on Climate Change (IPCC), 2013. Climate Change 2013: The Physical Science Basis. Working Group I Contribution to the Fifth Assessment Report of the Intergovernmental Panel on Climate Change. Available at: http://www.ipcc.ch/report/ar5/wg1/ (accessed February, 2014).

Melecky, M., Raddatz, C., 2011. How Do Governments Respond after Catastrophes? Natural-Disaster Shocks and the Fiscal Stance. The World Bank Group. Policy Research Working Papers.

Mendelsohn, R., 2011a. The Impact of Climate Change on Hurricane Damages in the United States. The World Bank. Policy Research Working Paper Series: 5561.

Mendelsohn, R., 2011b. The Impact of Climate Change on Global Tropical Storm Damages. The World Bank. Policy Research Working Paper Series: 5562.

National Research Council, 2010. Advancing the Science of Climate Change. The National Academies Press, Washington.

Nicholls, R.J., Marinova, N., Lowe, J.A., Brown, S., Vellinga, P., de Gusmao, D., Hinkel, J., Tol, R.S.J., 2011. Sea-level rise and its possible impacts given a "Beyond 4 degrees C world" in the twenty-first century. Phil. Trans. R. Soc. A 369 (1934), 161–181.

Nelson, D.B., 1991. Conditional heteroskedasticity in asset returns: a new approach. Econometrica 59 (2), 347–370.

Nordhaus, W.D., 2010. The economics of hurricanes and implications of global warming. Clim. Change Econ. 1 (1), 1–20.

Ramiah, V., 2013. Effects of the Boxing Day tsunami on the world capital markets. Rev. Quant. Finance Account. 40 (2), 383–401.

Ramiah, V., Regan-Beasley, J., Moosa, I. The Black Friday Effect, Advances in Investment Analysis and Portfolio Management, forthcoming.

Vergano, L., Umgiesser, G., Nunes, P.A.L.D., 2010. An economic assessment of the impacts of the MOSE barriers on Venice port activities. Transp. Res. Part D 15, 343–349.

Wang, L., Kutan, A.M., 2013. The impact of natural disasters on stock markets: evidence from Japan and the US. Comp. Econ. Stud. 55 (4), 672–686.

Worthington, A., 2008. The impact of natural events and disasters on the Australian stock market: a GARCH-M analysis of storms, floods, cyclones, earthquakes and bushfires. Glob. Bus. Econ. Rev. 10 (1), 1–10.

Worthington, A., Valadkhani, A., 2004. Measuring the impact of natural disasters on capital markets: an empirical application using intervention analysis. Appl. Econ. 36 (19), 2177–2186.

Worthington, A., Valadkhani, A., 2005. Catastrophic shocks and capital markets: a comparative analysis by disaster and sector. Glob. Econ. Rev. 34 (3), 331–344.

CHAPTER 6

Corporate Social Responsibility and Macroeconomic Uncertainty

Abraham Lioui
Department of Finance, EDHEC Business School, Nice, France

Contents

6.1 INTRODUCTION

In recent years, firms' activities related to corporate social responsibility (CSR) have come under increased scrutiny by the academic community. Until now the main objective assigned to firms has been to maximize shareholders' wealth. However, this approach can be too restrictive and may be dangerous if, in order to maximize shareholders' wealth, firms end up harming its other stakeholders (by polluting, exploiting the environment, or discriminating among employees, etc.). Thus, CSR, a priori, is not necessarily incompatible with the objective of increasing firm market valuation. Jensen (2002) stated this as follows:

> The choice of value maximization as the corporate scorecard must be complemented by a corporate vision, strategy and tactics that unite participants in the organization in its struggle for dominance in its competitive arena.

Among the research questions addressed was: does CSR impact firm value directly or a variable positively related to firm value? Bae et al. (2011) and Verwijmeren and Derwall (2010), for example, showed that firms with a good track record of employee well-being have, in general, a lower leverage (the ratio of debt to equity). Since the latter is negatively related to firm value, good employee treatment increases firms' valuation. El Ghoul et al. (2011) widened this finding by showing that firms with responsible practices related to employees, environment, and products experience a lower cost of capital and thus higher valuation. On the other hand, Bird et al. (2007) has shown that certain CSR strengths may have a negative impact on firm valuation. Fisher-Vanden and Thorburn (2011) recently showed that voluntary responsible activities, such as voluntary membership in environment programs, result in a price decline. The empirical evidence as to the impact of CSR on firm value is thus mixed, at best. In a meta-analysis of 167 studies covering the period 1972–2007, Margolis et al. (2007) conclude that, while positive, the overall effect of CSR on firm performance is small at best. Horvathova's (2010) meta-analysis extends to 2008–2009, a particularly complicated period for the worldwide economy. She shows that, using the findings from 37 studies, the empirical evidence is still inconclusive as to the sign of the relation between CSR and financial performance: half of firms find that the impact is positive while the rest document either a negative or an insignificant impact.

Several studies have suggested some explanations for the contradictory results obtained by the empirical literature. Margolis et al. (2007) raise the problem of aggregation as a potential source of heterogeneity in the empirical findings: while a broad indicator of CSR delivers a small to inexistent impact of CSR on firm performance, the association between financial performance and CSR is stronger for specific dimensions of CSR such as employees and community. Recent results in Bird et al. (2007) and Drusch and Lioui (2010) confirm this notion. Horvathova (2010) points out that the heterogeneity in the empirical methods is an important determinant of the difference in results. King and Lenox (2001) argue that empirical tests of the effect of environmental performance on financial performance fail to distinguish whether the firms operate in cleaner industries or whether they adopt cleaner technologies. Chatterji et al. (2009) suggest that the problem may lie with the data. They assess the extent to which the KLD database, extensively used in the literature, really measures what it is supposed to measure. Meanwhile, Sharfman (1996) validates the data, but only for the small sample available at that time, and Chatterji et al. (2009) were unable to validate all the data although they validated a substantial fraction of it. Rahman and Post (2012) extend this argument and Lioui and Sharma (2012) argue that it is of great importance to account simultaneously for direct and indirect effects. Finally, Vanhamme and Grobben (2009) warn against the instrumentalization of CSR by firms. They show that the use of CSR in firms' crisis time is more effective for firms that have a long CSR history than for those that have a short one. Similarly, Goss and Roberts (2011) show that borrowers engaging in CSR activities experience inferior credit market conditions.

Our purpose in this chapter is to show that market conditions are very important and condition the impact of CSR on firm value. Most of the studies, using mainly panel data on firms' CSR, do control in their empirical investigation for several firms' characteristics. However, no paper, to the best of our knowledge, has accounted for the particular business conditions that prevailed in the market at the time of the CSR. Macroeconomic uncertainty related to economic fundamentals is time varying, peaking during recessions. An important dispersion in beliefs on the part of market participants at a given point in time is also likely to increase the uncertainty surrounding the impact of CSR on firm value. We plan to investigate this issue and offer another possible explanation of the puzzling evidence regarding the relation between CSR and firm value.

This chapter is organized as follows. In Section 6.2, we motivate our research question and set up our hypothesis. In Section 6.3, we describe the data and in Section 6.4, we present our main findings. We then offer some concluding remarks.

6.2 CSR AND MACROECONOMIC UNCERTAINTY

In their seminal contribution, Bansal and Yaron (2004) forcefully argued that macroeconomic uncertainty is a key determinant of risk premia and thus asset prices. The flood of papers that followed only reinforced the sentiment among researchers and policy makers that uncertainty shocks are likely to play an important role in shaping the business cycle, and thus determining the duration of recessions and expansions. Bloom (2009) recently offered a rigorous model to apprehend the impact of macroeconomic uncertainty on key economic variables. Across the board, he shows that this variable has a strong and significant impact on interest rates, productivity, and economic growth. Since these fundamentals are key to stock market equilibrium, asset prices are also likely to be strongly impacted by such uncertainty. Bansal and Yaron (2004) and Bonomo et al. (2011) showed that the key systematic factor in stock markets is clearly macroeconomic uncertainty.

Macroeconomic uncertainty matters mainly because, as Kurz (1994) discussed, high uncertainty usually causes economic agents to have more dispersed beliefs. Bloom (2009) also showed that macroeconomic uncertainty is strongly correlated with a measure of the dispersion across macroeconomic forecasters over their predictions for future gross domestic product. When macroeconomics is proxied by stock market index volatility, it is a stylized fact that this measure strongly correlates with dispersions among forecasters of inflation and growth.

A CSR event in a period of high macroeconomic uncertainty is also a CSR event in a period of high dispersion of beliefs, both at firm level and also at the macroeconomic wide level. Hence there is an increased uncertainty as to the potential impact of the CSR event on firm valuation. Our conviction is that accounting for market conditions is part of the explanation for the mixed findings relative to the impact of CSR on firm valuation.

Hence our research question:

Hypothesis: High macroeconomic uncertainty increases the uncertainty surrounding the impact of CSR on firm value.

We detail hereafter the data we use to investigate this issue.

6.3 DATA

Several methodologies have been followed in the literature to demonstrate the impact of CSR on firm valuation (see Horvathova, 2010). Recently, a panel data analysis using the KLD database has clearly been a dominant approach. Unfortunately, to account for market conditions, this approach is not necessarily appropriate. This is because, by construction, market conditions are the same for a cross section of firms/CSR activities. As a consequence, a time-series approach seems better suited since it allows one to generate a time series of the indicator of macroeconomic uncertainty and hence a sample that covers different market conditions. Therefore we have opted to use an event study that will allow us to generate a time series of the impact of CSR events concurrently with a time series of the corresponding market conditions.

6.3.1 Events Data

In this study, we use the sample data compiled by Drusch and Lioui (2010). This sample, used to assess the impact of CSR in the French market, is made up of 148 actions taken by French firms in the field of CSR. CSR events were selected based on how well they fitted the definition of CSR and its triple bottom line approach. All CSR events needed to fall into one of the three categories: economic, social, or environmental, and should be beyond financial aspects with the objective of improving social, economic, or environmental aspects of society. In the end, there were 148 events selected, performed by 31 distinct firms from various industries. The minimum number of CSR events for a firm is 1 and the maximum 12 (Société Générale) which is still less than 10% of the sample. For each event, date 0 is the date of the CSR event, or, more precisely, it is the announcement date for the event, that is, the date stated on the press release.

This sample therefore represents a wide range of business activities in France. The events themselves are of various types. Some events are environmental in nature; for example, Accor became involved in the "Plant for the Planet" program, which consists of reforesting forests, thanks to savings made by laundry services in the group's hotels. BNP Paribas initiated a socially responsible action by offering €145,000 to various associations and projects benefitting the Marseilles area. This donation was directed toward social and professional inclusion, the apprenticeship tax, and vacations for disadvantaged children, among other things. On the other hand, we do not have many examples of economic-related CSR actions. Companies seem to prefer investing in environmental and socially related activities in order to make a clear distinction from their own financial activity. The only event that would approach this type is the setting up by Credit Agricole in 2008 of the "Grameen—Credit Agricole Microfinance Foundation" and the endowment of the fund with €50 million.

The 31 companies are included in the Compagnie des Agents de Change (CAC) 40 index, except for Aéroports de Paris (ADP). We mainly included CAC 40 firms in this study for specific reasons. First, as these firms are among the biggest French firms, they are visible in the market and their actions might have a more significant impact on investor reactions than those of smaller firms. Second, CAC 40 firms are the ones investing most heavily in CSR activities and disclosing the largest CSR events and involvement. This observation results from an in-depth analysis of all CSR-related information publicly available for French firms in general.

The French market is suitable for addressing the issues at hand for several reasons. Our sample includes the years 2007 and 2008, during which there was considerable market turmoil. In the French market, there was no bank/investment company bankruptcy during the crisis, unlike on the US or the UK stock exchanges. Moreover, listed French firms must disclose social and environmental reports together with their annual reports. This practice was originated by the NRE law[1] of 2001 and was implemented for the first time in 2003. The stock exchange is thus well aware of CSR activities on the part of French companies. Finally, French CSR has developed very quickly during the last decade, as revealed by the recent increase in the number of French companies listed in a socially responsible investment (SRI) index. French firms are also listed in international CSR rankings such as the "global 100 most sustainable corporations in the world." There were eight[2] French firms in the 2009 ranking and six[3] in 2008. The great majority of these companies belong to the CAC 40 index.

6.3.2 Cumulative Abnormal Return

We use a standard[4] event study methodology to estimate the abnormal return of the stocks following the events. For our baseline case, we use a window of 5 days, meaning plus and minus 2 trading days around the event. We use the market model to describe the data generating process of the stock returns, namely:

$$R_{i,t+1} = \alpha_i + \beta_i R_{M,t+1} + \varepsilon_{i,t+1} \tag{6.1}$$

where $R_{i,t+1}$ and $R_{M,t+1}$ are the return on the stock and the return on the market, respectively. In our setting, given that the vast majority of the stocks belong to the

---

[1] *Nouvelles Régulations Economiques* (New Economic Regulations).

[2] Accor, Air France-KLM, Credit Agricole SA, Danone, Lafarge SA, L'Oreal, Michelin, Saint-Gobain.

[3] Accor, Air France-KLM, Credit Agricole SA, Lafarge SA, L'Oreal, Société Générale.

[4] For details, see Campbell et al. (1997), Beccheti et al. (2012), and Fisher-Vanden and Thorburn (2011).

CAC 40, the latter is the natural market index to use in the regressions. β_i is the stock's beta measuring its systematic risk and α_i should be equal to interest $*(1 - \beta_i)$ under no mispricing, that is, assuming the market model to be the true asset pricing model. A 3-month trading period is used to estimate the parameters of the model. Abnormal return for day h in the event window is measured as:

$$AR_{i,h} = R_{i,t+h} - \widehat{\alpha}_i - \widehat{\beta}_i R_{M,t+h} \tag{6.2}$$

where $\widehat{\alpha}_i$ and $\widehat{\beta}_i$ are the estimates of α_i and β_i using Eqn (6.1) on a 3-month period preceding the event window. Therefore, using the market model, a proper account is given to the risk of the stock under consideration.

The cumulative abnormal return (CAR) of one event is defined as:

$$\text{CAR}_i = \frac{1}{5} \sum_{h=1}^{5} AR_{i,h} \tag{6.3}$$

This variable is on a daily basis. Although the market model has been extensively used in event studies of CSR as reported in the meta-analysis of Margolis et al. (2007), some authors[5] advocate the use of more sophisticated models. The two main extensions are additional factors and/or accounting for the heteroscedasticity of the market model residuals (concretely, using a Generalized Autoregressive Conditional Heteroskedasticity, GARCH, model). Fortunately, the market model works quite well and has a good fit for the return data, as will be confirmed below. More sophisticated models do not, in general, change the qualitative results from a market model, as shown recently by Beccheti et al. (2012). We do check for the robustness of our results for different window lengths.

6.3.2.1 Conditional Volatility of CAR (Vol)

The key variable of the current study is the conditional volatility of the CAR. For each event we compute its CAR, which enables us to do a time series for the CAR (148 observations since we have 148 events). To filter the conditional volatility of CAR, we use a parsimonious GARCH(1,1) specification such that:

$$
\begin{aligned}
\text{CAR}_i &= C + \eta_i \\
\eta_i &= \sigma_i \varepsilon_i \\
\sigma_i^2 &= \kappa + \delta_\sigma \sigma_{i-1}^2 + \delta_\eta \eta_{i-i}^2
\end{aligned}
\tag{6.4}
$$

where C and κ are constants, and δ_σ and δ_η are the loadings of the volatility on its own lag and the lag of the residual. ε is an i.i.d. sequence of standard normal random variables. More sophisticated data generating processes could be used, although the simple

---

[5] See Yamagushi (2008) and Ramiah et al. (2015).

model above captures relatively well the behavior of the CAR series in our context. The series that we use as our dependent variable for this chapter is σ_i.

6.3.3 Capturing Market Conditions

The most challenging task is to capture the ongoing market conditions at the time of the CSR-related event. Since our focus in this study was on the uncertainty surrounding the impact of a CSR event, it seemed to us natural to first capture the market conditions by some measure of the uncertainty in the economy. Our first measure is thus simply the volatility of the market index used in the event study during the event window. Stock market aggregate volatility has been extensively used as a proxy for uncertainty, even at the firm level, by Leahy and Whited (1996), Bloom et al. (2007), and Bloom (2009). More precisely, we take the return on the index during the 3 months prior to the event as well as the days of the window and compute the standard deviation of the index return. Like CAR, it is on a daily basis. We call this variable *Market*.

We use two additional measures of market conditions. The first one is still related to the uncertainty in the stock market but it is forward looking, unlike *Market*, which is essentially backward looking. We use the index of implied volatility from index options as published for the US market, namely the VIX index. This index is built based on the implied volatility extracted from the cross section of traded options. It is true that our CSR data are related to France and it would be better to have a local forward looking measure. Unfortunately, such a measure does not exist for our whole sample period for France. Fortunately, both US and European stock exchanges are well integrated, which makes it fairly reasonable to assume that VIX also reflects the views in the French stock market. We call this variable *VIX*.

The last index that we use to capture market conditions is the index of consumer sentiment (ICS). It is published monthly by the University of Michigan and Thomson Reuters. It aims to measure the consumer's attitude and sentiment *vis-à-vis* the economy. It has been normalized to have a value of 100 in December 1964 and we retained the index level during the month of the event. Once again, this a US-based measure, but a casual observation of the financial journals and Web sites in France shows that this index in particular is observed by stock market participants since it gives a good sense of the business cycle situation in the US, which impacts substantially on the views of market participants. We call this variable *ICS*.

6.3.4 Summary Statistics

Table 6.1 reports the summary statistics of the events. In Panel A we report the number of events per year. Panel B shows the number of events per type of CSR. We isolated CSR events related to environmental protection (Environment), help for the professional integration of young or handicapped people (Insertion), health (Health), sport (Sport),

Table 6.1 Summary Statistics of the Events

Panel A

Year	#Events
2003	1
2004	2
2005	15
2006	25
2007	27
2008	48
2009	30

Panel B

Event Type	#Events
Environment	33
Insertion	17
Health	16
Sport	6
Index	6
Suburbs	10
Others	60

Panel C

	CAR	Vol
Mean	0.01%	0.87%
Standard	0.87%	0.14%
Skewness	−0.154	0.602
Kurtosis	3.790	3.097
Minimum	−3.10%	0.65%
Maximum	2.37%	1.31%
Auto	0.153	0.882

inclusion or exclusion from a SRI index (Index), and help to suburbs (Suburbs). Panel C shows the summary statistics for the CAR and the volatility of CAR (Vol). To compute CAR, the market model is used as the benchmark in which the CAC 40 index acts as proxy for the market. The model is estimated using daily data and the event window is 5 days, that is, 2 days before and 2 days after the day of the event. For each stock, 3 months of trading data (58 observations) have been collected before this time period to run a standard market model regression with the stock return on the left-hand side and the market return as the independent variable. For each day of the event study window an abnormal return is computed, defined as the difference between the realized return and the predicted return using the parameter estimates from the regression. The CAR is the average daily abnormal return over the 5-day window. The conditional

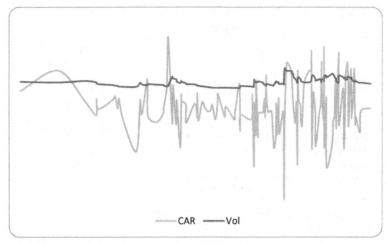

Figure 6.1 The cumulative abnormal return (CAR) and its corresponding volatility (Vol).

volatility of CAR is computed using a GARCH(1,1) model for the data-generating process of CAR. The data contain 148 distinct events involving 31 distinct companies and span the period from September 2003 to September 2009.

Table 6.1 summarizes the properties of the sample data as well as the dependent variable of our empirical analysis in the next section. As reported in Panel A, the number of relevant CSR events increased substantially from 1 or 2 at the beginning of the sample period to 48 in 2008 and remained at 30 in 2009, at the paroxysm of the financial crisis[6]. Panel B details the types of event. This classification has been suggested by Drusch and Lioui (2010) since the decomposition used in the KLD database is better suited to US companies, while there are some dimensions that matter more for French companies.

Panel C reports the details for the characteristics of the distribution of the CAR series. Average CAR is only 0.01%, which is very low (even before accounting for transaction costs), although on a daily basis. The standard deviation is substantial, 0.87% on a daily basis with a minimum for the CAR at −3.10% and a maximum at 2.37%. As one can see from Figure 6.1, the range of the CAR widened substantially at the end of the sample, that is, in the years of substantial market turmoil. The conditional volatility seems to have been relatively high throughout the sample and not only during the crisis period. It ranges from 0.65% to 1.31% with an average of 0.87% on a daily basis. It was positively skewed (0.602) in our sample period, which must be related to the particular market conditions in the sample end. Finally, of interest is the persistence of the volatility series, since the autoregressive coefficient is 0.882, which is relatively high, and this will need to be taken into account in our empirical investigation below.

---

[6] The Kyoto Protocol, adopted in 1997, entered into force in February 2005. This could be one possible explanation for the substantial increase in the number of the events by the end of our sample period.

Table 6.2 Summary Statistics of Indicators of Market Conditions

Panel A

	VIX	Market	ICS
Mean	25.67%	1.58%	73.92
Standard	14.07%	0.90%	12.87
Skewness	1.215	1.372	0.03
Kurtosis	3.845	4.426	1.67
Minimum	10.08%	0.63%	55.30
Maximum	71.80%	4.32%	96.90
Auto	0.967	0.979	0.96

Panel B

	VIX	Market	ICS
VIX	1		
Market	0.96	1	
ICS	−0.76	−0.73	1

In Table 6.2, we report the summary statistics of three indicators of market conditions as used in the text. VIX stands for an index of the implied volatility in the S&P 100 options with around 30 days to maturity. It is expressed on a yearly basis and computed as the average of the VIX values during the 5 days of the event window. *Market* is the volatility of the daily market return during the period of the event. It is expressed on a daily basis like CAR. ICS is published monthly by the University of Michigan and Thomson Reuters. It aims at measuring consumers' attitude and sentiment *vis-à-vis* the economy. It has been normalized to have a value of 100 in December 1964 and we retained the index level during the month of the event.

Figure 6.2 summarizes the properties of the sample data for our three measures of market conditions. Among the noteworthy features are (1) the strong persistence of the series, which needs to be carefully accounted for in the inferences and (2) the relatively high correlation between the three measures. This is comforting since this makes it likely that these measures help to capture a similar phenomenon. Hopefully the correlation is still not perfect and therefore there is no redundancy in looking at the results for the three variables. The two volatility measures are positively related (this is a standard result for papers comparing realized with implied volatilities in the options markets). They are both negatively correlated with ICS, since a low index implies a low market sentiment, usually related to high market uncertainty. As expected, these indicators peak in 2008, during which the VIX, for example, reached levels never seen before (80% implied volatility on an annual basis).

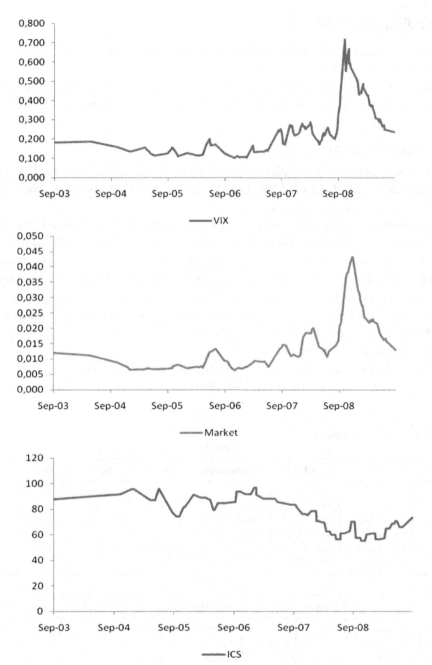

Figure 6.2 Indicators of market conditions.

6.4 EMPIRICAL FINDINGS

Our purpose, as stated previously, is to assess the impact of macroeconomic uncertainty in financial markets on the uncertainty related to the impact of CSR events on firms' market valuations. We systematically use, as the dependent variable, the volatility of CAR. We first present the results for our benchmark case and then some robustness checks.

6.4.1 Main Findings

In our benchmark case the explanatory variables are one of the indicators of market conditions as described above (Market, VIX, and ICS). The results of our benchmark investigation are reported in Table 6.3.

We report the results of a regression in which the dependent variable is the conditional volatility of the CAR. The dependent variable is one of the indicators of market conditions. In Panel B we added, as an independent variable, the lagged conditional volatility. For each regression we report three tests of the presence of a unit root in the residual of the regressions. Augmented Dickey Fuller 1 (ADF1) assumes an autoregressive model for the residual, ADF2 assumes an autoregressive model with drift, and ADF3 assumes an autoregressive model with drift and trend stationary. The critical value for the t statistic at 1% confidence is −3.44.

In Panel A, one can see that all three variables turn out to have a statistically significant impact on CAR's volatility. *Market* and VIX have a positive impact, meaning that a high

Table 6.3 Regression Analysis

Panel A

		Constant	Independent	Adj. R^2	ADF1	ADF2	ADF3
VIX	Coefficient	0.01	0.0050	0.25	−3.71	−4.15	−4.03
	t statistic	35.72	7.10				
Market	Coefficient	0.01	0.0866	0.31	−3.80	−4.23	−3.95
	t statistic	37.81	8.19				
ICS	Coefficient	0.01	−0.0001	0.46	−4.63	−5.15	−5.00
	t statistic	28.71	−11.24				

Panel B

		Constant	Lagged	Independent	Adj. R^2	ADF1	ADF2	ADF3
VIX	Coefficient	0.00	0.83	0.0010	0.78	−10.72	−8.51	−6.66
	t statistic	3.43	18.82	2.36				
Market	Coefficient	0.00	0.82	0.0189	0.79	−10.72	−8.54	−6.64
	t statistic	3.66	17.90	2.70				
ICS	Coefficient	0.00	0.79	−0.00001	0.78	−10.50	−8.43	−6.68
	t statistic	3.52	14.72	−2.51				

ADF, Augmented Dickey Fuller; Adj., adjusted.

market realized volatility or forward volatility will both substantially increase the volatility of CAR. This in turn renders extremely uncertain the impact of CSR on firm value. The impact of ICS is negative since bad consumer sentiment means a lower index, which increases the volatility and hence the negative relationship. All indexes thus impact in the same direction on the uncertainty surrounding the impact of CSR. The difference in the coefficients' size needs to be interpreted in the light of the difference in scale between the three indicators. For example, the coefficient on the market volatility is 0.08, meaning that an absolute daily increase in market volatility of 0.01 will increase the volatility of CAR by 0.08. Given that the average CAR is only 0.0001, this means that the impact of *Market* is economically very substantial. VIX is on yearly basis and hence the relatively lower size and, finally, ICS is on basis 100, which explains the even lower size of the coefficient.

Overall, Panel A delivers the main message of this chapter: market conditions do matter for the impact of CSR on firm value; bad market conditions are likely to increase the uncertainty of the impact of CSR on firm value; this increase in uncertainty is sizable.

Unit root tests presented in the last column for different specifications suggest that the behavior of the residuals of the model are reasonable, although only slightly in some cases. Hence it is important to control for the persistence of conditional volatility as has been done in Panel B.

In Panel B we thus augmented the regression by the lagged value of the CAR's volatility. As expected, this lagged value turns out to be extremely significant. Unit root tests in the last column strongly reject the null of a unit root in the residuals. Although extremely conservative, the indicators of market conditions turn out to still be statistically significant, although the size of their impact is lower now.

This benchmark case already shows the main point of this chapter: market conditions are extremely important for the impact of CSR on the market value of firms conducting CSR.

6.4.2 Robustness Checks

We performed various robustness checks on the previous findings, in several directions. The first was to control for the window length. As advocated by previous literature, this length may be of importance. The results are gathered in Table 6.4 and Figure 6.3.

We report the results of a regression in which the dependent variable is the conditional volatility of the CAR. The independent variables are the lagged volatility, the event type and an indicator of the market conditions, that is VIX in Panel A, market volatility in Panel B, and ICS in Panel C. For each regression we report three tests of the presence of a unit root in the residual of the regressions. ADF1 assumes an autoregressive model for the residual, ADF2 assumes an autoregressive model with drift, and ADF3 assumes an autoregressive model with drift and trend stationary.

Table 6.4 Robustness Checks: Changing the Event Window

Panel A

	Window [0, 1]		Window [0, 2]		Window [−1, 1]	
	Constant	VIX	Constant	VIX	Constant	VIX
Coefficient	0.014	0.004	0.009	0.008	0.011	0.002
t statistic	52.965	4.031	30.414	7.853	76.785	4.730
Adj. R^2	0.095		0.294		0.128	
ADF1	−2.493		−2.813		−2.530	
ADF2	−2.418		−2.698		−2.510	
ADF3	−2.387		−2.729		−2.523	
	Constant	Market	Constant	Market	Constant	Market
Coefficient	0.014	0.065	0.009	0.141	0.011	0.038
t statistic	55.051	4.693	32.182	8.857	79.653	5.077
Adj. R^2	0.126		0.347		0.145	
ADF1	−2.514		−2.786		−2.496	
ADF2	−2.383		−2.685		−2.454	
ADF3	−2.306		−2.564		−2.436	
	Constant	ICS	Constant	ICS	Constant	ICS
Coefficient	0.022	0.000	0.021	0.000	0.015	0.000
t statistic	38.720	−11.951	31.025	−14.075	53.461	−13.411
Adj. R^2	0.493		0.574		0.551	
ADF1	−3.665		−3.964		−3.673	
ADF2	−3.415		−4.027		−3.558	
ADF3	−3.445		−4.016		−3.611	

Panel B

Window [0, 1]

	Constant	Lagged	VIX
Coefficient	0.001	0.923	0.000
t statistic	2.328	29.679	1.139
Adj. R^2	0.872		
ADF1	−12.635		
ADF2	−8.890		
ADF3	−6.636		
	Constant	Lagged	Market
Coefficient	0.001	0.919	0.008
t statistic	2.422	29.161	1.422
Adj. R^2	0.873		
ADF1	−12.673		
ADF2	−8.914		
ADF3	−6.624		
	Constant	Lagged	ICS
Coefficient	0.003	0.875	−0.00001
t statistic	2.809	20.994	−1.993
Adj. R^2	0.874		
ADF1	−12.387		
ADF2	−8.706		
ADF3	−6.559		

Window [0, 2]

	Constant	Lagged	VIX
Coefficient	0.001	0.908	0.001
t statistic	2.421	27.730	1.974
Adj. R^2	0.888		
ADF1	−12.405		
ADF2	−8.934		
ADF3	−6.652		
	Constant	Lagged	Market
Coefficient	0.001	0.894	0.020
t statistic	2.680	26.734	2.530
Adj. R^2	0.890		
ADF1	−12.457		
ADF2	−8.992		
ADF3	−6.641		
	Constant	Lagged	ICS
Coefficient	0.003	0.860	−0.00002
t statistic	3.094	20.384	−2.544
Adj. R^2	0.890		
ADF1	−12.146		
ADF2	−8.901		
ADF3	−6.773		

Window [−1, 1]

	Constant	Lagged	VIX
Coefficient	0.001	0.934	0.000
t statistic	2.212	32.389	1.133
Adj. R^2	0.894		
ADF1	−12.407		
ADF2	−8.867		
ADF3	−6.091		
	Constant	Lagged	Market
Coefficient	0.001	0.930	0.004
t statistic	2.308	32.118	1.438
Adj. R^2	0.895		
ADF1	−12.440		
ADF2	−8.887		
ADF3	−6.081		
	Constant	Lagged	ICS
Coefficient	0.002	0.880	−0.00001
t statistic	2.920	22.042	−2.206
Adj. R^2	0.897		
ADF1	−12.130		
ADF2	−8.684		
ADF3	−6.067		

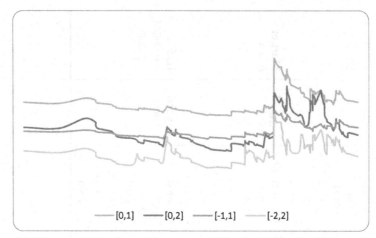

Figure 6.3 The cumulative abnormal return conditional volatility for different windows.

First of all, while the size of the conditional volatility does depend upon the window's size, the time series behavior of the conditional volatility is more or less the same as shown in Figure 6.3. Interestingly, the smaller the event's window, the greater the conditional volatility. In Panel A of Table 6.4 we present the results of the regression analysis when only the dependent variable is included in the regression. It turns out that, for all the window lengths, the three indicators of market condition have the right sign and are statistically strongly significant, confirming in this the previous results obtained with the wide window. When adding the lag of the conditional volatility, the findings are slightly less clear cut. When using the ICS as a measure of market conditions, the impact is always negative and statistically significant, even when one controls for the persistency of the conditional volatility. For VIX and *Market*, it seems that their significance depends on the window length, although the direction of the impact is the expected one (positive).

The second robustness check we performed is related to the particular type of CSR. As advocated by Bird et al. (2007) and Drusch and Lioui (2010), CSR event type is likely to matter for the impact of CSR on firm value. The results are reported in Table 6.5.

We report the results of a regression where the dependent variable is the conditional volatility of the CAR. The independent variables are the lagged volatility, the event type (*Type*), and an indicator of the market conditions, that is VIX in Panel A, Market volatility in Panel B, and ICS in Panel C. All the coefficients have been multiplied by 100 for readability. For each regression we report three tests of the presence of a unit root in the residual of the regressions. ADF1 assumes an autoregressive model for the residual, ADF2 assumes an autoregressive model with drift, and ADF3 assumes an autoregressive model with drift and trend stationary. The critical value for the t statistic at 1% confidence is -3.44.

Table 6.5 Robustness Checks: Accounting for CSR Event Type

Panel A: VIX

		Constant	Lagged	Type	Independent	Adj. R^2	ADF1	ADF2	ADF3
Environment	Coefficient	0.118	83.199	0.003	0.102	0.78	−10.71	−8.52	−6.69
	t statistic	3.38	18.75	0.21	2.33				
Health	Coefficient	0.120	82.950	0.010	0.104	0.78	−10.82	−8.52	−6.67
	t statistic	3.44	18.65	0.56	2.38				
Sport	Coefficient	0.119	83.214	−0.012	0.103	0.78	−10.77	−8.50	−6.67
	t statistic	3.42	18.77	−0.46	2.36				
Suburbs	Coefficient	0.118	83.192	−0.042	0.116	0.79	−10.71	−8.50	−6.80
	t statistic	3.44	19.01	−1.98	2.65				
Insertion	Coefficient	0.121	83.139	−0.010	0.101	0.78	−10.68	−8.48	−6.66
	t statistic	3.46	18.77	−0.61	2.32				
Crisis 2007	Coefficient	0.136	80.705	0.022	0.064	0.78	−10.65	−8.50	−6.67
	t statistic	3.72	17.05	1.42	1.26				
Crisis 2008	Coefficient	0.116	83.363	−0.006	0.116	0.78	−10.76	−8.55	−6.72
	t statistic	3.14	18.58	−0.29	1.80				

Panel B: Market

		Constant	Lagged	Type	Independent	Adj. R^2	ADF1	ADF2	ADF3
Environment	Coefficient	0.128	81.620	0.004	1.890	0.78	−10.71	−8.56	−6.69
	t statistic	3.61	17.85	0.32	2.68				
Health	Coefficient	0.129	81.422	0.009	1.906	0.78	−10.81	−8.54	−6.65
	t statistic	3.66	17.76	0.52	2.71				
Sport	Coefficient	0.129	81.642	−0.013	1.901	0.78	−10.78	−8.54	−6.66
	t statistic	3.66	17.86	−0.48	2.70				
Suburbs	Coefficient	0.129	81.468	−0.044	2.131	0.79	−10.71	−8.54	−6.81
	t statistic	3.72	18.07	−2.08	3.03				
Insertion	Coefficient	0.130	81.630	−0.009	1.861	0.78	−10.69	−8.52	−6.65
	t statistic	3.68	17.86	−0.51	2.64				
Crisis 2007	Coefficient	0.143	79.384	0.021	1.388	0.79	−10.65	−8.53	−6.67
	t statistic	3.91	16.44	1.36	1.75				
Crisis 2008	Coefficient	0.125	81.813	−0.009	2.228	0.78	−10.78	−8.61	−6.73
	t statistic	3.42	17.82	−0.48	2.26				

Continued

Table 6.5 Robustness Checks: Accounting for CSR Event Type—cont'd

Panel C: ICS

		Constant	Lagged	Type	Independent	Adj. R²	ADF1	ADF2	ADF3
Environment	Coefficient	0.288	78.943	0.003	−0.001	0.78	−10.49	−8.44	−6.71
	t statistic	3.47	14.66	0.20	−2.48				
Health	Coefficient	0.291	78.694	0.009	−0.001	0.78	−10.59	−8.43	−6.69
	t statistic	3.53	14.61	0.51	−2.52				
Sport	Coefficient	0.290	78.930	−0.012	−0.001	0.78	−10.55	−8.43	−6.70
	t statistic	3.52	14.69	−0.46	−2.51				
Suburbs	Coefficient	0.305	78.650	−0.041	−0.002	0.79	−10.44	−8.40	−6.81
	t statistic	3.72	14.81	−1.92	−2.75				
Insertion	Coefficient	0.288	78.995	−0.008	−0.001	0.78	−10.47	−8.41	−6.69
	t statistic	3.48	14.69	−0.50	−2.44				
Crisis 2007	Coefficient	0.239	78.742	0.019	−0.001	0.78	−10.53	−8.45	−6.67
	t statistic	2.49	14.69	1.04	−1.16				
Crisis 2008	Coefficient	0.279	78.715	0.007	−0.001	0.78	−10.48	−8.39	−6.62
	t statistic	3.26	14.61	0.45	−2.02				

ADF, Augmented Dickey Fuller; Adj., adjusted.

Across the board, the three indicators happen to be significant with the right sign in the vast majority of cases. Interestingly, when the uncertainty surrounding the impact of CSR is concerned, the CSR event type seems to be of little importance, if any. The "suburb" type happens to be the most important one with a negative impact on the uncertainty.

6.5 CONCLUSION

The findings of this chapter have highlighted the importance of the business environment in assessing the outcome of any CSR action of a firm, although such an action has a positive connotation a priori. When there is too much noise in a market as a result of increased dispersion in beliefs, a CSR action may simply be useless, since the market is anyway unable to have good foresight on the future prospects of the economy. If the impact of CSR on firm value is of primary importance for the responsible firm, then the findings in this chapter would suggest postponing such activities until markets are less turbulent. So does it pay to be good in bad times? Our answer would be in good times, nothing guarantees that a positive CSR action will impact positively on firm value, as a result of heterogeneous beliefs; in bad times, when the dispersion in beliefs is even greater, a positive impact on firm value is even less likely to happen.

ACKNOWLEDGMENTS

I would like to thank Laetitia Drusch for valuable research assistance. Discussions on related topics with Michelle Sisto and Zenu Sharma, and comments from Vikash Ramiah were extremely helpful.

REFERENCES

Bae, K.H., Kang, J.K., Wang, J., 2011. Employee treatment and firm leverage: a test of the stakeholder theory of capital structure. J. Financ. Econ. 100 (1), 130–153.
Bansal, R., Yaron, A., 2004. Risks for the long run: a potential resolution of asset pricing puzzles. J. Finance 59 (4), 1481–1510.
Beccheti, L., Ciciretti, R., Hassan, I., Nada, J., 2012. Corporate social responsibility and shareholder's value. J. Bus. Res. 65 (11), 1628–1635.
Bird, R., Hall, A., Momente, F., Reggiani, F., 2007. What corporate social responsibility activities are valued by the market? J. Bus. Ethics 76 (2), 189–206.
Bloom, N., 2009. The impact of uncertainty shocks. Econometrica 77 (3), 623–685.
Bloom, N., Bond, S., Van Reenen, J., 2007. Uncertainty and investment dynamics. Rev. Econ. Stud. 74 (2), 391–415.
Bonomo, M., Garcia, R., Meddahi, N., Tedongap, R., 2011. Generalized disappointment aversion, long-run volatility risk and asset prices. Rev. Financ. Stud. 24 (1), 82–122.
Chatterji, A., Levine, D., Toffel, M., 2009. How well do social ratings actually measure corporate social responsibility? J. Econ. Manag. Strategy 18 (1), 125–169.
Campbell, J., Lo, A., Mackinlay, C., 1997. The Econometrics of Financial Markets. Princeton University Press.
Drusch, L., Lioui, A., 2010. French corporate social responsibility: which dimension pays more? Bankers, Mark. Investors 109, 47–55 (November–December).

El Ghoul, S., Guedhami, O., Kwoh, C., Mishra, D., 2011. Does corporate social responsibility affect the cost of capital? J. Bank. Financ. 35 (9), 2388–2406.

Fisher-Vanden, K., Thorburn, K., 2011. Voluntary corporate environmental initiatives and shareholder wealth. J. Environ. Econ. Manag. 62 (3), 430–445.

Goss, A., Roberts, G., 2011. The impact of CSR on the cost of bank loans. J. Bank. Finance 35 (7), 1794–1810.

Horvathova, E., 2010. Does environmental performance affect financial performance? A meta-analysis. Ecol. Econ. 70 (1), 52–59.

Jensen, M., 2002. Value maximization and the corporate objective function. Bus. Ethics Q. 12 (1), 235–256.

King, A., Lenox, M., 2001. Does it *really* pay to be green? An empirical study of firm environmental and financial performance. J. Ind. Ecol. 5 (1), 105–106.

Kurz, M., 1994. On the structure and diversity of rational beliefs. Econ. Theory 4 (6), 877–900.

Leahy, J., Whited, T., 1996. The effects of uncertainty on investment: some stylized facts. J. Money Credit Bank. 28 (1), 64–83.

Lioui, A., Sharma, Z., 2012. Environmental corporate social responsibility and financial performance: disentangling direct and indirect effects. Ecol. Econ. 78 (June), 100–111.

Margolis, J., Elfenbein, H., Walsh, J., 2007. In: Does it Pay to Be Good ? A Meta-analysis and Redirection of Research on the Relationship between Corporate Social and Financial Performance. Working Paper. Harvard Business School, Boston, USA.

Rahman, N., Post, C., 2012. Measurement issues in environmental corporate social responsibility (ECSR): toward a transparent, reliable, and construct valid instrument. J. Bus. Ethics 105 (3), 307–319.

Ramiah, V., Pham, A., Nguyen, H., Moosa, I., 2015. Are european environmental regulations excessive? In: Financial Markets & Corporate Governance Conference, 2015.

Sharfman, M., 1996. The construct validity of the Kinder, Lydenberg & Domini social performance ratings data. J. Bus. Ethics 15 (3), 287–296.

Vanhamme, J., Grobben, B., 2009. "Too good to be true!". The effectiveness of CSR history in countering negative publicity. J. Bus. Ethics 85 (2), 273–283.

Verwijmeren, P., Derwall, J., 2010. Employee well-being, firm leverage and bankruptcy risk. J. Bank. Financ. 34 (5), 956–964.

Yamagushi, K., 2008. Reexamination of stock Price reaction to environmental performance: a GARCH application. Ecol. Econ. 68 (1–2), 345–352.

CHAPTER 7

Public Value of Environmental Investments: A Conceptual Outlook on the Management of Normatively Determined Risks

Camillo von Müller[1], Steven A. Brieger[1,2]
[1]Center for Leadership and Values in Society, University of St. Gallen, St. Gallen, Switzerland; [2]Institute of Corporate Development, Leuphana University of Lüneburg, Lüneburg, Germany

Contents

7.1 INTRODUCTION

This chapter discusses current and historic developments that exist with regard to holistic approaches toward the identification, evaluation, and risk management of externalities. If green concerns are widespread in postmaterialistic economically developed societies (Welzel, 2013), it is to assume that neither the fields of economics nor management can confine themselves to questions that focus on material living standards. That does not necessarily imply that the academic disciplines of "green finance," "environmental finance," "sustainable finance," "green economics," or "alternative management" necessarily conflict with the materialistic outlook of traditional mainstream economics, management, or finance. Rather, demand for academic contributions is particularly high, if theories help investors to "do good by doing well" (Brainerd et al., 2013). The objective of this chapter is to identify current theory developments and their historic underpinnings that fulfill investor demands for explaining and systemizing the ecological and monetary aspects of "green finance." In this context, the chapter focuses on a risk management perspective that conceptually applies evidence and theory of current public value research to the task of hedging normative risks such as legislative changes and reputational risks.

Handbook of Environmental and Sustainable Finance
ISBN 978-0-12-803615-0

In order to achieve this objective, the chapter proceeds as follows: Section 7.2 highlights the growing relevance of environmentally conscious investments in Austrian, German, and Swiss markets that form the economic and cultural context in which the public value approach discussed in this chapter has been developed. Section 7.3 reviews historic underpinnings of these theories focusing also on the classical argument of the diversion of private and social marginal costs that trigger the need for valuation approaches which include but are not limited to information provided by market outcomes. Section 7.4 discusses public value theory as conceptual approach that fulfills this demand by relying on subjective criteria of evaluation that goes beyond market-based processes of evaluation. Section 7.5 reviews empirical results of public value measurement provided by the Public Value Atlas, a recent research project developed at the University of St. Gallen. If public value ratings translate into different risk profiles, then investing into companies that produce high public values can be viewed as strategy to hedge risks that stem from potential perceived public value violations such as reputational risks. Section 7.6 highlights the link between reputational risks and public value. Section 7.7 reviews public value investments (PVI) as hedging tools under the framework of option pricing theory paying special attention to the issue of reputational risks. Section 7.8 summarizes main results and limitations, offering conclusions for practitioners in form of five propositions.

7.2 THE RISE OF ENVIRONMENTAL INVESTMENTS

Green investments, or in other terms, "socially responsible investing," "environmental, governmental and social investing (ESG)," "sustainable finance," or "environmental finance" have become increasingly popular since the 1980s, when environmental issues and problems came more into focus of public debates. Yet, the history of so-called responsible investment reaches back several hundred years. In biblical times, directives on how to invest ethically were laid down in Jewish laws. Christian, Islamic, and Jewish theologies have tended to support investments that agree directly with their beliefs. Catholic edicts and Quakers avoided investing in enterprises that profit from goods designed to enslave or kill fellow human beings. The Quakers have never condoned investing in war or slavery. Since the 1960s, investors have begun to care more systematically about environmental issues. However, it was not before the 1990s that the dangers of environmental pollution led to the founding of first ecological green funds in Europe (Mermod and Idowu, 2014). Since then, green investments, where investors do not base their decisions on cash flow statements alone but also take into account the social and ecological aspects of their investments, have become increasingly popular (Leahy, 2008). Buzz words that characterize investment areas in this field are terms such as "green technology," "sustainable living," "renewable energy," and "green construction."

Furthermore, wherever investors put their money into companies and industries that are responsible for external effects, which harm the environment (Nkambule and Blignaut, 2012), compensation claims may linger so that the financing of such companies can be a costly exercise, resulting in lower long-lasting profits and stock values. Also, for this reason, investors are interested in creating a "cleaner environment while also delivering competitive rates of returns" (Mermod and Idowu, 2014, p. 328). Moreover, the personal factor should not be underestimated. Many investors support environmental investment strategies due to own ethical, social, religious, and other preferences they share.

In the German-speaking area, the volume of the overall market is now valued at 134.5 billion Euro (Dittrich et al., 2014). According to the Sustainable Investment Market Report 2014 published by *Forum Nachhaltige Geldanlagen* (FNG), the *Forum of Sustainable Money Investments*, sustainable investment funds and mandates grow very strongly, in Austria by 29%, in Germany and Switzerland by 17%, which means that in all three countries the sustainable investment market has grown at a higher rate than the conventional investment market. In all three countries, institutional investors attach great importance to the integration of ESG criteria into their investment decision-making and ownership practices. Claudia Tober, FNG's Executive Director, points out that

> *[i]n Switzerland and Austria, corporate pension funds are the most frequent investors in sustainable investment solutions, while in Germany religious institutions and charitable organizations account for the lion's share of sustainable investments. But private investors also have a significant market share, of 41 per cent in Switzerland, 25 per cent in Germany and 14 per cent in Austria.*
>
> **Forum Nachhaltige Geldanlagen e.V. (May 7, 2014).**

In analogy to above-described market growth, mutual funds could increase their sustainability investments over the last years, resulting in a 45 billion Euro volume in 2014.

Meanwhile, all three countries occupy leading places in the global ecosystem of environmental and sustainable finance (Knoepfel and Imbert, 2013). This development is driven by their favorable country ratings. Sustainable rating agencies, which rate countries by using several criteria, covering from performance in environmental issues, climate change and energy, production and consumption as well as human rights and fundamental freedoms, give regularly Austria, Germany, and Switzerland very good class testimonies that ultimately exert an influence on further investment decisions made by investors (Oekom, 2014).

If the experiences of the last years can be extrapolated into the future, it seems fair to assume that investments that account for environmental, ethical, and social issues will grow in importance not only in countries such as Austria, Germany, and Switzerland, but will also develop from a niche market to mainstream worldwide. In fact, the hypothesis that responsible investment strategies such as ESG will play an increased role in the future is reconfirmed by value shifts, and increasing concerns with regard to the various environmental risks and challenges that exist as of today. If market development continues as described above it seems reasonable to assume that in the future, the incorporation of environmental as well as social and governmental risks will become—even more than today—an integral task of diligent investment management. In this context, the question how to integrate various forms of business, financial, environmental, and political risks into a common evaluative framework is of fundamental relevance to both academics and practitioners alike. The subsequent paragraphs thus present a historic overview and current account of integrated approaches toward the evaluation of internal and external risks and values.

7.3 HISTORICAL BACKGROUNDS

In 1913, the first international conference for environmental protection was held in Bern, Switzerland. Among the leading heads who had planned and directed the get-together of 17 nations from all over the globe was Basel-born scientist Paul Sarasin. Since then, Switzerland—and in particular Basel—has been playing a decisive role as pioneer in the international protection of the environment (Schefold, 2006). In the second half of the twentieth century, this role has been reflected in the theories of Swiss economists and management scholars and their incorporation of environmental aspects into their research. In his 1955 study on the economics of nuclear energy, economist and former president of Basel University, Edgar Salin (1955) quotes an influential 1819 article of *Cölnische Zeitung*, a leading German daily newspaper in the late eighteenth and early nineteenth century, that discusses the disadvantages of public gas-lighting. The author of the article quoted by Salin (1955) summarizes several arguments that contradict the installation of public gas lamps, such as theological reasoning (public illumination violates the divine order of nocturnal darkness), legal arguments (public lightning "taxes" those who prefer natural dimness), medical aspects (illuminated streets invite for late strolls that increase the risks of catching a cold), philosophical considerations (drunkards and lovebirds will be led to excesses in early morning hours), forensic calculations (the public illumination may dare thieves and burglars to undertake additional ventures), economic logic (an increased gas production demands additional import of coal that weaken national prosperity), and last but not least patriotic emotions (illumination ceremonies at national holidays will lose in appeal if constant illumination is publicly available) (Salin, 1955, pp. 9—10). Salin (1955) refers to *Cölnische Zeitung* so as to illustrate that processes of economic production and energy consumption have been viewed in context of the externalities argument right from the beginning of modern industrialization.

Salin's (1955) observations that environmental considerations have been established long ago within the realm of economic thought and analysis are backed by Walter A. Jöhr (1990), economist and former head of the University of St. Gallen in Switzerland. Jöhr (1990, p. 73) begins his historic account of ecologically conscious economics with Thomas Malthus' theories on the limits to demographic growth. He focuses on Malthus' observation that the human population tends to grow more rapidly than the stock of resources needed for its preservation. As Jöhr (1990) points out, Malthus views the fact that population growth exceeds resource growth as direct explanation for the reason why

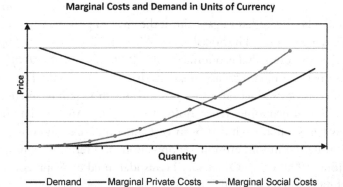

Figure 7.1 A textbook representation of the divergence of marginal private and social costs.

humanity is suffering from famine and poverty. The remedy to overcome this problem is technical progress that can raise resource growth above the growth rate of the population. Salin (1955) and Jöhr (1990) further agree on the fact that classical economists were aware of the growth-limiting nature of scarce (environmental) resources since in their theories, scarcity either translated into higher transport costs and/or increased the prices of raw materials.

With regard to the twentieth century development of ecological economics, Jöhr (1990, p. 74) identifies the observations of Cambridge economist Arthur Pigou on divergences between private and social marginal costs in competitive markets as important milestone. Figure 7.1 contains a textbook graphical representation of the divergence of marginal private and marginal social costs. From the viewpoint of producing firms, output should take place where the (market) demand curve and the marginal private costs (mpc) intersect. From the viewpoint of society, production should take place at the parameters determined by the intersection of the demand curve and the marginal social costs (msc) curve. Under the assumption of Pigou that msc > mpc, firms produce quantities above the social optimum.

The logic inherent to the social costs problem as represented by Pigou is the assumption that situations exist in which firms are not confronted with the true costs of capital that they employ for production. Even though the means of production of a firm can be determined in objective and physical terms, the question in how far these means enter the firm's cost function is resolved in normative terms depending not only on the question of the existence of property rights, and similar issues of regulation (i.e., usage fees, taxes, etc.) but also on how far consumers exercise their "voice options" (Hirschman, 1970) to fore the respective firm to change its behavior.

Among the most prolific ecological economists of the early and mid-twentieth century, who discussed the physical and normative aspects of the usage of natural resources in competitive markets, is Karl W. Kapp, whom the *Journal of Economic Issues* identifies in a 1976 obituary as "institutionalist scholar in the great tradition of European scholarship," and who succeeded Edgar Salin at the chair of political economy at the University of Basel. Kapp's work "The Social Costs of Private Enterprise" belongs to the standard literature on environmental economics (Jöhr, 1990, p. 75; JEI (1976)). In this book, Kapp identifies the limits of competitive markets that exist due to the fact that the full inclusion of total costs of production into a given firm's production function is solved in normative, and not in economic terms (Jöhr, 1990, p.75). Among the costs that Kapp (2000) views as being often excluded from cost functions, are environmental damages, such as air and water pollution, erosion, deforestation, and terminal consumption of energy sources (Jöhr, 1990, p.75). Other social costs identified by Kapp comprise of health problems, technical unemployment, and wrongly targeted scientific endeavors (Jöhr, 1990, p.75). Perry et al. (2014) recently reconfirmed Kapp's (2000) analysis with regard to the existence of incomplete costs functions by discussing fiscal strategies in selected countries in order to ensure that global energy prices reflect health and environmental costs of fuel use.

Given the normative and thus flexible nature of the question in how far legal environments and conscious consumers may induce firms to incorporate externalities into their cost functions, investors who aim at minimizing risks associated with this question are in need of tools that help them to gain a systematic outlook on the risks and values that a firm produces for its various stakeholders, as well as for society in general.

7.4 PUBLIC VALUE THEORY AND NORMATIVE VALUE CREATION

If Kapp's (2000) observations are true and market-based modes of production rest on nonmarket resources than the question arises, in how far nonmarket forces exist, that normatively induce firms to compensate the usage of these resources. For investors, this question is of particular relevance in transitional periods, i.e., when regimes of property rights (including the tax system) are subject to change so that firms are forced to incorporate previously public resources into their cost functions (e.g., in the form of taxes, or direct costs due to "enclosures," or ex-post compensation through compensation claims).

Viewed from this angle, "green" investments can be identified as strategies to manage "enclosure risks" by prompting investors to select firms that deliver positive returns without relying on the free exploitation of public resources. Hence, green investors "need a … [good] sense of society's expectations. They need … [analytic] tools to … constantly monitor how things evolve over time. In particular, a common framework is needed that allows for a language that is robust enough to systematically give different world views and value systems" (Meynhardt et al., 2014, p. 5). A framework that may fulfill these requirements is public value theory (see Meynhardt, 2009; Bozeman, 2007).

According to Meynhardt, public value is the combined view of what a society evaluates as being valuable which can be summarized by the phrase "public value is what the public values." What the public values, depends on the basic needs of its members. Meynhardt (2009) follows psychologist Epstein, who formulates four basic human needs—the basic need for positive self-evaluation, for maximizing pleasure and avoiding pain, for gaining control and coherence over one's conceptual system, and for forming positive relationships. These needs allow for the formulation of a multidimensional framework that accounts for market and nonmarket values of a given subject that is being evaluated. These dimensions include a moral—ethical dimension, a political—social dimension, a utilitarian—instrumental dimension, and a hedonistic—aesthetical dimension. Meynhardt and Bartholomes (2011) have confirmed the existence and relevance of all four dimensions empirically. According to Meynhardt (2009), entities such as a private company, an investment firm, a public administration, or an NGO create value along above-described dimensions. In other words, subjects such as customers, employees, and other stakeholders evaluate the products and services delivered by these entities along these dimensions. Markets may—but do not necessarily—reflect the outcome of these evaluations. Neither do these evaluations need to be consistent. For example, as consumers, individuals may purchase cheap airline tickets the price of which does not account for the externalities produced by airline companies, while they may vote at the same time for policies that induce environmental taxes as those proposed by Parry et al. (2014). In this situation, relying on the information reflected in ticket prices is not sufficient for aviation firms (and their investors) so as to estimate the probability of changes in their cost structures due to alternating future legal environments.

In order to arrive at a more detailed and complete picture with regard to the probability of changes in a firm's cost function due to changing normative environments, organizations need to ask themselves if and how they deliver public value to society: by behaving in a decent manner (moral—ethical dimension), by contributing to societal cohesion (political—social dimension), by raising the quality of life of human beings (hedonistic—aesthetical dimension), and/or by performing well in its core business (utilitarian—instrumental dimension).

This task is complex, since human needs differ individually and culturally. Some individuals show a preference for societal cohesion, favoring equality, justice, and fairness, others have a focus on individual liberties. In some world regions, in-group thinking and collectivism are more pronounced, while people living in individualist societies stress that individual liberty and responsibility are important prerequisites of modern society, if not individual freedom per se. Hedonic needs are distributed differently as well. Consequently, cultural values impact public values—the idea about what makes life worth living and how companies and organizations must behave to serve to the common good, are not equally distributed among human beings living in different cultural background.

Environmental investments correspond to all four basic needs of public value: they create financial returns (thus responding to the utilitarian—instrumental dimension of

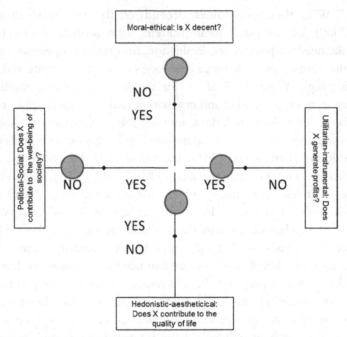

Figure 7.2 Evaluating public value. The framework represents a four-dimensional evaluation of a given good, service, organization, or action (X). For example, in a hypothetical evaluation outcome, subjects may believe that a ticket rebate of an aviation carrier may be as strategy that generates value along utilitarian—instrumental dimension (i.e., it generates profits for the company and earnings for its stockholders), and the hedonistic—aesthetical dimensions (it allows consumers to purchase airline tickets at lower costs), while conflicting with moral—ethical considerations (that may be due to high environmental costs, associated with this kind of strategy) and political—social dimensions (e.g., if the company has to cut employment benefits, or is relocating in order to refinance lower fares). The framework above represents a simplified evaluation of this kind of example. For a full discussion of public value valuations in form of a scorecard see Meynhardt et al. (2014).

public value), and often integrate moral—ethical dimensions into investment decisions (e.g., by aiming to minimize externalities), while also offering the promise of bettering the world (e.g., by reducing pollution growth), or at least society (and thus scoring high in political—social public value dimensions as well). Figure 7.2 depicts this kind of multidimensional evaluation in form of a simplified framework that rests on a duopolistic form of evaluation, i.e., either a distinct good or service of a given institution creates values along one of the dimensions identified by Meynhardt (2009) (marked by the word "YES"), or it does not ("NO").

In opposition to objective criteria that are often the base of investment decisions, the public value approach described above is based upon subjective evaluations: Public value is created when individuals evaluate how their experience of community and society are influenced by others.

Therefore, investors, who would like to take into account normative aspects of their investments such as regulative environments or reputational risks, can refer to public value theory in order to analyze risks from a multidimensional point of view. For example, an investment in a company, that pays high dividends, is creating public value along the utilitarian—instrumental dimension of public value. If this company, however, does not contribute to the quality of life of the society or if the company acts indecent by polluting the environment, public values are destroyed at the same time. In other words, an investment that—if evaluated on basis of the balance sheet of the given firm—may seem to be attractive, can at the same time bear severe risks due to potential compensation claims or punitive damages that result from violations of assets not on the balance sheet of the given firm (for a discussion on the effect of environmental fines on stock price performance c.f. Lorraine et al., 2004). This example shows that company performance of each public value dimension must be put into context. This is possible since public value is measured at the level of individuals, which means that it is adaptable to various circumstances and situations. We perceive of public value evaluations as being value-relativistic.

However, while public value theory may offer the base for holistic frameworks of evaluation, a major disadvantage of its approach is the difficulty to measure public value creation. Contrary to data based on objective criteria, that are usually easy to get from research and statistical institutes, the subjective public value approach requires evaluations by individuals the results of which are difficult and only with great effort to obtain. An attempt to overcome this obstacle is the design and compilation of special databases such as the Swiss Public Value Atlas that will be discussed in the next section.

7.5 EMPIRICAL EVIDENCE ON PUBLIC VALUE MEASUREMENT: THE PUBLIC VALUE ATLAS

The Public Value Atlas has been implemented first at the University of St. Gallen, Switzerland, in 2014, and is to be further developed at the universities of St. Gallen and Lüneburg (Leuphana), Germany. It is the first empirical attempt to measure and determine the public value contribution of organization and companies on a broad scale. Currently, the Public Value Atlas focuses on Swiss companies and other organizations in Switzerland. Its database contains a representative sample (N = 4483) of the Swiss population. Sample members were asked in an online survey to evaluate the public value of major Swiss organizations. This procedure is consistent with the psychological approach of public value described above. This type of measurement has the advantage that criteria of evaluations are not externally given but determined by individuals and their preferences on what is valuable and what is less valuable.

In order to obtain comparable results and to be able to aggregate data, validated single-item measures were used for each of the four public value dimension discussed above. Respondents were asked to indicate on a 1 (full disagreement) to 6 (full agreement) Likert scale whether they agreed with the statements to each organization (Table 7.1).

Table 7.1 Items for Measuring Public Value in Public Value Atlas

Dimension	Item
	Organization/Industry XY...
Moral—ethical	...behaves in a decent manner.
Hedonistic—aesthetical	...contributes to the quality of life in Switzerland.
Utilitarian—instrumental	...does good work in its core business.
Political—social	...contributes to social cohesion in Switzerland.

Results were calculated first by summing up the single dimension categories—moral—ethical, political—social, utilitarian—instrumental, and hedonistic—aesthetical—and second by dividing the sums by the total number of the four dimensions.

The results of the study are documented in detail on the project Web page www.gemeinwohlatlas.ch. Public institutions as well as cooperatives obtained top scores in the ranking. This result can be viewed as being consistent with the fact that the former are typically perceived to operate under objective functions that cater not only to the needs of specific status groups such as consumers, employees, and owners, but also to those of a broader public. The Public Value Atlas also contains information with regard to values perceived by respondents to be created by companies with exposure toward environmental issues, such as nuclear energy, agribusiness, and commodity trading. Here, the data reveal that in particular hedonistic—aesthetical (does the respective organization contribute to the quality of life?) and moral—ethical (is it decent?) dimensions were evaluated more negatively in comparison to other companies (see www.gemeinwohl.ch/atlas).

The Public Value Atlas offers information based on subjective evaluations, i.e., it does not reflect "objective" or "tangible" values produced. Rather, its results rest on psychologically based assessment of perceived values. In how far this assessment translates into financial values, has to be discussed yet. However, contrary to ex-post evaluations of corporate environmental performance, public value offers a status quo estimate not only of past but also of current and potential future value performance of organizations (Meynhardt et al., 2014). Even without a concrete estimate of the link between financial and public value performance, this tool can be used for identifying and systemizing normative risks, such as reputational risks. This approach will be discussed in the next paragraphs.

7.6 PUBLIC VALUE AND REPUTATIONAL RISKS

Globalization, stakeholder activism, and the influence of stakeholder activism have increased the relevance of reputation management more than ever (Fombrun and Van Riel, 2004). Corporate reputation, which Fombrun et al. (2002, p. 87) defines as a "cognitive representation of a company's actions and results that crystallizes the firm's ability to deliver valued to its stakeholders," plays more and more an essential role in investor's decision makings. Reputation is formed by the public's expectation or estimation

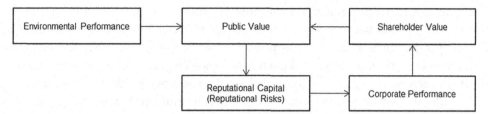

Figure 7.3 Effect of public value on shareholder value.

of attributes and characteristics that are ascribed to a company on the basis of past actions. Hence, reputation captures the assessment of a company, a collective judgment by a whole range of different social stakeholders. This includes the assessment of a company's environmental impact, in addition to its social and economic effects. Environmental externalities may affect consumer support of companies and a single accident can damage the financial performance of firms significantly. In short, when the reputation and thus legitimation of a company is challenged, it can lose access to essential resources that are needed in business operations (Barnett et al., 2006).

Because reputation is a construct of collective knowledge, recognition, and emotional reactions toward a company, thus it shares essential characteristics with the concept of public value. Public value can be seen as a signal from the public that reconfirms existing support or at least the nonexistence of disagreement of the value proposition of a given entity.

If managed in the right way, a high public value allows a company to build up reputational capital, which in turn can minimize reputational risks. Firms may be able to exploit this capital, e.g., by increasing the sale of products and services or reducing transaction costs, so that increases in reputational capital can lead to higher corporate performances and thus positively affect shareholder value. Since shareholder value is part of public value, we can say that the existence of public value fosters further public value creation (Figure 7.3).

The psychology-based approach of public value theory defines value creation as mental process: a value exists only if it is in someone's mind (Meynhardt, 2009). A high public value score that rests on the public rating of a given firm thus signals to investors that the firm is perceived to produce high value to the public. Investors with a focus on a firm's public value thus obtain information on the reputational risks and opportunities that the firm is confronted with along the four need-based dimensions described above.

7.7 HEDGING NORMATIVE RISKS BY BUYING A PUT-ON PUBLIC VALUE

"Existing empirical research confirms that there is a positive relationship between reputation and financial performance" of a firm (Roberts and Dowling, 2002). From an investors' point of view, information on the public value of a firm can thus be a relevant factor in the investment decision. Moreover, the present section also discusses how PVI can be viewed as instruments to hedge reputational risks that are based upon potential shifts not in the behavior of the firm but in public evaluations. Shifts in public preferences can result in regulative risks—such as the introduction of new taxes, e.g., on fuel, air, and water pollution—as well as reputational risks, where stakeholders "tax" companies for producing negative externalities, e.g., by no longer buying the respective products as consumers.

In order to discuss how PVI may be regarded as instruments for the hedging of such normatively determined risks, we think of the stocks of two companies that are evaluated differently in terms of public value but otherwise display identical financial characteristics. As a consequence of this assumption, both firms will behave differently once taxes on public value violations are introduced. Under this assumption, share prices (S) of the company that is violating public values will decrease relative to share prices of the other company since additional "taxes" will squeeze expected profits and diminish respective cash flows of the violating firm, while the firm that does not violate public values will be unaffected by the new tax and so will be its share price (S*) (c.f. Lorraine et al., 2004; who discuss the negative impact of fines on stock market performance). Under the assumptions of the current model, the share prices of the two firms are (at least) identical, so that we can write $S_{t0}/S_{t0*} = 1$. In an alternative scenario, taxes are introduced on public value violations. In this scenario, the model determines share price relations such that $S_{t1}*/S_{t1} < 1$, where t0 is the period before introduction of a new tax on public value violations; and t1 is the period after introduction of a new tax on public value violations.

Conceptually, this form of PVI can be viewed as strategy to hedge risks based on normative perceptions by purchasing a put option. The exercise price of this option is determined by the minimum share price of the public value complying firm ($S*_m$) in a world where only public value risks persist. The value of the PVI is determined in opportunity costs, i.e., buying an otherwise identical stock that just deviates with regard to public value risk exposure. Once the value of this stock decreases due to public value violations of the second firm (cf. paragraph above), PVI increase in value. Figure 7.4 represents a graphical illustration of this at the moment of the introduction of a new tax on public value violations.

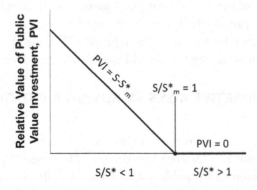

Figure 7.4 Public value investments (PVI) as Strategy to hedge normative (regulative and reputational) risks.

The conceptual discussion of the current section thus proposes public value evaluations as frameworks that may help investors not only to pick stocks in order to enhance portfolio performance by managing reputational risks (Robert and Dowling, 2002), but also to actively manage downside risks due to potential normative changes that translate into new regulations, or shifted consumer preferences. Viewed from this angle, environmental investments represent strategies of public value hedging.

7.8 CONCLUSION

The present article developed his observations against the background of a growing relevance of environmental issues in general (Welzel, 2013), as well as in corporate contexts. Within the setting of financial markets, this growth is being reflected in the development of environmental investments from a niche to mainstream as it is taking place currently.

Against this development, we discussed public value theory as a tool that helps investors to manage normatively determined risks associated with changes in the legislative environment, or the shifting of investor preferences. In a brief historic outlook we showed how environmental concerns have been subject to change within the development of economic and management theory in the past. Because public value is created when individuals evaluate how organizations and their behaviors influence their experiences of community and society, we argue that public value theory is suitable to offer systematic explanations for how environmental and reputational risks translate into a firm's production function. Due to the subjective and value-relativistic nature of processes of public value creation, this theory is apt to offer insights across diverse backgrounds of individual and cultural values, attributes, and preferences. Finally, by examining the empirical results of the Public Value Atlas from the perspective of reputational risk management, and by reviewing the mechanics of environmental investments as public value hedging strategy, we offer conceptual approaches of how investors are able to reduce risks due to environmental and reputational considerations. Our findings are constrained by various limitations such as (1) empirical difficulties of obtaining quantitative data on public values; (2) the lack of long-term or international data on public value creation that allows for general conclusions beyond distinctly defined contexts; (3) the fact that in light of limitation (1) and limitation (2) our public value approaches toward risk management are based on conceptual, not empirical, reasoning.

Given these limitations our observations and findings translate into the following five propositions:

- From the beginning of industrialization, externalities associated with energy consumption have been subject to economic and political debates. However, today—more than ever—these debates are of relevance to investors in the financial markets.

- **Proposition 1:** Based on current data on Austrian, Swiss, and German financial markets, we expect environmentally conscious investments to grow both in relevance as well as in volumes over the coming years.
- From the viewpoint of investors, above-named risks are determined by objective criteria such as the actual externalities produced by a given firm, as well as by normative decisions by society and its members that determine if and to what degree firms have to incorporate externalities into their production functions.
  - **Proposition 2:** We hence reason that environmentally conscious investors should care for both objectively and normatively defined externalities associated with their investments. Public value theory offers a conceptual approach that helps investors to fulfill this task.
- We observe that normatively induced inclusions of externalities into a firm's production function may be subject to change reflected in flexible legal environments and shifts of consumer preferences.
  - **Proposition 3:** We hence argue that environmentally conscious investors are in need of tools that help them to identify and account for risks associated with the fact that they are operating in dynamic value environments. These dynamics can be discussed under the framework of public value theory as well.
- In context of normatively determined investment risks we identify reputational risk management as task of diligent environmentally conscious investors.
  - **Proposition 4:** Reputational risks materialize in the minds of people: only when people perceive of values to be created, or destroyed a firm's reputation is enhanced or decreased. In order to account for this fact, reputational risk management strategy should be based on psychologically grounded frameworks.
- Empirical evidence suggests that financial firm performance is associated with environmental firm performance.
  - **Proposition 5:** Investors may perceive of environmental conscious investments as buying a put-on public value. PVI can thus be regarded as strategy to manage normatively determined risks.

REFERENCES

Barnett, M., Jermier, J., Lafferty, B.A., 2006. Corporate reputation: the definitional landscape. Corp. Reput. Rev. 9 (1), 26–38.

Bozeman, B., 2007. Public Values and Public Interest: Counterbalancing Economic Individualism. Georgetown University Press, Washington.

Brainerd, M., Campbell, J., Davis, R., 2013. Doing Well by Doing Good: A Leader's Guide. McKinsey Quarterly. September. Available at: http://www.mckinsey.com/insights/social_sector/doing_well_by_doing_good_a_leaders_guide as of 2015-02-08.

Dittrich, S., Kunzlmann, J., Tober, C., Vögele, G., 2014. Marktbericht Nachhaltige Geldanlagen 2014: Deutschland, Österreich und die Schweiz. Available at: www.forum-ng.org/images/stories/Publikationen/FNG_Marktbericht2014_Web.pdf.

Fombrun, C.J., van Riel, C., 2004. Fame and Fortune: How Successful Companies Build Winning Reputations. Pearson Education, New York.

Fombrun, C.J., Gardberg, N.A., Barnett, M.L., 2002. Opportunity platforms and safety nets: corporate citizenship and reputational risk. Bus. Soc. Rev. 105 (1), 85–106.

Forum Nachhaltige Geldanlagen e.V, May 7, 2014. Sustainable Investments in Germany, Austria and Switzerland Continue to Grow—Exclusion of Cluster Munitions and Anti-personnel Mines Becomes Standard Practice. Germany, Berlin.

Hirschman, A.O., 1970. Exit, Voice, and Loyalty. Responses to Decline in Firms, Organizations, and States, Cambridge (MA). Harvard University Press, Cambridge.

Jöhr, W.A., 1990. Bedrohte Umwelt. Die Nationalökonomie vor neuen Aufgaben. In: Binswanger, H.C., et al. (Eds.), Idem: Der Auftrag der Nationalökonomie. Ausgewählte Schriften. Mohr (Paul Siebeck), Tübingen, pp. 44–69.

JEI (Journal of Economic Issues). (1976). Obituary. Karl William Kapp. 1910–1976, 23 (9), 1.

Kapp, K.W., 2000. The Social Costs of Business Enterprise. First Published under the Title the Social Costs of Private Enterprise. Reprint of the 1963-second edition. Nottingham (UK) Spokesman, Russell House, Lyme Regis, Dorset.

Knoepfel, I., Imbert, D., 2013. Mapping Sustainable Finance in Switzerland. Available at: http://www.sfgeneva.org/doc/150213_MAPPING.pdf.

Leahy, J., 2008. Socially responsible investing. The quest for financial return and social good. Account. Irel. 40 (6), 47–49.

Lorraine, N.H., Collision, D.J., Power, D.M., 2004. An analysis of the stock market impact of environmental performance information. Account. Forum 28 (1), 7–26.

Mermod, A.Y., Idowu, S.O., 2014. Investing peacefully: a global overview of socially responsible investing. In: Mermod, A.Y., Idowu, S.O. (Eds.), Corporate Social Responsibility in the Global Business World. Springer, pp. 325–355.

Meynhardt, T., 2009. Public value inside: what is public value creation? Int. J. Public Adm. 32 (3–4), 192–219.

Meynhardt, T., Bartholomes, S., 2011. (De)Composing public value: in search of basic dimensions and common ground. Int. Public Manag. J. 14 (3), 284–308.

Meynhardt, T., Gomez, P., Schweizer, M.T., 2014. The public value scorecard: what makes an organization valuable to society? Performance 6 (1), 2–9.

Nkambule, N.P., Blignaut, J.N., 2012. The external costs of coal mining: the case of collieries supplying kusile powerstation. J. Energy South. Afr. 23 (4), 83–93.

Oekom, 2014. Taking Stock of Sustainability Performance in Corporate Management. Available at: www.oekom-research.com/homepage/english/oekom_CR_Review_2014_EN.pdf.

Parry, I., Heine, D., Lis, E., Li, S., 2014. Getting Energy Prices Right. From Principle to Practice. International Monetary Fund. Available at: http://dx.doi.org/10.5089/9781484388570.071.

Roberts, P.W., Dowling, G.R., 2002. Corporate reputation and sustained superior financial performance. Strategic Manag. J. 23 (12), 1077–1093.

Salin, E., 1955. Ökonomik der Atomkraft. Vor Einer Neuen Etappe der Industriellen Revolution. Im Sigilum-Verlag, Köln-Marienburg.

Schefold, B., 2006. Umweltökonomie. Die Entstehung einer Fachdisziplin vor dem Hintergrund von Weltuntergangsängsten, Kapitalismuskritik und Methodenkontroversen zur neoklassischen Theorie, in: der Gestaltungsanspruch der Wissenschaft. Aufbruch und Ernüchterung in den Rechts-, Sozial- und Wirtschaftswissenschaften auf dem Weg von den 1960er zu den 1980er Jahren. In: Acham, K., Nörr, K.W., Schefold, B. (Eds.), Aus den Arbeitskreisen "Methoden der Geisteswissenschaften" der Fritz Thyssen Stiftung. Stuttgart: Steiner 2006, 252-255, 296-299, 504-538.

Welzel, C., 2013. Freedom rising: human empowerment and the quest for emancipation. Cambridge University Press, Cambridge.

CHAPTER 8

What Holds Back Eco-innovations? A "Green Growth Diagnostics" Approach

Harald Sander
Technische Hochschule Köln, Cologne, Germany and Maastricht School of Management, Limburg, Netherlands

Contents

8.1 INTRODUCTION

Greening growth faces two major challenges: limiting the emission of pollutants, especially greenhouse gases, and dealing with limited natural resources in a sustainable way. This paper argues that greening the economy requires green innovations. Without such eco-innovations economic growth will sooner or later hit the boundaries of these environmental constraints.

Eco-innovations offer win—win solutions and are therefore at the heart of any green growth strategy that aims at reconciling economic growth and environmental protection, such as the Organization for Economic Cooperation and Development (OECD) Green Growth Strategy (OECD, 2011a). However, innovations require up-front investments and thus finance. But investing in eco-innovations requires even more difficult decision making as the uncertainty about their profitability is often much

Handbook of Environmental and Sustainable Finance
ISBN 978-0-12-803615-0

higher than for traditional investments. Investors have to deal with a plethora of market distortions, in particular environmental and research and development (R&D) externalities. Against this background it is important to enable policy makers to deal efficiently with these market distortions to ensure an appropriate level eco-innovations and green investments. However, often this is not the case, thus creating additional uncertainties about whether and how policy makers will eventually address these market distortions.

This paper intends to clarify the concept of eco-innovation for greening growth. In particular it will provide a systematic framework to analyze "what holds back eco-innovations." In this respect it reviews "Green Growth Diagnostics (GGD)," a tool for greening the economy recently proposed by OECD (2011c). It will be shown that the tool is a useful, though not perfect, instrument to prioritize business and policy action in order to remove the most binding constraints to eco-innovations. However, it is also argued that, given its limitations, its main value lies in structuring and enabling a collaborative discovery process among all stakeholders when obstacles to eco-innovation ought to be removed.

The paper starts with a discussion of the importance of eco-innovation for greening growth in Section 8.2. Section 8.3 provides a structured discussion of potential barriers to eco-innovation. Section 8.4 discusses recently proposed tools to weigh and identify the most binding constraint(s) to eco-innovations with a view on prioritizing country- and time-specific policy responses. Section 8.5 concludes.

8.2 ECO-INNOVATION FOR GREENING GROWTH

8.2.1 Innovations as Enabler for Greening Growth

In 2011, the OECD has launched its green growth initiative. By green growth they mean "…fostering economic growth and development while ensuring that natural assets continue to provide the resources and environmental services on which our well-being relies." (OECD, 2011a, p. 9).

Economic growth uses scarce resources. Therefore green growth requires decoupling from resource use. While absolute decoupling is understood as economic growth and/or a higher per capita gross domestic product (GDP) with less use of resources, relative decoupling signifies a reduction of resource use relative to per capita GDP. According to the so-called Environmental Kuznets Curve (EKC) absolute decoupling occurs eventually beyond a certain per capita GDP level.[1] There is, however, mixed evidence on decoupling. When we look, for example, on different type of pollutants,

---

[1] The EKC was introduced to the literature by Grossman and Krueger (1991) in a study on the environmental impact of North American Free Trade Area (NAFTA). The concept fits best for emissions of pollutants. For an early critical review of the concept, see, e.g., Stern et al. (1996).

we find relative and absolute decoupling in high-income countries for NO_x (comprising both nitric oxide and nitrogen dioxide, which both contribute to smog) and SO_2 (sulfur dioxide, which contributes to acid rain). In contrast, no, or at best, mixed evidence is found for decoupling with respect to CO_2 (carbon dioxide). In many developing economies even a relative decoupling for NO_x and SO_2 is often not found at all.[2]

Theoretical as well as a simple guesswork suggests that decoupling depends on (1) the spatial closeness of negative effects (do you "feel" the air pollution?), (2) the time distance to the effect, (3) time preferences (which are often lower with a lower per capita income), and (4) the cost of avoiding negative effects (e.g., abatement costs). Frankel (2009) argues that most pollution, e.g., resulting from emissions of NO_x and SO_2 is external to the polluter. While the desire for cleaner air increases with higher income, to become effective, the popular will must be translated into policy action. As such, adoption of higher environmental standards and thus decoupling beyond a certain peak requires effective government regulation that responds to the popular will. This reasoning also explains why decoupling with respect to CO_2 emissions is lagging behind. Additionally and different from other environmental challenges, the effects of global warming are highly uncertain effects that may or may not occur in the future. And if such effects do happen they may or may not affect the polluter, thus involving a clear free rider problem. This leads to a second and related argument for pushing eco-innovations: they can provide win—win solution and may get therefore adopted already for economic reasons.

As policies matter, it is also clear that the EKC should be understood as an empirical concept to analyze the actual growth-environment nexus in different countries, regions, or even globally, rather than a universal law. Especially developing countries, when enabled to adopt new clean technologies with a short time lag, can leapfrog the EKC.[3] Eco-innovation can therefore be understood and illustrated as an instrument that pushes down the EKC, thus allowing a higher per capita GDP without increasing pollution or resource use in either relative or absolute terms. Figure 8.1 shows the EKC before and after eco-innovation. The move from A to C instead to B illustrates the case of absolute decoupling. Or as OECD (2011b, p. 9) puts it: "Existing production technology and consumer behavior can only be expected to produce positive outcomes up to a point; a frontier, beyond which depleting natural capital has negative consequences for overall growth. By pushing the frontier outward, innovation can help to decouple growth from natural capital depletion. …Innovation is therefore the key in enabling green and growth to go hand in hand."

---

[2] For example, Frankel and Rose (2005) estimate the peak of the EKC for SO_2 at a per capita GDP of $5770, a level many developing countries have not passed.

[3] For an early critical review of the EKC in this respect, see Stern (2004).

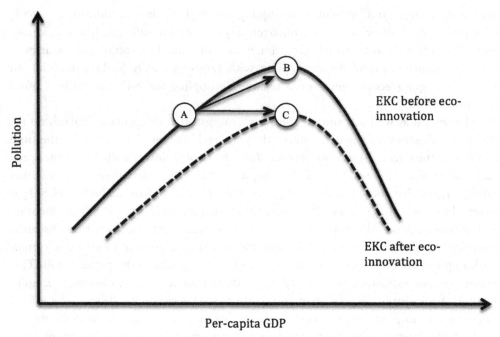

Figure 8.1 The environmental Kuznets curve (EKC), eco-innovation, and decoupling.

8.2.2 The Concept of Eco-innovation

OECD (2009, p. 40) defines eco-innovations as "the implementation of new, or significantly improved, products (goods and services), processes, marketing methods, organisational structures and institutional arrangements which, with or without intent, lead to environmental improvements compared to relevant alternatives." Eco-innovation is therefore a broad concept, which is not confined to technological innovations or technical improvements in resource efficiency, but it also covers social innovations whenever they—intentionally or unintentionally—improve resource efficiency and/or reduce pollution at any given level of economic development.[4]

Technological eco-innovations typically range from end-of-the-pipe pollution control technologies over cleaner production approaches that modify products and/or production methods to systematic environmental management and monitoring system, often dubbed as "eco-efficiency." However, introducing systematic green management systems already relies—at least partly—on nontechnological organizational and social

---

[4] For an early discussion of the concept of eco-innovation, see Rennings (2000). For more recent overviews and critical discussions, see Arundel and Kemp (2009), Wuppertal Institute for Climate, Environment and Energy (2009), Ekins (2010), Machiba (2010), and Walz et al. (2011).

innovations (see Machiba, 2010). According to OECD (2009) the latter are even more important when it comes to implementing "green supply chain management," "closed-loop production," or an "industrial ecology" with integrated systems of production or product service systems (PSS). Such social innovation may even require changes in norms and values in a society or—a bit less ambitious—changes in preferences. This can be illustrated by PSS, which deliver product functionality rather than the product itself. Popular examples are cloud computing rather than private computers, which change both, the way we use resources and the level of efficiency with which we use them. Another example for PSS is providing mobility solutions rather than private cars. Likewise, leasing systems, where the customers are provided with product services while the producer retains ownership of the product, allow for a (more) closed-loop production. The automobile industry but in particular the electronics industry is the most prominent examples where a "share economy" (Weitzman, 1984; Rifkin, 2014) can proliferate eco-innovations.

By adopting a systems perspective, the concept of eco-innovation goes clearly beyond a purely technological point of view on innovations. As such, it is also clear, that is even more difficult to trigger such innovation as they have not only to overcome traditional and well-known obstacles to R&D, but also market failures in environmental markets as well as inertia in values, norms, and behavior.

8.3 WHAT HOLDS BACK ECO-INNOVATIONS?

8.3.1 A Conceptual Frame

When eco-innovations offer win—win solutions, why are they then adopted as such a low level? A large part of the answer lies in the fact that they have to deal with two different types of market failures: those that hold back adopting environmental friendly product and processes and those that hold back innovations. Another part of the answer relates to the limited will in the political sphere to correct these market failures. In this section, I will provide an overview of potential barriers to eco-innovation before discussing ways to weigh these barriers against each other with respect to their relevance.

Eco-innovation, just like any other innovation start from the "idea"—the *invention*—to "bringing it to the show room"—the *innovation*. To be successful they need to be adopted broadly—the *diffusion* (Popp, 2011a). However, diffusion is often seen as the major bottleneck for eco-innovation. For example, Gerarden et al. (2015) find low adoption rates in the energy sector even when more efficient technologies would pay off.

This suggests that a good starting point for analyzing barriers to a broad adoption of eco-innovations is to differentiate whether or not they pay off. And if they pay off, why they are then not broadly adopted? If eco-innovations would have sufficiently high social (private and public) returns and if these returns would accrue to the innovator there would be no need to worry about a lack of eco-innovations. Private investors would sufficiently engage in it. Problems occur when either the returns are too low, or appropriability is not ensured, or both. In Section 8.4 this distinction will be a key for identifying binding constraint(s) to eco-innovations. Before moving there, the rationale and meaning of each problem is discussed below in more detail.

8.3.2 Low Appropriability

8.3.2.1 Market Failures

Imagine an eco-innovation worthwhile from a societal point of view, i.e., the sum of private and social benefits would exceed the sum of private and social costs. From environmental economics it is well known that when external effects exist, private actors will not be charged for social costs they impose on the society and not be rewarded for social benefits they create. As such, there would be too little eco-innovation because innovators cannot appropriate the benefits. But when it comes to innovation rather than to standard environmental externalities, we face a second externality with respect to R&D, which is also well known from the R&D literature: fast imitation and high adoption rates reduce private returns to innovations. What makes eco-innovations special is that they face both externalities, the so-called "double externality problem" (Jaffe et al., 2005). This problem can keep eco-innovations at a level far below the social optimum. While some eco-innovation may get adopted because of secondary benefits, e.g., fuel-efficient cars if the (discounted) savings in fuel exceed their higher costs, many others are hold back by double market failures.

The likely simultaneous occurrence of environmental externalities, which often limit the market size, and R&D market failures, which limit innovation activity and diffusion, requires policy action on both sides. If environmental damage associated with the production or consumption of a good is not reflected in the market price there is too much production and consumption of that good and too much environmental damage at a too low price. Moreover, markets for alternative goods or production processes might be underdeveloped or even nonexistent. Environmental economists would usually then recommend the "internalization" of external effects by environmental policies, such as a Pigou tax.

R&D market failures limit innovation activity and diffusion because R&D has in principle a public good nature unless protected by intellectual property rights. However, rewarding for R&D by means of patents provides an incentive to innovate makes eco-innovations also more expensive and reduces diffusion. R&D subsidies geared toward "clean" innovations are then needed.

Double market failure is the key constraint on eco-innovations and it is this double externality problem, which calls for a double policy response to trigger eco-innovation: an environmental policy, which internalizes external effects to create a market, and a technology policy to promote technology development and diffusion. Each single policy action might be a necessary condition for unleashing eco-innovation but neither is sufficient when undertaken in isolation. This reasoning, which appears to be an emerging consensus view among environmental economists, has recently received a theoretical foundation from a paper by Acemoglu et al. (2012). These authors have developed a model with "dirty" and "clean" inputs which are sufficiently substitutable and show that, e.g., with respect to limiting emissions of CO_2, an optimal policy response would involve both carbon taxes and research subsidies, thus addressing the double market failure problem.

Of course, other market failures may occur too, among which barriers to entry and other forms of imperfect competition are the most common ones. Another relevant market failure is network externalities, which refer to cases in which eco-innovations would not be profitable for an individual actor but only pay off if all—or at least a critical mass—of economic actors would adapt the eco-innovation, be it as consumer or as producer. Network externalities are especially important for eco-innovations with a high social innovation component, such as a share economy. And last but not least, other informational imperfections and capital market failures can hold back eco-innovation too.

8.3.2.2 Governance Failures

Not only market failures but also governance failures,[5] such as corruption, a low institutional quality and macroeconomic instability can constitute serious obstacles to obtain full rewards from eco-innovation, especially when property right is incomplete or wrongly assigned. With respect to environmental regulation, however, history and effective interest groups are often successful in pressuring for "perverse subsidies," i.e., subsidies to dirty energies such as coal. Given the organized power of incumbents which represent old, but well-organized industries, these activities often receive preference over new and cleaner activities as represented by eco-innovations that lack a similar negotiating power.

8.3.3 Low Economic Returns

Appropriability of returns, however, may not be the only—or even not the overwhelming—problem, especially for developing countries. Imagine that after perfect corrections of all market failures all due returns accrue to the investor: even then it could be that eco-innovation would not be triggered and adopted because the resulting returns would be considered too low. Of course, "too low" is relative: relative to "dirty innovations"

---

[5] I do not include "failures" to correct market failures here under the category of governance failures, but rather discussed these "failures" in Section 8.4 as an issue of (insufficient) policy implementation.

and relative to finance (opportunity) costs. But in respect to the latter this would be a general problem of an economy in triggering all types of investment and innovation, e.g., because of an ineffective financial system with low savings, high lending rates and high interest spreads, and limited access to international finance. While this would require general policy initiatives, with respect to eco-innovation we need therefore to focus on economic returns that remain lower relative to other innovations and investments even after correcting for market and governance failures. Two major groups of reasons why this can occur should be mentioned.

First, a lack of social resources, in particular traditional norms and values that do not (fully) accept new products, thus resulting in low adoption rates by consumers. But also production processes in general often show strong habit inertia, which can easily give traditional products and process a competitive edge. A study by Aghion et al. (2010) suggests that the innovation activity of an industry can depend on the industry's patent history. The authors study the patent history of the automobile industry by distinguishing "clean" and "dirty" patents. To make their point, they first establish a significant effect of higher fuel prices on pushing the industry toward cleaner innovation. This is what one would expect from theory and also from an internalization of external effects, which would make fossil fuels more expensive. However, and most importantly, the authors also show that companies with a history of "dirty patents" are less engaged in "clean patents" even after controlling for the price internalization effects. The lesson is that internalization of external effects is eventually not sufficient to promote eco-innovation to a socially optimal level. If an industry shows substantial path dependence, this makes a strong case for policies other than those directed at external effect internalization.

Second, the economy may simply lack the complementary economic resources necessary for successful adoption and diffusion of eco-innovations. This may concern the usual factors like infrastructure and the human capital, but also lack of finance when, for example, the banking system discriminates against nontraditional innovations or start-ups.

Finally access to green technology can be limited. Especially for developing countries access to appropriate green technology may be one of the most important limiting factors. The old approach of growing first and cleaning up later, where a movement along the EKC has been viewed as a "normal" development process is becoming increasing under pressure as it leads to high social and environmental costs, which are increasingly causing political pressures within emerging economies, especially when the immediate (intragenerational) costs, e.g., of air pollution, are getting extremely high, as, e.g., in Beijing. Because direct benefits from greening growth are often directly felt (for example, clean drinking water, smoke-reduced cooking stoves) there is a strong demand for eco-innovations, which may help to leapfrog the EKC when and if politicians react to popular will. This is less so with respect to long-term (intergenerational) costs of irreversible damage for future growth and prosperity as the costs for present generation depends on time preferences, but often these irreversible damages have cross-border regional and global (external) effects. As such there is an additional external pressure on emerging economies to leapfrog and push down the EKC.

However, greening growth in developing countries requires either technology transfer with a related development of an "absorptive capacity," or the development of own R&D capabilities, eventually adapted to local circumstances. And advanced countries need to pay attention to innovate adapted technologies for developing countries, especially when their home market is limited with respect to these technologies. It is also noteworthy that the share of flows of climate technologies of all flows from OECD to non-OECD countries has not only increased fast from the 1990 onward, but it has also outpaced other technology flows since the late 1990s (Dechezlepêtre et al., 2011).

More recently, there is fast-growing and substantial R&D in (some) developing countries, in particular China and India, especially with respect to adapting technologies. Lower technological as well as a lower regulatory distance matters, for example, in the automobile industry where, according to Dechezlepêtre, Perkins and Neumeyer (2012, p. 4) "…countries are more likely to receive newly-innovated technologies from source countries whose regulatory standards are 'closer' to their own." Moreover, developing country technologies may be more appropriate in terms of factor-proportions required in developing countries and thus easier to adapt to local circumstances. Finally, in many cases, developing countries may require additional finance for acquiring clean technologies.[6]

In sum, the lack of complementary resources, especially the access to adapted technologies in developing countries can be a substantial barrier to eco-innovation and diffusion.

8.4 IDENTIFYING AND PRIORITIZING BARRIERS TO ECO-INNOVATION

8.4.1 Prioritizing Policies toward Eco-innovation

Eco-innovations and the subsequent adoption of green technologies have to overcome multiple market failures, governance failures, behavioral barriers, and—not least—lacking complementary resources. The crucial issue is therefore whether policy makers take sufficient measures to ensure that eco-innovations and subsequent investments in green technologies take place at an appropriate scale. If economic resources would be endless and the political reform capital unlimited, there would be no need for prioritizing. But limited economic, financial, and political resources suggest that unleashing green growth by means of eco-innovations requires some effort to identify the most

---

[6] See, e.g., Popp (2011b) for an analysis of international technology transfer and the "clean development mechanism."

Table 8.1 Prioritizing Green Growth Strategies: The Hallegatte et al. (2011) Framework

	Low Immediate Benefits	High Immediate Benefits
Low irreversibility and inertia	**Low priority**, e.g., lower carbon, higher cost energy supply	**Medium priority**, e.g., lower carbon, lower cost energy supply
High irreversibility and inertia	**Medium priority**, e.g., costal zone and natural area protection	**High priority**, e.g., public urban transport

Source: Own simplified version based on Table 1 in Hallegatte et al. (2012).

pressing policy areas and to prioritize policy instruments. How can this be done and how helpful are currently discussed tools to prioritize?

To start with, studies on eco-innovations often attempt to deliver a comprehensive analysis of all technical, economic, natural, and social drivers and barriers of green growth, e.g., the so-called "fishbone approach" (see Wuppertal Institute for Climate, Environment and Energy, 2009). While this is a valuable first step to review all barriers and drivers of eco-innovation, such an analysis gives an equal weight to all dimensions and leaves therefore the policy maker with the problem of prioritizing.

In a recent study, Hallegatte et al. (2011) argue that policy makers should focus on areas where both, the net immediate benefits and the risks of irreversibility are high, as illustrated in Table 8.1.

Two limitations should be borne in mind when applying the Hallegatte et al. framework to eco-innovation. First, high-immediate-benefit policy areas may emerge as pressing but they may actually use little additional political reform capital as these areas, such as high urban pollution, are close to and felt by the people. By contrast, areas like global warming are eventually downplayed, although eco-innovation could eventually provide win–win solutions—not least by technological leapfrogging. Thus there may be a danger of "doing too little" when economically and politically more eco-innovation-friendly reforms are possible. The second concern is that the approach focuses on prioritizing policy areas but does not help prioritizing policy instruments within certain policy areas.

OECD (2011b, p. 46) gets closer to prioritizing policy instruments in a more general way by recommending "…a policy environment based on core "framework conditions" — sound macroeconomic policy, competition, openness to international trade and investment, adequate and effective protection and enforcement of intellectual property rights, efficient tax and financial systems — is a fundamental building block of any effective (green) growth strategy and allows innovation to thrive." But are all listed "framework conditions" really supportive for eco-innovations, especially in developing countries? Critics of such statements fear a "green Washington consensus" with "one size fits all" receipts and caution whether all countries can or even should address all "framework conditions" simultaneously.

The latter concern is based on a critique of the so-called Washington consensus approach with its uniform "presumptive" approach. Rather than assuming that all problems should be solved at once, Hausman et al. (2008) have argued that a careful "growth diagnostics" (GD) approach should first try to identify the most "binding constraints" on economic growth and then to target the area where there is the "biggest bang for the reform buck." In other words, one should prioritize by means of GD. It is interesting to note that with the launch of the green growth strategy in 2011, the OECD has also proposed "GGD" as a tool for prioritizing green growth policies and for identifying the (most) binding constraints to green growth.

8.4.2 Green Growth Diagnostics: A Useful Tool for Prioritizing?

GGD is based on the GD approach. It is helpful to review GD briefly first in order to understand why GGD has been proposed. Hausman et al. (2008) motivate their GD with the famous quote from Tolstoy's Anna Karenina "Happy families are all alike; every unhappy family is unhappy in its own way." The argument is that each country is different and held back by its "own" binding constraint(s). This binding constraint varies from country to country and can change over time. As GD should help to identify country- and time-specific policy instruments, it is essentially a heuristic instrument for prioritizing policies when economic resources are scarce. As such it can also serve as an instrument for policy dialogue when political reform capital is limited.

Figure 8.2 shows a slightly simplified version, the GD decision tree. It starts from the question why investment levels are (too) low. And at the first node, the suggested answer

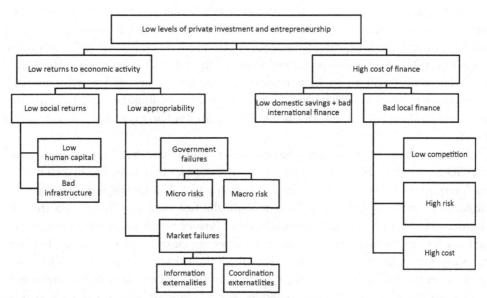

Figure 8.2 The growth diagnostics decision tree. *(Source: Simplified own version based on Hausman et al. (2008).)*

is that it is either because economic activities have low returns or—in case that returns are sufficiently high—finance is binding. Once it is found out whether finance or low returns are responsible, one can go deeper into the decision tree to identify the binding constraints. It is needless to say that several recursive cycles of analysis may be needed before one converges on a narrative, which gives a good diagnosis of the problem and can form the base for a "therapy." Moreover, over time the binding constraints may change. In fact, they should change if a policy has successfully removed a binding constraint. As such GD suggests permanent analyses and dialogues on growth policies and the political reform agenda.

It is now easy to see, why it is tempting to adapt GD to the green growth problem and to suggest a GGD. Five points can be highlighted:

- GGD should help to prioritize by identifying binding constraints to green growth.
- GGD should promote the economic use of scarce economic and social reform resources.
- GGD is not a presumptive approach, but allows country- and time-specific analysis and policy recommendations.
- GGD can thus facilitate an open dialogue among all stakeholders.
- GGD proposes continuous dialogue processes as binding constraints change over time.

OECD (2011c, p. 6) also highlights the crucial role of a country's level of development and the country- and time-specific nature of binding constraints: "The importance of constraints to green growth will vary according to level of development, socio-economic context, and existing economic and environmental policy settings. Low human capital or inadequate infrastructure will tend to be associated with lower levels of economic development (though not exclusively). Rectifying these constraints will be of high priority and perhaps a precondition to resolving many other constraints. Where human capital is relatively abundant and infrastructure relatively well-supplied, the focus should first be on resolving government and market failure."

While all this is welcome, there are unfortunately also some drawbacks when adapting the GD approach to the green growth problem. First, identifying the binding constraints to "green growth" or "green activities" is less clear-cut than identifying overall growth obstacles (as in the case of GD). To start with, the first node in GD decision tree, the distinction between low returns and high cost of finance is only useful for all investments not just for "green investments" (unless the latter are clearly discriminated). Therefore, for analyzing "what holds back green activities," the OECD decision tree concentrates entirely on the left-hand side of the GD decision tree and asks whether the returns to "green activities" are too low or whether investors cannot appropriate such returns. While this is in line with the reasoning given in Section 8.3, the problem emerges when going down further the OECD decision tree as it becomes visible that there is substantial overlap among the various different categories in the decision tree, as OECD (2011c, p. 6) concedes: "[T]he categories of constraints described in Figure 8.1 [here reproduced in as Figure 8.A1 in the appendix] are not entirely separable." This makes the approach much less practical and useful as compared to the relative successful GD approach.

I have argued earlier (Sander, 2011) that the approach could be considered as a more generic one directed to analyze binding constraints for eco-innovation in general and have proposed the following GGD decision tree, which is a slightly revised version of the OECD one. I concentrate here more narrowly on what holds back "eco-innovation" with a view on reducing overlap between the categories, in particular, by clearly allocating the double market failure problem to the "low appropriability" category. Figure 8.3 shows the proposed decision tree, which closely follows the line of arguments on what holds back eco-innovation laid out in Section 8.3.

The first issue would then be—as in OECD (2011c)—to determine whether low returns to eco-innovations or limits to appropriate these returns are holding back such activities. Clearly, this would already constitute a formidable research effort. But as a first step, it is logical to follow the literature that highlights the importance of double externalities and try to identify whether or not adequate environmental and technological

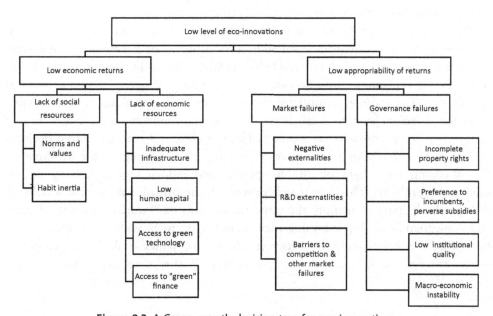

Figure 8.3 A Green growth decision tree for eco-innovations.

policies are in place to address double market failures.[7] Of course, other market failures as well as governance failure need to be cross-checked as well. In a second step, one could then check whether eco-innovations are suffering from low economic returns even after accounting for market and governance failures. This should then be followed by step three: identifying country-specific binding constraints and appropriate policies to reduce/remove these constraints.

Unfortunately what sounds logical and easy on paper can turn out to be difficult in practice: identifying which constraint is really binding. This is, however, often simply the result of the complex nature of eco-innovations. In fact, De Serres et al. (2010, p. 2) argue that because "...environmental damage often result from several interacting market failures, an appropriate policy response will in many cases involve a mix of complementary instruments." Moreover, GGD does not offer many clear-cut differential diagnostic tools whether, e.g., a too little eco-tax is binding or the missing inadequate infrastructure. Finally, every sector or environmental problem may be so different from the other that the GGD decision tree must eventually be retailored for the specificity of each problem. This may explain the low adoption rates so far (but some currently running GGD projects can be found already).

So is the instrument useless and only old wine in new bottles? In fact, despite the discussed problems, GGD could indeed be a useful instrument to prioritize policy instruments when applied to eco-innovations at the level of industries or with respect to certain environmental challenges. However, and unlike the successful GD approach, it needs much more efforts to be tailored to the problem at hand. But why then doing a GGD exercise in practice rather than sticking to the old tools?

First, a tailor-made GGD can be helpful as a guideline and "directory" to an increasingly complex environmental indicator "jungle." In a complex world with highly uncertain outcomes, GGD can help prioritizing by facilitating a policy dialogue for triggering eco-innovations. Second, when economic resources as well as political reform capital are scarce, identifying the most efficient, promising, and acceptable ways for greening growth is crucial to unleash eco-innovation, even if a mix of complementary instruments are needed. Third, a clear and well-structured decision tree helps to identify opportunity costs of using or not using policy instruments, thus shedding more light on the complexity and hidden cost of solving or not solving the problem. This way it will provide the informational base for a political dialogue by helping to identify the opportunity costs of "business as usual." Finally and consequently, its highest value may therefore be to provide an instrument for facilitating the policy dialogue between researchers, policy makers, the corporate sector, the finance sector, and the civil society. Rather than expecting a unique

---

[7] This is exactly what the study by Aghion et al. (2010) has done for eco-innovations in the automobile industry.

outcome out of a GGD exercise, one should view it as continuous process of trial and error, which can help developing and shaping a "green industrial policy" (Rodrik, 2014).

8.5 CONCLUSION

Eco-innovations in a broad sense, comprising technological as well as social innovations, promise to reconcile economic growth and environmental sustainability. Unfortunately, there is still too little of these good things. Multiple market failures and governance failures can hold them back, but also behavioral effects and inertia play an important role,[8] and in particular but not exclusively in developing countries lacking complementary economic and financial resources can be binding.

While many if not most of these problems will occur in reality, the question arises, which of these constraints are the most binding and should thus be addressed first to get the biggest effects. When economic resources and political reform capital are both limited it is not useful to address all problems at once, but rather to prioritize. The GGD approach attempts to offer a framework for identifying these constraints. However, it turns out that the tool is less useful and clear-cut than its paragon "GD." In fact, identifying binding constraints remains a formidable task and the GGD decision tree may need to be adapted to the environmental challenges at hand. However, this cannot be blamed on the tool, as the reason for this is the complexity of the problem itself.

Rather than expecting a ready-made tool, researchers and practitioners in eco-innovation should view it as a framework and a hopefully well-organized "inventory list" of potentially binding constraints which can help to evaluate and discuss policies alternatives for promoting eco-innovation. The particular value of the diagnostics approach is therefore not so much that it offers a new methodology, but that it provides a framework for a policy dialogue among all stakeholders in greening growth.

REFERENCES

Acemoglu, D., Aghion, P., Bursztyn, L., Hemous, D., 2012. The environment and directed technical change. Am. Econ. Rev. 102 (1), 131–166.

Aghion, P., Dewatripont, M., Du, L., Harrison, A., Legros, P., 2010. Industrial Policy and Competition. Working Paper. Harvard, Cambridge, MA.

Arundel, A., Kemp, R., 2009. Measuring Eco-innovation. UNU-Merit Working Paper Series #2009-017. UNU-Merit, Maastricht.

Dechezleprêtre, A., Glachant, M., Hascic, I., Johnstone, N., Ménière, Y., 2011. Invention and transfer of climate change mitigation technologies on a global scale: a study drawing on patent data. Rev. Environ. Econ. Policy 5 (1), 109–130.

---

[8] Gerarden et al. (2015) named (1) market failures, (2) behavioral effects, and (3) modeling flaws as major reason for the energy-efficiency gap.

Dechezleprêtre, A., Perkins, R., Neumeyer, E., 2012. Regulatory Distance and the Transfer of New Environmentally Sound Technologies: Evidence from the Automobile Sector. Grantham Research Institute on Climate Change and the Environment Working Paper No. 73. London. School of Economics, London.

De Serres, A., Murtin, A., Nicoletti, G., 2010. A Framework for Assessing Green Growth Policies. Economics Department Working Paper No. 774. OECD, Paris.

Ekins, P., 2010. Eco-innovation for environmental sustainability: concepts, progress and policies. Int. Econ. Econ. Policy 7, 267—290.

Frankel, J.A., 2009. Environmental Effects of International Trade. HKS Faculty Research Working Paper Series RWP09-006. Harvard, Cambridge, MA.

Frankel, J.A., Rose, A., 2005. Is trade good or bad for the environment? sorting out the causality. Rev. Econ. Stat. 87 (1), 85—91.

Gerarden, T.D., Newell, R.G., Stavins, R.N., 2015. Assessing the Energy-Efficiency Gap. Harvard Environmental Economics Program, Cambridge, MA.

Grossman, G.M., Krueger, A.B., 1991. Environmental Impacts of a North American Free Trade Agreement. National Bureau of Economic Research Working Paper 3914. NBER, Cambridge, MA.

Hallegatte, S., Heal, G., Fay, M., Tréguer, D., 2011. From Growth to Green Growth: A Framework. NBER Working Paper 17841. National Bureau of Economic Research, Cambridge, MA.

Hallegatte, S., Heal, G., Fay, M., Tréguer, D., March 24, 2012. From Growth to Green Growth. http://www.voxeu.org/article/growth-green-growth.

Hausman, R., Velasco, A., Rodrik, D., 2008. Growth diagnostics. In: Stiglitz, J., Serra, N. (Eds.), The Washington Consensus Reconsidered: Towards a New Global Governance. Oxford University Press, New York, pp. 324—354.

Jaffe, A.B., Newell, R.G., Stavins, R.N., 2005. A tale of two market failures: technology and environmental policy. Ecol. Econ. 54 (2), 164—174.

Machiba, T., 2010. Eco-innovation for enabling resource efficiency and green growth: development of an analytical framework and preliminary analysis of industry and policy practices. Int. Econ. Econ. Policy 7 (2), 357—370.

OECD, 2009. Eco-innovation in Industry. Enabling Green Growth. OECD, Paris.

OECD, 2011a. Towards Green Growth. OECD, Paris.

OECD, 2011b. Fostering Innovation for Green Growth. OECD, Paris.

OECD, 2011c. Tools for Delivering Green Growth. OECD, Paris.

Popp, D., 2011a. The Role of Technological Change in Green Growth. World Bank Policy Research Paper 6239: Washington, DC.

Popp, D., 2011b. International technology transfer, climate change, and the clean development mechanism. Rev. Environ. Econ. Policy 5 (1), 131—152.

Rennings, K., 2000. Redefining innovation: eco-innovation research and the contribution from ecological economics. J. Ecol. Econ. 32 (3), 319—332.

Rifkin, J., 2014. The Zero Marginal Cost Society: The Internet of Things, the Collaborative Commons, and the Eclipse of Capitalism. Palgrave Macmillan, New York.

Rodrik, D., 2014. Green Industrial Policy. Oxford Rev. Econ. Pol. 30 (3), 469—491.

Sander, H., 2011. The Use and the Usefulness of OECD's Green Growth Diagnostics. Institute of Global Business and Society Working Paper 2011-03. Cologne University of Applied Sciences, Cologne.

Stern, D.I., Common, M.S., Barbier, E.B., 1996. Economic growth and environmental degradation: the environmental kuznets curve and sustainable development. World Dev. 24 (7), 1151—1160.

Stern, D.I., 2004. The rise and fall of the environmental kuznets curve. World Dev. 32 (8), 1419—1439.

Walz, R., Köhler, J., Maarscheider-Weidemann, F., 2011. Global Eco-innovation, Economic Impacts and Competitiveness in Environmental Technologies. Fraunhofer ISI, Karlsruhe, 31 October.

Weitzman, M., 1984. The Share Economy: Conquering Stagflation. Harvard Publication Press, Boston.

Wuppertal Institute for Climate, Environment and Energy, 2009. Eco-innovation — Putting the EU on the Path to a Resources and Energy Efficient Economy (Wuppertal).

APPENDIX 1

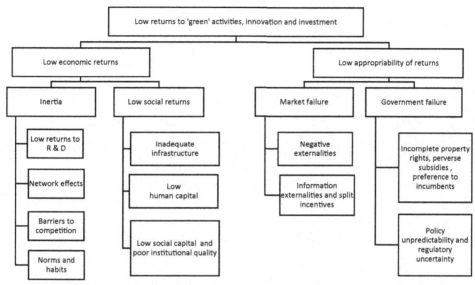

Figure 8.A1 The OECD green growth diagnostics decision tree. *(Source: OECD (2011c).)*

CHAPTER 9

Trade Openness and CO_2 Emission: Evidence from a SIDS

Raja Vinesh Sannassee, Boopen Seetanah
Department of Finance & Accounting, Faculty of Law & Management, University of Mauritius, Reduit, Mauritius

Contents

9.1 INTRODUCTION

International trade is often recognized as an important engine of economic growth, where the benefits of trade liberalization give access to a broader variety of goods and services on the market, easier access to foreign technologies which in turn enables local firms to boost up their efficiency in terms of production and resource allocation. However, over the last decade or so, the sustainability of such economic growth has gained significant prominence in the literature with countries favoring policies and measures which are conducive to more sustained growth targets. Nevertheless, very often it seems that the trade-off between economic growth and environment protection is rarely maintained and very often economic growth is associated with a decline in environmental quality (Hanley et al., 2001). From an economic point of view, the comparative advantage propounded by Ricardo argues that the freer trade is, the greater the propensity of countries to specialize in the production of goods where they hold a comparative advantage. As a result, they will turn to lower cost of production but at the same time provide more goods at relatively lower prices. On the other hand, Hanley et al. (2001) argue that the negative environmental consequences of increased output from trade liberalization have in many instances outweighed the gains from increased income.

In this regard, one branch of the existing literature has argued that it is very likely that developed nations would specialize in capital of human intensive activities which are likely to be less emission intensive. As a result, one could posit that trade may therefore result in increased pollution in developing countries due to the increased production of these emission intensive goods in these countries.[1]

However, the aforementioned studies have overwhelmingly focused on developed country cases with empirical evidences on developing countries, particularly African economies and SIDS, being very scarce. The present study thus attempts to supplement the existing literature by bringing new evidences for the case of a SIDS African economy namely Mauritius. Mauritius poses as a good case study as it is one of the best performers and most open economy of the African continent, albeit it is totally devoid of any natural resources.

Mauritius is often regarded as one of the richest and most successful African economies, with its Gross Domestic Product (GDP) rising from US$260 in 1968—year of independence—to a current figure approximating US$8500, with the economy growing almost consistently at an average of around 5% yearly for the last quarter century. Interestingly, the success of the Mauritian economy can largely be attributed to its policy of trade openness, given its small domestic market. The traditional ingredients of growth have been sugar, textile, and tourism (Khadaroo and Seetanah, 2007; Seetanah and Khadaroo, 2009), with the economy also diversifying into financial services and information and computer technologies, keeping the traditional sector base. The prominence of trade is fundamentally due to three factors: (1) preferential access to trade, (2) conducive environment for investment, and (3) exogenous factors.[2]

However, rising standards of living have been accompanied by an increased demand for energy resources and in this regard, Mauritius has developed a strong reliance on imported fossil fuels for its energy needs close to 82%, and this dependency is increasing. The carbon dioxide emissions associated with the burning of fossil fuels are also on the rise, and Mauritius has a per capita carbon dioxide emission around 3 tons (Central Statistics Office, 2013). Additionally, Mauritius also faces the inherent environmental vulnerabilities associated with SIDS (Durbarry and Seetanah, 2015), which include a small land area, susceptibility to natural disasters, geographical isolation, limited natural resources and sensitive ecosystems, and limited human and institutional capacity. As a result of its economic success, the standard of living on Mauritius has increased with significant changes in consumption patterns. The new demands placed on an already limited natural resource base are resulting in increased environmental problems, such as road congestion, water scarcity, land and coastal degradation, inadequate waste disposal, and pollution.

---

[1] Studies by Grossman and Krueger (1991), Lucas et al. (1992), Wyckoff and Roop (1994), Suri and Chapman (1998), Anderson et al. (2011) have thoroughly discussed and tested this proposition; albeit their findings depict mixed results.

[2] Examples of exogenous factors include the Multifiber Agreement and the AGOA which has led to massive influx of FDI in the garment and textile industry.

As mentioned previously, the present study attempts to delineate the impact of openness on CO_2 emission for the case of Mauritius over the period 1976—2013.[3] In the presence of both I(0) and I(1) data series and to account for the possibility of dynamism in the link, we adopted an ARDL (autoregressive distributed lag model) approach to model the determinant of CO_2 emission with a particular focus on trade openness. The results are expected to supplement the scarce literature on the relationship between trade and environment in developing country cases and island states.

The rest of the chapter is structured as follows: Section 9.2 provides a brief literature review while Section 9.3 dwells in the methodological framework and analysis of the results and Section 9.4 concludes.

9.2 LITERATURE REVIEW

9.2.1 Theoretical Review

Extensive research and large theory-based approaches have been proposed on the relationship between trade liberalization and the environment. For instance, using the Heckscher—Ohlin type model, Antweiler et al. (2001) and López et al. (2007) have argued that whenever countries open up to trade, the effects may be decomposed into three broad categories, namely *scale, technique*, and *composition* effects.

The scale effect states that with trade liberalization, there is a trade-off between greater productivity which results in greater output and environmental degradation (Appiah-Konadu, 2013). The *technique effect,* on the other hand, relates to the use of environmental friendly production techniques which would allow for an improvement in the environmental quality of products during a country's trade liberalizing phase (Appiah-Konadu, 2013). From an economic point of view, environmental products are classified as normal goods indicating that with a further increase in individual's income, higher environmental quality products are prioritized (Onder, 2012). This is justified by the induced demand function for more environmental quality goods but which also call for more environmental regulations, standards, and protection since income has risen. Finally, Grossmann and Krueger (1995) suggest that the *composition effect* represents the channel through which the pollution haven hypothesis (PHH) impact on the environment (Copeland and Taylor, 2004). Basically, the composition effect relates to the changes in the structure of the economy following trade liberalization whereby countries normally specialize as per their comparative advantage. Thus, it analyses the extent to which trade openness fosters the adoption of new technologies or it shifts the economy toward the tertiary industry. Thus, the composition effect depicts the shifts in the level of carbon

---

[3] The total CO_2 emission for 2013 for Mauritius was 4118 kt and the CO_2 emission per capita was 3.125 for the same year.

emissions due to changes in the shares of different goods in the aggregate production of the countries (Onder, 2012).

Other authors have analyzed the theoretical interplay between trade liberalization and the environment through a discussion of the relationship between trade and CO_2 emission. In this regard, the environmental Kuznets curve (EKC) (1995) has become the fundamental economic theory which explains the link between economic growth and environmental degradation. Kijima et al. (2010) defines the EKC as "the hypothesis that the relationship between environmental degradation and per capita income demonstrates an inverted-U shape nature" and it depicts the long-run relationship between economic growth and environmental degradation. More importantly, the EKC hypothesis summarizes an essentially dynamic process of change; as the income of an economy grows over time, its emission level of carbon dioxide increases at first, then it reaches a peak and finally starts to decline after a threshold level of income has been crossed. In short, it may be perceived as a representation of the natural process of economic development from a clean agrarian economy to a polluting industrial economy, and, finally, to a clean service economy (Arrow et al., 1995).

In a similar vein, the PHH, which also attempts to uncover the possible link between trade and CO_2 emission, propounds that governments in most developing countries avoid imposing strict environmental standards on local companies so as to boost their competitiveness in global markets. Accordingly, the PHH is based on the differences in the flexibility of environmental regulations between developed and developing countries. Eskeland and Harrison (2003) argue that environmental regulations can shift polluting activities involved in the production of tradable commodities to developing countries. With trade liberalization, pollution-intensive companies in developed countries, where regulations are stringent, move toward developing countries with poor environmental regulations. In the long run, therefore, developing countries have a comparative advantage in pollution-intensive production while developed countries specialize in clean production.

Given the above analogy, Esty (2001) argues that trade liberalization may be problematic since the economic pressures to remain competitive may generate a "race toward the bottom" whereby countries may be encouraged to decrease environmental standards in order to remain attractive as low-cost markets for manufacturing. Additionally, in situations where trade liberalization improves the mobility of capital and production, developed countries may be forced to sacrifice environmental protection to remain competitive which may unfortunately lead to short-term economic gains only.

Finally, "the Porter Hypothesis," developed by Porter (1991) and Porter and van der Linde (1995), has also gained prominence as a result of its stringent proposed environmental regulations and trade standards which have proved beneficial both to the environment and to the competitiveness element of firms (Appiah-Konadu, 2013). The basic underpinning behind the hypothesis is the attempt to counteract the trade-off relationship as propounded by the standard view of free trade and the environment (Rutqvist, 2009). Porter argues that through the application of stringent environmental policies termed as the "go green" concept, firms and nations alike may benefit through the emanation of new competitive advantages. This may be achieved in the following manner. Firstly, regulations and trade standards will enable companies to become aware of alternative and efficient production methods. Secondly, because of the restrictive nature of such regulations, firms which have already accounted for change will benefit, and this constitutes an initial comparative advantage over those firms which are to implement changes for the first time. In a nutshell, the Porter Hypothesis propounds that if firms are compelled to reduce their pollution emanation and adopt flexible ways which will enable them to innovate and to adopt measures which are regulations compliant, they may benefit from "innovation offsets" as a result of the rise in resource productivity thereby leading to improved comparative advantage.

9.2.2 Empirical Review

Since the early 1990s, there has been a growing body of empirical literature on the economic growth, trade openness, and pollution nexus for countries with different development stages. However, one could argue that the empirical findings have displayed mixed results at best. Although several studies have demonstrated the positive interplay between trade openness and wealth generation, they have nevertheless also displayed varying resulting impacts on the environment (Grossman and Krueger, 1993; Copeland and Taylor, 2004; Kahuthu, 2006).

For instance, Halicioglu (2009) examined the dynamic causal relationship between CO_2 emission, energy consumption, income, and foreign trade for Turkey using time-series data for the period 1960–2005. The results showed that income was the most significant variable explaining CO_2 emissions in Turkey, followed by energy consumption and foreign trade.

More recently, Onoja et al. (2014) conducted an empirical investigation of the role played by trade openness on CO_2 emissions in Africa for the period spanning 1960–2010. The econometric results indicated that GDP growth rate and trade openness were the major long-run and short-run determinants of greenhouse gas emissions on the African continent. Interestingly, their findings also provided support to the Kuznet's environmental curve theory. Similarly, other authors who have uncovered a positive relationship between trade and CO_2 emissions include, among others, Khalil and Inam (2006) for Pakistan and Chebbi et al. (2010) for Tunisia.

On the other hand, Managi et al.'s study (2009) focused on trade and income as endogenous variables and investigated the relationship between trade openness, economic development, and environmental quality using the instrumental variables technique and their study displayed mixed results for the OECD and non-OECD sample of countries. Specifically, trade was found to be beneficial to the environment in OECD countries but having a detrimental impact on the environment in the case of non-OECD countries. The authors further pointed out that the impact was large in the long run while the short run showed a smaller impact.

Additionally, irrespective of the above, one could argue that the impact of trade on the environment, as suggested by the theoretical underpinning, would ultimately depend upon a country's comparative advantage, its environmental protection regulations, and the development of trade, respectively. This is why most of the empirical studies have used parametric models such as quadratic or cubic functions to test for the presence of the EKC hypothesis and that of other theories.

In this regard, many authors have focused their attention on the delineating relationship between income and pollution which provides support to the EKC model. Existence of such a relationship was found in various studies namely those by Grossman and Krueger (1993, 1995), Suri and Chapman (1998), and Cole (2004).

Similarly, McCarney and Adamowicz (2006) used panel data from 143 countries to examine the effect of free trade on the environment for the years 1970—2000. Their methodology consisted of building a random effect model to assess the impact of trade and to uncover whether an EKC may be observed in the case of these countries. The results displayed two very important findings. Firstly, *trade* was found to have a positive and significant impact, implying that as trade openness increases, CO_2 emissions also increase, *ceteris paribus*. Secondly, a negative coefficient was noted on the *haven* which captured the combined effects of GDP per capita and openness to trade. Being statistically significant, it indicated a possible EKC path since a reduction in CO_2 emissions could be observed at higher levels of income per capita.

Loi (2012), in his study of six Asian countries for the period of 1980—2006, analyzed the interrelationship between trade liberalization and environmental degradation and the results provided no support to the EKC hypothesis. However, it supported the PHH, with the latter indicating that a greater propensity to liberalize trade will lead to a subsequent increase in the level of CO_2 emissions and energy consumption in these East Asian countries.

9.3 METHODOLOGY AND ANALYSIS

We use a multivariate model in which CO_2 emissions are a function of trade openness and other related explanatory variables (the different determinants on the carbon dioxide emissions in Mauritius). Our conceptual model is largely based on the earlier work of Seetanah and Sannassee (2011), amended for the sake of this study with the inclusion of a trade openness proxy. Thus we claim

$$CO_2 = f(\text{IVTGDP, GDP, SER, POP, VEHICLES, TRADE}) \quad (9.1)$$

where IVTGDP is the investment ratio, GDP is a measure of the output level, SER (secondary enrollment ratio) is a proxy for the education level and human capital, POP is the population level, and VEHICLES is the number of vehicles on road. The rationale for their inclusion can be found in Seetanah and Sannassee (2011). Of interest to us is the TRADE variable which is measured by the ratio of export plus import on GDP. The theoretical link between trade and environmental degradation has been discussed both theoretically and empirically in the earlier section. We use annual data on the total fossil fuels per capita carbon dioxide emissions and real GDP for the period of 1976–2013. Data on per capita carbon dioxide emissions, stemming from fossil fuels burning, cement manufacture, and gas flaring, are extracted from the CDIAC (Carbon dioxide Information Analysis Center) database. Per capita emissions are expressed in metric tons of carbon. The data exclude emissions from land use and agriculture (including deforestation). GDP, SER, POP, IVTDGP, and VEHICLES data are gathered from the Central Statistical Office.

9.3.1 The Econometric Model

Taking logs (on both sides) of the above equation (for more meaningful interpretation in terms of rate of change) and denoting the lowercase variables as the natural log of the respective uppercase variable results in the following:

$$co2 = \beta_0 + \beta_1 ivtgdp + \beta_2 gdp + \beta_3 ser + \beta_4 pop + \beta_5 vehicles + \beta_6 trade + \varepsilon \qquad (9.2)$$

The disturbance term ε is the deviation from the above relationship.

Before considering the appropriate framework for the econometric model, we investigate the univariate properties of all data series and determine the degree to which they are integrated. Both the Augmented Dickey-Fuller (ADF) (Dickey and Fuller, 1979) and Phillips–Perron (PP) (Phillips and Perron 1988) unit-roots tests are employed for this purpose and the variables are found to be I(1) except the trade variable. In the presence of a mix of I(1) and I(0) variables, the testing and estimation procedure, advanced in Pesaran et al. (1996) and Pesaran and Shin (1999) to examine the existence of a long-term relationship (cointegration), is employed in our analysis. Unlike other cointegration approaches such as the Johansen's (1988) maximum likelihood technique, the ARDL technique does not require the variables in the model to be integrated to the same order and is thus considered to be appropriate in our case.

The error correction version of the ARDL model (see Pesaran et al., 1996) is given by:

$$\Delta co2 = \beta_0 + \sum_{i=1}^{n} b_i \Delta co2_{t-i} + \sum_{i=1}^{n} c_i \Delta ivtgdp_{t-i} + \sum_{i}^{n} d_i \Delta gdp_{t-1} + \sum_{i=1}^{n} e_i \Delta ser_{t-i}$$

$$+ \sum_{i=1}^{n} f_i \Delta pop_{t-i} + \sum_{i=1}^{n} g_i \Delta vehicles_{t-i} + \sum_{i=1}^{n} h_i \Delta trade_{t-i} + \delta_1 co2_{t-1} + \delta_2 ivtgdp_{t-1}$$

$$+ \delta_3 gdp_{t-i} + \delta_4 ser_{t-1} + \delta_5 pop + \delta_6 vehicles_{t-1} + \delta_7 trade_{t-1} + \varepsilon_t$$

$$(9.3)$$

For the econometric model, the hypothesis which is tested is the null of "non-existence of the long-run relationship" and is defined by

$$H_0: \delta_1 = \delta_2 = \delta_3 = \delta_4 = \delta_5 = \delta_6 = \delta_7$$

And the alternative hypothesis is

$$H_1: \delta_1 \neq 0, \quad \delta_2 \neq 0, \quad \delta_3 \neq 0, \quad \delta_4 \neq 0, \quad \delta_5 \neq 0, \quad \delta_6 \neq 0, \quad \delta_7 \neq 0$$

The recommended statistic is the F statistics for the joint significance of δ_1, δ_2, δ_3, δ_4, δ_5, δ_6, and δ_7. Computation of such F statistics requires running the following regression

$$\Delta co2 = \beta_0 + b\Delta co2_{t-1} + c\Delta ivtgdp_{t-1} + d\Delta gdp_{t-1} + e\Delta ser_{t-1} + f\Delta pop_{t-1}$$
$$+ g\Delta vehicles_{t-1} + h\Delta trade_{t-1} + \varepsilon_t$$

and a variable addition test is subsequently done by including the following:

$$\delta_1 co2_{t-1} + \delta_2 ivtgdp_{t-1} + \delta_3 gdp_{t-1} + \delta_4 ser_{t-1} + \delta_5 pop_{t-1} + \delta_6 vehicles_{t-1} + \delta_7 trade_{t-1}$$

It should however be noted that the distribution of the F statistics is nonstandard, irrespective of whether regressors are I(0) or I(1). Pesaran et al. (1996) have tabulated the appropriate critical values[4] for different number of regressors. The F statistics in the present case equates 5.34 and exceeds the upper bound of the critical value band. We thus reject the null hypothesis of no long-run relationship between the variables irrespective of their order. The test results thus suggest that there is a long-run relationship between the variables and can thus be specified in an error correction model (ECM) as well.

9.3.2 Estimation Results of the ARDL

Given that the specification is cointegrated, the unrestricted error correction representation of the ARDL model is given by Eqn (9.3).

We then proceeded to estimate the coefficients of the long-run relations and the associated ECM using the ARDL approach. The order of the distributed lag on the dependent variable was selected by the SBC and was equal to 1.

The Bayesian Information Criterion or Schwarz Criterion (SBC) criteria select the ARDL (1, 0, 0, 0, 0, 0, 1) for our model. The long-run estimated coefficients are shown in Table 9.1.

The results from Table 9.1 demonstrate that trade has contributed positively and significantly to raising CO_2 emission in the country in the long run. Additionally, a

---

[4] If the computed F statistics falls outside the band, a conclusive decision can be made without needing to know whether the underlying variables are I(0) or I(1). If it falls with the band (critical), the result of the inference is inconclusive and depends on whether the underlying variables are I(0) or I(1). If need be a unit root test may be carried out.

Table 9.1 Estimated Long-run Coefficients Based on ARDL Approach

Regressor	Coefficient (SBC 1, 0, 0, 0, 0, 0, 1)	t-ratio
ivtgdp	0.53***	2.26
gdp	0.69***	3.11
ser	−0.28**	−2.31
pop	0.35***	2.27
vehicles	0.78***	2.41
trade	0.59**	2.12
Constant	2.14**	2.38

ARDL (1, 0, 0, 0, 0, 0, 1) selected based on Schwarz Bayesian Criterion, CO_{2t} is the dependent variable.
Dependent variable is *co2*.
The selected ARDL model passes the standard diagnostic tests (serial correlation, functional form, normality, and heteroscedasticity).
Note: ** and *** indicate 5% and 1% significance level, respectively.

1% increase in trade openness is associated with a 0.60% increase in CO_2 emission. Such a result is consistent with the findings of Onoja et al. (2014) for the case of Africa, Khalil and Inam (2006) for Pakistan, and Chebbi et al. (2010) for Tunisia.

Such results for Mauritius were indeed expected in view of the prevalence of the manufacturing sector as a major contributor to growth for a period spanning more than 20 years since the early 1980s. Mauritius, being an island totally devoid of natural resources, has laid an overt emphasis on fostering trade openness, which has led to expanded export figures, most notably in the garment and textile industry, aided by favorable external conditions such as preferential access to the EU and US markets, which have served to attract FDI from East Asian NICs. Unsurprisingly, such a trend has been accompanied by a concomitant increase in the use of fossil fuel, and hence the resulting increase in CO_2 emission.[5]

Furthermore, and as expected, *vehicles*, *gdp*, *invgdp*, and *pop* are found to be positively related to environmental degradation while the level of education (SER) is found to be negatively related to CO_2 emission. Such a finding is unsurprising given that the higher the level of education, the greater the propensity for people to be better informed about potential environmental risks and damages and as such they would be more willing to support measures aimed at protecting the environment and at promulgating sustainable development (SD) (e.g., Torgler and Garcia-Valinas, 2007). In addition, a higher level of education is likely to increase the capacity of society to innovate technologically and accelerate alternatives to fossil fuels for the energy needed to support economic growth.

[5] We augmented the original econometric specification to add a dummy variable (taking the value of 0 before 1995 and 1 post 1995) to take into account the changes in the garment regulations (with the advent of the ATC in 1995 whereby China could export quota free) and our estimations remain more or less the same as the one reported.

Table 9.2 Error Correction Representation for the Selected ARDL Model

Regressor	Coefficient (SBC 1,0,0,0,0,0,1)	t-ratio
Δ ivt	0.17*	1.88
Δ gdp	0.22**	2.21
Δ ser	−0.08	−1.34
Δ pop	0.11	1.39
Δ vehicles	0.46*	1.89
Δ trade	0.23**	2.53
constant	2.33*	1.84
ecm_{t-1}	−0.11***	−3.62
R square	0.76	
DW	2.07	

Dependent variable is co2.
Note: *, **, *** indicate 10%, 5%, and 1% significance level, respectively.

A relatively higher coefficient associated with vehicles suggests that over the past four decades, an increase in the fleet of vehicles on our roads has had a significant detrimental effect on environmental quality, and this result, it may be argued, provides support to the widely held view that, among various greenhouse gases, CO_2 emissions through the combustion of fossil fuels (e.g., coal, petroleum, and natural gas) seem to be the major contributor of global warming. Such results are in line with the findings of Seetanah and Sannassee (2011).

We proceed to the estimation of the ECM associated with our long-run estimates using SBC. The results are reported in the table below.

The findings from Table 9.2, where estimates of the ECM are presented, suggest that trade has also had a positive and significant effect on CO_2 in the short run. The coefficients for the other explanatory variables also have the expected sign and significance. Moreover, the coefficient of the ECM of the selected ARDL (1,0,0,0,0,0,1) is negative and highly significant at 1% level. This confirms the existence of a stable long-run relationship and points to a long-run cointegration relationship between the variables. The ECM represents the speed of adjustment to restore equilibrium in the dynamic model following a disturbance. The coefficient of the ECM approximates −0.11 in the system and this implies that a deviation from the long-run equilibrium following a short-run shock is corrected by approximately 11% after each year and it requires some 10 years for the model or system to restore equilibrium in long run.

9.4 CONCLUSION

This chapter has sought to investigate the link between trade openness and the environment in the case of Mauritius for the period 1976–2013 using an econometric approach to model CO_2 emissions. In the presence of both I(0) and I(1) variables, an ARDL has been employed and the results have shown that trade has served to increase the level of CO_2 in the country in both the long run and short run. In this regard, the results demonstrate that a 1% increase in trade openness is accompanied by an upshot of 0.60% in CO_2 emission. The results are consistent with the findings of Onoja et al. (2014), Khalil and Inam (2006), and Chebbi et al. (2010) for developing country cases. The error correction framework has also confirmed similar results in general for the short run.

Such results for Mauritius can be explained by the fact that the manufacturing sector has been a major contributor to growth since the early 1980s. Given that the island is totally devoid of natural resources, successive governments have had but no other alternative than to adopt an outward looking approach which has led to very commendable export figures most notably in the garment and textile industry. In this regard, one should not underplay the crucial importance of favorable external conditions such as preferential access to the EU and US markets, which have served to attract FDI from East Asian NICs.

The findings also delineated the prevalence of trade liberalization on CO_2 emission. However, as postulated by the PHH, one would expect a future decline in such emission numbers as Mauritius moves up the ladder of industrialization and after Mauritius surpasses its peak level of CO_2 as per the EKC.[6] In addition, over the last few years, there has been an increasing number of Mauritian garment firms delocalizing their operations from Mauritius and moving into cheaper location sites such as Madagascar and Bangladesh (in line with the PHH) and one could expect that such a trend will serve to curb, although not expected to decrease in the short term, the increasing trend of CO_2 emission.

Nevertheless, various measures and actions have been taken by the island's different governments over the years to try mitigate the potential negative impact of CO_2 emission on the environment. In this regard, the Government of Mauritius has committed themselves to implementing the international recommendations on SD which recognize the specific vulnerabilities of SIDS, such as the 1992 Agenda 21, the 1994 Barbados Programme of Action for SIDS, the 2002 Johannesburg Plan of implementation (JPOI), the 2005 Mauritius Strategy (MSI) for SIDS, and the 2012 Rio+20 "*The Future We Want.*" In a similar vein, the National Environmental Policy of 2007 also outlines a series of thematic policy objectives and strategies to address environmental challenges. Sectoral policies have, in addition, been developed across various thematic areas including the likes of land, biodiversity, forests, water and wastewater, solid waste, coastal zone management, and tourism and energy among others, to help mitigate the detrimental impact of environmental degradation and CO_2 emission.

---

[6] Mauritius is yet to reach its peak level of CO_2 emission as propounded by the EKC as highlighted in the study by Seetanah and Sannassee (2011).

REFERENCES

Anderson, M.J., Crist, T.O., Chase, J.M., Vellend, M., Inouye, B.D., Freestone, A.L., Sanders, N.J., Cornell, H.V., Comita, L.S., Davies, K.F., Harrison, S.P., Kraft, N.J., Stegen, J.C., Swenson, N.G., 2011. Navigating the multiple meanings of beta diversity: a roadmap for the practicing ecologist. Ecol. Lett. 14 (1), 19–28.

Antweiler, W., Copeland, B.R., Taylor, M.S., 2001. Is free trade good for the environment? Am. Econ. Rev. 91 (4), 877–908.

Appiah-Konadu, P., 2013. The Effect of Trade Liberalisation on the Environment: A Case Study of Ghana. MPhil Thesis. University of Ghana, Ghana.

Arrow, K., Bolin, B., Costanza, R., Dasgupta, P., Folke, C., Holling, C.S., Jansson, B.-O., Levin, S., Maler, K.-G., Perrings, C.A., Pimentel, D., 1995. Economic growth, carrying capacity, and the environment. Science 268, 520–521.

Chebbi, H.E., Olarreaga, M., Zitouna, H., 2010. Trade Openness and CO_2 Emissions in Tunisia. Working Paper No. 518. Faculty of Economic Science and Management of Nabeul (FSEGN), Tunisia.

Cole, M.A., 2004. Trade, the pollution heaven hypothesis and the environmental Kuznets curve: examining the linkages. Ecol. Econ. 48 (1), 71–81.

Copeland, B.R., Taylor, M.S., 2004. Trade, growth, and the environment. J. Econ. Lit. 42 (1), 7–71.

Dickey, D.A., Fuller, W.A., 1979. Distribution of the estimators for autoregressive time series with a unit root. J. Am. Stat. Assoc. 74 (366), 427–431.

Durbarry, R., Seetanah, B., 2015. The impact of long haul destinations on carbon emissions: the case of Mauritius. J. Hosp. Mark. Manag. 24 (4).

Eskeland, G.S., Harrison, A.E., 2003. Moving to greener pastures? Multinationals and the pollution Haven hypothesis. J. Dev. Econ. 70 (1), 1–23.

Esty, D.C., 2001. Bridging the trade-environment Divide. J. Econ. Perspect. 15 (3), 113–130.

Grossman, G.M., Krueger, A.B., 1991. Environmental Impacts of a North American Free Trade Agreement. National Bureau of Economic Research Working Paper Working Paper 3914, NBER, Cambridge, MA.

Grossman, G.M., Krueger, A.B., 1993. Environmental Impacts of a North American Free Trade Agreement. In the Mexico-US Free Trade Agreement. P. M. Garber. MIT Press, Cambridge.

Grossman, G.M., Krueger, A.B., 1995. Economic growth and the environment. Q. J. Econ. 110 (2), 353–377.

Halicioglu, F., 2009. An econometric study of CO_2 emissions, energy consumption, income and foreign trade in Turkey. Energ. Policy 37 (3), 1156–1164.

Hanley, N., Shrogen, F., White, B., 2001. An Introduction to Environmental Economics. Oxford University Press, London.

Johansen, S., 1988. Statistical analysis of co-integration vectors. J. Econ. Dyn. Control 12 (2–3), 231–254.

Kahuthu, A., 2006. Economic growth and environmental degradation in a global context. Environ. Dev. Sustain. 8 (1), 55–68.

Khalil, S., Inam, Z., 2006. Is trade good for environment? A unit root cointegration analysis. Pak. Dev. Rev. 45 (4), 1187–1196.

Kijima, M., Nishide, K., Ohyama, A., 2010. Economic models for the environmental Kuznets curve: a survey. J. Econ. Dyn. Control 34 (7), 1187–1201.

Khadaroo, A.J., Seetanah, B., 2007. Does transport infrastructure matter in overall tourism development? Evidence from a sample of island economies. Tour. Econ. 13 (4), 675–684.

Loi, N., 2012. The Impact of Trade Liberalization on the Environment in Some East Asian Countries: An Empirical Study. Available Online: http://www.cerdi.org/uploads/sfCmsContent/html/323/NhuyenDuy.pdf.

López, R., Galinato, G., Islam, A., 2007. Government Expenditures and Air Pollution. Working Paper. University of Maryland, US.

Lucas, G., Wheeler, N., Hettige, R., 1992. The Inflexion Point of Manufacture Industries: International Trade and Environment. World Bank Discussion Paper. No. 148, Washington, DC.

Managi, S., Hibiki, A., Tsurumi, T., 2009. Does trade openness improve environmental quality? J. Environ. Econ. Manag. 58 (3), 346—363.

McCarney, G., Adamowicz, V., 2006. The Effects of Trade Liberalisation of the Environment: An Empirical Study. International Association of Agricultural Economists Annual Meeting, August 12—18, Queensland, Australia.

Onder, H., 2012. Trade and Climate Change: An Analytical Review of Key Issues. Economic Premise, Poverty Reduction and Economic Management Network (PREM). World Bank. No.86.

Onoja, A.O., Ajie, E.N., Achike, A.I., 2014. Econometric analysis of influences of trade openness, economic growth and urbanization on greenhouse gas emission in Africa (1960—2010). J. Econ. Sustain. Dev. 5 (10), 14—23.

Pesaran, M.H., Shin, Y., Smith, R.P., 1996. Testing for the Existence of a Long-run Relationship. DAE Working Paper, No. 9622.

Pesaran, M.H., Shin, Y., 1999. An autoregressive distributed lag modelling approach to cointegration analysis. In: Strom, S. (Ed.), Econometrics and Economic Theory in the 20th Century, the Ragnar Frisch Centennial Symposium. Cambridge University Press, Cambridge.

Phillips, P., Perrone, P., 1988. Testing for a unit root in time series regression. Biometrica 75, 335—346.

Porter, M.E., 1991. Towards a dynamic theory of strategy. Strategic Manage. J. 12 (Winter Special Issue), 95—117.

Porter, M.E., Van der Linde, C., 1995. Toward a new conception of the environment-competitiveness relationship. J. Econ. Perspect. 9 (4), 97—118.

Rutqvist, J., 2009. Porter or Pollution Haven? An Analysis of the Dynamics of Competitiveness and Environmental Regulations. Harvard.

Seetanah, B., Khadaroo, A.J., 2009. An analysis of the relationship between transport capital and tourism development in a dynamic framework. Tour. Econ. 15 (4), 785—803.

Seetanah, B., Sannassee, V., 2011. On the Relationship between CO_2 Emissions and Economic Growth: The Mauritian Experience. Paper Presented at CSAE Centre. Oxford University, Oxford.

Suri, V., Chapman, D., 1998. Economic growth, trade and the energy: implications for the environmental Kuznets curve. Ecol. Econ. 25 (2), 195—208.

Torgler, B., Garcia-Valinas, M.A., 2007. The determinants of individuals', attitude towards preventing environmental damage. Ecol. Econ. 63, 536—552.

Wyckoff, A.W., Roop, J.M., 1994. The embodiment of carbon in imports of manufactured products: implications for international agreements on Greenhouse Gas Emissions. Energ. Policy 22, 187—194.

CHAPTER 10

Will TAFTA Be Good or Bad for the Environment?

Dhimitri Qirjo, Robert Christopherson
Department of Economics & Finance, School of Business & Economics, The State University of New York at Plattsburgh, Plattsburgh, NY, USA

Contents

10.1 INTRODUCTION

The North American Free Trade Agreement (NAFTA) has generated a long debate, before and after its establishment, in regard to its consequences for the environment. NAFTA was unique in its relationship to the environment since it was the first international major trade agreement that included environmental provisions in a side agreement (The North American Agreement on Environmental Cooperation). Today, there are ongoing negotiations between officials of the US and EU for the creation of another major common free trade area, the Trans-Atlantic Free Trade Agreement (TAFTA). In this chapter, we investigate the potential effects of TAFTA on pollution emissions in a typical TAFTA member country.

The debate over the effect of trade liberalization on environmental outcomes has received much attention. Early work in this literature identifies three main channels in which trade liberalization can affect pollution emissions. These are known in the literature as the scale, technique, and composition effects. Holding all other factors constant, the scale effect refers to the increase in pollution emissions, because of the increase in national aggregate production, due to bigger markets as a result of trade liberalization. To understand the technique effect, one has to consider environmental quality as a normal good. Thus, trade liberalization increases income per capita, and hence the

Handbook of Environmental and Sustainable Finance
ISBN 978-0-12-803615-0

179

demand for environmentally clean goods. Moreover, free trade increases the accessibility of cleaner technologies especially for developing countries. Consequently, ceteris paribus, national pollution emissions fall under trade liberalization via the technique effect. The composition effect is related to the empirical evidence that relative capital-intensive goods are pollution-intensive goods. Thus, trade liberalization generates economic growth, which in turn generates physical capital accumulation and increase the national physical capital. The latter will increase the overall production of the capital-intensive goods and therefore increase national pollution emissions. Combining the composition effect with the Heckscher–Ohlin theory, it is easy to conclude that trade liberalization (TAFTA), will increase pollution emissions in a relatively capital-abundant country (such as the US or Luxembourg).

The recent literature has focused more on the composition effect of international trade, attempting to identify sources related to a country's comparative advantage. One school of thought suggests that trade liberalization will increase pollution in low-income countries due to the existence of relatively lax environmental regulations, giving them comparative advantage in dirty goods, and decrease pollution in high-income countries with comparative advantage in clean goods. This is known as the pollution haven effect. Despite the existence of robust theoretical models consistent with the pollution haven argument, there is limited empirical evidence to support it. Another school of thought identifies comparative advantage according to the Heckscher–Ohlin theory, where the rich, capital-abundant country has comparative advantage over the dirty, capital-intensive good. This is called the factor endowment effect. This latter argument has been substantiated with more empirical support than the pollution haven hypothesis.

However, the most recent literature as represented by Antweiler et al. (2001) focuses on identifying the comparative advantage in a country using both opposing forces (the pollution haven and the factor endowment). For example, according to the pollution haven argument, the implementation of TAFTA should decrease (increase) pollution emissions in the relatively rich (poor) US (Estonia) because of relatively strict environmental regulations, as compared to an average TAFTA member. On the other hand, according to the factor endowment argument, it should increase (decrease), the United States' (Estonia's) pollution emissions, due to the intensification of its national production, of the capital-intensive (labor-intensive) goods. Thus, the implementation of TAFTA, at least theoretically, would reduce the pollution emissions in the US (Estonia) if the pollution haven effect dominates (is being dominated by) the factor endowment effect. Moreover, in a relatively labor-abundant (capital-abundant) and rich (poor) country, such as Ireland or/and Sweden (Spain), pollution emissions should be theoretically reduced after the implementation of TAFTA due to pollution haven and factor endowment arguments. See Figures 10.1 and 10.2 (in Section 10.3 of this chapter).

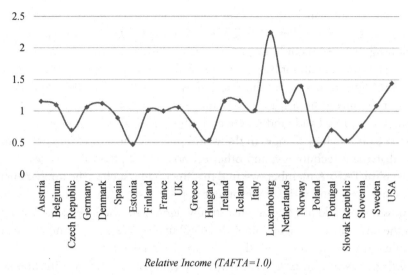

Relative Income (TAFTA=1.0)

Figure 10.1 Relative income for 24 TAFTA members over 1995–2010.

In this chapter, we follow the works of Antweiler et al. (2001) and Cole (2003) in order to examine the environmental consequences of the implementation of TAFTA. Using data over the 1995–2010 time period, for 23 European Union member countries and the United States, we empirically investigate the role of TAFTA implementation on per capita pollution emissions as measured by four air pollutants: carbon dioxide, greenhouse gases (GHGs), nitrogen oxides, and sulfur oxides; and a general pollutant such as municipal waste (MW).

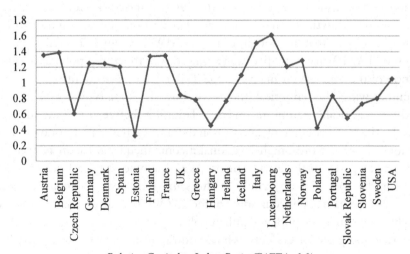

Relative Capital to Labor Ratio (TAFTA=1.0)

Figure 10.2 Relative capital abundance in 24 TAFTA country members over 1995–2010.

By exploring our panel data set, we develop four econometric models following closely the works of Antweiler et al. (2001) and Cole (2003), in order to first examine the relationship between economic growth and pollution, second to evaluate the effects of income gains brought about by income inequality and government effectiveness (GE) on pollution, third to investigate the direct composition effect of growth on pollution, and finally to evaluate the impact of income gains as a result of trade liberalization on pollution. We use fixed and random effect methods to deal with unobservable variables in our regressions, such as changes in the world prices of dirty or clean goods, improvements in abatement technology, and other economic and physical variables.

For a typical TAFTA member, we find that freer trade, such as the implementation of TAFTA, could be good for the environment in the cases of carbon dioxide and nitrogen oxides emissions. We do not find any statistically significant evidence of the impact of free trade on the environment for the other three pollutants. We do not find any statistically significant evidence consistent with the pollution haven or/and factor endowment arguments, with the exception of GHGs, where we find robust evidence in support of the pollution haven hypothesis. Thus, we believe a relatively poor member of TAFTA, such as the Czech Republic will see a raise in GHGs per capita emissions in response to the implementation of TAFTA. In contrast, the US as a relatively rich member of TAFTA will observe a reduction of GHGs per capita emissions.

For both GHGs and CO_2, when we account for trade variables we find robust empirical evidence of the existence of an inverse U-shaped relationship between economic growth and pollution known as the environmental Kuznets curve (EKC). All the countries in our sample find themselves on the right-hand side of the peak of the EKC, meaning that higher economic growth, due to TAFTA, may help reduce GHGs and CO_2 emissions per capita. Moreover, we show that for a typical TAFTA member country, growth as a result of trade liberalization does not appear to improve the techniques of production enough, to reduce MW per capita. Therefore, the implementation of TAFTA may raise per capita emissions of these two pollutants, due to the domination of the scale effects in an average TAFTA member country.

In contrast with other empirical studies we show that for a typical TAFTA member, there exists a statistically significant inverse relationship between income inequality, as measured by the GINI coefficient, and carbon dioxide, GHGs, and sulfur oxides per capita emissions. In other words, higher national income inequality for an average TAFTA member may be good for the environment. As an example, the US has improved its air quality over the past 15 years, while at the same time seen an increase in income inequality, this too has happened in some of the 23 OECD countries, but not universally. One might surmise that the push and pull of politics and economics are at play. While the wealthy favor and push for tax cuts, which contribute to income inequality, both the wealthy and the poor favor better environmental standards. We know that environmental quality is generally considered a normal good. Therefore, an increase of the income for the wealthy due to favorably tax cut policies may encourage them to provide political support toward stringent environmental regulations. While the wealthy can avoid living in relatively more polluted zones, the poor are often located closer to industrial areas and experience more pollution. However, enforcing more stringent environmental regulations helps everyone.

Moreover, we find another surprising result over the relationship of a country's direct composition effect of growth, as measured by its capital to labor ratio, and pollution. Our estimates show that higher capital abundance, for a typical TAFTA member, is associated with a reduction of carbon dioxide, GHGs, sulfur dioxide per capita emissions, and MW per capita. Previous studies have found a positive relationship between a country's capital abundance and its pollution emissions, linking capital-intensive goods with dirty goods.

When analyzed carefully, we observe that employment levels and labor force participation rates have fallen in many of the 24 countries in our data set, especially during the 2003–2005 period and more dramatically from 2008 to 2010. Additionally, as indicated by the literature in developed economics, we also observe an aging population phenomenon in many countries in our data set, indicating a further decline in labor force participation rate. Therefore, we infer that the negative relationship between the capital to labor ratio and four of our pollutants is most likely due to this reduction of the labor force, of an average TAFTA member.

The rest of this chapter is organized as follows. In Section 10.2, we provide a theoretical and empirical literature review of the relationship between trade liberalization and environmental quality. In Section 10.3, we describe our data set. In Section 10.4, we present the functional form of our four estimating equations. In Section 10.5, we discuss our empirical methodology. In Section 10.6, we present our empirical findings using the help of five tables. Finally, Section 10.7 concludes the chapter.

10.2 LITERATURE REVIEW

There is a large and burgeoning literature over the role of international trade on pollution emissions. See, for example, Anderson and Blackhurst (1992), Grossman and Krueger (1991, 1993, 1995), Chichilnisky (1994), Copeland and Taylor (1994, 1995), Antweiler et al. (2001), Cole (2003), Mani and Cunha (2011), and Onder (2012) to name a few. Copeland and Taylor (2004) provide an excellent comprehensive review on the international trade and environmental literature. They claim that economists, started to pay much more attention to the breakdown of the trade consequences on the environment into three main channels (scale, technique, and composition channel), after the pioneering work of Grossman and Krueger (1993). Much of the work on the relationship between trade and pollution, use the Grossman and Krueger (1993) technique, especially when evaluating the environmental impact of NAFTA.

The creation of common free market area, such as NAFTA, EU (which is also a custom and monetary union), or TAFTA, brings international markets closer, and therefore, intensifies the volume of international trade among them. This is translated into an increase in production, drives the prices of traded goods down, which in turn cause an increase in consumption for individuals living in these areas. As a result, this process increases national per capita income, but also increases pollution levels mainly due to the increase of national aggregate production. This phenomenon is known as the scale effect in the environmental economics literature.

At the same time, trade liberalization can cause a change in production methods used in the production process, mainly via two channels. First, since environmental quality is generally considered to be a normal good, an increase in per capita income, due to trade liberalization, changes consumer preferences toward demanding better environmental quality. This forces the governments to implement better quality environmental policies, such as increasing taxes on dirtier goods or providing tax breaks to firms that develop cleaner technologies. Second, like a technological improvement in an economy is not only beneficial to that sector, but it may also be transferable; trade liberalization could create positive technological spillover effects. This is especially seen in developing countries, where domestic firms might take advantage by accessing cleaner available technologies. The above two channels are used to identify the so-called technique effect in the literature.

In addition to the scale and technique effects, trade liberalization can also have composition effects. This effect refers to changes in pollution levels as a result of changes in the relative shares of different goods in the aggregate national production. This is consistent with the notion of comparative advantage in international trade observed from either a Ricardian or/and Heckscher—Ohlin prospective. In other words, liberalized trade forces a country to produce more of the goods they are relatively better at (can produce it at relatively cheaper cost) and let other goods be produced in foreign countries. This changes the composition of national aggregate production, where each country increases the production of goods where they enjoy a comparative advantage, and therefore, exports them and import the other set of goods. Consequently, if a country has a comparative advantage in goods that are produced with relatively cleaner technologies, trade liberalization would expand these cleaner industries, and result in lower pollution levels.

A set of papers provide empirical evidence that the relative pollution-intensive goods are more likely to be capital-intensive goods. For instance, see Mani and Wheeler (1997), Antweiler et al. (1998, 2001), Cole and Elliot (2003). Therefore, according to the Heckscher—Ohlin theory of international trade (which states that relatively capital-abundant countries export, relatively capital-intensive goods, to relatively labor-abundant countries and import the relatively labor-intensive goods from labor-abundant countries), trade liberalization should increase pollution levels in capital-abundant countries and decrease national pollution levels in labor-abundant countries.

Focusing on the changes of the above three effects on sulfur dioxide (SO_2), due to the implementation of NAFTA, Grossman and Krueger (1993) find that SO_2 emission levels raise when national income per capita increases in low-income countries. However, they fall when per capita income raises in high-income countries, providing a unique relationship between trade liberalization and pollution levels. Studying Mexico, they claim that NAFTA produces an inverted U-shaped EKC between economic growth (as measured by real gross domestic product (GDP) per capita) and SO_2 emission levels, with a turning point (TP) of approximately 5000 US dollars (in 1985 US dollars). Since Mexico had a higher GDP per capita, NAFTA should have brought about a decrease in Mexico's SO_2 emission levels.

One branch of theoretical and empirical work on the relationship between international trade and pollution examines what is known as the pollution haven hypothesis. See, for example, Grossman and Krueger (1995) or Copeland and Taylor (1995) as two classical theoretical papers in line with this hypothesis. The pollution haven effect hypothesis states that the economic growth due to trade liberalization will make countries increase their national production levels according to their comparative advantage. Thus, for a relatively low-income country, participation in a common market area, such as TAFTA, is translated into a rise of national production levels of the traded goods that are produced relatively cheaper. For example, a relatively low-income country could have a comparative advantage in a dirty industry, due to the existence of relatively lax environmental regulations, or relatively noneffective implementation policies toward clean industries. Several early studies in this literature, such as Pething (1976), Sierbert et al. (1980), McGuire (1982), Baumol and Oates (1988), Copeland and Taylor (1994), and Chichilnisky (1994), provide theoretical models, where the cost of relatively less clean goods are cheaper in countries with no pollution policies. Chichilnisky (1994) and Copeland and Taylor (1994, 1995), in particular, provide theoretically robust and simple models in the support of the pollution haven hypothesis. Thus, trade liberalization in line with this hypothesis would increase production of relatively dirty goods in relative low-income per capita countries as compared to relatively rich income per capita countries. Consequently, at least theoretically, due to income per capita differences between both countries, the US should expect a reduction of pollution levels due to its participation in TAFTA, while Hungary should expect an increase in pollution levels. However, we do not find many empirical papers in support of the pollution haven argument. For example, Grossman and Krueger (1993) evaluate the impact of NAFTA on SO_2 emission levels. In the Mexico—United States case, they arrive at the conclusion that the impact of NAFTA on SO_2 emission levels do not likely come from the existence of lax environmental regulations in Mexico. Jaffe et al. (1995) conclude that there is little empirical support for the pollution haven hypothesis, where he lists only a few studies that contradict his conclusion, such as Low and Yeats (1992), Ratnayake (1998), Birdsall and Wheeler (1992), and Lucas et al. (1992). Recent papers such as Antweiler et al. (2001), Levinson and Taylor (2002), Cole (2003), Ederington and Minier (2003) find some empirical evidence in support of pollution haven hypothesis.[1]

There is another part of the literature in agreement with the Heckscher—Ohlin theory of international trade that claims that a reduction of international trade costs would intensify the exports of relatively capital-intensive (labor-intensive) goods, produced mainly in capital (labor)-abundant countries, toward labor (capital)-abundant countries. As indicated earlier the relatively pollution-intensive goods are more likely to be relatively capital-intensive goods, we would expect an increase of national pollution levels in capital-abundant countries (the US or/and Italy) and a fall of national pollution levels (Slovenia or/and Estonia). In the trade and environmental literature, this is referred to as the factor endowment effect.

Grossman and Krueger (1993), using data on SO_2, provide empirical evidence, especially for the case of Mexico—United States that increased trade due to NAFTA, is more likely consistent with the factor endowment hypothesis, indicating an increase (decrease) in SO_2 emission levels in the United States (Mexico) since the US is a relatively capital-abundant country as compared to Mexico. For more empirical evidence in support of the factor endowment effect, see Walter (1973), Tobey (1990), Low and Yeats (1992), Grossman and Krueger (1993), Jaffe et al. (1995), Xu (1999), Cole and Rayner (2000), and Antweiler et al. (2001).

Recent literature on trade and environmental impacts claims that previous empirical studies have found conflicting evidence not only because of data limitation or/and quality, but mainly because they were trying to establish a unique relationship between international trade and national pollution emissions across all countries. However, a country's comparative advantage, as shown above, could result from the pollution haven or/and the factor endowment argument. Both these effects, for a relatively rich and capital-intensive country, produce pollution levels that move in opposite directions, and may even cancel out. Thus, the implementation of TAFTA, at least theoretically, would reduce the pollution emissions in the US (Poland), if the pollution haven effect dominates (is being dominated by) the factor endowment effect. This result is in line with the theoretical pioneering paper of Antweiler et al. (2001), who were the first to

---

[1] In this chapter we do not distinguish between the pollution haven effect and the pollution haven hypothesis, but there is a difference between them. The pollution haven effect states that there would be a relocation of plant production of relatively dirty firms toward countries with lax environmental regulations. In other words, the implementation of TAFTA according to the pollution haven effect should relocate the production of relatively clean (dirty) industries toward the relatively high per capita income side of the Atlantic, such as the US (Portugal). The pollution haven hypothesis claims that trade liberalization will increase production of the relatively dirty (clean) goods in the relatively rich (poor) countries. Thus, in line with the pollution haven hypothesis, the implementation of TAFTA will increase production of the clean (dirty) goods in the US (Portugal).

provide a theoretical model for explaining the existence of earlier contradicted empirical findings of the pollution haven and factor endowment arguments. They break down the composition effects of international trade into factor endowment and pollution haven arguments and show empirically that trade liberalization lowers pollution emissions.

10.3 DATA SOURCES

We have data for 24 countries over 16 years from 1995 to 2010. Countries included are the following: Austria, Belgium, the Czech Republic, Germany, Denmark, Estonia, Spain, Finland, France, the United Kingdom, Greece, Hungary, Ireland, Iceland, Italy, Luxembourg, the Netherlands, Norway, Poland, Portugal, Sweden, Slovenia, Slovakia and the United States of America. We present our variables, their unit of measurements, and their data sources in Table 10.1.

We denote sulfur oxides and nitrogen oxides with SO_x and NO_x respectively in our models. (SO_x) and (NO_x) are both the by-products of the energy transformation process and when converted into energy, these pollutants are released into the environment. In the air these substances are turned into acidifying agents, often called "acid rain," and on the ground these pollutants cause both soil and water acidification. The OECD factbook (2014) reports that over the past 25 years, we have seen a steady decline in both (SO_x) and (NO_x) emissions, due to technological improvements, energy conservation, stronger environmental regulations, and a switch to more nonfossil fuel energy sources. However, this improvement in OECD countries has been offset in other parts of the world, where economic growth has results in increased fossil fuel use.

Municipal waste is denoted as MW in our models. The OECD factbook (2014) defines MW as, "the waste collected and treated by or for municipalities. It covers waste from households, including bulky waste, waste from commerce and trade, office build-ings, small businesses, yard and gardens, etc...." We use the kilograms of MW generated per capita to see the effect on the environment of human activity.

We denote carbon dioxide as CO_2 in our models. Carbon dioxide is probably the most discussed and often cited indicator of climate change and global warming. Like (SO_x) and (NO_x), this too is generated from the use of fossil fuels, which are typically used to produce energy. However, another significant source of carbon emissions is from the cement manufacturing process.[2] The fossil fuels that create the most carbon dioxide are coal, oil, and natural gas, World Bank (2011). Carbon dioxide is also a significant contributor to ocean acidification; as it dissolves into water, carbonic acid is created. The burning of wood and waste also contributes to carbon emissions,

---

[2] World Bank (2001) available online at http://databank.worldbank.org/data/views/reports/metadataview.aspx#.

Table 10.1 Data Sources

Variable	Source	Unit of Measurement
$E(SO_x)$, $E(NO_x)$, $E(MW)$	OECD (2013b)	Kilogram per capita
$E(CO_2)$	World Bank (2011)	Metric tons per capita
$E(GHG)$	OECD (2013a)	Thousand kilogram per capita
Real GDP per capita (I)	PENN World Table 8.0	Current PPPs 2005 US $
Capital to labor ratios (KL)	PENN World Table 8.0	Current PPPs 2005 US $
Trade intensity (TI)	World Bank (2011)	Current US $
GINI coefficients	OECD (2013c), CIA, Eurostat	Percentage (0−100)
Government effectiveness	Kaufmann et al. (2010)	$-2.5 \leq GE \geq 2.5$

PPP, Purchasing Power Parity.

so eliminating fossil fuels would not eliminate this problem, but would greatly reduce its impact.

While carbon dioxide is the largest contributor to GHGs, it is only one of several poisonous gases emitted into our atmosphere; others include methane, nitrous oxide, chlorofluorocarbons, hydrofluorocarbons, perfluorocarbons, and sulfur hexafluoride, OECD Factbook (2014). We use these six to show total gross emission of GHGs, but convert them to per capita terms. These gases can have significant impacts on human health, global warming, ecosystems, volatile weather, and economic output. Efforts are underway to cut GHGs emissions, and the US and China have recently pledged to significantly cut GHGs in the years ahead. Table 10.3 discusses our results and establishes a link between GHGs and economic growth.

We use real GDP per capita and the capital to labor ratio, taking data from the PENN Table 8.0 (available publicly online at: http://www.ggdc.net/pwt). We denote these variables I and KL respectively in our model and both are measured in current PPPs 2005 US dollars. For statistical discretions and analysis of the data included in the PENN Table 8.0, see Feenstra et al. (2013). We measure the real GDP per capita (which we simply call income per capita throughout this chapter) by dividing a country's expenditure-side real GDP by its population. When calculating the relative real GDP per capita variable, denoted by RI, we divide each country's real GDP per capita by the TAFTA average real GDP per capita, for the 24 countries included in our sample. We illustrate the relative income per capita of each country in our data set, see Figure 10.1, where the relatively rich countries are the ones with higher than 1.00 relative incomes per capita.

We measure the capital to labor ratio of a country by dividing the physical capital stock with its labor force. We measure the national labor force by incorporating, the national persons engaged variable, with the national working hours variable and the education index (the latter index is from Barro/Lee data set as shown in the PENN World Tables). Similarly, we calculate the relative capital to labor ratio variable, denoted

Table 10.2 Estimation Results for Carbon Dioxide

Variable	Random Effects			Fixed Effects		
	10.2	10.3	10.4	10.2	10.3	10.4
Constant	10.6***	11.39***	11.31***	10.72***	11.58***	11.54***
I	−0.00006	0.00004	0.00017**	−0.00005	0.00006	0.0002***
I²	9.9e-10	−1.55e-10	−2.54e-09	4.25e-10	−6.7e-10	−4.14e-09**
I³			2.51e-14			4.8e-14**
GINI	−0.055***	−0.048**	−0.056***	−0.058***	−0.049**	−0.039**
GE	0.88***		0.8***	0.8***		
KL		−0.058**	−0.0619**		−0.06***	−0.066**
(KL)²		0.0001	0.0001		0.0002	0.0001
T			−0.0217**			−0.0269***
T(RKL)			0.00001			0.004
T(RKL)²			0.0035			0.005
T(RI)			−0.004			0.006
T(RI)²			0.004			−0.005
R²	0.308	0.363	0.399	0.258	0.305	0.325
F-test	1779***	1867***	1653***	14.29***	18.01***	13.74***
BP/LM test						
Hausman (fixed vs random)			6.15			
TP						23,973

Table 10.3 Estimation Results for Greenhouse Gas Emissions

Variable	Random Effects			Fixed Effects		
	10.2	10.3	10.4	10.2	10.3	10.4
Constant		14.95***		14.24***	15.33***	15.15***
I	9.09e-06	0.000136*	0.0002***	2.85e-05	0.0002**	0.00025***
I^2	-2.73e-09	-3.96e-09**	-4.18e-09*	-3.5e-09**	-4.7e-09***	-6.0e-09***
I^3	3.6e-14***	3.9e-14***	2.47e-14	4.1e-14***	4.4e-14***	5.17e-14**
GINI	-0.069***	-0.059***	-0.059***	-0.072***	-0.06***	-0.058***
GE	1.1***			0.914***		
KL		-0.0628**	-0.0651**		-0.0667**	-0.0684**
$(KL)^2$		1.7e-5	3.24e-05		0.00004	2.1e-05
T			-0.0014			-0.0025
T(RKL)			0.0144			0.0147
$T(RKL)^2$			-0.0047			-0.005
T(RI)			-0.04195*			-0.0329
$T(RI)^2$			0.0206			0.0117
R^2	0.162	0.297	0.307	0.167	0.299	0.315
F-test	1669***	1676***	1677.36***	13.68***	13.78***	13.36***
BP/LM test	65.5****	62.17***				
Hausman (fixed vs random)			16.52*			
TP		17,222	25,646			20,833

by *RKL*, by dividing each country's capital to labor ratio by the TAFTA average capital to labor ratio. Note that we calculate the TAFTA average capital to labor ratio (TAFTA average income per capita) by first averaging the yearly capital to labor ratios (incomes per capita), across all countries available in our data set and then averaging capital to labor ratios (incomes per capita), over the 1995–2010 time period. We demonstrate the relative capital abundance of every country in our data set in Figure 10.2, where the relatively capital-abundant (labor-abundant) countries are the ones with higher (lower) than 1.00 relative capital to labor ratio.

As in Antweiler et al. (2001), we call the volume of trade divided by the value of the GDP $\left(\frac{X+M}{GDP}\right)$. Trade intensity is denoted by T in our model. All the data for trade intensity are obtained from the World Bank Indicators (2011).

We measure a country's income inequality with the help of the GINI index. We denote the GINI coefficient with GINI in our model. All the data for the GINI index in the US, Iceland 2010, Ireland 2009, Austria 2010, Belgium 2010, France 2010, Luxembourg 2010, Portugal 2010, and Sweden 2010 are obtained from OECD (2013). The rest of the GINI indexes are obtained from CIA Factbook and Eurostat.[3]

We denote government effectiveness with the variable GE in our model. Kaufmann et al. (2010) define GE as "capturing perceptions of the quality of public services, the quality of the civil service and the degree of its independence from political pressures, the quality of policy formulation and implementation, and the credibility of the government's commitment to such policies." GE data are available at www.govindicators.org.

10.4 FOUR ESTIMATING EQUATIONS

Our models predict per capita emission levels and data for all pollutants are in per capita emission level terms. Subscripts t and i are used to indicate year (1995–2010) and country, respectively. $E(Z_{it})$ is used to denote per capita emission levels of pollutants, where Z_{it} denotes the specific pollutant, such as carbon dioxide, GHGs, MW, nitrogen oxides, and sulfur oxides, or $Z_{it} \in [CO_{2_{it}}, GHG_{it}, MW_{it}, NO_{x_{it}}, SO_{x_{it}}]$. In construction of our four econometric models, we follow closely the work of Antweiler et al. (2001), and Cole (2003).

---

[3] The following GINI index data are obtained from CIA World Factbook: Iceland 2006, Austria 2007, Belgium 2005, Finland 2008, Germany 2006, Hungary 2009, Luxembourg 2008, Norway 2008, Portugal 2007, and Spain 2005, available online at https://www.cia.gov/library/publications/the-world-factbook/fields/2172.html. The rest of the data for the GINI index are taken from Eurostat (available publicly online at http://ec.europa.eu/eurostat/tgm/table.do?tab=table&language=en&pcode=tessi190).

First, we investigate the relationship between per capita emission levels of each pollutant and per capita income levels in each country. We do this in order to investigate the existence of the EKC in the absence of direct composition effect of growth and trade variables. This represents our model 10.1 and it is presented as follows:

$$E(Z_{it}) = \theta_i + \xi_t + \beta_1 I_{it} + \beta_2 I_{it}^2 + \beta_3 I_{it}^3 + \varepsilon_{it} \qquad (10.1)$$

I denotes per capita income levels for each year (t) in each region (i). The square of income per capita is included, in order to investigate the existence of an inverted U-shaped EKC. Moreover we include the cubic income per capita, to examine the impact of very high-income per capita levels, on pollution. θ_i denotes the country-specific constant term, while ξ_t denotes the time-specific constant term. Since our pollution data are in term of per capita emission levels for each region, it is not possible to separate the scale and technique effects of growth. Therefore, the per capita income variable is used as a direct measure of both scale and technique effects.

Model 10.2: Adding Income Inequality and GE

In order to measure the political economy effect of growth on pollution we add a measurement of (1) national inequality as represented by the GINI coefficient and (2) government effectiveness denoted by GE, to the above model and call this model 10.2:

$$E(Z_{it}) = \theta_i + \xi_t + \beta_1 I_{it} + \beta_2 I_{it}^2 + \beta_3 I_{it}^3 + \beta_4 GINI_{it} + \beta_5 GE_{it} + \varepsilon_{it} \qquad (10.2)$$

The slopes of GINI and GE are used to measure the political economy effect of growth on pollution in both regions. Other studies have also included political economy variable in addition to income variables when measuring the relationship between growth and pollution. For example, Cole (2003) and Torras and Boyce (1998) use the GINI coefficient and the literacy rate as political economy variables in their models. Due to data limitation, we have substituted the literacy rate for the government efficiency variable, taken from Kaufmann et al. (2010).

Model 10.3: Adding the Capital to Labor Ratio

To measure the direct composition effect of growth, or the importance of capital abundance, we add the per capita, capital to labor ratio denoted by KL, and its square denoted by $(KL)^2$ to model 10.2. The squared capital to labor ratio is included in model 10.3, to account for the diminishing effect of capital accumulation at the margin. This is shown in model 10.3 as follows:

$$E(Z_{it}) = \theta_i + \xi_t + \beta_1 I_{it} + \beta_2 I_{it}^2 + \beta_3 I_{it}^3 + \beta_4 GINI_{it} + \beta_5 GE_{it} + \beta_6 KL_{it} + \beta_7 (KL)_{it}^2 + \varepsilon_{it} \qquad (10.3)$$

Model 10.4: Adding the Trade Variables

In order to measure the effect of trade liberalization on pollution, due to the possible implementation of TAFTA, we add three trade explanatory variables to model 10.3. More specifically, we add to the right-hand side of model 10.3: (1) the trade variable, T as a measurement of trade intensity; (2) the interaction of the trade intensity with relative capital to labor ratio denoted by $T(RKL)$, in order to detect the factor endowment argument; (3) the interaction of the trade intensity with the squared relative capital to labor ratio denoted by $T(RKL)^2$, in order to account for diminishing factor endowment effect, at the margin; (4) the interaction of the trade intensity with relative per capita income, denoted by $T(RI)$, in order to investigate the pollution haven argument; and (5) the interaction of the trade intensity with squared relative per capita income, denoted by $T(RI)^2$, in order to account for diminishing pollution haven effect at the margin. We measure all above effects with the help of the model 10.4 as follows.

$$
\begin{aligned}
E(Z_{it}) = {}& \theta_i + \xi_t + \beta_1 I_{it} + \beta_2 I_{it}^2 + \beta_3 I_{it}^3 + \beta_4 GINI_{it} + \beta_5 GE_{it} + \beta_6 KL_{it} \\
& + \beta_7 (KL)_{it}^2 + \beta_8 T + \beta_9 [T(RKL)]_{it} + \beta_{10}\left[T(RKL)^2\right]_{it} \\
& + \beta_{11}[T(RI)]_{it} + \beta_{12}\left[T(RI)^2\right]_{it} + \varepsilon_{it}
\end{aligned}
\tag{10.4}
$$

In the above equation, the slopes of β_9 and β_{11} are used to measure the factor endowment and the pollution haven argument, respectively. Theoretically, according to the factor endowment effect, TAFTA should produce a positive (negative) sign of β_9 for a relatively capital (labor)-abundant country, as compared to the average capital abundance of all countries in our data set. For example, since the US and Luxembourg are relatively capital-abundant countries while Poland and Slovenia are labor-abundant countries (see Figure 10.2 in Section 10.3 of this chapter), and pollution-intensive goods are considered to be capital intensive, the US and Luxembourg should export capital-intensive goods to Poland and Slovenia. Consequently, the implementation of TAFTA will increase pollution levels in the US and Luxembourg but decrease pollution in Poland and Slovenia. In terms of our regression if β_9 is positive, this means that for an average capital-abundant member of TAFTA, trade liberalization will increase its pollution emissions via the factor endowment channel.

On the other hand, along the lines of the pollution haven argument, TAFTA should produce a negative (positive) sign of β_{11} for the US and Luxembourg (Poland and Slovenia) because relatively richer countries design and implement stringent environmental regulations (see, for example, Dasgupta et al., 1995) and the US and Luxemburg (Poland and Slovenia) are relatively richer (poorer) countries. See Figure 10.1 in Section 10.3 of this chapter. Therefore, according to model 10.4, a negative sign of β_{11} means that for a relatively high-income per capita TAFTA member country, trade liberalization should decrease pollution emissions.

The overall impact of trade is measured by the trade intensity variable (T). Theoretically, we expect to find that TAFTA would reduce per capita pollution levels in the US, if the raise of per capita pollution levels due to the factor endowment effect is outweighed by the fall of per capita pollution levels in the US, due to the pollution haven effect. In line with Antweiler et al. (2001), we theoretically expect that TAFTA will definitely increase per capita pollution emissions for a relatively capital-abundant and poor country (such as Spain in our data set) because the factor endowment hypothesis goes in the same direction with the pollution haven argument. Similarly, TAFTA should theoretically decrease per capita pollution emissions for a relatively labor-abundant and rich country (such as Ireland and Sweden in our data set) again, because the factor endowment and the pollution haven move in the same direction.

10.5 EMPIRICAL METHODOLOGY

In all tables of the next section, we present estimates using both random and fixed effects. For all pollutants, we found that ordinary least squares (OLS) estimators are biased as compared to random effects, using Breusch—Pagan Lagrange Multiplier (BP/LM) test. The null hypothesis for the above test is that there are no significant differences across units. In other words, there are no panel effects. In all regressions for all pollutants, we reject the null hypothesis with a 99% confidence level (see row 17 of each table in the next section, for the χ^2 value and its statistically significance) and conclude that there are significant differences across TAFTA members in our data set; therefore we cannot run simple OLS regressions.

We perform a Hausman test in order to decide if random or fixed effects specification is preferred. The null hypothesis for this test is that the random disturbance term is not correlated with the regressors. In other words, if we reject the null hypothesis, fixed effects is the preferred method, otherwise random effects indicate better estimators. For more on this test, see Chapter 9 of Greene (2008). We present the result of this test on the 18th row of each table. However, on certain occasions the Hausman test of fixed versus random effects provides negative χ^2 values, and therefore, in those cases is not an appropriate test for deciding which method is better. This explains the missing values of χ^2 in some of our models. Moreover, even in the cases when we clearly reject or do not reject the null hypothesis, we present both estimates. Keep in mind that the random effects method uses both intercepts: θ_i, ξ_t as parts of the random error term. Theoretically, the random effects are more efficient because they are generally not related to observations in our sample. On the other hand, the fixed effects specification uses the above intercepts as parameters of our regressions conditional over time. In most cases, as indicated in next section tables, the preferred method is the fixed effects (we reject the null hypothesis). However, in these cases, we are cautious when interpreting the results from fixed effects (especially when they differ in sign and statistically significance with random effects) because they limit the cross–sectional variation.

10.6 EMPIRICAL RESULTS

The empirical results are presented by the various types of pollutants. We first present Table 10.2, which provides the estimation results for carbon dioxide per capita emissions (CO_2), for the 24 TAFTA members included. We should emphasize that each table reports the estimation results using random effects of models 10.2, 10.3, and 10.4 in the second, third, and fourth columns, respectively and the estimation results of the same models using fixed effects are reported in the fifth, sixth, and seventh columns, respectively. Note that model 10.1 results are not presented in any tables, because we did not find any statistical significance of the relationship between per capita emissions (for three of our pollutants: carbon dioxide, MW, and GHGs) and per capita income. For two pollutants, nitrogen and sulfur oxide, we find a statistically significant relationship, but the sign and significance is the same with model 10.2.

In every model, for all pollutants we find negative and statistically significant estimators for the GINI index, indicating that as inequality increases pollution emissions per capita falls (with the exception of MW and NO_x, where GINI estimators are negative but not significant). This result is surprising and it is inconsistent with previous findings. For example, Torras and Boyce (1998) show that a rise in income inequality is associated with an increase of the concentrations of two measures of water pollution, sulfur dioxide, and particulate matter. Especially in developed countries, such as the majority in our sample, it should be easier for citizens to organize and force their governments to develop and implement better environmental friendly laws. However, we question this argument given there may be more complex political and economic forces behind the relationship between national income inequality and pollution. For example, we know that Germany, Denmark, the UK, Ireland, and the US are good examples, where we have seen an increase in their income inequality and a decrease in GHGs and CO_2 per capita over the past 15 years. We argue that it might be possible that while the rich favor tax cut policies, which contributes to income inequality, both the rich and the poor favor better environmental standards. Since the rich could benefit more from tax cut policies and the environmental quality is considered a normal good, they may support stringent environmental regulations. While the rich can avoid living in relatively polluted areas, the poor are often located closer to industrial areas and experience more pollution. Therefore, enforcing more stringent environmental regulations is Pareto optimal for all members of the society.

However, the other political economy variable representing the GE is not reported in any of our tables for model 10.4 (and in most cases for model 10.3), because it appears to be insignificant for all pollutants. In the last row of all tables presented in this section, we report the TP of the inverted U-shaped EKC (which we denoted by TP) only for statistically significant estimators of I, I^2 at the 90% significance level, or higher.

We find a negative and empirically robust relationship between capital abundance, as measured by capital to labor ratio denoted by KL, and per capita emissions of CO_2, GHGs, and MW (and for NO_x using model 10.3). For the other pollutant, KL is still negative, but not statistically significant. This is a surprising result since is not consistent with the theoretical literature and other empirical studies, which show that capital-abundant countries produce more dirty goods. For example, Antweiler et al. (2001), Cole and Elliot (2003), Mani and Wheeler (1997), among others, provide theoretical and empirical evidence of a positive relationship between national capital abundance and pollution. However, taking a closer look in our data, we noticed that employment levels and labor force participation rates have fallen in most of our 24 countries, especially during 2003—2005, but more severely during the great recession of 2008—2010 that has affected mainly the US, the UK, and the Southern European countries. Moreover, consistent with the literature in development economics, we observe an aging population phenomenon in many countries, over the 1995—2010 period, signifying a deeper fall in their labor force participation rate. Consequently, we believe that the negative relationship between capital to labor ratio and four of our pollutants is less likely originating from capital accumulation in TAFTA members during this specific time period, but most likely due to the shrinkage of the labor force in an average TAFTA member. We denote by *, **, *** significance at 90%, 95%, and 90% confidence level, respectively for all the results presented in all tables. To conserve space in each table, we do not present the standard errors, t-statistics, or p-values, although they are available upon request.

Model 10.2 is worth noting that when we add the political economy variables, on the right-hand side of model 10.1, we find statistically significant evidence of a positive relationship between carbon dioxide emissions per capita and GE. In other words, ceteris paribus, a one unit yearly increase of GE, for an average TAFTA member in our sample, produces an increase of approximately 0.88 metric tons of CO_2 per capita emissions in a year. This implies that GE affects per capita emissions of CO_2 in a counterintuitive manner, since we may assume that especially in the case of developed countries, such as the US and EU members, more effective are the governments of these countries' higher quality, and stricter environmental regulations should be implemented there. Model 10.2, as all other models, indicates that income inequality as measured by the GINI index is good for the reduction of CO_2 per capita emissions. The estimators of GINI index are all negative and statistically significant.

Both models 10.3 and 10.4 indicate a statistically significant and negative coefficient for the direct measurement of the composition effect. It shows that, ceteris paribus, a one dollar increase of an average TAFTA member country's capital to labor ratio in a year will decrease CO_2 by approximately 0.06 metric tons per capita.

Using fixed effects, when we include the trade variable as presented in model 10.4, we find statistical significant evidence consistent with an inverted U-shaped EKC, for the relationship between per capita income and per capita CO_2 emissions. This means that TAFTA will initially increase CO_2 per capita emissions, via its growth effect (where initially the scale effect dominates the technique effect) and then eventually as income per capita increases further (see the slope of I^2), CO_2 per capita emissions fall (meaning that the technique effects eventually dominates the scale effect). However, for extremely high-income levels (as indicated by the coefficient of I^3), CO_2 per capita emission levels rise again when per capita income increases. We find the TP for each statistically robust inverted U-shaped relationship between per capita income and per capita CO_2 emissions and reported them in the last row of Table 10.2. Since all our data on CO_2 (as for all other pollutants) are not in concentrations, but in the form of CO_2 emissions per capita, it is impossible to separate scale and technique effects.

However, using random effects, with the help of model 10.4, we find a positive and monotonic relationship between per capita income and CO_2 emissions per capita, indicating that scale dominates the technique effect for all countries. This is because the statistically significant estimator of income per capita is positive, while the coefficient of I^2 is insignificant. However, at 86% level of significance I^2 becomes statistically significant and is negative, meaning that there is an inverted U-shaped relationship between per capita income and CO_2 emissions per capita consistent with fixed effect estimation as explained above.

We perform the Hausman fixed versus random test (as reported in row 18, column 4 in this case) and concluded that we cannot reject the null hypothesis, meaning that the random effects are preferred. Therefore, one may conclude that growth may help increase CO_2 per capita emissions in a typical TAFTA member. However, the slope of I^2 is negative and just misses the threshold of being statistical significance at 90% level of significance according to the random effects estimates. This leads us to believe that higher-income per capita may help reduce CO_2 per capita emissions. However, at extremely high-income per capita, we observe an increase on the CO_2 per capita emissions.

Model 10.4 includes the trade intensity and relative trade variables as independent variable in our regression, but does not provide any statistically significant evidence of the factor endowment or/and pollution haven arguments for CO_2 per capita. However, freer trade seems to reduce CO_2 emissions per capita as measured by the factor intensity variable. Thus, a 1% yearly increase of a national volume of trade, divided by its real GDP level, decreases CO_2 approximately by 0.02 metric kg per capita in a year. Thus, model 10.4 provides robust empirical evidence that the implementation of TAFTA may decrease CO_2 per capita emissions via the raise of the trade intensity. Moreover, the signs of $T(RI)$ and $T(RKL)$ are consistent with the pollution haven and factor endowment hypotheses, but they are both insignificant. However, the inclusions of the trade variables, as illustrated by model 10.4, increase the value of R^2.

In Table 10.3, we present the estimation results for GHGs per capita. Consistent with the results of Table 10.2, national income inequality results in a reduction of GHGs per capita. All GINI estimators are negative and statistically significant. Both models 10.3 and 10.4 show a negative and statistically significant relationship between national capital abundance and GHGs per capita. However, most importantly inclusion of trade variables provides statistical significant evidence, consistent with an inverted U-shaped EKC, indicating the scale effect dominates the technique effect, but then eventually the roles are reversed. However, for extremely high levels of growth, GHGs per capita increase with a rise of national per capita income.

In contrast with CO_2, the inclusion of trade variables does not provide any statistically significant evidence of the relationship between trade liberalization and GHGs per capita. However, the signs of the important trade variables: T, $T(RI)$, and $T(RKL)$ are consistent with the pollution haven and factor endowment hypotheses, but they are all insignificant with the exception of the pollution haven effect measured from the random effects method. The negative sign of $T(RI)$, indicates that a relatively poor (rich) country member of TAFTA, such as Greece (Belgium), will see an increase (decrease) in GHGs per capita as a consequence of TAFTA. Therefore, Greece will have comparative advantage in the dirty goods, and therefore, it will increase the aggregate production levels of the dirty industries. The sign of the trade intensity coefficient is negative indicating a reduction of GHGs in a typical TAFTA member country. Again, the value of R^2 increases with the inclusion of capital accumulation and trade variables in our regressions.

Table 10.4 illustrates the estimation results for MW per capita. In contrast with the results of the previous two tables, all GINI estimators are insignificant, but still negative. Consistent with CO_2 and GHGs, models 10.3 and 10.4 show a negative and statistically significant relationship between the direct composition effect of growth (KL) and MW per capita emissions. In contrast with GHGs, the inclusion of trade variables provides statistically significant evidence consistent with a positive line of EKC, in the case of MW per capita emissions, indicating that scale dominates the technique effect for a typical TAFTA member. This result stems from the fact that the statistically significant estimator of income per capita is positive, while the coefficient of I^2 is insignificant. Therefore, growth as a result of trade liberalization, such as TAFTA, does not appear to reduce MW per capita in their production process.

Consistent with GHGs, the inclusion of trade variables does not provide any statistically significant evidence of the relationship between trade liberalization and MW per capita. Moreover, we do not find any statistically significant empirical evidence for the existence of the pollution haven or/and factor endowment effects. However, the sign of $T(RI)$ is consistent with the pollution haven argument. In contrast to the previously

Table 10.4 Estimation Results for Municipal Waste

Variable	Random Effects			Fixed Effects		
	10.2	10.3	10.4	10.2	10.3	10.4
Constant	387.68	418.9***	398.9***	434.58***	473.8***	445.06
I	0.0023	0.0084**	0.009**	0.0015	0.008**	0.0050
I²	6.43e-08	-3.92e-08	-1.19e-08	7.07e-08	-4.18e-08	1.0e-07
I³						-2.33e-12*
GINI	-0.754	-0.770	-0.814	-1.226	-1.124	-1.363
GE	26.157*			15.628		
KL		-5.109***	-3.539**		-5.625***	-3.592**
(KL)²		0.0446***	0.0336**		0.0476***	0.0347***
T			0.373			0.906
T(RKL)			-1.222			-1.642
T(RKL)²			0.409			0.489
T(RI)			-0.573			-0.616
T(RI)²			0.574			0.789
R²	0.524	0.489	0.474	0.502	0.407	0.24
F-test				8.65***	9.03***	6.27***
BP/LM test	1286***	1263***	1161***			
Hausman (fixed vs random)	8.62**	10.61**	12.3			
TP						

analyzed pollutants, the value of R^2 decreases with the inclusion of capital abundance and trade variables in these regressions.

Table 10.5 presents the estimation results for per capita nitrogen oxides emissions NO_x. In line with MW, but in contrast with CO_2 and GHGs, all GINI estimators are insignificant, but again negative. Quite the opposite with all the previously analyzed pollutants, in the case of NO_x, all models show statistically insignificant estimators of capital abundance (*KL*). In contrast with all the other pollutants, the inclusion of trade and capital abundance variables provides statistically insignificant evidence for income and square income per capita variables. Therefore, growth as a result of trade liberalization, such as TAFTA, is not significantly related with NO_x emissions per capita. However, using model 10.2, it appears that growth monotonically reduces NO_x per capita emissions, but eventually for extremely high levels of income, growth increases NO_x per capita emissions. This relationship stands for models 10.3 and 10.4, but becomes insignificant.

In line with GHGs and CO_2, the inclusion of trade variables does not provide any statistically significant evidence of the existence of the pollution haven effect. Consistent with CO_2, freer trade produces a negative and statistically significant estimator of the trade intensity variable. In particular, holding all the other factors constant, a yearly increase of a national trade intensity by 1% decreases NO_x per capita emissions of an average country member, by approximately 0.15 kg per capita a year. However, we find statistically significant empirical evidence inconsistent with the factor endowment hypothesis. This is indicated by the negative sign of *T(RKL)* estimator, which shows that a relatively capital-abundant country member of TAFTA will see a decrease in NO_x per capita emissions with the implementation of TAFTA. In line with CO_2 and GHGs, the value of R^2 increases with the inclusion of capital abundance and trade variables in our regressions.

Table 10.6 shows the estimation results for sulfur oxides per capita emissions (SO_x). All models suggest, statistically significant evidence consistent with a U-shaped EKC, for the relationship between per capita income and per capita SO_x emissions. Since the TP of the U-shaped relationship are relatively high, for per capita sulfur oxides emissions to fall, they must have been risen in the past, indicating a possible wave shape of the EKC. Therefore, this relationship is still consistent with literature of EKC.

In line with CO_2 and GHGs, all models indicate a negative and statistically significant relationship between national income inequality and SO_x per capita emissions. In other words, a yearly increase in national income inequality by 1% will approximately reduce SO_x by 0.8 per capita. Consistent with CO_2, GHGs, and MW, model 10.3 indicates a statistically robust and negative relationship between capital abundance and per capita SO_x emissions. However, this relationship loses statistical significance when we introduce our trade variables.

Table 10.5 Estimation Results for Nitrogen Oxides

Variable	Random Effects			Fixed Effects		
	10.2	10.3	10.4	10.2	10.3	10.4
Constant	54.99**	60.36	59.56***	60.14***	64.43***	61.389***
I	−0.001*	−0.0004	−8.62e−05	−0.0009*	−0.0002	−7.8e−05
I^2	−4.58e−09	−7.39e−09	−9.37e−09	−1.15e−08	−1.49e−08	−9.24e−09
I^3	1.81e−13*	1.75e−13*		2.32e−13**	2.26e−13**	
GINI	−0.194	−0.141	−0.0842	0.233	−0.154	−0.123
GE	8.44			5.89***		
KL		−0.151	−0.170		−0.234	−0.161
$(KL)^2$		−0.002	−0.0018		−0.0018	−0.002
T			−0.1887***			−0.154**
T(RKL)			−0.216**			−0.238*
T(RKL)^2			0.169***			0.180**
T(RI)			0.201			0.156
T(RI)^2			0.0307			0.0518
R^2	0.294	0.397	0.533	0.301	0.404	0.534
F-test				30.52***	34.24***	33.22***
BP/LM test	1309***	1294***	1459***			
Hausman (fixed vs random)	30.01***	70.01***				
TP						

Table 10.6 Estimation Results for Sulfur Oxides

Variable	Random Effects			Fixed Effects		
	10.2	10.3	10.4	10.2	10.3	10.4
Constant		165.82***	172.60***	173.50***	179.80***	188.94
I	−0.0081***	−0.0071***	−0.0085***	−0.0083***	−0.0072***	−0.0077***
I²	1.66e−07***	1.5e−07***	1.6e−07***	1.6e−7***	1.48e−07***	1.2e−07***
I³	−1.07e−12***	−9.7e−13***	−8.3e−13***	−1.1e−12***	−9.7e−13***	−2.1e−13
GINI	−0.874**	−0.874**	−0.7869746**	−1.117***	−1.106***	−1.026***
GE	4.748			2.865		
KL		−0.941**	−0.441		−0.984**	−0.491
(KL)²		0.0083*	0.0049		0.0087**	0.004
T			−0.212			−0.069
T(RKL)			−0.417			−0.505
T(RKL)²			0.0596			0.101
T(RI)			0.980**			0.949**
T(RI)²			−0.390**			−0.517***
R²	0.240	0.250	0.289	0.241	0.251	0.059
F-test				22.51***	16.89***	12.63***
BP/LM test	1444***	1327***	860.62***			
Hausman (fixed vs random)	10.28**	10.9**				
TP						

Consistent with GHGs and MW, model 10.4 does not provide any statistically significant evidence of a relationship between trade liberalization and SO_x per capita. Moreover, we do not find any statistically significant empirical evidence for the existence of the factor endowment effects. However, we do find statistically significant evidence in contrast with the pollution haven hypothesis as indicated by the positive sign of $T(RI)$. In line with most of the previously analyzed pollutants, looking at the random effects the value of R^2 increases with the inclusion of capital abundance and trade variables in our regressions.

10.7 CONCLUSION

In this chapter we empirically investigate the effects of trade liberalization, such as the implementation of TAFTA, on several pollutants: carbon dioxide, GHGs, MW, nitrogen oxides, and sulfur oxides. We use cross-sectional panel data for 23 EU members and the US, over the 1995–2010 time period. We find that for members of TAFTA, trade liberalization, conditional on country characteristics, has a small but statistically significant impact on carbon dioxides and nitrogen oxides per capita emissions. In particular, holding other factors constant, a 1% increase in trade volume per real GDP of a typical TAFTA member may help reduce per capita emission of carbon dioxide by approximately 0.002 metric tons. Similarly, this increase in trade may help reduce per capita emissions of nitrogen oxides by approximately 0.15 kg. We do not find any statistically significant evidence consistent with the pollution haven or/and factor endowment arguments, with the exception of GHGs, where we find robust evidence of the existence of the pollution haven hypothesis. Consequently, a rich member of TAFTA, such as Norway will see a fall in GHGs per capita emissions in response to the implementation of TAFTA.

Moreover, the inclusion of trade variables as explanatory variables seems to add much in our analysis when evaluating the impact of growth on pollution. We show that only after their inclusion in our regressions, we were able to find statistically significant estimators, of income per capita consistent with the ECK, for GHGs and CO_2 per capita emissions. Thus, for an average TAFTA member, for GHGs and CO_2, when we add trade variables, the scale effect dominates the technique effect, but is eventually reversed at higher-income per capita. Therefore, for a usual TAFTA member, we find statistically significant evidence (when we include the trade variables), indicating that growth could assist in designing and implementing more efficient regulations, in order to help reduce per capita emissions of GHGs and CO_2. However, in the case of MW, we find robust empirical evidence that economic growth tends to increase MW per capita.

In contrast with other studies in the literature, this chapter also provides robust empirical evidence that income per capita earnings originating from capital to labor ratio tend to reduce per capita emissions of carbon dioxide, GHGs, MW, and sulfur dioxide.

This is a counterintuitive result since previous literature finds convincing evidence that dirty goods are relatively more capital intensive. Thus, an increase of capital accumulation will decrease the price of capital, and therefore will raise its aggregate production of capital-intensive goods, increasing national pollution emissions. We believe that our result is different from previous findings since it stems from a number of variables, but all related to a consistent declining of the labor force in almost all countries in our data set, especially during 2003–2005 and 2008–2010 time periods. This leads us to suspect that the negative relationship between the capital to labor ratio and our four pollutants does not originate from capital accumulation, but it may be related to the fall of the labor force in the majority of the countries in our data set (mainly the US, the UK, and Southern European countries) due to the double-dip recessions and population aging. Therefore, while our result is inconsistent with what previous researchers have found, it is empirically valid and consistent for countries in our data set over the 1995–2010 time period.

No robust evidence is found to support the argument that better government efficiency could help reduce per capita emissions for our pollutants, with the exception of CO_2, where we find statistically significant evidence quite opposite with this argument. However, in this chapter, we also report a surprising finding quite opposite to that of previous empirical studies on the relationship between the national distribution of power as measured by the GINI index and national emissions levels. We believe both political and economic factors are at play, with economic factors creating a more unequal distribution of income and political variables improving environment standards. We find statistically significant evidence of a negative relationship between a representative TAFTA member's income inequality and its carbon dioxide, GHGs, and sulfur oxides per capita emissions. Moreover, for the other two pollutants we find the same, but not statistically significant impact of income inequality on pollution. Therefore, we conclude that for a standard TAFTA member, higher-income inequality may prove to be good for the environment.

REFERENCES

Anderson, K., Blackhurst, R., 1992. The Greening of World Trade Issues. Harvester Wheatsheaf, New York.

Antweiler, W., Copeland, B.R., Taylor, M.S., 1998. Is Free Trade Good for the Environment? NBER, Cambridge, MA. Paper 6707.

Antweiler, W., Copeland, B.R., Taylor, M.S., 2001. Is free trade good for the environment? Am. Econ. Rev. 91 (4), 877–908.

Baumol, W.J., Oates, W.E., 1988. The Theory of Environmental Policy. Cambridge University Press, Cambridge, MA.

Birdsall, N., Wheeler, D., 1992. Trade policy and industrial pollution in Latin America: where are the pollution havens? In: Low, P. (Ed.), International Trade and the Environment. World Bank, Washington, DC, pp. 159–167. World Bank Discuss. Paper 159.

Chichilnisky, C., 1994. North-South trade and the global environment. Am. Econ. Rev. 84 (4), 851–874.

Cole, M.A., 2003. Development, trade, and the environment: how robust is the environmental Kuznets curve? Environ. Dev. Econ. 8 (4), 557–580.

Cole, M.A., Rayner, A.J., 2000. The uruguay round and air pollution: estimating the composition, scale and technique effects of trade liberalization. J. Int. Trade Econ. Dev. 9 (3), 343–358.

Cole, M.A., Elliott, R.J.R., 2003. Determining the trade-environment composition effect: the role of capital, labor and environmental regulations. J. Environ. Econ. Manag. 46 (3), 363–383.

Copeland, B.R., Taylor, M.S., 1994. North-South trade and the environment. Q. J. Econ. 109 (3), 755–787.

Copeland, B.R., Taylor, M.S., 1995. Trade and transboundary pollution. Am. Econ. Rev. 85 (4), 716–737.

Copeland, B.R., Taylor, M.S., 2004. Trade, growth, and the environment. J. Econ. Lit. 42 (1), 7–71.

Dasgupta, S., Mody, A., Roy, S., Wheeler, D., 1995. Environmental Regulation and Development: A Cross-country Empirical Analysis. Working Paper 1448. Policy Research Department, The World Bank.

Ederington, J., Minier, J., 2003. Is environmental policy a secondary trade barrier? An empirical analysis. J. Can. Econ. 36 (1), 137–154.

Feenstra, R.C., Inklaar, R., Timmer, M.P., 2013. The Next Generation of PENN World Tables. Mimeo. University of California, Davis and University of Groningen.

Greene, W.H., 2008. Econometric Analysis, seventh ed. Prentice Hall, Upper Saddle River.

Grossman, G.M., Krueger, A.B., 1991. Environmental Impacts of a North American Free Trade Agreement. National Bureau of Economic Research Working Paper 3914, Cambridge, MA.

Grossman, G.M., Krueger, A.B., 1993. Environmental impacts of a North American free trade agreement. In: Garber, P.M. (Ed.), The US-Mexico Free Trade Agreement. MIT Press, Cambridge, MA, pp. 13–56.

Grossman, G.M., Krueger, A.B., 1995. Economic growth and the environment. Q. J. Econ. 110 (2), 353–377.

Jaffe, A.B., Peterson, S.R., Portney, P.R., Stavins, R.N., 1995. Environmental regulation and the competitiveness of us manufacturing: what does the evidence tell us? J. Econ. Lit. 33 (1), 132–163.

Kaufmann, D., Kraay, A., Mastruzzi, M., 2010. The Worldwide Governance Indicators: A Summary of Methodology, Data and Analytical Issues. World Bank Policy Research. Working Paper, 5430.

Levinson, A., Taylor, M.S., 2002. Trade and the Environment: Unmasking the Pollution Haven Effect. Mimeo, Georgetown University.

Low, P., Yeats, A., 1992. Do dirty' industries migrate. In: Low, P. (Ed.), International Trade and the Environment. World Bank, Washington, DC, pp. 89–104. Discussion Paper 159.

Lucas, R.E.B., Wheeler, D., Hettige, H., 1992. Economic development and environmental regulation and the international migration of toxic industrial pollution: 1960–1988. In: Low, P. (Ed.), International Trade and the Environment. World Bank, Washington, DC, pp. 67–86. Discussion Paper 159.

Mani, M., Wheeler, D., 1997. In Search of Pollution Havens? Dirty Industry Migration in the World Economy. World Bank, Working Paper 16.

Mani, M., Cunha, B., 2011. Dr-cafta and the Environment. Poverty Reduction and Economic Management Unit. World Bank.

McGuire, M.C., 1982. Regulation, factor rewards, and international trade. J. Public Econ. 17 (3), 335–354.

OECD, 2013a. Environmental Data Compendium 2013. OECD, Paris.

OECD, 2013b. Emissions of Air Pollutants. OECD Environmental Statistics (Database).

OECD, 2013c. OECD Social and Welfare Statistics (Database).

OECD Factbook, 2014. Economics, Environmental and Social Statistics.

Onder, H., 2012. Trade and climate change: an analytical review of key issues. PREM, World Bank Econ. Premise 86, 1–8.

Pethig, R., 1976. Pollution, welfare, and environmental policy in the theory of comparative advantage. J. Environ. Econ. Manag. 2 (1), 160–169.

Ratnayake, R., 1998. Do stringent environmental regulations reduce international competitiveness? evidence from an inter-industry analysis. Int. J. Econ. Bus. 5 (1), 77–96.

Siebert, H., Eichberger, J., Gronych, R., Pethig, R., 1980. Trade and the Environment: A Theoretical Enquiry. Elsevier, Amsterdam, North-Holland.

Tobey, J.A., 1990. The effects of domestic environmental policies on patterns of world trade: an empirical test. Kyklos 43 (2), 191–209.

Torras, M., Boyce, 1998. Income, inequality and pollution: a reassessment of the environmental kuznets curve. Ecol. Econ. 25 (3), 147–160.

Walter, I., 1973. The pollution content of American trade. Econ. Inq 11 (1), 61–70.

World Bank, 2011. World Development Indicators 2011. CD-ROM, the World Bank, Washington, DC.

Xu, X., 1999. Do stringent environmental regulations reduce the international competitiveness of environmentally sensitive goods? A global perspective. World Dev. 27 (7), 1215–1226.

CHAPTER 11

Feminism, Environmental Economics, and Accountability

Tehmina Khan
RMIT University, Melbourne, VIC, Australia

Contents

11.1 INTRODUCTION

The basic principles of ecofeminism require the addressing of the key national and global economics' concerns including the ones created by the simplistic economics formula of inputs required for production being capital, land, and labor to produce outputs. This limited consideration of inputs, according to Henderson (1984) needs to be replaced by the new conceptualization of minimal entropy society with revised key inputs that are required and that cannot be excluded from the equation including capital, resources and knowledge.

Ecofeminism principles are based around nature being the central consideration for preservation and protection, requiring efficient use of natural resources, asking for the consideration of nurturing and community growth and development as important priorities and indicators of success (Henderson, 1984) rather than the conventional economic

GDP measures which have been criticized for their lack of consideration of comprehensive performance, output, and impacts at the national level (Stockhammer et al., 1997). The conventional GDP measures are considered as inadequate and unreliable measures of social welfare (Van Den Bergh, 2009). This is a brief consideration of ecofeminism which has been provided here to establish that there is a strong link between environmental economics and ecofeminism and that these principles define the basic premise of environmental economics which entails broader environmental and societal oriented considerations and which encompasses a significant departure from conventional economics' considerations.

11.2 IMPACT OF FEMINISM IN ECONOMICS

The impact of feminism in economics has been considered as of minor in nature, including the presence of feminist scholarship in the field of economics (Albelda, 1995). Multiple implications are provided due to this factor. These include less emphasis on female-oriented research, for example, environmental economics research (as it has ecofeminist characteristics), the invisibility of the contributions of female economists and researchers and severely biased depictions (if not complete exclusions of any representations) of the roles and contributions of females in households and societies in economics literature (Hewitson, 2001). Gender differences in the discipline of Economics are disappearing (Kahn, 1993) in relation to PhD attainments by females in economics.

Over the recent decades, the number of women attaining PhDs in economics has increased, there has not been a proportionate increase in their representation as assistant and associate professors in academia (Ginther and Kahn, 2004, 2006). Slightly better results are observed in the area of environmental economics as a higher percentage of women have been found to be present at the associate/assistant professor levels, although the number of publications has been lesser and the number of citations to articles written by female environmental economists has been lower in comparison to the male economist academics (Bhattacharjee et al., 2007). More recently, Hamermesh (2012) has found a sharp increase in the fraction of female authors of economics' journal articles, yet the female share of tenure-stream faculty at institutions that grant PhDs remains dramatically below the female share of authorship in the United States (Scott and Siegfried, 2012).

It appears as if female economists' contributions are increasing in Economics' scholarship, research, and policy development, yet the level of contributions and recognition of their contributions is still not on par to that of male economists. Only 10.5% of the full professors at PhD granting institutions in the United States are females (Scott and Siegfried, 2012). Motivated by the small fraction of female economists' contributions that are recognized through promotion in the field of economics and the link between environmental economics and feminism, the contributions of female environmental economists in research and in governments are considered and presented.

11.3 FEMALE ECONOMISTS' CONTRIBUTIONS IN ENVIRONMENTAL ECONOMICS

The following list of female economists who have contributed in the field of environmental economics was attained from the "Encyclopedia of Women in Today's world" (2011): *Esther Duflo, Elinor Ostrom, Anny Ratnawati, Salsiah Alisjahbana, and Dr Shamshad Akhtar.*

An analysis of the female economists' academic and professional contributions in the field of environmental economics is presented below. Please note that this is not a comprehensive representation of the economists' detailed profiles, biography, or contributions especially outside the field of environmental economics.

11.3.1 Esther Duflo

(The initial source of the data is Professor Esther Duflo's CV dated September 2014 as presented on the Massachusetts Institute of Technology's Web site Duflo (2014).)

Professor Esther Duflo works in the Department of Economics at Massachusetts Institute of Technology. She has received multiple awards recognizing her research including the Elaine Bennett Prize for research in 2003, the John Bates Clark Model in 2010, and Erna Hamburger Prize in 2014, to mention a few. As at 24th of October 2014, there are 22,069 citations listed in Google Scholar to Duflo's work.

Professor Duflo is the Abdul Latif Jameel Professor of Poverty Alleviation and Development Economics at MIT and is the director of the Abdul Latif Jameel Poverty Action Lab at MIT. She is the editor/co-editor/associate editor of seven highly ranked Economics journals. She is a member of the President's Global Development Council, Director of the Development Program at the Centre for Economic Policy Research, Research Associate at the National Bureau of Economic Research, and Board Member of the Bureau for Research and Economic Analysis of Development. She has published more than 50 articles, 4 books and has multiple working papers since the late 1990s.

Her primary research area is social economics, she has addressed key global economic issues including poverty, social systems in developing countries including India, women's empowerment, education in developing countries; she has undertaken substantial research in environmental economics as well. Her key contributions in environmental economics include water economics-related research including her published article in 2012 in American Economic Journal: *Economic Policy*, which is a top-ranked economics journal. As at October 23rd, 2014, this article has been cited 84 times, as identified in Google Scholar. Her "Nudging Farmers to Use Fertilizer: Evidence from Kenya" was published in 2011 in *American Economic Review* which is a highly ranked Economics journal. As at October 23rd, 2014, this article has been cited 40 times. The main point of discussion and analysis in this article is the implementation of a program that encourages farmers in Western Kenya to undertake timely purchase and use of fertilizer (Duflo et al., 2009). This research would come under the umbrella of benefits' assessments of environmental decision making (a category of environmental economics and management identified by Callan and Thomas, 2012). Fertilizer use efficiency, under which the correct timing of fertilizer use is a crucial factor for maximizing output (food production) and minimizing environmental impacts (Zu and Chen, 2002; Randall and Mulla, 2001) is a key consideration in Duflo et al.'s (2009) article.

The next set of articles that Duflo has written that incorporates environmental economics issues are "Cooking Stoves, Indoor Air Pollution and Respiratory Health in Rural Orissa, India" and "Indoor Air Pollution, Health and Economic Well-being" (Duflo et al., 2008a,b). Duflo et al. have addressed the issue of indoor environmental pollution and impact on health of women and children in India due to regular open-fire cooking in low-income households. The second article has provided a broader picture of the use of cheap and highly polluting fuel for cooking in multiple economically developing countries, as well as the policy impacts requiring the adoption of less-polluting cooking equipment in multiple countries. These articles have been cited 37 and 83 times, respectively as at 24th of October 2014.

Another article that has been written on the same topic that of indoor air pollution considers the adoption of low-polluting stoves in Orissa, India and the psychological barriers against sustained use of low-polluting stoves (for more than a year) and includes the evaluation of related health impacts, was published in 2012. Duflo is the second author on this article. This article has been cited 71 times as at 24th of October 2014. An example of benefit-cost analysis in environmental decision making is provided in Duflo and Pande's (2005) "Dams," which is a cost-benefit analysis of the construction of dams in India. Although it focused more on social benefits and costs, geographical considerations and impacts on local topography were provided. This article has been cited 232 times as at 24th of October 2014. Duflo et al. (2013) work can be considered seminal from the environmental auditing research perspective. This research encompassed a field and experimental study in which the structure of audit fees was altered in Gujarat in India. Duflo et al. (2013) found that prior to this study, environmental auditors were paid by the businesses and this resulted in false audit opinions and under reporting of emissions.

The experiment undertaken by Duflo et al. (2013) encompassed the creation of a central pool for audit fees and a secondary check of auditor's readings by technical agencies. Duffy et al. found that this system resulted in more accurate reporting as well as reduced pollution caused by the businesses. The implications from this article for the auditing profession are substantial, and Duflo et al. (2013) have proved that audit fees paid by the business being audited create a conflict of interest and that the findings from the project support a centralized pool for payments for audits in order to promote better quality audits. The function of the regulator is reenforced in this research.

11.3.1.1 Duffy's Work and Accountability for the Environment

Majority of Duffy's work has focused on social rather than environmental issues. Nevertheless her research described above provides evidence of impacting work in the area of environmental economics as well. Duffy has provided a multidisciplinary perspective in her work; she has taken into consideration behavioral, health, political, global, and accounting impacts beyond the economics' perspectives that are visible as policies' impacts analyses and econometric modeling.

11.3.2 (The Late) Professor Elinor Ostrom (1933—2012)

Ostrom is the first woman to be awarded the Nobel Prize for Economic Sciences. "She was senior research director of the Vincent and Elinor Ostrom Workshop in Political Theory and Policy Analysis, Distinguished Professor and Arthur F. Bentley Professor of Political Science in the College of Arts and Sciences, and professor in the School of Public and Environmental Affairs" (Indiana University, 2014). Ostrom received the Nobel Prize for her research in promoting and proving the role that all members of society can play in the equitable and sustainable sharing of "Commons" (natural resources available for all members of society). She tested the theory of "The Tragedy of the Commons" which implies that private ownership is the only means to protect the finite resources of the planet (On the Commons, OTC, 2011). She proved with various field studies that community ownership can promote conservation and protection of land from over utilization.

The main aspects of Ostrom's principles for managing Commons are strongly linked to the principles of accountability. There are two main aspects of accountability: responsibility and answerability. Ostrom clearly described the requirement for boundaries, rules for governance of common goods, as well as flexibility in relation to the rules and responsibilities as determined by the stakeholders (community members) impacted by the rules of governance of the Commons. Ostrom through her research stressed that there needs to be respect for the rules and responsibilities established by the communities, by outside authorities, and not just sanctions against the violators of the rules that have been established externally. The other principles that Ostrom proposed and promoted in her research are to utilize low-cost mechanisms for dispute resolution and to develop a sophisticated system of tiered responsibilities and governance requirements (OTC, 2011).

Going through Ostrom's complete achievements will not be a justified enough representation over here. She was a highly distinguished economist whose contributions have been substantial and that have had a major impact in the field of economics. The focus here is predominantly on her contributions in environmental economics. In relation to general statistics on Ostrom's work as at 24th of October 2014, the total citations to Ostrom's work were 84,361 at that particular date. The publication details are attained from Ostrom's 2013 curriculum vitae as available on the Indiana University Web site (Source: Indiana University, 2013).

In relation to her impact in environmental economics, Ostrom was on the advisory board of the Centre for the Study of Institutions, Population, and Environmental Change in 2010–2011. She was a member of the Resilience Alliance in 2000, on the Board of the Stockholm Resilience Centre in 2007, on the senior advisory group of the MacArthur Foundation, on "Advancing Conservation in a Social Context" in 2006, was a member of the University of Michigan, School of Natural Resources and Environment, External Advisory Committee for Strategic Assessment, from 2005 to 2007. She was a member of the editorial team of 23 Economics' journals including the *Ecological Economics, Ecology and Society, Global Environmental Change,* and *Global Environmental Politics.* Her multiple funded projects in the area of environmental economics are included but are not exclusive to: National Science Foundation, "Biocomplexity in linked Bioecological-Human Systems: Agent-Based Models of land use decisions and emergent land use patterns in forested regions," Swedish International Development Cooperation Agency, "A study of how aid, incentives, and sustainability are related," United Nations Development Program, "A Proposal to Monitor and Assess the Parks and People Project in the Nepal Terai." MacArthur Foundation, "The International Forestry Resources and Institutions Research and Training Program in Madagascar and the Eastern Himalayas," National Science Foundation, "A Proposal to Support a Centre for the Study of Institutions, Population, and Environmental Change," Ford Foundation's "The International Forestry Resources and Institutions Research and Training Program." Food and Agriculture Organization of the United Nations, Forests, Trees, and People Programme, "Collaboration on Communal Management: The International Forestry Resources and Institutions (IFRI) Database." Ford Foundation's "Nepal Irrigation Institutions and Systems Database," National Science Foundation's "The Role of Institutions in the Survival and Efficiency of Common-Pool Resources," U.S. Geological Survey's "The Comparative Performance of Institutional Arrangements for Groundwater Resources," U.S. AID's "Institutions and Common-Pool Resources in the Third World: What Works?", to mention a few.

Her seminal work in the form of books in the area of environmental economics includes (note that this list is not at all inclusive in any way and that the point here with these representations of Ostrom's work is to demonstrate the degree of influence that Ostrom has had in environmental economics).

The Future of the Commons: Beyond Market Failure and Government Regulation (Ostrom et al., 2012), in which Ostrom has applied the liberal tradition of political economy to promote free choice and self-management among communities to address natural resources' problems without the involvement of governments and other external bodies.

In *Property in Land and Other Resources,* Cole and Ostrom (2012) provided detailed discussions of the nature of different types of property rights as well as the impact of the exercise of different types of property rights and systems on the use of scarce natural resources are provided, with the use of case studies in multiple world regions.

Improving Irrigation in Asia: Sustainable Performance of an Innovative Intervention in Nepal (Ostrom et al., 2011) The main focus in this publication was on self-governance for water management of irrigation systems using a longitudinal case study in Nepal. A detailed analysis of the multistakeholder engagement within a community setting in Nepal that has worked well integrated with a local government initiative of irrigation systems that was managed adequately by the farmers, which promoted efficiency in the use of the water resources for farming purposes, was presented.

Seeing the Forest and the Trees: Human-Environment Interactions in Forest Ecosystems (Moran and Ostrom, 2005) In this text the varied nature of anthropocentric impacts on world forests is provided, and the unique nature of the impacts relating to the density of human population and its ability to conserve or to destroy nature was investigated. Case studies have been used; methods for forests monitoring and management have been considered and presented; and a rich theoretical and conceptual framework for human-environment interactions has been provided.

Asian Irrigation in Transition: Responding to Challenges (Shivakoti et al., 2005): This book provides a detailed case study approach using multiple geographical regions including India and Thailand as examples of water resources' issues and water resources' management as well as a detailed analysis of water irrigation (and systems) in Asia. The chapter written by Vermillion et al. (2005) in this book addressed the topic of "empowerment with accountability" resulting from the irrigation reforms in Asia. The main points discussed in this research refer to the changed approach of empowerment with accountability which refers to the assigning of greater responsibility to multiple parties including water users (the farmers) and the changing role of the government as a regulator and auditor as well as greater answerability for management and governance over the water resources by water users and members of Water Use Associations.

In *The Drama of the Commons* (Stonich et al., 2002), there are multiple aspects of management of natural resources that were addressed. These include a chapter on psychological considerations in relation to effective governance and accountability for natural resources. Other topics addressed in the book include natural resources' management, cooperation for natural resources management, property management and environmental protection, systems and institutional setups for the management of the commons, and a recap of the outcomes and impacts of the research undertaken in this area over the decades.

In *Improving Irrigation Governance and Management in Nepal* (Shivakoti and Ostrom, 2002) and in *Reformulating the Commons*, Ostrom (2001) shed light on Commons management and irrigations systems' economics analyses in two different geographical contexts, once again emphasizing the effectiveness and efficiency associated with self-management, governance, and accountability frameworks implemented by the farmers.

Other books authored (coauthored) by Ostrom include:

Institutions, Ecosystems, and Sustainability (Costanza et al., 2001),

Local Commons and Global Interdependence: Heterogeneity and Cooperation in Two Domains (Ostrom and Keohane, 1994),

Rules, Games, and Common-Pool Resources (Ostrom et al., 1994),

Institutional Incentives and Sustainable Development: Infrastructure Policies in Perspective (Ostrom et al., 1993),

Crafting Institutions for Self-Governing Irrigation Systems (Ostrom, 1992): In this publication, institutional economics was addressed. Multiple geographical regions have been used to present case studies on environmental and institutional management.

Ostrom's journal articles are briefly described. The following is not an exhaustive list. The articles most relevant to environmental economics have been selected and are by no means a complete representation. A brief description of the articles that are not directly related to the issues of the Commons and Irrigation Systems is provided. The purpose of undertaking this exercise is to shed light on the diversity of research that was undertaken by Ostrom in the field of environmental economics. Her research goes beyond the consideration of the Commons and the economics around irrigation systems in multiple geographical regions of the world. Conference proceedings have been excluded from the analysis as most of the proceedings are not available (the proceedings are mostly provided to conference attendants and are not available through Google Scholar).

In the article "Can Communities Plan, Grow and Sustainably Harvest from Forests?" (Ghate et al., 2013), the findings from experimental settings implemented in the form of games to attain data on attitudes and treatments of forests in a state of India have been presented. The main findings of the article are that communities rely heavily on forests for multiple purposes including firewood, building, and fuel. Although, communication between participants was found to generate better initiatives for sustainable forestry management, more efficient use of scarce resources and reduced deforestation was found as actions undertaken as a result of communications between participants. The two indigenous communities' participants demonstrated a preference for noncommercialized nature of harvesting.

In "Planetary Opportunities: A Social Contract for Global Change Science to Contribute to a Sustainable Future" (DeFries et al., 2012), Ostrom with multiple other authors presented the argument for addressing the problems associated with population growth in urban areas of the world. A multidisciplinary research approach that should focus on practically addressing the world problems associated with environmental damage, land management that reduces environmental impacts and biodiversity loss, recycling of phosphorus in agricultural use, a combined approach between global concerns and local action that would provide practical solutions to meet increasing needs while reducing environmental impacts was promoted in this article. Multiple geographical examples were provided.

In "Nested Externalities and Polycentric Institutions: Must We Wait for Global Solutions to Climate Change before Taking Actions at Other Scales?" Ostrom (2012) discussed the issues associated with considering global emissions' reduction as a public good that would benefit a large number of human beings from a global approach. In place of a global approach, Ostrom provided support for a polycentric approach that involves the consideration of the microimpacts created by smaller units that promote better action for the physical environment. Action taken at the organizational level can result in collective global impact, Ostrom proposed in this article. Ostrom provided multiple examples of government action taken in the US at the state level for reducing emissions and in Europe at the government level for sustainable fishery. Ostrom emphasized that local action among communities for environmental protection should be encouraged and supported.

In "Institutions for Managing Ecosystem Services" (Allen et al., 2012), the community management of common-pool resources that provide an opposing view to the tragedy of the Commons has been presented, with multiple examples of successful natural resources' management, so long as effective communication is implemented within the communities. In "Reconnecting to the Biosphere" (Folke et al., 2011), global stewardship for promoting sustainability is discussed with considerations, including examples of changes in governance undertaken and required for natural resources' management that addresses environmental and physical boundaries including the provision of natural capital. A new social contract that requires sustainability supporting expectations, responsibilities, and changed actions on behalf of governments and citizens of countries is called for.

The article "A Multimethod Approach to Study the Governance of Social-Ecological Systems" (Janssen et al., 2011) addressed a microapproach to natural resources management. The additional feature for effective communal communication that is identified through field experiments is trust. The article findings were that trust and communication together promote better resources' management results.

In "The Challenge of Forest Diagnostics" (Nagendra and Ostrom, 2011), the analyses and judgments provided by forest users and compared to statistical analyses of forest change, for example, forest regeneration were collated. It was found that judgmental analyses may be relevant in single time comparison instances but for more accurate analyses statistical techniques are required.

In "Moving beyond Panaceas: A Multi-Tiered Diagnostic Approach for Social-Ecological Analysis," Ostrom and Cox (2010) have addressed the "panacea problem" of over simplification to tackle ecological problems. Government and private ownerships were discussed with the consideration of social-ecological frameworks that involve the identification and analysis of various interactions between actors and systems that create impacts including environmental impacts. In this article the study of the institutional, systems and actors' interactions and impacts were encouraged to be taken into consideration for the purpose of more effective (in relation to, for example, reducing environmental impacts' goals) policy development and implementation.

In "Analyzing Collective Action" (Ostrom, 2010a), a number of structural variables have been presented as derived from game theory and other theoretical frameworks to proffer the case for collective action without the involvement or influence of external parties, for example, government, resulting in meaningful action for reduced environmental impacts in community settings.

In "Polycentric Systems for Coping with Collective Action and Global Environmental Change," Ostrom (2010b) promoted polycentric approaches toward addressing environmental problems rather than the centralized, government driven (collective action) approach. She stressed on the need for the involvement of multiple actors who can implement different types of actions for positive environmental changes, for example, to reduce carbon emissions by undertaking localized action.

In "A Polycentric Approach for Coping with Climate Change" (Ostrom, 2009b), Ostrom shed light on the weaknesses associated with programs such as REDD that according to Ostrom encourage free riding. In place of centralized programs, Ostrom has encouraged a polycentric approach encompassing multiple levels of responsibilities and involvement in environmental projects and experimental approaches adopted and tested in multiple micronetworks, with assessments of emissions reductions' goals planned resulting in implementations at microlevels.

In "Connectivity and the Governance of Multilevel Social-Ecological Systems: The Role of Social Capital" Brondizio et al. (2012), using the example of Xingu Indigenous Park in Brazil as the setting for a case study, identified the complexities associated with ecosystems' governance and management. The key focus in the article was on social capital which is created by institutional involvement in local networks and through interconnectivity between community groups in order to promote environmental and biodiversity protection in local settings.

The focus in the book chapter "The Contribution of Community Institutions to Environmental Problem-Solving" (Ostrom, 2009a) was once again on the consideration of polycentric theory and its key principles. The criticism of collective action including the problem of free riding has been provided. Examples of small scale institutional setups and success in relation to implementation of goals were provided in the US context.

In "A General Framework for Analysing Sustainability of Social-Ecological Systems," Ostrom (2007) presented a detailed outline of a societal system with nested subsystems such as resource systems, governance systems, measures, and outcomes as components of complex socioecological systems as a framework for application to multiple microsettings in order to evaluate the interactions between the subsystems and the impacts of the subsystems on each other.

"Top-Down Solutions: Looking Up from East Africa's Rangelands" (Mwangi and Ostrom, 2009a) was a look at the Maasai pastoralists in Kenya. The past history and the current practices were evaluated. The nature of change in institutionalization from self-management to colonization, followed by government intervention and the decline in the region's biodiversity and the negative impacts on the environment when self-management has been replaced with corporatization were evaluated and presented. This article provides a comprehensive depiction of multiple policy impacts on land, biodiversity, and management using one region as an example with consideration of multiple stages of influence. It supports the notion of self-management without external including government intervention for better natural resources' management and biodiversity conservation.

In "A Century of Institutions and Ecology in East Africa's Rangelands: Linking Institutional Robustness with Ecological Resilience of Kenya's Maasailand" (Mwangi and Ostrom, 2009b), the focus was on institutional setup and the impact on ecology. The Maasai people were considered and their institutional setup was compared to the government and colonial setup. It was found that indigenous management works best in relation to land use and conservation.

In Institutions, Ostrom (2008) has compared multiple types of governance systems for common-pool natural resources. She recommended a multiple tiered approach for resources' management rather than reliance on one type of system. She used the Gordon model of fishery bioeconomics to support her argument. She provided examples of government-controlled and privately managed resources and the weaknesses associated with them.

"Insights into linking Forests, Trees, and People: From the Air, on the Ground, and in the Lab" (Ostrom and Nagendra, 2006): This piece of research supported a multilevel governance and accountability structure that allows greater involvement of community members in decision making. According to the authors, this setup works effectively in relation to resources' management and conservation compared to strict enforcements by governments without deep community involvement.

In "Diversity and Resilience of Social-Ecological Systems" (Norberg et al., 2008), Ostrom and coauthors critically considered the negative impacts of the short term profit maximization focus on environmental damage and biodiversity loss. The development of sustainable socioecological systems was discussed. In relation to this, the concept of resilience and its related factors including self-organization was considered. Diversity among species and the related ecosystems functioning was evaluated. The concept of diversity from the biological sciences was applied to institutional diversity. With the example of the fishing industry the concept of complementarization as opposed to diversity was discussed, the human preference for homogeneity as opposed to diversity as demonstrated in agricultural practices and the lack of recognition for diversity conservation in biodiversity were assessed. A community level stakeholder approach that supports micro, regional diversity in decision making for environmental protection purposes was proposed.

In "Implications of Leasehold and Community Forestry for Poverty Alleviation" (Karmacharya et al., 2007), the focus was on forestry leasing and management in the context of Nepal. The main factors that were found to stand against effective forest management in the cases of forest leasing to economically disadvantaged families include lack of long-term returns to the communities for land care and rehabilitation of the degraded region and social conflicts (the resolution of which was crucial for the success of the project). It was found that if forestry leasing and management programs are managed well, including in relation to the allocation of funding and allocation of fraction of the income attained from the sale of forest products to the least privileged members of the groups, greater benefits can be attained in relation to better forest management and sustainable use of the land over the long term.

In "Decentralization and Community-Based Forestry: Learning from Experience" (Agrawal and Ostrom, 2006), the concept of decentralization and environmental resources was discussed with a critical perspective of centralization and its lack of impact was outlined. It was found that the primary factors that local actors can play in relation to decentralized forestry management include taking measures for resources' protection, implementing and monitoring required compliance, and undertaking and enforcing sanctions against noncompliance. There is a view in the article that in some instances governments "pretend" to decentralize but that this is more of an act rather than real action. Multiple case studies from different regions including Asia, Africa, and Latin America were presented with a discussion of actions and impacts.

"Fifteen Years of Empirical Research on Collective Action in Natural Resource Management: Struggling to Build Large-N Databases Based on Qualitative Research" (Poteete and Ostrom, 2008) This article was a comparative study of multiple data collection techniques relating to research on natural resources' management. Multiple qualitative and quantitative methods were compared including the case study approach, field-based empirical studies, large N-studies, and secondary data analysis (of data derived from data bases) approach. A main finding presented in the article is the lack of cross-country research that Ostrom and coauthor have suggested is of concern as lack of such research in environmental economics represents a barrier in relation to comparison and has national and global policy implications.

In "Coupled Human and Natural Systems" (Liu et al., 2007), a case study approach was adopted to demonstrate the nature of coupling between nature and human systems in multiple regions from economically developed and developing countries of the word. Multiple variables that relate exclusively to human actions, and to nature and combined (that is the impact of human actions on nature variables) were considered. The impacts of micro- and macroeconomic factors including the impacts of markets and government policies were examined, especially in relation to impacts that are created over vast geographical regions. Changes in biodiversity were compared to changes in community related factors, for example, scatter plots are shown of changes in birds' population in a region of the US linked to changes in the number of human dwellings in the region. The authors have stated that there are different types of impacts caused by different spatial or temporal considerations that add to the complexity of consideration of coupling between human and nature systems in different regions of the world.

In "Robustness of Social-Ecological Systems to Spatial and Temporal Variability" (Janssen et al., 2007), a historical context was adopted to consider the setup relating to land management and transaction costs in medieval Europe. The problems associated with private ownership and boundaries' enforcement on resources' management in the UK and India were discussed. The impacts of changes in governance on socioecological systems were considered. Policy intervention that has caused lesser accountability and consideration for natural resources' management was criticized as a deterrent against optimal resources' governance and management.

In "The Globalization of Socio-Ecological Systems: An Agenda for Scientific Research" (Young et al., 2006), the impacts of multiple globalization factors on human–environment or socioecological systems have been considered. Factors that were considered include global environmental changes, urbanization, connectedness, multinational corporations, and their impacts on the complexity of networks and on regional biodiversity (loss). Detailed studies of resilience, vulnerability, and adaptability in multiple socioecological systems as a result of global impacts were called for.

In "Political Science and Conservation Biology: A Dialog of the Deaf" (Agrawal and Ostrom, 2006), the lack of engagement of political scientists with biodiversity conservation was presented as a problem associated with lack of biodiversity considerations as priorities in the field of political and policy economics. The benefits of addressing biodiversity loss issues and problems in this area of research were presented and a call was made to understand the impacts of government policies on ecological systems and the setting of rules for long-term sustainability of crucial and sensitive ecosystems.

In "15 Incentives Affecting Land Use Decisions of Nonindustrial Private Forest Landowners," York et al. (2005) have presented the different types of policy tools that impact private forest owners' behaviors are considered. Tools include tax policies, costs' sharing, easement, and certification programs. The potential impact of these policies was considered in the context of the United States. Other factors, for example, the opportunities for massive economic gains that can result from selling land for urban development were mentioned. This research serves as an introduction to the types of policies that exist for private forests' owners to encourage forest protection rather than destruction caused, for example, by property development on forest land.

In "Conserving the World's Forests: Are Protected Areas the Only Way?" Hayes and Ostrom (2005) discuss the effectiveness relating to safeguarding goals of protected world forest areas. The main issue discussed relates to the monitoring and control mechanisms that are not implemented in a large number of the protected areas. The economic and social costs associated with maintaining protected areas are analyzed. Multiple examples and studies are compared to conclude that government-driven forest protection may not necessarily be the only means to promote biodiversity conservation. On the contrary, protected forestry regions face multiple imminent dangers associated with encroachment, illegal logging, fires, and biodiversity destruction, especially if the local communities see the protected regions as a threat to their livelihood or if they interfere in their ways of living. Greater stakeholder involvement and assigning of responsibility and accountability to communities is proposed for better forestry and biodiversity conservation.

In "Local Enforcement and Better Forests" (Gibson et al., 2005), quantitative data analysis is applied to data attained from the International Forestry and Resources' Institutions database in order to assess the impact of rules enforcement on forest conditions. It is found that consistent and regular rules enforcement is extremely important for forest areas' protection and conservation.

In "Multi-Level Governance and Resilience of Social-Ecological Systems" (Ostrom and Janssen, 2005), a historical perspective of economic growth in developing countries and the impacts of financial aid injected into the economically developing regions have been discussed. Policies relating to environmental conservation are critically analyzed. Weaknesses associated with government-controlled resources and the lack of monitoring and control are evaluated; the inadequacy associated with the misguided expectation of altruistic measures to be implemented by governments is analyzed. The problems associated with optimal policy strategies are presented. The resilience-connectivity model is shown to demonstrate the multiple stages of policy implementation, and different considerations/approaches relating to governance that are required at different stages of the cycle are discussed. Actions that are required to deal with multiple disturbances or external factors have been assessed.

"Legal and Political Conditions of Water Resource Development" (Ostrom and Ostrom, 1972) is a combined consideration of water engineering and water economics with organizational cogitations, including gaming capabilities involving costs and benefits, capabilities, and limitations. Water rights and collective enterprise have been discussed with examples from a few states in the US.

In "Ecological Systems and Multi-Tier Human Organization," Moran et al. (2002) have shed light on multiple types of decision making from the perspectives of heuristics and the level of knowledge and understandings. The types of decision making are considered in the context of nature and land management (ecosystems) at five levels of governance: international, national, regional, communities (groups), and individuals and households levels. Multiple factors at each level are assessed in relation to impacts.

In "The Study of Human-Ecological Systems in the Laboratory" (Ostrom et al., 2002), three assumptions relating to contemporary natural resource policy are argued to be wrong that: (1) resource users are short-term users who require immediate gains, (2) they are without norms and values (including lacking inclinations toward accountability for natural resources), and (3) objective, distant interference is capable of implementing changed behaviors and better actions for the protection of natural resources; resource governance has to be driven centrally. Through experimental results, individuals are found to be fallible, but they are observed to have the ability to learn from mistakes and adaptation and are not found to be myopic, rather they are found to have the capability to exhibit norms (rules) following in order to control impacts on natural resources.

In "Property-Rights Regimes and Natural Resources: A Conceptual Analysis" (Schlager and Ostrom, 1992), the multiple types of property rights as relevant to control over natural resources have been described. These include de facto and de jure property rights and their implications; the right of exclusion is considered in detail. A case study approach (using the example of the Maine lobster industry in the United States) is used to present the application of the conceptualization of property rights to the lobsters' fishing industry in Maine.

11.3.2.1 Accountability and Ostrom

The main focus in Ostrom's research was on self-governance, self-management, and the exercise of accountability from regulators for the Commons management, resources' efficiency of use, and responsibility for sharing of natural resources. The following works have clearly encompassed community-driven accountability for the governance of Commons considerations by Ostrom.

In "Community Driven development," Ostrom and coauthors (Dongier et al., 2008) emphasized empowerment of community members of society for forestry management. The authors have provided evidence of successful land rehabilitation societal initiatives undertaken and implemented with efficiency as a result of community involvement and acceptance of responsibility for the effective management of the Commons. The initiatives described and discussed include community-organized irrigation systems, group-based microfinanced projects, with embedded responsibility for answerability to other group members for the success of the project involving natural resources management including forests' management. Dongier et al. (2008) found indications of positive impacts on land rehabilitation, water management, and conservation as a result of these community initiatives.

The crucial work that NGO's can play in community empowerment is stressed including NGO-initiated community-driven development projects. The benefits of such projects are described to include the establishment of an enabling environment, responsible investments matched with demand, participatory mechanisms and stakeholder engagement, social and gender inclusion, capacity building of community-based organizations, responsibility for facilitating community access to required information, monitoring and evaluation, flexibility, allowance for scaling up, and an exit strategy. A supportive legal and regulatory environment that promotes accountability in the relevant government departments is emphasized.

Ostrom's research carries with it a long-lasting impact in academia and in relation to policy considerations by different levels of governments. Her substantial contributions in environmental economics including the applied research and its findings are invaluable, especially in the context of community-based Commons management policies and decentralization.

11.3.3 Anny Ratnawati

Anny Ratnawati was a lecturer at the Bogor Agriculture Institute, Indonesia. She is now the deputy finance minister in Indonesia. Anny's masters and PhD are in macroeconomics from the Bogor Agriculture Institute. Previously, she was the director general of budgeting in the Finance Ministry (The Jakarta Post, 2010).

In 2007, the Ministry of Finance in Indonesia initiated an interministry working group to direct efforts and research toward the High Level Event on Climate Change, organized for the Ministers in Finance as well as the 13th session of the Conference of the Parties that was held in Bali in 2007 under United Nations Framework Convention on Climate Change (UNFCCC). The working group comprised of multiple senior officials from the President's office, the Ministry of Environment, the Ministry of Foreign Affairs, and the coordinating Ministry of Economic Affairs. Anny Ratnawati was a regular participant and representative of the Ministry of Finance of Indonesia. As a result of this initiative in 2008, Indonesia published its plan for integrating climate change adaptation and mitigation initiatives into its national plan and budgeting and the National Council on Climate Change was formulated. The report that was prepared by the working group in which Ratnawati played a prominent role focused on the fiscal, macroeconomic, and policy factors that at that time were identified as key considerations for mitigating Indonesia's carbon footprint.

The key points and findings of this report were as follows (Source: Fiscal Policy Office, Indonesia, 2008).

Climate change is a key economic development and planning issue for the government of Indonesia. Timely planning and response is considered as being of key economic benefit for the country. In the long term, Indonesia's fossil fuel emissions are considered as a contributory factor to the worsening carbon footprint, in addition to deforestation, burning, and peat land degradation. The forestry and energy sectors have faced major policy failure and governance problems. Weak forestry governance is accepted to be a hindrance against investment, access to international markets, and a factor for increased costs. Although there are multiple renewable energy and efficiency options available, lack of investment in this area is acting as a hindrance against meaningful progress. There are international financial incentives' programs available that can be utilized resulting in better forestry management and reduced carbon footprint, as well as decreased biodiversity destruction. These international opportunities include Climate Investment Funds, global carbon markets, the Forest Carbon Partnership Facility and private sector partic-ipation. Indonesia, at that time, saw participation in the United Nations Collaborative Program on Reducing Emissions from Deforestation and Forest Degradation (REDD) in Developing Countries as a strategic (financial) opportunity to attain large amounts of revenue for reducing deforestation and land degradation, as well as for reducing carbon emissions. The country was identified as expecting to attain approxi-mately US$1 billion per annum but the condition as expected and imposed was to demonstrate verified and successful efforts to reduce forest destruction and degradation. The report considered the "right" implementation of fiscal policy for better forestry management and reduced forest loss that was required to focus more on better forestry governance and accountability.

An evaluation of the complexity associated with energy use and the lack of support for renewable energy and reduced carbon options were provided in the report. The fac-tor of higher costs from increasing emissions in economically developing countries was discussed. At the same the lack of financial resources to undertake initiatives for reducing carbon emissions in countries like Indonesia was mentioned. Five main factors that the finance departments of governments can influence for climate change actions were iden-tified as follows: International flows of investments and funds, national investment and financial sector policies, fiscal policy instruments including taxes and subsidies, budgetary, and expenditure policies and regulation.

The main sources of financing beyond government funding, which was identified as being inadequate, were outlined as follows: international financial institutions' funding, development banks funding, carbon markets that involve payments from companies, and governments for reduced carbon emissions funding. It was established that the demand per annum for reduced carbon emissions by 2030 could range from US$25 billion to US$100 billion. The connection between international carbon reduction initiatives, including market initiatives and the policy implementation, and changes required at mul-tiple levels including better governance and accountability was established and presented.

From this brief analysis of the key report that was issued in 2008 by the Ministry of Finance, which included major contributions from Anny Ratnawati, it can be seen that this milestone initiative acted as a major step toward further action for reduced defores-tation and biodiversity destruction in Indonesian forests. Since 2008, the REDD project has achieved some outcomes including: implementation of training in field measurement of carbon inventory in 2012, as well as development and installation of a specialized data-base for Indonesia's national forest assessment and monitoring. Training has been provided in forest and carbon inventory implementation in Palu in 2012. Further training and education (through workshops and dialog with local communities) in land rehabil-itation has been provided in Central Sulawesi. 2011 was mostly focused on planning around methods of implementation of REDD. Multiple stakeholder engagements were undertaken at the higher, national government levels. 2010 was a main focus on the development and deployment of funding through REDD for the country (UN–REDD, 2012).

Currently, the project is in the phase of undertaking further consultations with the Ministry of Forestry to implement measures for improved governance in the sector. At present, deforestation rate has doubled in the country. The primary issue identified is the lack of verifiable monitoring data from the Ministry of Forestry. From the financing perspective only a small amount has been received for the education and awareness and capacity building project (Lang, 2014). It is not the purpose in this chapter to critically evaluate the progress of REDD initiatives in Indonesia. What needs to be brought to attention is the contribution that Anny Ratnawati has made in bringing to global attention the economic destruction that is being caused of Indonesia's forests due to lack of governance and accountability (Khan, 2014) on behalf of multiple parties including government officials, illegal loggers, and palm oil plantation businesses. The report produced in 2008 and the Forum held of finance ministers have brought to light the complex problems that economically developing countries like Indonesia face in the form of regional barriers and complexities including lack of education. The report has brought to global attention the dire consequences of lack of action to mitigate climate change impacts and the problems of lack of clarification of property rights. Lack of microgovernance and grass roots changes required to promote greater governance and accountability for forest protection and damage mitigation have been brought to surface as major issues.

11.3.4 Professor Armida Salsiah Alisjahbana

Professor Alisjahbana is the Indonesian State Minister for National Development Planning and Head of National Development Planning Agency of Indonesia. She is a professor in the Faculty of Economics at Padjadjaran University in Bandung, Indonesia. Before taking on the role of minister, Alisjahbana was involved in multiple projects with the United Nations and the Australian Agency for International Development (AusAid) (National Graduate Institute for Policy Studies, 2012). She attained her PhD in Economics from the University of Washington.

As far as her contribution in the area of environmental economics is concerned, Armida has published the book titled "Green accounting and sustainable development in Indonesia" in 2004; she has published articles in the field of environmental economics as well. Her key works in the area include: in her 2004 coauthored article (Alisjahbana and Yusuf, 2004a), Alisjahbana has discussed the country specific issues and negative economic factors that have hindered progress toward sustainable development. She has evaluated the concepts of weak and strong sustainability from the perspective of savings and genuine savings. She has provided the conceptual outline of genuine national savings that should incorporate measures of deprecation of nonrenewable natural resources, depreciation of renewable natural resources including the consideration of excess felling which requires a calculation of the amount of round wood produced in excess of its sustainable growth.

Other factors that need to be considered in order to determine genuine national savings include environmental degradation expressed as EDL which is defined as environmental degradation from local and global pollution, multiple characteristics of costs relating to nonrenewable resources' use, unit cost of emission abatement of the pollutant, volume of pollutant emitted per unit of output produced by the manufacturing sector (pollution intensity), and marginal social cost of CO_2 emission.

The problems associated with unreliable data attained from the Ministry of Forestry have been discussed. She applied Hamilton's and Lutz' (1996) and Hamilton and Ward's (1997) modeling and methodology to Indonesian data from international sources. The main conclusion that was reached as a result of the application of this model was that Indonesia's economy has not been sustainable. The trends relating to wealth per capita changes and use of natural resources changes were linked with economic policy changes over the periods of 1980–1997 (the preeconomic crisis period) and 1997–2000. The connection between a faster depletion of natural resources (oil and gas) and the economic crisis after 1997 was established. The changes in resource rent during this period were discussed.

Empirical evidence was provided of increase in natural resources' depletion as a result of proliferation of poverty during the economic crisis period. The impact of the crisis on the value of the Indonesian rupiah and the augmented earnings from resources' exports in foreign currency (converted to larger sums in local currency) was emphasized as a key influencing factor on rapid natural resources depletion during those years. It was recommended in the article that the country should adopt natural resources' management policies including property rights legal requirements, royalties, concessions, zoning, and regulation of natural resources management and commitments for important environmental protection expenditure measures; the policies that would support optimal resources' extraction path.

Her coauthored book chapter (Resosudarmo et al., 2011) addressed the energy and climate changes in Indonesia. The efficient supply of energy to the economically disadvantaged fractions of the Indonesian society was addressed. The issues associated with the energy policy, Indonesia becoming a net importer of oil from 2006, the rising energy prices and environmental degradation issues have been discussed in detail. The political setup and agenda has been discussed from the perspective of a historical consideration of the Suharto government compared to the democratic setup, with the associated desired goals and objectives, from decentralization to economic centrism. The importance of greater distribution of earnings to the regions where they were generated, rather than to Jakarta is analyzed, together with the associated problems of corruption, political stabilization more recently, and increased energy use; energy inefficiency due to lack of technologies and cleaner fuel sources have been critically analyzed. The reduced oil production, coupled with increased consumption, caused the government to focus more on coal (implying more mining) and natural gas (large resource reserves have been found in Indonesia) consumption.

Alisjahbana and coauthors have discussed the environmental degradation impacts of the differential energy policy of Indonesia and Indonesia's recent commitment to reduce GHG emissions with and without international financial assistance by 2020. The issues associated with energy concessions in Indonesia have been examined in detail. Other works published by Alisjahbana, with coauthor Yusuf include a book on green accounting and sustainable development in Indonesia (2004b) and an article that discusses the key concepts of green accounting applied and critically analyzed from the green economics perspective (2004b).

The concepts from green accounting are considered including Green GDP, Genuine Savings, and Change in Wealth Per Capita. The use of green GDP is cautioned against as the indicator could be misleading. Other indicators such as genuine savings have been proposed as better measures of sustainable development and performance. The concept of green accounting has been discussed in relation to the changes that have occurred in Indonesia including the decentralization that has allowed more autonomy to regional governments and sectors. The economic principles behind weak and strong sustainability were evaluated. The concept of ecodomestic product which measures the output or income net of depreciation of natural capital and environmental costs was discussed in the context of Indonesia. The main weakness associated with Green GDP that may not have the capability to indicate weak sustainability has been outlined with the following modeling $YG = Y - R$ where Y is the conventional GDP and R is resource depletion, each variable is a function of time which is expanded on to result in: $GyG = gY - r.gR/1 - r$, $dGYG/dr = gY - gR/(1 - r)2$, growth of green GDP being a function of growth of GDP, resource depletion and relative resource dependence.

As growth of resource depletion equals the growth of GDP, then Green GDP will equal the growth of GDP. The authors emphasize that because of this weakness in the model, in instances of rapid natural resources depletion, Green GDP will equal economic, conventional GDP, and in this sense will not be a reliable indicator of a country's sustainability performance. The ratio of resource depletion to GDP (r) is presented as a better indicator of sustainability performance. It is considered in the contexts of multiple regions of Indonesia, for example, East Kalimantan and Riau, both of which are resource rich regions, with Kalimantan facing more resource depletion than Riau. Alternative measurements such as Genuine Savings and Change in Wealth Per Capita that can utilize existing data that are required for the System of Integrated Environmental and Economic Accounting (SEEA)'s Green GDP are presented as better and more accurate depictions of weak sustainability.

Professor Alisjahbana is an expert in the field of environmental economics, from the perspective of macrosustainability accounting measures. She has shed light on the weaknesses associated with the use of conventional GDP measures and Environmental and Economic Accounting measures. She has proposed genuine national savings models as better depictions of national sustainable development or lack thereof.

11.3.5 Dr Shamshad Akhtar

Dr Shamshad Akhtar is the Under-Secretary-General of the United Nations and Tenth Executive Secretary of the Economic and Social Commission for Asia and the Pacific. She is the United Nations Sherpa for the G-20. She has served as the Vice President of the Middle East and North African region of the World Bank. In her role at the United Nations she has been involved in sustainable financing strategy initiatives of the United Nations. Dr Akhtar was a US Fulbright Fellow at Harvard University; she attained her PhD in Economics from Scotland's Paisley College of Technology and a Master in Development Economics from the University of Sussex in the UK (United Nations Economic and Social Commission for Asia and the Pacific, ESCAP, 2014).

Akhtar has contributed immensely in the area of sustainable development in her capacity as the Under-Secretary-General of the United Nations. An example is her speech at the Dialogue on Strengthening Connectivity Partnership in November 2014. Her focus in the speech was on effective negotiations between small and large economies; consensus and agreement in relation to regional connectivity. She outlined the main works of ESCAP including the development of transportation networks across Asia as well as information highways and networks. She linked these developments with economic concepts of greater economies of scale and with the environmental economics considerations of reduced carbon footprints through better-connected transportation options; resulting from the funding and building of sophisticated, regionally connected roads and railway networks; with the involvement of multiple countries including China, Thailand, Vietnam, Russian Federation, Mongolia, and through information highways, for example, the Asian Information Super Highway (Akhtar, 2014a). Other areas of environmental economics that Akhtar has addressed include (these topics were addressed at speeches delivered between 2009 and 2011 when Akhtar was the Vice President of the Middle East and African region of the World Bank): food security, safety nets for poor people as a barrier against price rises, efficient management of water and arable land, supply chain efficiency, agricultural risk management, investment and policy measures to reduce inefficiencies, investment in "critical" infrastructure including better connectivity, for example, through port facilities and strategic grain reserves. She addressed the contribution of the World Bank to promote Agricultural and Rural Development in the MENA (Middle East and North African) region (Akhtar, 2011a).

In the speech delivered in March 2011 at the First Arab Development Symposium, Akhtar (2011b) described multiple factors that have impacted international food prices and have caused increased vulnerabilities among the poor, specifically in the Arab region. The factors that Akhtar discussed, that have impacted increasing demand for food include growth in populations and incomes, limited water resources, decrease in renewable water resources, complications associated with water resource management, especially from the cross-country perspective, change in rainfall patterns as a result of climate change, emphasis on greater agricultural water productivity, energy prices, total factor productivity, among other factors. Akhtar (2011b) discussed the vulnerabilities of the key Arab regions that face high exposures to food prices and food quantity risks including Yemen, Jordan, Djibouti, Lebanon, Iraq, and Tunisia. She emphasized the importance of adequate stock levels and policy development that would act as positive measures toward food security and pricing risk mitigation in the most vulnerable Arab regions. She discussed the detailed short-term and long-term responses of the World Bank to address the Food Crisis. These include rapid response grant funding including the Food Crisis Response Program and the International Development Association (IDA) Crisis Response Window. She highlighted the World Bank's facilitation of long-term programs in agriculture and food security in IDA countries through the Global Agriculture and Food Security Program (GAFSP), worth millions of dollars. Regular programs including lending, policy consulting, and technical advice for long-term initiatives to support agricultural productivity and long-term resilience of supply chains have been described as well. She shed light on the World Bank investment in technology, R & D, and improved agricultural water management. She stressed on improved weather forecasting and monitoring, food security risk management, faster food distributions, effective social safety nets including nutritional outcomes, and support of microfarming (as a global food security solution).

She has given an account of the World Bank's research on import supply chains and the proposed improvements to these in order to increase efficiencies in transportation, logistics, and storage. She provided a summary of the food security policy adoption for three different types of Arab regions: fiscal deficit countries that are highly dependent on cereal imports, countries with fiscal deficits that are not highly dependent on cereal imports, and countries with fiscal surpluses that are highly dependent on cereal imports.

Her speech of 2010 at the International Forum on Human Development (Akhtar, 2010a) was a focus on Morocco. Akhtar emphasized World Bank's role in sustainable development and economic growth in Morocco. She mentioned her visit to the rural areas of Morocco and her observations relating to social mobilization, social services as well as domestic water availability. In the speech, she discussed multiple socioeconomic factors including literacy, mortality rates and income sources for women including the manual production of Argan oil from Argan nuts harvested from the plains. She provided examples from India and Brazil of successful lending projects to women and self-management on behalf of women that has resulted in reduced exploitation and greater success in microfunding and community development, specifically targeted toward women. In this speech, Akhtar provided the connection between the economic development of the disadvantaged members of an economically suffering society and the use of natural resources (Argan Nuts) for the economic betterment of the specific fraction of the Moroccan rural society, the women of rural Morocco.

In her speech at the Dubai School of Government (2010b), Akhtar focused on the contributions of the World Bank in reducing spatial disparities in regions of the Arab World. She presented multiple examples including the World Bank's Tunisia's Northwest Mountainous and Forestry Areas Development Project. The aim of the project was to improve the socioeconomic conditions of the specific regions including Tunisia in the Middle East and North Africa region. The goal of the project was to ensure the sustainable management of the regions' natural resources. From an environmental economics perspective, the project focused on providing equipment and training in relation to improved forestry management, improved crop production, livestock husbandry, small-scale irrigation schemes, improvement of environmental and natural resources' management, soil and water conservation, pastures and rangelands underutilization, and improvements in local infrastructure (The World Bank, 2014).

11.3.5.1 United Nations ESCAP

Akhtar is the Executive Secretary of the United Nations Economic and Social Commission for Asia and the Pacific (ESCAP). It comprises of 53 member states and 9 associate members. The main goal of ESCAP is to address the key challenges faced by this wide region including from an environmental economics perspective: sustainable development, environment and development, social development, and disaster risk reduction. In the 2014 (August) policy statement, Akhtar discussed the sustainability development challenges faced by the Asian and Pacific regions including the promotion of economic growth within the limits of sustainable growth. She discussed the post-2015 Sustainable Development agenda, specifically for the Asian-Pacific region. The important concepts of monitoring and accountability were emphasized for discussion at the Asia Pacific Forum for Sustainable Development. Akhtar analyzed the link between sustainable development goals, financing of the goals, the Climate Finance Summit, and other international conventions dealing with sustainable development. She provided the key points of the Global Sustainability Agenda as follows:

Balanced social, economic, and environmental developments with a key focus on interdependencies and linkages, coordination of sustainable development agenda across the member states and their ministries with a crucial focus on reporting, governance and accountability, serious considerations of financing sustainable development supporting technologies including cleaner technologies, knowledge sharing across countries and sharing of experiences for learning and advancement with multiple facets of sustainable development, development of sophisticated databases and national and regional reporting (to serve the important accountability requirements), and the promotion of the involvement of the private sector through sponsoring of the development of innovative and sustainable development technologies, implementations, and advancements.

She stressed the key initiatives planned to be implemented by ESCAP to promote the goals for sustainable development including cooperation from member states to implement an effective program delivery strategy as well as the integration of intergovernmental analytical and knowledge management strategies. Implementation of tailored sustainable development strategies in the most economically vulnerable countries in the region was emphasized. She stressed on the utilization of crucial partnerships with other key bodies for example the Association of Southeast Asian Nations (ASEAN) to strategize and implement key areas for sustainable development including technology transfer, sustainable agriculture, and crucial capacity building (Akhtar, 2014b).

11.3.5.2 Contributions of Dr Shamshad Akhtar, Summary

It can be seen from the key contributions of Dr Akhtar in the capacity of the Vice President of the Middle East and North African region of the World Bank and as the Executive Secretary of ESCAP that she has provided immense degree of service at an executive level to environmentally supporting national and international initiatives, encompassing social, environmental, and economic developments. She has advocated sustainable development, financing of projects that are not only supportive of the economic development of the most disadvantaged members of society (the world's poorest) but projects that have involved the advancement of environmental protection and conservation as well as sustainable use of natural resources.

There are multiple examples as discussed above of her representations of the United Nations initiatives, for example, the promotion of efficient transportation networks, information networks, agricultural efficiencies, multiple projects and funding to promote food security in high risk countries, funding of projects that have and will support natural resources' management, for example, efficient agricultural practices and efficient water asset use. Although Dr Akhtar has presented the projects undertaken by the institution that she represents, her serving in this capacity represents her substantial contributions in promoting sustainable development in key regions that have most of the world's population. These regions face the most serious issues associated with Climate Change, for example, reduction in water resource, issues associated with lack of reporting and accountability in relation to natural resources' use, and the management and allocation of the scarce resources. She has emphasized the promotion and implementation of crucial projects that support the economic advancement, education, and betterment of women among the most underprivileged and disadvantaged fractions of economically developing societies. Her work is an example of an ecofeminist approach and its promotion at national levels, combined with the advancement of female rights to income, to self-reliance and to education in the highest risk countries of the world.

11.3.6 Upcoming Female Environmental Economists

The three emerging environmental economists have been selected based on their substantial contributions to environmental economics research, priority areas. The three economists have more than 20 publications in the field of environmental economics. They have been involved in multiple environmental policy impact and evaluation projects and have been associated with large funding national and international bodies such as the World Bank and Organization for Economic Co-operation and Development (OECD).

11.3.6.1 Dr Phoebe Koundouri

The source of the information provided here on Dr Koundouri is from the London School of Economics and Political Science Web site. Dr Koundouri is a visiting senior research fellow at the Grantham Research Institute. She is an associate professor at the Department of International and European Economics Studies at Athens University of Economics and Business and the director of the Research Team on Socio-Economic and Environmental Sustainability (RESEES). She is highly cited and is an associate editor and editorial board member of 12 academic journals. Dr Koundouri has been involved in multiple natural resources' management projects funded by the World Bank, OECD, European Bank of Reconstruction and Development, United Nations, NATO, WHO, the European Commission, and other national and international organizations. Dr Koundouri attained her PhD in Economics from the University of Cambridge.

Dr Koundouri's research has been published in top economics' journals including the *Journal of Environmental Economics and Management, the American Journal of Agricultural Economics, Oxford Economic Papers, Land Economics,* and *Water Resources Research.* She is ranked among the top 5% authors in IDEAS. Dr Koundouri's main research areas include natural resources economics and econometrics (water, energy, biodiversity and ecosystem services, land use, and air), pollution, climate change and intergenerational issues, uncertainty, risk, natural disasters, agricultural economics, game theory and the environment, interdisciplinary approaches to natural resources management and policy, and philosophy applied to environmental and resource economics (London School of Economics and Political Science, 2015).

11.3.6.2 Dr Celine Nauges

Dr Celine Nauges is an Australian Research Future Fellow with the School of Economics at the University of Queensland, Australia. Dr Nauges attained her PhD in Economics from the University of Toulouse, France, in 1999. Her main research areas relating to environmental economics include microeconometrics applied to agriculture, natural resources, environmental management, urban water management in developed and developing countries, and risk assessment of agricultural practices. She has more than 30 published academic journal articles and multiple working papers. She has been involved in numerous grant and consultancy projects relating to environmental economics including consultancies for the OECD on environmental policy and water saving, environmental policy and organic food consumption, household behavior and environmental policies consultancies for the World Bank on irrigation practices in Jordan, energy efficiency of Brazilian water utilities, a project on assessing the economies of scale and density in water supply and sewerage service provision, water and sanitation policies in South West Sri Lanka, to mention a few.

Dr Nauges serves on the editorial boards of the Environmental and Resource Economics, Review of Agricultural and Environmental Studies and Water Resources and Economics. She is an advisory board member of South Asian Network for Development and Environmental Economics (University of Queensland Australia, 2015). It is important to note that Dr Nauges is an Australian Research Council (ARC) Future Fellow and her contract is dated from 2011 to 2015. The Future Fellowships program implemented by the Australian Government, ARC attracts midcareer researchers for 4-year fellowships to undertake research relating to areas of national importance (ARC, 2015).

11.3.6.3 Associate Professor Sarah Wheeler

Associate Professor Sarah Wheeler is a Future Fellow at the University of South Australia, Business School. She has more than 80 peer-reviewed publications in the areas of organic farming, irrigated farming, water markets, and water scarcity. She is the associate editor of the *Australian Journal of Agricultural and Resource Economics* and of *Water Resources and Economics*. She has attained multiple grants including the prestigious ARC grants relating to environmental economics priority research areas in the areas of agricultural economics, water economics, pollutant inventory, irrigation, water productivity in multiple African regions, water trade, climate change, and irrigation adaptability in the Murray–Darling basin in Australia, water markets, and national water initiatives (environmental policies research), to mention a few.

11.4 CONCLUDING COMMENTS

The works of five prominent and three emerging female economists in the field of environmental economics, research, and policy have been presented as a display of the substantial contributions of these economists. This chapter has provided evidence that female economists are capable of making significant additions in the field of environmental economics and that their work has earned them high regard and senior positions in organizations and in faculties.

Multiple examples of different types of ecofeminist works that have been undertaken and that can be reapplied and further developed in multiple contexts from an environmental research perspective have been provided. There is immense potential in the further development of ecofeminism and environmental economics' research, as well as women's representation in environmental economics research and policy setting and implementation.

Less than 4 years ago, Dolado et al. (2012), based on data from 2005, presented their findings that the most popular subtopics among female economists were Health, Education, and Welfare as well as Labor and Demographic Economics. Mathematical Economics, Agricultural, Natural Resource, and Environmental Economics were found to be the least popular subtopics among female economists. At the same time they found a decreasing gap in relation to gender segregation in fields such as Environmental Economics.

The prominent research undertaken by the female economists as discussed in this article strongly supports the decreasing gender gap in agricultural and environmental economics. Their research entails a wide range of theoretical, methodological, and empirical undertakings. The research undertakings of these economists are highly relevant and critical in relation to policy development, change, and implementation in multiple contexts including in multiple regions, especially the developing regions of the world.

Our world's natural resources are limited; the long-term survival of the human race depends on the health of our planet. The detailed study of the relationship between political economy and the environment, biodiversity, and ecosystems is the core notion of environmental economics. Female academics in environmental economics will continue to make key contributions toward sustainable development through national and international policy development and implementation. It is one of the most important research agendas that we have to look forward toward for promoting greater accountability and national and international change for the betterment of our planet.

REFERENCES

Agrawal, A., Ostrom, E., 2006. Political science and conservation biology: a dialog of the deaf. Conserv. Biol. 20 (3), 681–682.

Akhtar, S., 2010a. In: Human Development and Redistribution of Wealth in Morocco Presented at the International Forum on Human Development, Agadir. Available at: http://web.worldbank.org/WBSITE/EXTERNAL/COUNTRIES/MENAEXT/0,contentMDK:22751185~menuPK:247615~pagePK:2865106~piPK:2865128~theSitePK:256299,00.html (accessed 02.11.14.).

Akhtar, S., 2010b. In: Poor Places, Thriving People: How the Middle East and North Africa Can Rise above Spatial Disparities, Opening Remarks Presented at the Flagship Launch. Dubai School of Government. Available at: http://web.worldbank.org/WBSITE/EXTERNAL/COUNTRIES/MENAEXT/0,contentMDK:22616119~menuPK:247603~pagePK:2865106~piPK:2865128~theSitePK:256299,00.html (accessed 02.11.14.).

Akhtar, S., 2011a. Unfolding Developments in the Arab World and its Implications for the Region. Available at: http://web.worldbank.org/WBSITE/EXTERNAL/COUNTRIES/MENAEXT/0,contentMDK:22865708~pagePK:146736~piPK:146830~theSitePK:256299,00.html (accessed 10.11.14.).

Akhtar, S., 2011b. Food Security in the Arab World: Price Volatility and Vulnerabilities and the World Bank Response. Available at: http://web.worldbank.org/WBSITE/EXTERNAL/NEWS/0,contentMDK:22864816~menuPK:34474~pagePK:34370~piPK:42770~theSitePK:4607,00.html (accessed 06.11.14.).

Akhtar, S., 2014a. Asia-Pacific Regional Partnerships for Connectivity and Sustainable Development. Available at: http://www.unescap.org/speeches/asia-pacific-regional-partnerships-connectivity-and-sustainable-development (accessed 10.11.14.).

Akhtar, S., 2014b. 2014 Policy Statement by the Executive Secretary Presented at the Seventieth Commission Session. Available at: http://www.unescap.org/speeches/2014-policy-statement-executive-secretary (accessed 11.11.14.).

Albelda, R., 1995. The impact of feminism in economics: beyond the pale? A discussion and survey results. J. Econ. Educ. 26 (3), 253–273.

Alisjahbana, A.S., Yusuf, A.A., 2004a. Assessing Indonesia's sustainable development: long-run trend, impact of the crisis, and adjustment during the recovery period. ASEAN Econ. Bull. 21 (3), 290–307.

Alisjahbana, A.S., Yusuf, A.A., 2004b. Green Accounting and Sustainable Development in Indonesia. Unpad Press, Bandung.

Allen, J.H., Duvander, J., Kubiszewski, I., Ostrom, E., 2012. Institutions for managing ecosystem services. Solutions 2 (6), 44–48 (November–December 2011).

Australian Research Council (ARC), 2015. Future Fellowships. Available at: http://www.arc.gov.au/ncgp/futurefel/future_default.htm (accessed 02.03.15.).

Bhattacharjee, S., Herriges, J.A., Kling, C.L., 2007. The status of women in environmental economics. Rev. Environ. Econ. Policy 1 (2), 212–227.

Brondizio, E.S., Ostrom, E., Young, O.R., 2012. Connectivity and the governance of multilevel socio-ecological systems: the role of social capital. In: Christophe, B., Pe'rez, R. (Eds.), Agro-Ressources Et Écosystèmes: Enjeux Sociétaux Et Pratiques Managériales. Publie' Avec Le Soutien, Du Conseil Regional Nord-Pas De Calais.

Callan, S., Thomas, J., 2012. Environmental Economics and Management: Theory, Policy, and Applications. South-Western Cengage Learning, Mason.

Cole, D., Ostrom, E., 2012. Property in Land and Other Resources. Lincoln Institute of Land Policy, Cambridge.

Costanza, R., Low, B., Ostrom, E., Wilson, J., 2001. Institutions, Ecosystems, and Sustainability. Lewis Publishers, Boca Raton.

DeFries, R.S., Ellis, E.C., Chapin III, F.S., Matson, P., Turner II, B., Agrawal, A., Crutzen, P., Field, C., Gleick, P., Kareiva, P., Lambin, E., Liverman, D., Ostrom, E., Sanchez, P., Syvitski, J., 2012. Planetary opportunities: a social contract for global change science to contribute to a sustainable future. Bioscience 62 (6), 603–606.

Dolado, J., Felgueroso, F., Almunia, M., 2012. Are men and women-economists evenly distributed across research fields? some new empirical evidence. SERIEs 3 (3), 367–393.

Dongier, P., Van Domelen, J., Ostrom, E., Ryan, A., Wakeman, W., Bebbington, A., Polski, M., 2008. Community-driven development. In: Poverty Reduction that Works (United Nations Development Programme). Earthscan, Oxon.

Duflo, E., 2014. Esther Duflo. Available at: http://economics.mit.edu/files/9094 (accessed October, 2014).

Duflo, E., Pande, R., 2005. Dams (No. W11711). National Bureau of Economic Research.

Duflo, E., Greenstone, M., Hanna, R., 2008a. Cooking stoves, indoor air pollution and respiratory health in rural Orissa. Econ. Polit. Wkly. 43 (32), 71e76.

Duflo, E., Greenstone, M., Hanna, R., 2008b. Indoor air pollution, health and economic well-being. SAPIENS Surv. Perspect. Integr. Environ. Soc. 1 (1), 6e16.

Duflo, E., Kremer, M., Robinson, J., 2009. Nudging farmers to use fertilizer: theory and experimental evidence from Kenya. Am. Econ. Rev. Am. Econ. Assoc. 101 (6), 2350–2390.

Duflo, E., Greenstone, M., Pande, R., Ryan, N., 2013. Truth-Telling By Third-Party Auditors And The Response Of Polluting Firms: Experimental Evidence From India (No. w19259). National Bureau of Economic Research.

Fiscal Policy Office Jakarta, 2008. Low Carbon Development Options for Indonesia. Available at: file://ntapprdfs01n01.rmit.internal/el1/e75321/850indonesialowcarbon.pdf (accessed 10.11.14.).

Folke, C., Jansson, Å., Rockström, J., Olsson, P., Carpenter, S.R., Chapin III, F.S., Crepin, A., Gretchen, D., Kjell, D., Ebbesson, J., Elmqvist, T., Galaz, V., Moberg, F., Nilsson, M., Osterblom, H., Ostrom, E., Persson, A., Peterson, G., Polasky, S., Steffen, W., Walker, B., Westley, F., 2011. Reconnecting to the biosphere. Ambio 40 (7), 719–738.

Ghate, R., Ghate, S., Ostrom, E., 2013. Can communities plan, grow and sustainably harvest from forests. Econ. Polit. Wkly. 48 (8), 59–67.

Gibson, C.C., Williams, J.T., Ostrom, E., 2005. Local enforcement and better forests. World Dev. 33 (2), 273–284.

Ginther, D.K., Kahn, S., 2004. Women in economics: moving up or falling off the academic career ladder? J. Econ. Perspect. 18, 193–214.

Ginther, D.K., Kahn, S., 2006. Does science promote women? Evidence from academia 1973–2001. In: Freeman, R., Goroff, D. (Eds.), Science and Engineering Careers in the United States: An Analysis of Markets and Employment. University of Chicago Press, Chicago.

Hamermesh, D.S., 2012. Six decades of top economics publishing: who and how? J. Econ. Lit. 51 (1), 162–172.

Hamilton, K., Lutz, E., 1996. Measuring sustainable development. In: Prepared for OECD Round Table on Sustainable Development, May 31, 2001. "Green National Accounts: Policy Uses and Empirical Experience". World Bank, Washington, DC. Environment Department Paper No. 39.

Hamilton, K., Ward, M., 1997. Greening the national accounts: valuation issues and policy uses. In: Uno, K., Bartelmus, P. (Eds.), Environmental Accounting in Theory and Practice. Kluwer, London.

Hayes, T., Ostrom, E., 2005. Conserving the world's forests: are protected areas the only way. Indiana Law Rev. 38, 595–618.

Henderson, H., 1984. The warp and the weft. The coming synthesis of eco-philosophy and eco-feminism. Dev. Seed Change (4), 64–70.

Hewitson, G., 2001. A Survey of Feminist Economics. Discussion Paper. La Trobe University, School of Economics and Commerce, Melbourne, Australia, pp. 1–52.

Indiana University, 2013. Elinor Ostrom, 1933–2012 Curriculum Vitae. Available at: http://www.indiana.edu/~workshop/people/lostromcv.htm (accessed 25.10.14.).

Indiana University, 2014. Elinor Ostrom, Nobel Laureate. Available at: http://elinorostrom.indiana.edu/ (accessed 24.10.14.).

Janssen, M.A., Anderies, J.M., Ostrom, E., 2007. Robustness of social-ecological systems to spatial and temporal variability. Soc. Nat. Resour. 20 (4), 307–322.

Janssen, M.A., Bousquet, F., Ostrom, E., 2011. A multi-method approach to study the governance of social-ecological systems. Nat. Sci. Soc. 19 (4), 382–394.

The Jakarta Post, 2010. Anny Ratnawati: Our Second Iron Woman. Available at: http://www.thejakartapost.com/news/2010/05/25/anny-ratnawati-our-second-iron-woman.html (accessed 10.11.14.).

Kahn, S., 1993. Gender differences in academic career paths of economists. Am. Econ. Rev. 83 (2), 52–56.

Karmacharya, M., Karna, B., Ostrom, E., 2007. Implications of leasehold and community forestry for poverty alleviation. In: Webb, E., Shivakoti, G. (Eds.), Decentralization, Forests and Rural Communities: Policy Outcomes in South and Southeast Asia. Sage Publications, New Delhi.

Khan, T., 2014. Kalimantan's biodiversity: developing accounting models to prevent its economic destruction. Acc. Aud. Acc. J. 27 (1), 150–182.

Lang, C., 2014. Are Norway's REDD Deals Reducing Deforestation? Available at: http://www.redd-monitor.org/2014/03/11/are-norways-redd-deals-reducing-Deforestation/ (accessed 30.10.14.).

Liu, L., Dietz, T., Carpenter, S., Folke, C., Alberti, M., Redman, C., Schneider, S., Ostrom, E., Pell, A., Lubchenco, E., Taylor, W., Ouyang, Z., Deadman, P., Kratz, T., Provencher, W., Liu, J., 2007. Complexity of coupled human and natural systems. Science 317 (5844), 1513–1516.

London School of Economics and Political Science, 2015. Dr Phoebe Koundouri. Available at: http://www.lse.ac.uk/GranthamInstitute/profile/phoebe-koundouri/ (accessed 10.01.15.).

Moran, E.F., Ostrom, E., Randolph, J.C., 2002. Ecological systems and multi-tier human organization. In: Kiel, D. (Ed.), Knowledge Management, Organizational Intelligence and Learning, and Complexity. Eoloss Publishers, Paris.

Moran, E.F., Ostrom, E., 2005. Preface and acknowledgements. In: Moran, E., Ostrom, E. (Eds.), Seeing the Forest and the Trees: Human-environment Interactions in Forest Ecosystems. MIT Press, Boston.

Mwangi, E., Ostrom, E., 2009a. Top-down solutions: looking up from east Africa's rangelands. Environ. Sci. Policy Sustain. Dev. 51 (1), 34–45.

Mwangi, E., Ostrom, E., 2009b. A century of institutions and ecology in east Africa's rangelands: linking institutional robustness with the ecological resilience of Kenya's Maasailand. In: Volker, B., Padmanabhan, M. (Eds.), Institutions and Sustainability. Springer, the Netherlands.

Nagendra, H., Ostrom, E., 2011. The challenge of forest diagnostics. Ecol. Soc. 16 (2), 20 [Online].

National Graduate Institute for Policy Studies, 2012. In: H.E. Dr. Armida S. Alisjahbana, Minister of National Development Planning and Head of National Development Planning Agency (BAPPENAS) of Indonesia: Special Symposium "Indonesian Economic Outlook. Available at: http://www.grips.ac.jp/en/oldevents/indonesian-economic-outlook/ (accessed 10.11.14.).

Norberg, J., Wilson, J., Walker, B., Ostrom, E., 2008. Diversity and resilience of social-ecological systems. In: Norberg, J., Cummings, G. (Eds.), Complexity Theory for a Sustainable Future. Columbia University Press, New York.

On The Commons, 2011. Elinor Ostrom's 8 Principles for Managing a Commons. Available at: http://www.onthecommons.org/magazine/elinor-ostroms-8-principles-managing-commmons (accessed 24.10.14.).

Ostrom, E., 1992. Crafting Institutions for Self-governing Irrigation Systems, International Center for Self-governance. ICS PRESS, Michigan.

Ostrom, E., 2001. Reformulating the commons. In: Burger, J., Ostrom, E., Norgaard, R., Policansky, D., Goldstein, B. (Eds.), Protecting the Commons: A Framework for Resource Management in the Americas. Island Press, Washington DC.

Ostrom, E., 2007. A general framework for analysing sustainability of socio-ecological systems. Science 274 (325), 419–422.

Ostrom, E., 2008. Institutions and the Environment. Econ. Aff. 28 (3), 24–31.

Ostrom, E., 2009a. The contribution of community institutions to environmental problem-solving. In: Breton, E., Brosio, G., Dalmazzone, S., Garrone, G. (Eds.), Governing the Environment: Salient Institutional Issues. Edward Elgar Publishing Ltd, Massachusetts.

Ostrom, E., 2009b. A Polycentric Approach for Coping with Climate Change. Policy Research Working Paper No. 5095. Background Paper to the 2010 World Development Report. The World Bank, Washington, DC.

Ostrom, E., 2010a. Analysing collective action. Agric. Econ. 41 (1), 155–166.

Ostrom, E., 2010b. Polycentric systems for coping with collective action and global environmental change. Global Environ. Change 20 (4), 550–557.

Ostrom, E., 2012. Nested externalities and polycentric institutions: must we wait for global solutions to climate change before taking actions at other scales? Econ. Theory 49 (2), 353–369.

Ostrom, V., Ostrom, E., 1972. Legal and political conditions of water resource development. Land Econ. 48 (1), 1–14.

Ostrom, E., Keohane, R., 1994. Local Commons and Global Interdependence, Heterogeneity and Cooperation in Two Domains. Sage Publications, London.

Ostrom, E., Janssen, M.A., 2005. Multi-level governance and resilience of social-ecological systems. In: Spoor, M. (Ed.), Globalisation, Poverty and Conflict. Springer, Rotterdam.

Ostrom, E., Nagendra, H., 2006. Insights on linking forests, trees, and people from the air, on the ground, and in the laboratory. Proc. Natl. Acad. Sci. 103 (51), 19224–19231.

Ostrom, E., Cox, M., 2010. Moving beyond panaceas: a multi-tiered diagnostic approach for social-ecological analysis. Environ. Conserv. 37 (04), 451–463.

Ostrom, E., Schroeder, L., Wynne, S., 1993. Institutional Incentives and Sustainable Development: Infrastructure Policies in Perspective. Westview Press, New York.

Ostrom, E., Gardner, R., Walker, J., 1994. Rules, Games, and Common-pool Resources. University Of Michigan Press, Ann Arbor.

Ostrom, E., Lam, W., Pradhan, P., 2011. Improving Irrigation in Asia: Sustainable Performance of an Innovative Intervention in Nepal. Edward Elgar Publishing, Northampton.

Ostrom, E., Gatzweiler, F.W., Judis, R., Hagedorn, K., 2002. The study of human-ecological systems in the laboratory. In: ACE Seminar on Sustainable Agriculture in Central and Eastern European Countries: The Environmental Effects of Transition and Needs for Change, Nitra, Slovakia, September 2002. Shaker Verlag GMBH, pp. 99–113.

Ostrom, E., Chang, C., Pennington, M., Tarko, V., 2012. Future of the Commons: Beyond Market Failure and Government Regulations. Institute of Economic Affairs, London.

Poteete, A.R., Ostrom, E., 2008. Fifteen years of empirical research on collective action in natural resource management: struggling to build large-N databases based on qualitative research. World Dev. 36 (1), 176—195.

Randall, G.W., Mulla, D.J., 2001. Nitrate nitrogen in surface waters as influenced by climatic conditions and agricultural practices. J. Environ. Qual. 30 (2), 337—344.

Resosudarmo, B., Alisjahbana, A., Nurdianto, D.A., 2011. Energy security in Indonesia. In: Anceschi, L., Symons, J. (Eds.), Energy Security in the Era of Climate Change: The Asia-Pacific Experience. Palgrave Macmillan, New York.

Schlager, E., Ostrom, E., 1992. Property-rights regimes and natural resources: a conceptual analysis. Land Econ. 68 (3), 249—262.

Scott, E., Siegfried, J., 2012. American economic association universal academic questionnaire summary statistics. Am. Econ. Rev. 102 (3), 631—634.

Shivakoti, G.P., Ostrom, E., 2002. Improving Irrigation Governance and Management in Nepal. ICS Press, Washington, DC.

Shivakoti, G., Vermillion, D., Lam, W., Ostrom, E., Pradhan, U., Yoder, R., 2005. Asian Irrigation in Transition: Responding to Challenges. SAGE Publications, New Delhi.

Stockhammer, E., Hochreiter, H., Obermayr, B., Steiner, K., 1997. The Index of Sustainable Economic Welfare (ISEW) as an alternative to GDP in measuring economic welfare. The results of the Austrian (revised) ISEW calculation 1955—1992. Ecol. Econ. 21 (1), 19—34.

Stonich, S., Stern, P.C., Dolsak, N., Dietz, T., Ostrom, E., Weber, E.U., 2002. The drama of the commons, introduction. In: Stonich, S., Stern, P.C., Dolsak, N., Dietz, T., Ostrom, E., Weber, E.U. (Eds.), The Drama of the Commons. National Academies Press, Washington, DC.

UN-REDD, 2012. Key Results and Achievements-programs in Indonesia. Available at: http://www.un-redd.org/key_results_achievements_indonesia/tabid/106623/default.aspx 9 (accessed 30.10.14.).

United Nations ESCAP, 2014. Biography of Executive Secretary Shamshad Akhtar. Available at: http://www.unescap.org/executive-secretary/biography (accessed 10.11.14.).

University of Queensland, 2015. "Nauges, Celine." Available at: http://www.uq.edu.au/economics/nauges-celine (accessed 10.2.15.).

Van Den Bergh, J.C., 2009. The GDP paradox. J. Econ. Psychol. 30 (2), 117—135.

Vermillion, D., Ostrom, E., Yoder, R., 2005. The future of irrigated agriculture in Asia: what the twenty-first century will require of policies, institutions and governance. In: Shivakoti, G., Vermillion, D., Lam, W., Ostrom, E., Pradhan, U., Yoder, R. (Eds.), Asian Irrigation in Transition. Sage Publications India Pvt Ltd, New Delhi.

The World Bank, 2014. Northwest Mountainous and Forestry Areas Development Project. Available at: http://www.worldbank.org/projects/p072317/northwest-mountainous-forestry-areas-development-project?lang=en (accessed 12.11.14.).

York, M., Janssen, M., Ostrom, E., 2005. 15. Incentives affecting land use decisions of nonindustrial private forest landowners. In: Dauvergne, P. (Ed.), Handbook of Global Environmental Politics. Edward Elgar Publishing Limited, Massachusetts.

Young, O.R., Berkhout, F., Gallopin, G.C., Janssen, M.A., Ostrom, E., Van Der Leeuw, S., 2006. The globalization of socio-ecological systems: an agenda for scientific research. Global Environ. Change 16 (3), 304—316.

Zhu, Z.L., Chen, D.L., 2002. Nitrogen fertilizer use in China—contributions to food production, impacts on the environment and best management strategies. Nutr. Cycl. Agro Ecosyst. 63 (2—3), 117—127.

SECTION 3

Environmental/Sustainable Finance

CHAPTER 12

Does National Culture Affect Attitudes toward Environment Friendly Practices?

Kartick Gupta, Ronald McIver
Centre for Applied Financial Studies (CAFS), School of Commerce, UniSA Business School, University of South Australia, Adelaide, SA, Australia

Contents

12.1 INTRODUCTION

Businesses activity creates incomes, growth, and, as an undesirable by-product, contributes to the many environmental problems that society faces. Recognition of its environmental impact suggests that business has a critical role to play in influencing and/or implementing changes to address these environmental issues (Senge, 2007). Not surprisingly, corporate environmental performance has been the topic matter of considerable scholarly interest over many decades (e.g., Russo and Fouts, 1997; Bansal and Gao, 2006; Ramiah et al., 2013). Regulatory standards and stakeholder expectations on the reporting of environmental performance have also evolved over this period (Franklin, 2008; Moosa and Ramiah, 2014), consistent with concerns over environmental health and safety in the 1970s, the rise in the 1990s of an increasingly influential environmental movement, and concerns over climate-related risks in the 2010s. Thus, it is now accepted that corporate environmental performance impacts financial analysts', investors', regulators', and other stakeholders' judgments regarding the value of individual businesses (Murninghan and Grant, 2013).

While stakeholder expectations have placed pressure on companies to improve their environmental performance, the interpretation of requirements and implementation of actions to achieve this is likely to differ between countries due to differences in culture (Park et al., 2007; Ho et al., 2012). This reflects culture's key role as a differentiator in cross-cultural ethics (Hofstede, 1985; Alas, 2006; Scholtens and Dam, 2007; Donleavy et al., 2008; Tan and Chow, 2009; Weaver, 2001). This includes ethical attitudes (Franke and Nadler, 2008), decision making (Srnka, 2004), judgments (Whipple and Swords, 1992), perceptions (Vitell and Paolillo, 2004), sensitivity (Simga-Mugan et al., 2005), and value systems (Ford et al., 2005).

In this chapter we study empirically the influence of national culture on firm-level attitudes toward environment friendly practices. We hypothesize that cross-country cultural differences are significant determinants in explaining differences in firm-level environmental performance. For example, people in countries with a long-term orientation will be more inclined to engage in environment friendly practices. Similarly, people in countries with high uncertainty avoidance will be more likely to take steps to minimize environmental spillovers/pollution from the firm's operations that may result in high fines or penalties. On average firms in countries displaying these cultural features will be expected to display higher levels of environmental performance.

To test our hypotheses we base measures of six cultural dimension factors on Hofstede et al. (2010).[1] Firm-level environment performance indices are constructed from the Thomson Reuters' Asset4 data set environmental indicators for 45 countries covering the period from 2002 to 2012. Following presentation of descriptive statistics, we employ a panel data regression model, with both year and industry fixed effects, to examine the relationship between cultural dimension factors and firm-level environmental performance.

We find statistically significant negative relationships between corporate environmental performance and the Power Distance and Masculinity versus Femininity dimensions of culture. This suggests that firms in societies that are more accepting of hierarchical structures where power is distributed unequally will on average, display lower levels of environmental performance. This holds in a similar way for firms that are located in societies that are more competitive, with a focus on achievement, assertiveness, and material reward. We also identify statistically significant positive relationships between corporate environmental performance and the Individualism versus Collectivism, Uncertainty Avoidance, Long-Term Orientation versus Short-Term Normative Orientation, and Indulgence versus Restraint dimensions of culture. Firms, on average, display higher levels of environmental performance in societies where the individual is expected to focus on their own (and immediate families) welfare, rather than that of the broader community; belief and behavior are more strongly guided by societal codes and expectations, leading to greater conformity in individual behavior and ideas; saving and investment in

---

[1] Hofstede's original four cultural dimensions are Power Distance, Masculinity versus Femininity, Individualism versus Collectivism, and Uncertainty Avoidance (Hofstede, 1980, 2001). These were augmented by a fifth dimension, Long-Term Orientation versus Short-Term Normative Orientation, presented in Hofstede and Hofstede (2005). A sixth dimension, Indulgence versus Restraint, is introduced in Hofstede et al. (2010). Each of the cultural dimensions, and their interpretation, are discussed in Section 12.2.

human capital (e.g., education) are prioritized to ensure preparedness for the future; and strict social norms restrain the free gratification of individual needs.

The rest of the chapter is as follows. We next provide a brief review of literature and develop our hypotheses. Following this, we discuss our data and methodology, including sample statistics and correlations between variables. Discussion of the regression results follows. We then conclude the chapter.

12.2 LITERATURE REVIEW

In the following we develop a brief overview of insights from a range of the extant literature on culture and environmental performance in order to develop testable hypotheses.

McCarty and Shrum (2001) propose that individuals' beliefs about environmental matters, and the likelihood that they will behave proenvironmentally, reflect more fundamental beliefs regarding their interaction with the world and other people. These fundamental beliefs, or value orientations, are held by individuals and their support adaptation to physical and social environments (Kluckhohn, 1951, cited in McCarty and Shrum, 2001, p. 94). National culture distinguishes members of one group of humans from another (Hofstede, 1980), and so is one of the fundamental determinants of differences between individuals from different cultural backgrounds. This is attributed mostly to the differences in shared value and belief systems, traditions, customs, etc., which act to constrain the range of available individual behaviors (Poortinga, 1992). This is not only true for individuals but also for organizations (Hofstede, 1985).

What constitutes environmentally responsible behavior may vary significantly across cultures. A mechanism potentially linking a society's culture and the environment is the impact of culture on normative ethical beliefs regarding what is morally correct behavior (Cohen and Nelson, 1994). Differences in the sets of beliefs, values, and morals, embedded in the national culture provide norms to individuals as to what are acceptable and unacceptable behaviors (Vitell et al., 1993; Ehrlich and Ehrlich, 2002). Members of societies evidencing high levels of social/human capital will balance their rights with their collective responsibilities; for example, stewardship of natural resources (Etzioni, 1995). Culturally embedded social and human capitals may therefore be seen as prerequisites for environmental (natural capital) improvement (Pretty and Ward, 2001). Therefore, not surprisingly, significant cross-cultural variability in people's attitudes about nature, and its conservation, are observed in practice (e.g., Kellert, 1996).

Perceptions of what acceptable business conduct is within a given society also reflect normative ethical beliefs which, in turn, form the basis for common business practice. Additionally, government regulation of business activity is impacted by cultural factors. These influence sovereign states' policies, the structures of public and private agencies that act to operationalize the policies, and the behaviors of the public officials responsible for implementing the policies (Gorham, 1997; Atzenhoffer, 2012).

Some prior research has focused on particular aspects of culture and their likely relationship to environmental outcomes. Cernea (1993), Elgin (1994), and Ward (1998) argue for the relationship between a society's environmental conditions and the level of connectedness among its members. The higher the level of connectedness, the more likely individuals are to care about public goods, including the environment. Kinsley (1995) claims a significant relationship between religious spirituality and the ecological condition of a society. Greater spirituality facilitates cooperation within society, reducing the transactions costs associated with management of public resources. In contrast, distrust reduces such cooperative efforts, suggesting lower levels of outcomes in the management of the environment and other public resources (Baland and Platteau, 1998). Given that much of this research is theoretical or anecdotal suggests a need to gain empirical support for or against these claims.

Hofstede's (1980, 2001) dimensions of national cultural values are the de facto standard in cross-cultural social science (including business and management) empirical research. Hofstede (2001) presents evidence validating his original four measures by presenting over 400 significant correlations between their index scores and data from other sources. Additionally, a wide range of studies support the association between Hofstede's cultural dimensions and corporate social performance (Shepherd et al., 2013).[2] Hofstede et al. (2010) six cultural dimensions (and the abbreviation for each of their respective measures) are Power Distance (PDI); Individualism versus Collectivism (IDV); Masculinity versus Femininity (MAS); Uncertainty Avoidance (UAI); Long-Term Orientation versus Short-Term Normative Orientation (LTO); and Indulgence versus Restraint (IVR). These cultural dimensions and their interpretation are presented in Table 12.1.

Although use of the Hofstede measures is common in the empirical literature that examines the relationship between cultural factors and environmental performance, this research has provided conflicting results. In some cases this may be due to limitations imposed by data choice and/or availability. For example, Halkos and Tzeremes (2013) focus on pollution emissions, and so capture only this limited aspect of

---

[2] While broadly used and accepted as proxies of national culture (see Minkov and Hofstede, 2011), Hofstede's original four cultural dimensions (and the two later additions) have been subject to both extensive analysis and criticism. These include (e.g., McSweeney, 2002; Smith et al., 2002) representativeness in sample composition; adequacy of sample sizes; and the methodological basis for Hofstede's development of the cultural dimensions. McSweeney (2002), in particular, questions the assumptions underlying Hofstede's claim of representativeness of the survey results from which the original four cultural dimensions were drawn. First, the cultural dimensions are based on two surveys capturing approximately 117,000 IBM staff, all sharing an "IBM culture." Second, as a number of country subsamples of less than 200 were used, the statistical validity of any inferences drawn from such samples may be low. Additionally, Smith et al. (2002) reject use of Hofstede's fifth dimension of culture, as the data on which it is based are derived from student survey results, and for a small sample of countries (23). This suggests a lack of comparability to the four original cultural dimensions. However, this earlier measure was replaced by one based on data from the World Values Survey, covering 93 countries and regions (Minkov and Hofstede, 2011).

Table 12.1 Hofstede's Six Dimensions of National Culture

Dimension	Interpretation
Power Distance Index (PDI)	This dimension expresses the degree to which the less powerful members of a society accept and expect that power is distributed unequally. The fundamental issue here is how a society handles inequalities among people. People in societies exhibiting a large degree of power distance accept a hierarchical order in which everybody has a place and which needs no further justification. In societies with low power distance, people strive to equalize the distribution of power and demand justification for inequalities of power.
Individualism versus Collectivism (IDV)	The high side of this dimension, called individualism, can be defined as a preference for a loosely knit social framework in which individuals are expected to take care of only themselves and their immediate families. Its opposite, collectivism, represents a preference for a tightly knit framework in society in which individuals can expect their relatives or members of a particular in-group to look after them in exchange for unquestioning loyalty. A society's position on this dimension is reflected in whether people's self-image is defined in terms of "I" or "we."
Masculinity versus Femininity (MAS)	The masculinity side of this dimension represents a preference in society for achievement, heroism, assertiveness, and material rewards for success. Society at large is more competitive. Its opposite, femininity, stands for a preference for cooperation, modesty, caring for the weak, and quality of life. Society at large is more consensus oriented. In the business context, Masculinity versus Femininity is sometimes also related to as "tough versus gender" cultures.
Uncertainty Avoidance Index (UAI)	The Uncertainty Avoidance dimension expresses the degree to which the members of a society feel uncomfortable with uncertainty and ambiguity. The fundamental issue here is how a society deals with the fact that the future can never be known: should we try to control the future or just let it happen? Countries exhibiting strong UAI maintain rigid codes of belief and behavior and are intolerant of unorthodox behavior and ideas. Weak UAI societies maintain a more relaxed attitude in which practice counts more than principles.
Long-Term Orientation versus Short-Term Normative Orientation (LTO)	Every society has to maintain some links with its own past while dealing with the challenges of the present and the future. Societies prioritize these two existential goals differently.

Continued

Table 12.1 Hofstede's Six Dimensions of National Culture—cont'd

Dimension	Interpretation
	Societies which score low on this dimension, for example, prefer to maintain time-honored traditions and norms while viewing societal change with suspicion. Those with a culture which scores high, on the other hand, take a more pragmatic approach: they encourage thrift and efforts in modern education as a way to prepare for the future. In the business context and in our country comparison tool this dimension is related to as "(short term) normative versus (long term) pragmatic" (PRA). In the academic environment the terminology monumentalism versus flexhumility is sometimes also used.
Indulgence versus Restraint (IVR)	Indulgence stands for a society that allows relatively free gratification of basic and natural human drives related to enjoying life and having fun. Restraint stands for a society that suppresses gratification of needs and regulates it by means of strict social norms.

Source: The Hofstede Centre, National Culture Dimensions. http://geert-hofstede.com/dimensions.html (accessed 19.12.14.).

environmental performance. In contrast, Husted (2005) and Park et al. (2007) evaluate the impact of cultural factors on environmental sustainability. Finally, Ringov and Zollo (2007) examine the role of cultural factors in determining corporate social and environmental performance.

Halkos and Tzeremes (2013) consider the impact of Hofstede's PDI, MAS, IDV, and UAI cultural dimensions on ecoefficiency. Ecoefficiency is defined in terms of countries' per capita emissions of carbon dioxide (CO_2) and sulfur dioxide (SO_2) relative to output measured as gross domestic product (GDP) per capita. Data envelopment analysis is applied to data for the year 2000. The empirical results indicate that countries that score highly on IDV and PDI achieve higher ecoefficiency levels. Countries with high levels of UAI achieve lower ecoefficiency levels.

Husted (2005) and Park et al. (2007) also use Hofstede's PDI, MAS, IDV, and UAI cultural dimensions to examine the role of cultural factors in environmental sustainability. Environmental sustainability is measured in terms of the World Economic Forum's Global Leaders of Tomorrow Environment Task Force Environmental Sustainability Index (ESI) (World Economic Forum, 2001). Husted (2005), who controls for both economic development and population growth, identifies that higher levels of IDV increase a country's social and institutional capacities for sustainability. Higher levels of PDI and MAS are associated with reductions in a country's social and institutional capacities for sustainability. In contrast, Park et al. (2007), who control for country levels of GDP and education, find that only two of Hofstede's dimensions of national culture consistently impact environmental performance. These are PDI and MAS, with higher levels of each reducing a country's ESI score. However, as Husted (2005) and Park et al., (2007) studies use country-level measures of sustainability, they are unable to provide insight into the impact of national cultural factors at the individual firm level, or to control for differences in environmental performance across industry sectors.

Ringov and Zollo (2007) use the Innovest Group's Intangible Value Assessment (IVA) score of firm social and environmental performance, allowing consideration of firm-level effects of national culture on IVA. In addition to Hofstede's PDI, MAS, IDV, and UAI cultural dimensions, Ringov and Zollo (2007) use the equivalent GLOBE cultural dimensions (House et al., 2004) for each. These are PDI, Collectivism, Gender Egalitarianism, and UAI, respectively. Controlling for firm size, growth, profitability, Tobin's Q, leverage, productivity, multinationality, and industry, they find that higher levels of PDI and MAS (lower levels of Gender Egalitarianism), are associated with poorer scores on social and environmental performance. However, Ringov and Zollo's more general focus on corporate social and environmental performance reduces comparability with studies focused on environmental sustainability and/or performance.

We suggest that additional empirical examination of the relationship between cultural dimensions and environmental outcomes. This reflects a number of considerations. First is the lack of agreement in the literature about the significance of a broad range of cultural factors' roles in influencing environmental outcomes. Second are limitations in the data used in a number of studies, especially where country-level rather than firm-level data on environmental performance has been used. Finally, is the ability to use the expanded set of six dimensions of national culture (Hofstede et al., 2010), rather than Hofstede's four original dimensions (Hofstede, 1980, 2001).

As discussed above, national culture influences the value and belief systems shared within each society. In particular, it impacts on normative ethical beliefs regarding what is morally correct behavior, and so on, collective attitudes and efforts to supporting the environment, and environmental outcomes. For example, a large number of individuals accrue small benefits from environmental improvements, such as pollution abatement, the sum total of which is argued to outweigh any costs incurred by a small number of polluters; the collective benefit from pollution reduction is positive. We suggest that as environmental sustainability and performance are usually collective activities, they may be seen to be governed by cultural factors. Higher levels of those cultural dimensions that support collective activity at the level of the society will be associated with improved environmental performance outcomes at the level of the firm.

Based on the cultural dimensions identified in Table 12.1, and the potential relationship between each factor and the propensity to engage in collective behavior, we develop the following hypotheses:

1. Firms based in cultures with a low PDI measure will evidence higher levels of environmental performance.

Societies where there is a lower degree of PDI are less likely to accept highly hierarchical structures in which everybody has a place and will be more likely to engage in collective environmentally focused activities. A negative coefficient is expected on this cultural dimension. As power distance increases, environmental performance will, on average, be lower.

2. Firms based in cultures that are highly collectivist will exhibit higher levels of environmental performance.

   Societies with a preference for a tightly knit framework, favoring collectivism, are more likely to engage in environmentally sustainable activities. The coefficient on the IDV cultural dimension is expected to be negative. As the degree of individualism in the society increases, environmental performance will, on average, fall.

3. Firms based in cultures that are more feminine will exhibit higher levels of environmental performance.

   Societies that are defined more strongly as feminine, will display a greater preference for cooperation. This suggests a greater likelihood of engaging in collectivist activity to enhance the quality of life, with the coefficient on the MAS cultural dimension being expected to be negative. As the degree of masculinity in the society increases, environmental performance will, on average, decrease.

4. Firms based in cultures that display a higher preference for UAI will exhibit higher levels of environmental performance.

   A preference for UAI suggests that members of a society feel uncomfortable with uncertainty and ambiguity. They are more likely to engage in behaviors that reduce the potential for future loss. For example, they are more likely to engage in pollution abatement behavior to avoid potential payment of compensation for damage caused by unabated pollution. The coefficient on this cultural dimension is expected to be positive. As the measured preference for uncertainty avoidance increases, environmental performance will, on average, increase.

5. Firms based in cultures with a bias toward long-term perspectives will display higher levels of environmental performance.

   Societies that display a greater long-term orientation are more likely to engage in collectivist behavior, including that providing long-term benefits in terms of environmental outcomes. This reflects that benefits gained from proenvironmental behaviors (e.g., an improved environment) will potentially, if at all, only be experienced in long term by the individual. The payoff may, therefore, be seen more as a collective payoff. A positive coefficient is expected on the LTO cultural dimension. As the measured preference toward long-term actions/perspectives increases, environmental performance will, on average, increase.

6. Firms located in cultures that display lower levels of tolerance for individual indulgence (i.e., higher levels of constraint) will display higher levels of environmental performance.

Societies that display higher levels of constraint are more likely to engage in collectivist behaviors. The coefficient on the IVR cultural dimension is expected to be positive. As the measured level of intolerance toward indulgence increases, environmental performance will, on average, increase.

12.3 METHODOLOGY AND DATA

The environmental performance data are derived from Thomson Reuters' Asset4 data set, which is extensively used within the global investment community in analyzing firm-level performance on environmental, social, and governance matters. An advantage of the Asset4 data is that it is constructed from publically available primary data, both sourced from the company and verified for quality and accuracy through extensive use of stock exchange filings, media reports, etc.

Table 12.2 provides the 38 questions of which the Asset4 environmental indicator set is comprised and the percentage of firms responding in the affirmative to each. The Asset4 data set provides binary information on each company's response to each of the 38 questions. As can be seen from Table 12.2, there is considerable variation in the percentage of firms that meet the requirements of the environmental performance outcome underlying each of the questions. For example, 70% of companies respond in the positive to having a policy for reducing the use of natural resources, and also for describing the processes in place through which their resource efficiency policies are implemented. However, only 5% of companies respond in the positive to commenting on their progress in achieving previously set (environmental) objectives.

To construct our environmental performance score we have retrieved information on each company's binary-coded response to each of the questions, and score 1 where the company affirms that it meets the requirement specified in the question and 0 otherwise. Consistent with a range of the extant literature (e.g., Aggarwal and Goodell, 2009), we next add all scores for each company and divide through by the total number of questions (38). By construction this gives an environmental performance index taking a value between 0 and 1, with 0 indicating no compliance in any area, and a 1 indicating 100% compliance with all 38 environmental performance requirements. Our data cover the 2002—2012 time period.

Table 12.3 presents index scores on each of the national cultural dimensions for which the Hofstede Centre collects data. Each index is constructed so as to have a range of 100 points, and scaled so as to generally achieve values between 0 and 100 (see Hofstede et al., 2008).

A brief examination of Table 12.3 highlights the considerable diversity, at the national level, in each of the cultural dimensions. It also shows the degree to which any particular country is consistent across each of the cultural dimensions in terms of its likely propensity

Table 12.2 Asset4 Environmental Indicators and Percentage of Firm's Complying

	Description	Percentage
1	Does the company have a policy for reducing the use of natural resources?	70
2	Does the company describe the implementation of its resource efficiency policy through the processes in place?	70
3	Does the company have a policy for reducing environmental emissions or its impacts on biodiversity?	56
4	Does the company report on initiatives to increase its energy efficiency overall?	51
5	Does the company describe the implementation of its emission reduction policy through the processes in place?	51
6	Does the company have a policy for maintaining an environmental management system?	50
7	Does the company report on initiatives to recycle, reduce, reuse, substitute, treat, or phase out total waste, hazardous waste, or wastewater?	47
8	Does the company monitor its resource efficiency performance?	47
9	Does the company describe the implementation of its resource efficiency policy through a public commitment from a senior management or board member?	44
10	Does the company have a policy to lessen the environmental impact of its supply chain?	41
11	Does the company report on partnerships or initiatives with specialized NGOs, industry organizations, governmental or supragovernmental organizations that focus on improving environmental issues?	37
12	Does the company have an environmental product innovation policy (ecodesign, life cycle assessment, dematerialization)?	37
13	Does the company report about take-back procedures and recycling programs to reduce the potential risks of products entering the environment? Or does the company report about product features and applications or services that will promote responsible, efficient, cost-effective, and environmentally preferable use?	32
14	Does the company use environmental criteria (ISO 14000, energy consumption, etc.) in the selection process of its suppliers or sourcing partners?	32
15	Is the company aware that climate change can represent commercial risks and/or opportunities?	31
16	Does the company report on initiatives to reduce the environmental impact of transportation of its products or its staff?	31
17	Does the company report on initiatives to use renewable energy sources?	29
18	Does the company set specific objectives to be achieved on emission reduction?	27
19	Does the company monitor its emission reduction performance?	25

Table 12.2 Asset4 Environmental Indicators and Percentage of Firm's Complying—cont'd

	Description	Percentage
20	Does the company report on initiatives to protect, restore, or reduce its impact on native ecosystems and species, biodiversity, protected and sensitive areas?	25
21	Does the company describe the implementation of its environmental product innovation policy?	24
22	Does the company describe, claim to have or mention the processes it uses to accomplish environmental product innovation?	24
23	Does the company report on its environmental expenditures or does the company report to make proactive environmental investments to reduce future risks or increase future opportunities?	24
24	Does the company report on initiatives to reuse or recycle water? Or does the company report on initiatives to reduce the amount of water used?	23
25	Does the company report on at least one product line or service that is designed to have positive effects on the environment or which is environmentally labeled and marketed?	23
26	Does the company set specific objectives to be achieved on resource efficiency?	21
27	Does the company show an initiative to reduce, reuse, recycle, substitute, phased out, or compensate CO_2 equivalents in the production process?	21
28	Does the company report or provide information on company-generated initiatives to restore the environment?	20
29	Does the company describe the implementation of its emission reduction policy through a public commitment from a senior management or board member?	20
30	Does the company have environmentally friendly or green sites or offices?	18
31	Does the company report on the concentration of production locations in order to limit the environmental impact during the production process? Or does the company report on its participation in any emissions trading initiative? Or does the company report on new production techniques to improve the global environmental impact (all emissions) during the production process?	18
32	Has the company received product awards with respect to environmental responsibility? Or does the company use product labels (e.g., Forestry Stewardship Council (FSC), Energy Star, Marine Stewardship Council (MSC)) indicating the environmental responsibility of its products?	13
33	Does the company report on specific products which are designed for reuse, recycling, or the reduction of environmental impacts?	11
34	Does the company develop products or technologies for use in the clean, renewable energy (such as wind, solar, hydro- and geothermal, and biomass power)?	10

Continued

Table 12.2 Asset4 Environmental Indicators and Percentage of Firm's Complying—cont'd

	Description	Percentage
35	Does the company report on initiatives to reduce, substitute, or phase out ozone-depleting (CFC-11 equivalents, chlorofluorocarbon) substances?	9
36	Does the company set specific objectives to be achieved on environmental product innovation?	8
37	Does the company report or show to be ready to end a partnership with a sourcing partner, if environmental criteria are not met?	8
38	Does the company comment on the results of previously set objectives?	5

to engage in collectivist behaviors. For example, Australia scores relatively high on IDV and low in terms of LTO. This suggests, in line with our hypotheses, a reduced likelihood of collective action on the environment. However, a relatively low score on PDI and relatively high score on IVR suggest that the propensity to engage in collectivist behaviors may be relatively high.

Tables 12.4 and 12.5 present summary statistics on the environmental index data, the dependent variable in the empirical modeling that follows. Table 12.4 presents statistics on environmental index scores by country, and the number of observations on which these country-level statistics are based, highlighting that the sample is comprised of 23,914 firm-year observations, with comprehensive coverage of firms in high-, middle-, and low-income countries. This includes firms located in both emerging market and transition economies (e.g., the BRIC economies). However, the sample is weighted toward firms located in higher income, western economies, consistent with expectations regarding the impact of stakeholder pressure to report on corporate social and environmental outcomes.

The mean scores in Table 12.4 suggest that sample firms in Morocco, Peru, Singapore, and China score relatively poorly in terms of responses to questions related to the Asset4 environmental performance indicators, while those in Spain, Finland, France, and Hungary score relatively high. Table 12.5 provides annual summary statistics across the data set, highlighting the trend increase in the propensity of firms to report positively regarding additional aspects of their environmental performance. Also highlighted in Tables 12.4 and 12.5 is the considerable variability in environmental index scores, both by year and by country.

In addition to the six dimensions of national culture used as independent variables in the empirical modeling in this chapter, we use a range of control variables. These include price-to-book value (PTBV); the percentage of closely held shares (Blockholding); the volatility of returns over the previous year (Vol.); illiquidity, measured as the daily ratio of absolute stock return to dollar volume (Illiquidity); the size of the firm measured in terms of the log value of its total assets (Asset); and the debt-equity ratio (Leverage).

Table 12.3 Sample Distribution

Country	PDI	IDV	MAS	UAI	LTO	IVR
Australia	36	90	61	51	21	71
Austria	11	55	79	70	60	63
Belgium	65	75	54	94	82	57
Brazil	69	38	49	76	44	59
Canada	39	80	52	48	36	68
Chile	63	23	28	86	31	68
China	80	20	66	30	87	24
Colombia	67	13	64	80	13	83
The Czech Republic	57	58	57	74	70	29
Denmark	18	74	16	23	35	70
Finland	33	63	26	59	38	57
France	68	71	43	86	63	48
Germany	35	67	66	65	83	40
Greece	60	35	57	112	45	50
Hong Kong	68	25	57	29	61	17
Hungary	46	80	88	82	58	31
India	77	48	56	40	51	26
Indonesia	78	14	46	48	62	38
Ireland	28	70	68	35	24	65
Israel	13	54	47	81	38	
Italy	50	76	70	75	61	30
Japan	54	46	95	92	88	42
Luxembourg	40	60	50	70	64	56
Malaysia	104	26	50	36	41	57
Mexico	81	30	69	82	24	97
Morocco	70	46	53	68	14	25
The Netherlands	38	80	14	53	67	68
New Zealand	22	79	58	49	33	75
Norway	31	69	8	50	35	55
Panama	95	11	44	86		
Peru	64	16	42	87	25	46
The Philippines	94	32	64	44	27	42
Poland	68	60	64	93	38	29
Portugal	63	27	31	104	28	33
Russia	93	39	36	95	81	20
Singapore	74	20	48	8	72	46
South Korea	60	18	39	85	100	29
Spain	57	51	42	86	48	44
Sweden	31	71	5	29	53	78
Switzerland	34	68	70	58	74	66
Taiwan	58	17	45	69	93	49
Thailand	64	20	34	64	32	45
Turkey	66	37	45	85	46	49
The United Kingdom	35	89	66	35	51	69
The United States	40	91	62	46	26	68

Table 12.4 Country-wise Environment Index

Country	Mean	Standard Deviation	Median	Observations
Australia	0.231	0.215	0.184	1566
Austria	0.315	0.219	0.342	143
Belgium	0.24	0.235	0.184	198
Brazil	0.389	0.227	0.447	167
Canada	0.208	0.202	0.132	1428
Chile	0.297	0.191	0.263	78
China	0.188	0.158	0.158	272
Colombia	0.227	0.206	0.158	25
The Czech Republic	0.412	0.143	0.368	14
Denmark	0.329	0.245	0.303	192
Finland	0.433	0.251	0.447	231
France	0.458	0.253	0.474	724
Germany	0.393	0.273	0.355	522
Greece	0.259	0.222	0.211	105
Hong Kong	0.239	0.229	0.184	370
Hungary	0.544	0.159	0.579	15
India	0.407	0.253	0.395	315
Indonesia	0.271	0.189	0.316	25
Ireland	0.251	0.252	0.158	223
Israel	0.236	0.213	0.171	48
Italy	0.297	0.297	0.237	385
Japan	0.396	0.284	0.395	3481
Luxembourg	0.371	0.264	0.316	37
Malaysia	0.271	0.178	0.237	144
Mexico	0.301	0.262	0.263	54
Morocco	0.121	0.139	0	13
The Netherlands	0.376	0.262	0.368	281
New Zealand	0.288	0.221	0.211	87
Norway	0.356	0.246	0.368	164
Panama	0.303	0.267	0.184	18
Peru	0.18	0.056	0.158	7
The Philippines	0.26	0.194	0.184	58
Poland	0.217	0.19	0.158	83
Portugal	0.38	0.252	0.421	100
Russia	0.27	0.212	0.263	123
Singapore	0.185	0.207	0.105	341
South Korea	0.362	0.28	0.289	236
Spain	0.417	0.26	0.421	367
Sweden	0.358	0.25	0.342	444
Switzerland	0.317	0.239	0.289	553
Taiwan	0.313	0.254	0.263	429
Thailand	0.35	0.249	0.316	83
Turkey	0.305	0.239	0.276	92
The United Kingdom	0.35	0.221	0.316	2508
The United States	0.228	0.254	0.132	7165
All countries	**0.298**	**0.259**	**0.237**	**23,914**

Table 12.5 Yearly Distribution of Environment Index (ENV Index)

Year	Mean	Standard Deviation	Median	Observations
2002	0.140	0.172	0.053	797
2003	0.152	0.170	0.105	1117
2004	0.148	0.169	0.079	1742
2005	0.169	0.179	0.132	1870
2006	0.224	0.213	0.184	1894
2007	0.326	0.247	0.316	1981
2008	0.349	0.269	0.316	2549
2009	0.338	0.268	0.316	2865
2010	0.346	0.259	0.316	3249
2011	0.372	0.273	0.342	3335
2012	0.392	0.277	0.368	2515

Each of these variables is common in studies of corporate governance and asset pricing, and can be argued to potentially impact firm-level action on and reporting of environmental performance. For example, given their greater resource base, larger firms are more likely to be able to engage in and report on multiple aspects of their environmental performance. Firms that have high levels of reported performance on other aspects of corporate social performance, risk management, etc., and which demand a higher market to book valuation, may also be more likely to report positively on environmental performance. Greater ownership control exercised in the presence of high levels of closely held shares may reduce the amount of resources devoted to environmental performance, as may high levels of leverage and relatively higher riskiness. Table 12.6 provides summary statistics for the dependent and each of the control variables. As the values of illiquidity ratio is low, we consider alternative illiquidity measure proposed by Lesmond et al. (1999), where a stock with a no change in price over a period of time is considered as illiquid. However, the results remain qualitatively similar.

To test the hypotheses discussed in this paper, we develop the following empirical model:

$$\text{ENV Index}_{it} = \alpha_i + \beta_1\text{PDI}_i + \beta_2\text{IDV}_i + \beta_3\text{MAS}_i + \beta_4\text{UAI}_i + \beta_5\text{LTO}_i + \beta_6\text{IVR}_i$$
$$+ \beta_7\text{PTBV}_{it} + \beta_8\text{Blockholding}_{it} + \beta_9\text{Illiquidity}_{it} + \beta_{10}\text{Vol.}_{it}$$
$$+ \beta_{11}\text{Asset}_{it} + \beta_{12}\text{Leverage}_{it} + \text{Year Dummy} + \text{Industry Dummy} + \varepsilon_{it}$$

ENV Index$_{it}$ is the value taken by the environmental index constructed for firm i in year t. PDI$_i$, IDV$_i$, MAS$_i$, UAI$_i$, LTO$_i$, and IVR$_i$ are the indices for PDI, IDV, MAS, UAI, LTO, and IVR, respectively, for the country in which firm i is located. As per Hofstede (2001) these values are considered to be relatively time invariant, and so only take a single value for each country. PTBV$_{it}$, Blockholding$_{it}$, Illiquidity$_{it}$,

Table 12.6 Summary Statistics of Main Variables

	Mean	Standard Deviation	Distribution				
			5th	25th	50th	75th	95th
Environmental Index	0.298	0.259	0.000	0.053	0.237	0.500	0.763
PTBV	2.617	2.345	0.660	1.210	1.920	3.130	7.020
Blockholding	0.249	0.232	0.002	0.031	0.187	0.408	0.701
Illiquidity	2.105E-08	2.286E-07	3.619E-12	4.170E-11	2.658E-10	1.228E-09	3.237E-08
Volatility	0.023	0.011	0.010	0.015	0.020	0.027	0.044
Asset	15.642	1.568	13.174	14.615	15.543	16.640	18.420
Leverage	0.183	0.153	0.000	0.051	0.161	0.278	0.473

Vol.$_{it}$, Asset$_{it}$, and Leverage$_{it}$, represent the values of the control variables for firm i in year t. As indicated in the regression model, we control for both year and industry fixed effects.

To consider the potential presence of multicollinearity between the independent variables in our model, we examine the Pearson correlation matrix (Table 12.7). None of the pairwise correlation values in our sample is greater than the critical value of 0.8, although several are relatively high (i.e., between IVR and PDI at 0.700, IVR and IDV at 0.752, and LTO and IDV at 0.716). Accordingly, we tentatively conclude that our model is likely to be sufficiently free of the symptoms of multicollinearity.

12.4 EMPIRICAL RESULTS

We report the main results of our regression model in Table 12.8. We use panel data fixed effect regression modeling, with robust standard errors. The use of fixed effects is in line with Gormley and Matsa (2014) who note that fixed effect estimators are superior to other measures in controlling for unobserved heterogeneity. We first regress the dependent variable ENV Index on each of the national cultural factors separately, also including both the control variables and accounting for year and industry fixed effects. The results of each of these separate regressions are reported in the columns 1 through 6, highlighting that, in isolation, a number of the cultural dimensions are not statistically significant in determining environmental outcomes. Column 7 presents the regression results using the full set of national cultural factors. When considered in combination, all six indices of national cultural dimension have statistically significant coefficients at the 1% level. In the following we focus on this final set of results.

In the case of the indices for PDI, MAS, UAI, LTO, and IVR, the sign of the coefficient is consistent with that predicted in its associated hypothesis. However, the coefficient on IDV, while statistically significant, has the opposite sign to that expected. With respect to the control variables, with the exception of Illiquidity, all are statistically significant at the 1% level.

We thus find support for the hypotheses that in societies where power distance is lower, femininity is higher, uncertainty avoidance is greater, there is a greater long-term orientation, and greater emphasis is placed on restraint, that firm-level environmental performance is, on average, higher. We argue that each of these factors adds to the pressure to act cooperatively, leading to a greater collective focus on environmental outcomes for societies displaying relatively high values in each, separately, of these cultural dimensions. In turn, these cultural norms are reflected in accepted business behavior, leading to firms within these societies to place a greater emphasis on achieving and reporting higher standards of environmental performance.

Table 12.7 Correlation Coefficients

	ENV Index	PDI	IDV	MAS	UAI	LTO	IVR	PTBV	Block-holding	Illiquidity	Volatility	Asset	Lever-age
ENV Index	1												
PDI	0.0666***	1											
IDV	−0.126***	−0.745***	1										
MAS	0.0471***	−0.0100	0.0333***	1									
UAI	0.174***	0.293***	−0.429***	0.330***	1								
LTO	0.207***	0.417***	−0.716***	0.253***	0.542***	1							
IVR	−0.138***	−0.700***	0.752***	−0.182***	−0.461***	−0.699***	1						
PTBV	−0.112***	−0.0649***	0.132***	−0.0638***	−0.158***	−0.164***	0.113***	1					
Blockholding	−0.0403***	0.419***	−0.419***	−0.0719***	0.165***	0.219***	−0.401***	0.0122*	1				
Illiquidity	−0.0284***	−0.0381***	0.0433***	−0.0157***	−0.00792	−0.0656***	0.0455***	−0.0298***	0.0697***	1			
Volatility	−0.0321***	−0.0137*	0.0134*	−0.0318***	−0.0295***	−0.0327***	−0.00264	−0.0707***	0.0481***	0.159***	1		
Asset	0.329***	0.120***	−0.108***	0.0593***	0.162***	0.158***	−0.193***	−0.239***	−0.0828***	−0.218***	−0.239***	1	
Leverage	0.0533***	−0.0766***	0.133***	−0.0526***	−0.0651***	−0.137***	0.106***	−0.0266***	−0.0739***	−0.0236***	−0.0475***	0.149***	1

Note: *$p < 0.1$, **$p < 0.05$, ***$p < 0.01$.

Table 12.8 Main Regressions

	ENV Index	ENV Index	ENV Index	ENV Index	ENV Index	ENV Index	ENV Index
PDI	−0.00132*** (−13.48)						−0.00174*** (−12.47)
IDV		0.00000643 (0.09)					0.000924*** (8.28)
MAS			0.000110 (1.61)				−0.000868*** (−11.48)
UAI				0.000905*** (14.26)			0.000937*** (11.98)
LTO					0.00116*** (19.71)		0.00206*** (23.10)
IVR						−0.0000411 (−0.39)	0.000564*** (3.36)
PTBV	0.00340*** (6.00)	0.00344*** (6.04)	0.00354*** (6.21)	0.00465*** (8.09)	0.00533*** (9.35)	0.00345*** (6.03)	0.00628*** (11.04)
Blockholding	−0.00431 (−0.73)	−0.0345*** (−5.61)	−0.0344*** (−6.14)	−0.0533*** (−9.30)	−0.0674*** (−11.53)	−0.0363*** (−5.93)	−0.0210*** (−3.41)
Illiquidity	9304.6* (1.73)	11072.4** (2.01)	11152.6** (2.02)	9336.6* (1.73)	9251.7* (1.67)	10955.0** (1.99)	3131.0 (0.59)
Volatility	−1.606*** (−9.87)	−1.555*** (−9.49)	−1.556*** (−9.50)	−1.541*** (−9.42)	−1.467*** (−9.01)	−1.557*** (−9.50)	−1.546*** (−9.65)
Asset	0.0859*** (86.15)	0.0837*** (83.78)	0.0836*** (84.04)	0.0816*** (81.15)	0.0806*** (80.13)	0.0836*** (82.19)	0.0829*** (82.80)
Leverage	−0.0753*** (−7.98)	−0.0653*** (−6.88)	−0.0640*** (−6.75)	−0.0575*** (−6.09)	−0.0381*** (−4.06)	−0.0652*** (−6.86)	−0.0560*** (−6.00)
Year fixed effects	Yes	Yes	Yes	Yes	Yes	Yes	Yes
Industry fixed effects	Yes	Yes	Yes	Yes	Yes	Yes	Yes
N	23,914	23,914	23,914	23,914	23,896	23,848	23,848
Adjusted R-square	0.462	0.458	0.459	0.463	0.468	0.459	0.486

Note: t statistics in parentheses: *$p < 0.1$, **$p < 0.05$, ***$p < 0.01$.

In the case of the result for individualism, we reject our initial hypothesis. This may reflect that the index for this cultural dimension captures an aspect of self-image, or belonging, rather than a dimension associated with collective behavior. The measured degree of individualism is related to the preference for a social framework in which individuals are expected to take care of themselves and their immediate families, versus a preference for a society in which unquestioning loyalty to a group will provide the individuals support from that group. However, this result does suggest that this aspect of national culture and its impact on environmental performance may warrant further examination.

In column 7 we combine the different cultural dimensions and run them in a single panel regression. The results reported in column 7 should be interpreted carefully due to high collinearity between different cultural dimensions. Overall, the combined cultural dimension results are largely consistent with individual cultural dimensions except Individuality, Masculinity, and Indulgence dimensions turning to statistically significant at the 1% level.

With respect to the control variables, we find that risk (Vol.), leverage, and the proportion of closely held shares reduce reported environmental performance. We hypothesize that this is a result of lower levels of the firm's resources being devoted to environmentally friendly activities. Increased firm size (as measured by the log value of total assets), and hence access to a greater array of resources, increases reported environmental performance at the level of the firm. The explanatory power of the regressions in columns 1–7 remains close to a reasonable 45%.

12.5 CONCLUSION

In this chapter we study empirically the influence of national cultural differences on firm-level attitudes toward environment friendly practices, as reflected in firm-level reports of environmental performance. Our central proposition is that as environmental sustainability and performance are usually collective activities, societal responses in this area may be seen to be governed by cultural factors. Differences in the sets of beliefs, values, and morals, embedded in national culture norms determine what are considered as acceptable and unacceptable behaviors for individuals and businesses, impact government regulation of business activity. Higher levels of those cultural dimensions that support collective activity at the level of the society will be associated with improved environmental performance outcomes at the level of the firm, leading to cross-country differences in firm-level environmental performance.

To test our hypotheses we construct firm-level environment performance indices from binary data related to 38 questions on corporate environmental performance outcomes. The binary data is derived from the Thomson Reuters' Asset4 data set, a contribution that allows the use of a relatively large data set with which to address our research questions. Our sample includes firms from 45 countries covering the period from 2002 to 2012, and is comprised of 23,914 firm-year observations. While the sample is weighted toward firms located in higher income, western economies, it has a comprehensive coverage of firms in high-, middle-, and low-income countries, including firms in emerging market and transition economies. We base measures of six cultural dimension factors—PDI, IDV, MAF, UAI, LTO, and IVR—on Hofstede et al. (2010).

Using a fixed effects panel data regression model in which we control for year and industry fixed effects, we find statistically significant relationships between corporate environmental performance and all six of Hofstede's dimensions of culture. We hypothesize that firms will evidence higher levels of environmental performance if based in cultures with low power distance, that are highly collectivist, that are more feminine, have a higher preference for uncertainty avoidance, are biased toward long-term perspectives, and display lower levels of tolerance for individual indulgence.

We conclude that national cultural variations are significant in determining firm-level corporate environmental performance. Our statistical results, which are based on a larger sample size than prior literature in the area (e.g., Husted, 2005; Park et al., 2007; and Ringov and Zollo, 2007), support the hypotheses that in societies where power distance is lower, femininity is higher, uncertainty avoidance is greater, there is a greater long-term orientation, and greater emphasis is placed on restraint, that firm-level environmental performance is, on average, higher. We also find that a greater number of the individual cultural factors are significant in explaining environmental outcomes than in some of these earlier studies. However, in the case of the result for individualism, while the coefficient is statistically significant, it has the opposite sign to that expected. Thus we reject our initial hypothesis about the impact of individualism, as measured by Hofstede et al. (2010), on environmental performance.

REFERENCES

Aggarwal, R., Goodell, J., 2009. Markets and institutions in financial intermediation: national characteristics as determinants. J. Bank. Financ. 33 (10), 1770–1780.

Alas, R., 2006. Ethics in countries with different cultural dimensions. J. Bus. Ethics 69 (3), 237–247.

Atzenhoffer, J.P., 2012. Could free-riders promote cooperation in the commons? Environ. Econ. Policy Stud. 14 (1), 85–101.

Baland, J.M., Platteau, J.P., 1998. Division of the commons: a partial assessment of the new institutional economics of land rights. Am. J. Agric. Econ. 80 (3), 644–650.

Bansal, P., Gao, J., 2006. Building the future by looking to the past. Organ. Environ. 19 (4), 458–478.

Cernea, M.M., 1993. The sociologist's approach to sustainable development. Financ. Dev. 30 (4), 11–13.

Cohen, D.V., Nelson, K., 1994. Multinational ethics programs: cases in corporate practice. In: Hoffman, W.M., Kamm, J.W., Frederick, R.E., Petry Jr., E.S. (Eds.), Emerging Global Business Ethics. Quorum Books, Westport.

Donleavy, G.D., Lam, K.-C.J., Ho, S.S., 2008. Does East meet West in business ethics: an introduction to the special issue. J. Bus. Ethics 79 (1–2), 1–8.

Elgin, D., 1994. Building a sustainable species-civilization. A challenge of culture and consciousness. Futures 26 (2), 234–245.

Ehrlich, P.R., Ehrlich, A.H., 2002. Population, development, and human natures. Environ. Dev. Econ. 7 (1), 158–170.

Etzioni, A., 1995. The Spirit of Community: Rights Responsibilities and the Communitarian Agenda. Fontana Press, London.

Ford, C.W., Nonis, S.A., Hudson, G.I., 2005. A cross-cultural comparison of value systems and consumer ethics. Cross Cult. Manag. Int. J. 12 (4), 36–50.

Franke, G.R., Scott Nadler, S., 2008. Culture, economic development, and national ethical attitudes. J. Bus. Res. 61 (3), 254–264.

Franklin, D., 2008. Just good business: a special report on corporate social responsibility. The Economist 386, 3–6 (January).

Gorham, E., 1997. Human impacts on ecosystems and landscapes. In: Iverson Nassauer, J. (Ed.), Placing Nature: Culture and Landscape Ecology. Island Press, Washington.

Gormley, T.A., Matsa, D.A., 2014. Common errors: how to (and not to) control for unobserved heterogeneity. Rev. Financ. Stud. 27 (2), 617–661.

Halkos, G.E., Tzeremes, N.G., 2013. National culture and eco-efficiency: an application of conditional partial nonparametric frontiers. Environ. Econ. Policy Stud. 15 (4), 423–441.

Ho, F.N., Wang, H.-M.D., Scott, J.V., 2012. A global analysis of corporate social performance: the effects of cultural and geographic environments. J. Bus. Ethics 107 (4), 423–433.

Hofstede, G., 1980. Culture's Consequences, International Differences in Work-related Values. Sage Publications, Beverly Hills.

Hofstede, G., 1985. The interaction between national and organizational value systems. J. Manag. Stud. 22 (4), 347–357.

Hofstede, G., 2001. Culture's Consequences: Comparing Values, Behaviors, Institutions and Organizations across Nations. Sage Publications, Inc., Thousand Oaks.

Hofstede, G., Hofstede, G.J., 2005. Cultures and Organizations: Software of the Mind, second ed. McGraw-Hill, New York.

Hofstede, G., Hofstede, G.J., Minkov, M., 2010. Cultures and Organizations: Software of the Mind, third ed. McGraw-Hill, New York.

Hofstede, G., Hofstede, G.J., Minkov, M., Vinken, H., 2008. Values Survey Module 2008: Manual, Release 08-01. Available at: http://geert-hofstede.com/dimensions.html (accessed 19.12.14.).

House, R.J., Hanges, P.J., Javidan, M., Gupta, V., 2004. Culture, Leadership, and Organizations: The GLOBE Study of 62 Societies. Sage Publications, Thousand Oaks.

Husted, B.W., 2005. Culture and ecology: a cross-national study of the determinants of environmental sustainability. Manag. Int. Rev. 45 (3), 349–371.

Kellert, S.R., 1996. The Value of Life. Island Press, Washington.

Kinsley, D.R., 1995. Ecology and Religion. Prentice Hall, Englewood Cliffs.

Kluckhohn, C., 1951. Values and value orientations in the theory of action: an exploration in definition and classification. In: Parsons, T., Shils, E. (Eds.), Toward a General Theory of Action. Harvard University Press, Cambridge, 388–433.

Lesmond, D., Ogden, J., Trzcinka, C., 1999. A new estimate of transaction costs. Rev. Financ. Stud. 12 (5), 1113–1141.

McCarty, J.A., Shrum, L.J., 2001. The influence of individualism, collectivism, and locus of control on environmental beliefs and behavior. J. Public Policy Mark. 20 (1), 93–104.

McSweeney, B., 2002. Hofstede's model of national cultural differences and their consequences: a triumph of faith-a failure of analysis. Hum. Relat. 55 (1), 89–118.

Minkov, M., Hofstede, G., 2011. The evolution of Hofstede's doctrine. Cross Cult. Manag. Int. J. 18 (1), 10–20.

Moosa, I., Ramiah, V., 2014. The Costs and Benefits of Environmental Regulation. Edward Elgar Publishing, Ltd., Cheltenham.

Murninghan, M., Grant, T., 2013. Corporate responsibility and the new "materiality". Corp. Board 34 (203), 12–17.

Park, H., Russell, C., Lee, J., 2007. National culture and environmental sustainability: a cross-national analysis. J. Econ. Financ. 31 (1), 104—121.

Poortinga, Y., 1992. Towards a conceptualization of culture for psychology. In: Iwawaki, S., Kashima, Y., Leung, K. (Eds.), Innovations in Cross-cultural Psychology. Swets and Zeitlinger, Amsterdam, the Netherlands.

Pretty, J., Ward, H., 2001. Social capital and the environment. World Dev. 29 (2), 209—227.

Ramiah, V., Martin, B., Moosa, I., 2013. How does the stock market react to the announcement of Green policies? J. Bank. Financ. 37 (5), 1747—1758.

Ringov, D., Zollo, M., 2007. Corporate responsibility from a socio-institutional perspective: the impact of national culture on corporate social performance. Corp. Gov. 7 (4), 476—485.

Russo, M.V., Fouts, P.A., 1997. A resource-based perspective on corporate environmental performance and profitability. Acad. Manag. J. 40 (3), 534—559.

Scholtens, B., Dam, L., 2007. Cultural values and international differences in business ethics. J. Bus. Ethics 75 (3), 273—284.

Senge, P.M., 2007. Waking the sleeping giant. Business as an agent for consumer understanding and responsible choice. J. Corp. Citizsh. 26, 25—27.

Shepherd, D.A., Patzelt, H., Baron, R.A., 2013. "I care about nature, but...": disengaging values in assessing opportunities that cause harm. Acad. Manag. J. 56 (5), 1251—1273.

Simga-Mugan, C., Daly, B.A., Onkal, D., Kavut, L., 2005. The influence of nationality and gender on ethical sensitivity: an application of the issue-contingent model. J. Bus. Ethics 57 (2), 139—159.

Smith, P.B., Peterson, M., Schwartz, S.H., 2002. Cultural values, sources of guidance, and their relevance to managerial behaviour: a 47-nation study. J. Cross-Cult. Psychol. 33 (2), 188—208.

Srnka, K.J., 2004. Culture's role in marketers' ethical decision making: an integrated theoretical framework. Acad. Mark. Sci. Rev. 12 (6), 1—32.

Tan, J., Chow, I.H.-S., 2009. Isolating cultural and national influence on value and ethics: a test of competing hypotheses. J. Bus. Ethics 88 (1), 197—210.

Vitell, S.J., Paolillo, J.G.P., 2004. A cross-cultural study of the antecedents of the perceived role of ethics and social responsibility. Bus. Ethics A Eur. Rev. 23 (2—3), 185—199.

Vitell, S.J., Nwachukwu, S.L., Barnes, J.H., 1993. The effects of culture on ethical decision-making: an application of Hofstede's typology. J. Bus. Ethics 12 (10), 753—760.

Ward, H., 1998. State, association, and community in a sustainable democratic polity: towards a Green associationalism. In: Coenen, F., Huitema, D., O'Toole, L.J. (Eds.), Participation and the Quality of Environmental Decision Making. Kluwer Academic Publishers, Dordrecht, the Netherlands.

Weaver, G.R., 2001. Ethics programs in global business: culture's role in managing ethics. J. Bus. Ethics 30 (1), 3—15.

Whipple, T.W., Swords, D.F., 1992. Business ethics judgments: a cross-cultural comparison. J. Bus. Ethics 11 (9), 671—678.

World Economic Forum, 2001. Environmental Sustainability Index. Global Leaders of Tomorrow Environment Task Force. World Economic Forum, Geneva, Switzerland.

CHAPTER 13

The Economic and Financial Effects of Environmental Regulation

Imad A. Moosa
School of Economics, Finance and Marketing, RMIT University, Melbourne, VIC, Australia

Contents

13.1 INTRODUCTION

Regulation in general is a form of government intervention in economic activity and interference with the working of the free market system. From a legislative point of view, regulation may be defined as a process of the promulgation, monitoring, and enforcement of rules. Environmental regulation is a formal response to the risk faced by the society posed by environmental changes. Economic activity requires the use of resources (both renewable and nonrenewable) extracted from the environment to produce goods and services—and by doing that, waste is produced and dumped on the environment. Lasting prosperity depends on protecting both the extractive potential and waste absorption capacity of the environment.

In thinking about how spending on environmental protection relates to future prosperity, we must first consider the yardsticks used to measure economic performance. Some of us tend to think that when we discharge pollutants as a by-product of economic activity, that discharge is "free" (and it is free in a way to dump a piece of asbestos in a public park). Those who think this way tend to view the cost of pollution control as a net cost, when in fact it is often repaid (sometimes many times over) by the costs avoided elsewhere in the economy and by preserving the environment at large. There are also those who think that economic growth is the only factor that maintains prosperity, which is a valid proposition if prosperity is measured only in terms of the monetary value of gross domestic product (GDP) or GDP per capita. But we know very well that a high GDP per capita does not compensate for breathing polluted air.

A typical argument against environmental regulation is that it kills jobs as factories close down because of the high cost of cleanup, or because of relocation to countries with weaker regulatory standards. However, the costs of environmental regulation can be offset as firms develop cheaper ways to cleanup pollutants. As new technologies are developed to cope with regulatory requirements, new jobs are created. In many cases it can be demonstrated that environmental regulation has resulted in far lower costs and job losses than business executives feared initially. The fear of environmental regulation is not confined to its effects on unemployment because it is typically blamed for macroeconomic mishaps in any shape or form. For example, Crandall (1980) argues that "it has become increasingly fashionable to attribute a myriad of our economic and social difficulties to excessive government regulation." Specifically, regulation is blamed for high inflation, slow GDP growth, declining productivity growth, currency depreciation, and even for the pessimism of entrepreneurs. These claims are invariably politically motivated and supported by no more than circumstantial evidence, which rests on the coincidence of the advent of environmental regulation and the stagflation of the 1970s.

The objective of this chapter is to consider the microeconomic, financial, and macroeconomic effects of environmental regulation. In the following sections we discuss in turn the effects of environmental regulation on the costs of production, plant location, productivity at the firm and sectoral level, stock prices and returns, employment, net exports, competitiveness, economic growth, and aggregate productivity. These are the perceived economic and financial effects of environmental regulation.

13.2 THE COSTS OF PRODUCTION

Firms incur costs in the production of marketable goods and services (commercial output), arising from payments to the factors of production (or inputs): labor, capital, and land (including minerals and energy). Compliance with environmental regulation requires the diversion of resources from, or to use additional resources over and above, what is used for the production of goods and services. The use of additional resources causes upward shifts in the cost curves, implying higher costs for the same level of output. Fixed costs rise when a firm installs equipment for the purpose of after-production cleanup, whereas variable costs go up because labor services are required to ensure compliance. It follows that firms operating in a country or a region with more stringent environmental regulation typically face higher costs of production compared with identical firms that operate in a country or a region with a lax attitude toward the control of pollution and other environmental issues. While this proposition sounds intuitive, there is no consensus view on the magnitude of the regulation-caused increase in production costs or indeed whether or not incurring these additional costs is worthwhile, at least from the perspective of the underlying firm. While estimates of the costs incurred by U.S. firms for the purpose of compliance with environmental regulation run into billions of dollars, some economists argue that the cost estimates may be significantly overstated, in the sense that gross costs may be far greater than net costs (that is, net of the benefits) because compliance may be productive (Berman and Bui, 1999).

Pollution control expenditure is typically difficult to identify. For example, if a new piece of equipment is bought to replace an existing piece (because the former is more efficient and produces less emission) a problem arises as to whether part or all of the cost of acquiring the new piece should be considered as pollution control expenditure (hence on compliance with environmental regulation). The issue here is to determine if a particular cost is incurred in the absence of environmental regulation. Furthermore, the allocation of managerial time devoted to compliance with environmental regulation is difficult to measure. This is a major problem that is encountered when the survey approach is used to estimate the costs of environmental regulation. In the US, for example, the Census Bureau's pollution abatement control expenditures (PACE) survey is designed to measure the capital expenditure and operating costs associated with environmental regulation. PACE has been criticized for potentially missing a large portion of the costs associated with environmental regulation. Firms typically fail to keep a special track of their expenditure on environmental protection, which means that these items have to be estimated somehow. The outcome is invariably inaccurate estimates.

The estimation of capital expenditure of the "end-of-line" type (such as scrubbers, filters, and precipitators) is rather an easy task, since these items can be readily recognized and because they are used for one purpose only: pollution control. However, when capital expenditure is of the "production process enhancement" type (such as the installation of new equipment that leads to improvement in productive efficiency and reduction in pollution) the task becomes much more difficult. If environmental protection is the primary reason for acquiring the equipment, the whole production process enhancement expenditure must be reported. If, on the other hand, the motivation is improved in productive efficiency, only the pollution control features added for compliance purposes must be reported.

Concerns are also raised with respect to operating expenses. While the salaries and wages of those involved in environmental tasks are rather easy to account for, this is not the case for a production team that spends a small but nonzero amount of time on these tasks. The same goes for the time spent by the management on environmental issues and compliance. Likewise, the cost of materials, parts, and components used as operating supplies for pollution control (or used for the repair and maintenance of pollution control equipment) may be easy to estimate, but this cannot be said of the incremental costs associated with the consumption of environmentally preferable materials and fuels. Apart from the potential underreporting of capital expenditure and operating costs, it is difficult to estimate other costs such as the adverse impact on output resulting from an outright suspension of production (caused by the installation of pollution control devices) or the loss of operational flexibility (as an outcome of regulatory compliance).

An alternative approach to the estimation of the costs incurred by firms for the purpose of regulatory compliance is the analytical cost function approach, which is used by Becker and Henderson (2001) to examine the cost that environmental regulation imposes on firms belonging to various industries. The basic results obtained by following this approach show that heavily regulated firms incur higher costs of production than less-regulated firms. Another factor that seems to play a role in determining the effect of regulation on the costs of production is the firm (or plant) age, as they find young firms to be more vulnerable when it comes to the costs imposed by regulation. By using the analytical cost function approach, the cost estimates obtained by Becker and Henderson (2001) turn out to be higher than those obtained by analyzing the PACE figures of regulated firms. They attribute this finding to the proposition that "PACE misses a substantial portion of environmental expenditures" and that the survey results "understate costs much more for plants in non-attainment areas than for those in attainment areas, narrowing the estimated gap between the two groups." Morgenstern et al. (1998) argue that while the estimated cost of environmental regulation (based largely on self-reported information) is often cited as an assessment of the burden of current regulatory efforts and a standard against which the associated benefits are measured, little is known about how well reported expenditures relate to the "true costs." They argue that the potential for both incidental savings and uncounted burdens means that actual costs could be either higher or lower than reported expenditures.

In its 2011 report to Congress, the Office of Management and Budget (OMB) estimated the total annual cost of major federal regulations in the US at an aggregate level in the $44–62 billion range (Office of Management and Budget, 2011). However, Sinclair and Vesey (2012) point out that the OMB admits that these cost estimates have their limitations, in part because they rely on agencies' ex ante estimates, and exclude any impacts that cannot be quantified or monetized with the information available. Moreover, these figures only capture regulations that are subject to OMB review under Executive Order 12866, and for which agencies estimate both costs and benefits. Another report, commissioned by the Small Business Administration, drew on a number of sources in an attempt to arrive at a more comprehensive measure of regulatory costs, estimating the total federal regulatory cost burden on businesses to be $1.75 trillion in 2008 (Crain and Crain, 2010). Sinclair and Vesey (2012) argue that without passing judgment on the accuracy of this estimate, it is clear that its dramatic departure from the OMB's estimates reveals that the true cost of the regulatory burden on the U.S. economy is uncertain and debatable.

There is no doubt that regulatory compliance adds to the costs of producing commercial output, hence affecting profitability and the level of production. And there are more than the direct costs of compliance, as some costs encompass the whole economy.

However, we must also bear in mind two observations pertaining to firm-specific costs and economy-wide costs. The first is that compliance with environmental regulation may enhance productive efficiency that makes the net costs of compliance lower than gross costs. The second is that we should take into account the economy-wide and society-wide benefits of regulation. We pay to consume goods and services without complaining about the costs involved in this consumption. Likewise, we should be prepared to pay for cleaner environment—this is like paying for health care.

13.3 PLANT LOCATION

The literature on plant location, as motivated by environmental regulation, deals with decisions on where to produce within one country and across countries. As far as environmental regulation is concerned, the underlying argument is that firms choose to produce in regions and countries where the cost of compliance is low—that is, where regulation is less stringent or not fully enforced and where regulatory standards are low.

Research on the effect of environmental regulation on plant location follows two approaches: the survey approach and the econometric approach. The results of surveys of business executives involved in plant location decisions generally lead to the conclusion that environmental regulation is not a major determinant of plant location (for example, Epping, 1986; Schmenner, 1982; Duerksen, 1983; Wintner, 1982; Stafford, 1985; Lyne, 1990). Levinson (1996) argues that the survey results are difficult to interpret because some surveys involve open-ended questions about the potential determinants of location, while others ask respondents to rank a preselected list of factors. He further points out that, even if they are consistently conducted, surveys may be of little value if the respondents, through intent or ignorance, misrepresent the true effects of environmental regulation on location choice. When it comes to environmental regulation, survey participants have every incentive for misrepresentation.

A large number of empirical studies have been conducted on the effect of environmental regulation on plant location. Crandall (1993) finds that environmental compliance costs, as measured by the Census Bureau, do not have a "measurable effect on the regional distribution of manufacturing employment." By using an empirical specification following McFadden's (1974) conditional logit model, Bartik (1988) examines the locations chosen by branch plants of the Fortune 500 companies during the period 1972—1978. The results lead him to support "the prevailing wisdom that environmental variables have only small effects on business locations." McConnell and Schwab (1990) use the same econometric approach to examine data on vehicle assembly plants, which (in the process of painting cars and trucks) emit volatile organic compounds that contribute to urban ozone (smog). They find significant coefficients only for those regions that are extremely far out of compliance. Friedman et al. (1992) use the conditional logit model and establishment-level data on the planned locations of foreign firms within the US. In one specification they include a variable that is similar to the one used by Crandall (1993) (that is, total state-wide pollution control capital expenditure per dollar of gross state product from manufacturing). The estimated coefficient turns out to be statistically insignificant, a result that may be explained in terms of the nature of the expenditure variable as a measure of statutory incidence (as it includes only direct capital expenditure without controlling for the states' industrial compositions).

The finding that environmental regulation does not affect plant location decisions is attributed by Becker and Henderson (2000) to some loopholes in the analysis. For example, many U.S. studies have been conducted at the state level, ignoring significant regulatory differences within states. Other studies lump together polluting and nonpolluting industries by examining industrial groupings such as "all manufacturing" or by using the categories of the two-digit standard industrial classification. Another loophole is that the proxies for environmental regulation are not based on the specific regulatory process—rather, they are indices of "green" activity in a particular location as indicated by congressional voting records, the existence of environmental laws and similar indicators. Some studies (for example, Bartik, 1988) focus on the pre-1978 period of regulation or use cross-sectional data or estimation methods that can be problematical (for example, McConnell and Schwab, 1990). Becker and Henderson (2000) investigate the effects of air quality regulation in the US on decisions pertaining to plant location, births, sizes, and investment patterns in major polluting industries and show that there has been significant relocation of polluting industries from more to less polluted regions (from attainment to nonattainment regions) to avoid more stringent regulation in more polluted regions. They also show that the timing of plant investments in regulated regions has been altered significantly. There is no reason, however, why all of the evidence for the neutrality of environmental regulation with respect to plant location should be discarded in favor of the results of Becker and Henderson, which show otherwise. We have a situation where we deal with an ideological issue with a tool (econometrics) that can be abused to produce the desired results.

The plant location issue has an international dimension. The choice between domestic and foreign plant locations, which is covered by the literature on foreign direct investment (FDI), has been considered by Jaffe et al. (1995). In choosing between domestic and foreign locations, firms consider a variety of factors including the market to be served by the plant, the quality of the labor force, foreign exchange risk, political risk, and the available infrastructure. Hence, it is rather difficult to isolate the effect of environmental regulation on the decision to choose between domestic and foreign locations. The literature on FDI hardly considers environmental regulation (or regulation in general) as a factor that determines FDI decisions. For example, Moosa (2002) argues that the choice between exporting (that is, choosing a domestic location) and FDI (choosing a foreign location) depends on profitability, opportunities for market growth, costs of production, and economies of scale. Moosa and Cardak (2006) list 14 determinants of

FDI, none of which is environmental regulation. Although it is not listed as an independent determining variable, environmental regulation can be thought of as playing a role via its effects on the costs of production, profitability, incentives, the cost of capital, and productivity. It may be even related to corruption, as corruption provides an opportunity to evade compliance, and to political risk as it pertains to changes in the "rules of the game." Leonard (1988) finds no evidence to indicate that U.S. firms have moved abroad for the purpose of avoiding U.S. environmental regulation. Wheeler and Mody (1992) conclude that multinational firms appear to base their FDI decisions primarily upon such factors as labor costs and access to markets, as well as the presence of a developed industrial base. In conclusion, the stringency or otherwise of environmental regulation may have ramifications for foreign plant location, but it is certainly a secondary factor. The literature on FDI supports this proposition.

13.4 PRODUCTIVITY AT THE FIRM AND SECTORAL LEVEL

Gray and Shadbegian (1993) refer to what they call the "mismeasurement effect" of environmental regulation on productivity, which results from the fact that measures of productivity do not distinguish between the inputs used for the production of commercial output and those used for the purpose of regulatory compliance. The use of observed productivity figures in empirical work involves measurement errors that lead to biased results. But even if compliance costs (that is, the amounts of factor inputs directed toward environmental regulation) are subtracted to correct for the mismeasurement effect, environmental regulation still exerts some impact on (true) productivity. Compliance-triggered changes in the production process may reduce productive efficiency because regulation enhances uncertainty, which complicates the decision-making process. On the other hand, one way in which environmental regulation could have a positive impact on productivity at the industry level is by forcing exceptionally inefficient plants to close down. To the degree that production is shifted to other plants with higher productivity, the productivity of the industry as a whole could actually improve.

The conflicting results produced by studies of the effects of environmental regulation on productivity can be attributed to selection bias and measurement errors. Selection bias may occur because plants that can most easily implement pollution reduction may actually choose to do so voluntarily, without being forced by regulators. Firms may choose to go green for reasons other than regulatory compliance, including changes in the production process that involve cleaner, more efficient technologies. If the effect of environmental regulation on economic outcomes is measured by looking at the relation between changes in productivity and pollution control expenditure, without taking into account the possibility that some firms may indulge in pollution control activities voluntarily, the effect of regulation on productivity will be underestimated. Measurement errors may also impart a bias because of the difficulty of quantifying expenditure (as we have seen) and productivity, which may be measured in a number of ways, not to mention the difference between observed and true productivity.

Berman and Bui (1999) produce results that reveal strong evidence indicating that regulation has induced higher environmental operating costs, but they point out that these costs had a transitory effect on the productivity of South Coast refineries. Greenstone et al. (2012) use detailed production data from nearly 1.2 million plant observations, drawn from the 1972–1993 annual survey of manufacturers, to estimate the effects of air quality regulation on manufacturing plants' total factor productivity. They find that among surviving polluting plants, stricter air quality regulations are associated with a roughly 2.6% decline in total factor productivity. Graff and Neidel (2011) examine the relation between ozone levels and farm worker productivity in the central valley of California, producing results indicating that lower levels of ozone are correlated with higher productivity and that even ozone levels well below current standards have significant and adverse effects on productivity.

On the other hand, there are those who believe that environmental regulation may have a positive effect on productivity because pollution is a waste. Although it is held by regulatory agencies, a large number of scholars subscribe to this view. For example, Porter and van der Linde (1995b) prescribe recommendations to help regulators design productivity-enhancing regulation and suggest a number of channels whereby environmental improvement has benefits for both processes and products. The Porter hypothesis postulates that environmental regulation may induce technological innovation to accomplish compliance (Porter, 1990, 1991). Believers in the Porter hypothesis are not only Al Gore and the Environmental Protection Agency (EPA) but also a large number of scholars in the field of environmental economics. For example, Lanjouw and Mody (1993) found that rising environmental compliance costs are associated with the patenting of new technologies with a lag of 1–2 years.

13.5 STOCK PRICES, RETURNS, AND RISK

In an efficient market, stock prices reflect all available information, including information on compliance with environmental regulation or taking the initiative to go green. When a regulatory announcement is made or a firm decides to adopt a new environmental policy, the information is disseminated to various stakeholders and the public in general. As the market processes the information, investors conclude that the announcement is good or bad for the underlying firm—these views will be reflected on prices and returns. For example, investors may conclude that a firm adopting a new environmental policy is likely to become less prone to environmental accidents and that it is well positioned to be in compliance with any new and more stringent environmental regulation. As a result, the firm is assigned a lower perceived risk in the form of a lower beta, which produces a lower cost of equity capital. With a lower level of perceived risk, investors seek a lower required rate of return to invest in the underlying firm. Since the required rate of return is the discount rate in the dividend discount model, a lower rate produces better stock valuation. On the other hand, the adoption of costly environmental policies and practices by a firm may be perceived (by investors) to be an unnecessary burden that will have a negative effect on the firm's financial indicators.

If the decision to go green is perceived in a positive manner, the adoption of a new environmental policy should, ceteris paribus, lead to higher stock prices. This is essentially the argument put forward by Feldman et al. (1996) who demonstrate the linkage between changes in environmental risk and a firm's stock price. They analyze a sample of 300 large public U.S. firms to find out if investment in environmental management leads to reduced risk, and if such risk reduction is valued by financial markets. Their findings suggest that investing in environmental management leads to substantial reduction in the perceived risk of a firm, with an accompanying increase in the stock price.

The technique of event study is used frequently to examine the response of stock prices and returns to firm-specific announcements pertaining to environmental management and performance. Hamilton (1995), White (1995), and Klassen and McLaughlin (1996) use event study methodology to demonstrate the following: (1) news of high levels of toxic emissions results in significant negative abnormal returns; (2) firms with strong environmental management practices produce higher returns (than firms with poor practices) after a major environmental disaster; and (3) environmental performance is associated with nontrivial positive abnormal returns. The first and second results indicate that investors expect firms to incur significant costs in the process of environmental cleanup, but these costs are lower for firms with superior environmental records. The third result suggests that recognition of environmental performance has a positive reputational effect, which boosts stock returns. Hamilton (1995) uses event study to investigate the effect of the pollution data released by the EPA in the June 1989 *Toxic Release Inventory* (TRI). The results indicate that the TRI data do provide news to investors. For firms that reported TRI data to the EPA, the average abnormal return on the day when the information was made public was significantly negative.

Ramiah et al. (2013) investigate the impact of 19 announcements of environmental regulation on the prices of stocks listed on the Australian Stock Exchange over the period 2005–2011. Using event study, they attempt to find out whether these announcements produce positive or negative abnormal returns. They use a standard event study approach that involves the estimation of the relation between the return on a particular stock and the return on a market portfolio over a period of time prior to the event under examination. By using this methodology, Ramiah et al. demonstrate that the Australian stock market was particularly sensitive to the announcement of the carbon pollution reduction scheme (CPRS). In a similar study, Ramiah et al. (forthcoming-b) conduct an empirical exercise to explore the effects of environmental regulation announcements on corporate performance in China when performance is represented by stock returns. By using event study, they obtain findings confirming that certain announcements of environmental regulation affect risk and return in the Chinese stock market and that some of the environmental policies are successful as indicated by the negative abnormal returns for the polluting industries. However, they come across some evidence for policy ineffectiveness. Similar results have been obtained by analyzing U.S., French, and British markets (Ramiah et al., 2015; Ramiah et al., forthcoming-a; Pham et al., 2014).

The evidence on environmental performance and corporate financial performance is mixed at best. Empirical studies reveal that the relation between environmental compliance and financial performance is not clear. A large number of studies suggest a positive relation between environmental compliance and financial performance (Spicer, 1978; Waddock and Graves, 1997; Schnietz and Epstien, 2005; Wahba, 2008; Hart, 1997; Porter and van der Linde, 1995a; Reinhardt, 1999; Dowell et al., 2000). On the other hand, some studies suggest a negative relation between environmental compliance and financial performance (Chen and Metcalf, 1980; Jagg and Freedman, 1992; Wagner et al., 2002). Some studies conclude that the relation between environmental compliance and financial performance is insignificant (Mahapatra, 1984; McWilliams and Siegel, 2000; Mill, 2006; Murray et al., 2006). The diversity of results is likely to be due to the ideologically motivated desire to produce results that confirm preconceived beliefs. Most of this research is not motivated by a quest for the truth but by a quest to prove a point.

One of the most important studies challenging the view that environmental expenditure has a negative effect on profitability is that of Feldman et al. (1996) who argue that when appropriately evaluated, expenditure on environmental management may be shown to provide substantial positive returns and a lasting value to the firm. Some of the benefits that would accrue are improved product quality, enhanced productivity, and ease of manufacturing and distribution. Likewise, McGuire et al. (1988) suggest that the explicit costs of environmental management are not only minimal but they also generate other management benefits, such as higher morale and enhanced productivity. Therefore, they suggest that although the costs of environmental management can be significant, other costs are reduced or revenues go up as a result of adopting environmentally friendly policies.

As far as the effect of environmental regulation on risk is concerned, we have already seen that a firm's environmental performance affects beta. In general terms, environmental regulation creates uncertainty with respect to costs and revenues, which makes profitability more volatile (and volatility means risk). However, Feldman et al. (1996) believe that sound environmental management leads to better short-term environmental performance as well as the prospect of further improvements in the future. This should consequently reduce firm-specific risk, which is valued by financial markets. By using real-world data on more than 300 of the largest U.S. public companies, they produce results that validate this hypothesis. They quantify changes in systematic risk for each firm over two time periods and relate these changes to a number of financial and environmental variables using multiple regression analysis. By following this approach, they manage to isolate and quantify the effects of several measures of environmental management and performance that turn out to be statistically significant.

Ramiah et al. (2013) examine changes in the risk associated with stock returns following the announcement of environmental regulation within the framework of event study. They argue that the environment in which businesses operate changes significantly with the adoption of new measures of environmental regulation, with unknown effect on systematic risk. Two major conclusions can be drawn from their work. First, an industry effect is evident in risk variation following the announcement of environmental regulation (the outcome can be positive, negative, or neutral). The second conclusion is that the majority of the sectors did not experience changes in overall short-term systematic risk. More or less similar results are obtained by Ramiah et al. (2015; forthcoming-b), Pham et al. (2014), and by Ramiah et al. (forthcoming-a).

13.6 EMPLOYMENT

One of the concerns raised by the opponents of environmental regulation is the envisaged adverse effect on employment. While some politicians and (right-wing) media make repeated claims that regulation is the "enemy of job creation," this claim is rarely backed by evidence. An article in the *Washington Post* states explicitly that "economists who have studied the matter say that there is little evidence that regulations cause massive job loss in the economy, and that rolling them [regulations] back would not lead to a boom in job creation" (Yang, 2011). Some evidence points to a lack of consumer demand (not regulation) as the primary obstacle to reducing unemployment. According to the Bureau for Labor Statistics (2011), among the mass layoffs recorded between 2008 and the first half of 2011, just 0.3% of employers attributed the layoff to regulation, compared to 34.6% claiming that business demand was to blame. On the other hand, a Gallup poll reported that 22% of small business owners consider "complying with government regulations" as the most important problem they are facing right now, while "consumer confidence" and "lack of consumer demand" were named by 15% and 12% of the respondents, respectively (Jacobe, 2011). Bruce Bartlett, a senior advisor in both the Reagan and George Bush I administrations, is quoted by Eberlly (2011) as suggesting that "no hard evidence has been offered for claims that regulation is the 'principal factor' holding back employment."

Business executives typically claim that strict environmental standards force them to close down factories or move them overseas (the plant location argument). Restrictions on the use of natural resource, such as limits on timber harvesting, are opposed by business leaders because they believe that these restrictions have adverse consequences for employment. Labor unionists also fear job losses if environmental regulation raises production costs or restrict supply. Repetto (1990) argues that many economists sub-scribe to a more sophisticated version of this argument, pointing out that diverting capital to invest in pollution control equipment, instead of capacity expansion or productivity improvement, limits employment growth over time. However, he disagrees with this view, suggesting that environmental protection actually creates more jobs than the jobs lost. For example, limits on the extraction of natural resources may threaten jobs in extractive industries, but the imposition of these limits saves or creates jobs in recrea-tion industries and in high-tech industries attracted to a high-quality environment. Another example is that environmental regulation that requires pollution control and leads to higher energy prices creates jobs in industries supplying pollution control and en-ergy conservation equipment and services (which is consistent with the narrow interpre-tation of the Porter hypothesis). Since these industries are more labor-intensive than the heavily polluting industries (such as energy, basic metals, and chemicals) it is plausible to suggest that expenditure on environmental protection creates jobs, even if it is at the expense of employment in polluting industries.

Economists tend to put forward widely diversified set of views on the effect of envi-ronmental regulation on employment. Greenstone et al. (2012) argue that the decline in U.S. manufacturing employment from 18 million (25.3% of total U.S. employment) in 1970 to 12 million (9% of total employment) in 2012 mirrors the introduction and expansion of U.S. environmental policy. This statement is nothing short of a travesty. The fall in manufacturing employment in the US is the result of a conscious decision to move away from manufacturing industry to services, particularly financial services, which free marketeers consider to be "natural," a phase of economic evolution. Liu (2005) quotes Alan Greenspan as saying, in a testimony to Congress, that "thinking jobs are better than doing jobs" and that "the US will keep higher-paying jobs in financial services, manage-ment, design, development, sales and distribution and let the emerging economies have the low-paying assembly line jobs in factories owned by US companies." One can justi-fiably ask the following question: How is it that making a car or even a pair of shoes involves less thinking than what a stock broker does?

We know by now that Greenspan has been proved wrong on almost everything he said when he was the boss at the Fed. The facts and figures relating to manufacturing employment and output tell a clear story. Manufacturing employment in the US reached a high level of 19.3 million in 1978—thereafter it declined persistently to stand at 11.6 million in 2010. Most of this decline occurred between 2000 and 2010 (from 17.2 to 11.6 million). Environmental regulation cannot explain such a drastic loss of manufacturing employment—a better explanation is the love affair with the financial sector, the belief that a modern economy should supply financial services rather than clothes and footwear.

The weakness and sluggishness of the U.S. economy in the aftermath of the global financial crisis have brought about calls for dismantling regulation to boost economic activity and reduce a stubbornly high unemployment rate. This may be thought of as a rather strange claim, given that the great recession was caused by the global financial crisis, not by environmental regulation. The opponents of regulation still argue that regulation is preventing sluggishness from turning into vibrancy. However, Shapiro and Irons (2011) dismiss calls for dismantling regulation to revive the economy by suggesting that concern about the negative consequences of regulation "should not lead to unjustified efforts to weaken government regulators and regulations" because "careful review of the available evidence indicates that regulations do not tend to significantly impede job creation." They suggest, to the contrary, that "an emphasis on deregulation can contribute to enormous economic dislocation." Furthermore, the great recession (and any recession) is a cyclical phenomenon whereas the effect of regulation is structural.

Berman and Bui (2001) contend that environmental regulation does not exert any effect on input prices if input markets are large and competitive. Based on the results obtained from a formal model, they conclude that the local air quality regulations introduced during the period 1979–1992 in the Los Angeles region were not responsible for a large decline in employment. In fact, they argue, "they [regulations] probably increased labour demand slightly." A possible explanation for their findings is that the regulation targeted capital-intensive industries with relatively little employment. To reconcile their findings with the claim that environmental regulation "costs jobs," they argue that employers may overestimate the job loss induced by regulation by confusing the firm's product demand curve with that of the industry.

A wide range of empirical studies have been conducted to examine the relation between environmental regulation and employment, including aggregate policy modeling (macroeconomic and general equilibrium), economy-wide microeconomic studies, industry-specific studies, and analyses of plant location and growth. Estimates of the economy-wide job impacts of environmental regulation are typically based on simulations of large macroeconomic and general equilibrium models. In a review article of macroeconomic modeling as applied to environmental regulation, Goodstein (1994) finds that seven of the nine studies he reviewed showed increases in employment, one showed a decrease and one produced mixed results. He concludes that "on balance, the available studies indicate that environmental spending... has probably led to a net increase in the number of jobs in the U.S. economy... (although) if it exists, this effect is not large."

Morgenstern et al. (2000) provide evidence indicating that "increased environmental spending does not cause a significant change in industry-level employment." They conclude that "EP [environmental protection], economic growth, and jobs creation are complementary and compatible" and that "investments in EP create jobs and displace jobs, but the net effect on employment is positive." Yan and Carr (2013) analyze the net employment growth and employment stability effects of the Clean Air Act on employment and find that (1) enhanced employment stability is associated with nonattainment regions for total suspended particulates (TSP) and 1-h ozone; (2) negative employment growth in TSP and carbon monoxide nonattainment regions; and (3) positive employment growth in 1-h ozone and sulfur dioxide nonattainment regions.

On the other hand, Beard et al. (2011) reach a peculiar conclusion by looking at the effect of on-budget regulatory agency spending on private sector employment in the US. They find that reducing the total budget of all U.S. federal regulatory agencies by 5% (or $2.8 billion) produces 1.2 million more private sector jobs each year and that firing one regulatory agency staff member creates 98 jobs in the private sector. These results sound ridiculous, most likely the product of extensive data mining motivated by an ideological antiregulation stance. Sinclair and Vesey (2012) use data sources and econometric methods similar to those employed in Beard et al. (2011), but they obtain very different results. While Beard et al. (2011) predict that reducing the regulators' budget by a small percentage would have a dramatic, positive impact on employment, Sinclair and Vesey (2012) cannot draw any definitive conclusions about the direction or size of the impact of the regulators' budget on employment. The results of Beard et al. cannot be explained on theoretical or intuitive grounds. How does the firing of a regulator create more jobs in the industries regulated by his former employer? For one thing, firing a regulator does not necessarily mean that regulation becomes lax. And cutting the budget of a regulatory agency may make regulators more aggressive and regulation more blunt.

The stylized facts do not seem to support the proposition that environmental regulation has a negative effect on employment or corporate profit for that matter. Moosa and Ramiah (2014) examine the level and growth rate of unemployment in the US over the period 1966–2012 and demonstrate that the level of employment has been experiencing a steady secular growth resulting from the growth of population and the economy. They do not observe any long-term effect of environmental regulation on the level of employment and identify several occasions on which the level of employment declined then recovered, exhibiting typical cyclical variation. For example, they contend that the behavior of employment cannot be attributed to environmental regulation because it is not that regulation was tightened in 1974 and relaxed in 1976, and nothing like that happened in 2007 and 2010 (these are years of recession and recovery). By analyzing cross-sectional data, they also show that the unemployment rate is unrelated to regulatory burden and negatively related to environmental performance, reaching the conclusion that countries with high environmental performance tend to have low unemployment rates. This does not necessarily mean that environmental performance leads to a decline in unemployment, but it means that employment is not negatively related to environmental performance.

13.7 NET EXPORTS

The available evidence on the effects of environmental regulation on net exports is inconclusive. Tobey (1990) uses a quantitative measure of national environmental stringency whereby the environmental policies of about 40 countries are sorted on a scale from 1 (strict) to 7 (tolerant) and finds that this index does not have a statistically significant effect on net exports. Tobey's results are consistent with those obtained by Walter (1982), Pearson (1987), and Leonard (1988). Likewise, Grossman and Krueger (1993) conclude that pollution control costs in the US have not affected imports from Mexico or activity in the Maquiladora sector along the US—Mexico border, thus casting doubt on the proposition that environmental regulation has significant adverse effects on net exports. Harris et al. (2000) conclude that, as soon as both the importing and exporting country-specific effects are taken into consideration, the relationship between stricter regulations and foreign trade becomes statistically insignificant. They suggest that environmental costs do not have a real impact, neither negative nor positive, on foreign trade.

Low and Yeats (1992) analyze trade patterns in the exports of "dirty" industries in different countries and find that during the period 1965—1988, the relative proportion of "dirty" product exports in various countries' total exports declined for the developed countries and went up significantly in Eastern Europe, Latin America, and West Asia. By using a Heckscher—Ohlin analysis of revealed comparative advantage based on export/import patterns for different sectors, they also find that developing countries enjoyed a rapidly increasing comparative advantage in pollution-intensive products during that period because their exports of dirty products grew more rapidly than exports generally. Maitra (2003) examines the relation between environmental regulation and net exports over the period 1967—1977 for 78 industrial categories after controlling for other explanatory variables. He finds no statistically significant relation between the two variables, but when he restricts the sample to manufacturing industry the relation turns out to be significantly negative. As in Kalt (1988), when the chemical industry is excluded, the relation becomes more strongly negative, a result that Jaffe et al. (1995) explain in terms of the strong export performance of this sector.

Some relevant work has been conducted on developing countries. Babiker (2009) examines the effect of environmental regulation on the petrochemical exports of Kuwait. While he fails to provide strong support for the proposition that environmental policy is used strategically in some export markets, he underscores the legitimacy of the concerns that strict environmental regulation in developing countries has a negative impact on their exports. He finds it ironic that domestic rather than international regulation poses more challenging questions to industry competitiveness. Based on China-ASEAN panel data in the period 1990—2007, Wu (2013) uses the gravity model to analyze the effect of the environmental regulation of ASEAN on China's exports. He uses energy consumption per capita and carbon dioxide emissions per capita as two proxies for the stringency of environmental regulation. His results reveal that China's exports are less sensitive to the environmental regulation of ASEAN member countries when the scale of exports is relatively small and that exports are more sensitive to ASEAN environmental regulation when "dirty products" cover a greater proportion. The evidence is indeed mixed and inconclusive.

13.8 COMPETITIVENESS

The loss of competitiveness is related to the loss of comparative advantage. The conventional wisdom is that environmental regulation has an adverse effect on competitiveness in international markets because it imposes significant costs and impedes productivity growth. The loss of competitiveness is reflected in declining exports, rising imports, and the tendency of regulated firms, particularly in pollution-intensive industries, to move to other countries. On the other hand, there is the opposite view that environmental regulation may boost international competitiveness. This proposition, which is associated mostly with the work of Porter (1991), is (understandably) widely endorsed by regulators. The EPA has for long concluded that environmental regulation induces "more cost-effective processes that reduce emissions and the overall cost of doing business" (Environmental Protection Agency, 1992). Esty and Porter (2002) suggest that a growing body of research indicates that competitiveness and environmental performance are compatible, if not mutually reinforcing. This is because low pollution and efficient energy indicate a highly productive use of resources.

Some in public policy circles would argue that the competitiveness of the economy as a whole can be enhanced by environmental regulation. It has been suggested that the imposition of strict environmental regulation induces innovation that can create lasting comparative advantage for domestic firms, if other countries eventually follow the lead to stricter regulation (see, for example, Gardiner, 1994). Barbera and McConnell (1990) found that lower production costs in the nonferrous metals industry were brought about by environmental regulation that led to the introduction of new, low-polluting, and efficient production practices. By forcing a reexamination of products and processes, regulation may induce an overall increase in the resources devoted to research, which would enhance competitiveness.

Stewart (1993) puts forward a different view by suggesting that the imposition of stringent environmental regulation may harm international competitiveness, even though most empirical studies have not established a strong causal association between regulation and competitiveness. The threat posed by regulation, the argument goes, is particularly significant in the US, due to the "exceptionally complex, burdensome and costly character of its regulatory and legal system." Yet Stewart admits that even if that were true, focusing solely on competitiveness is myopic and that the contribution of a cleaner environment and resource conservation to well-being must also be taken into account. What matters ultimately, he argues, is the broad overall performance of the economy, including the ecological and health benefits generated by environmental regulation. Nevertheless, issues of competitiveness and trade have high political visibility, particularly when they are manifested in plant closings and relocations.

International competitiveness is typically measured in terms of the real effective exchange rate, which depends on (1) the nominal exchange rate; (2) domestic inflation; and

(3) foreign inflation. Since a higher real exchange rate (a strong currency in real terms) implies lower competitiveness, environmental regulation will have an erosive effect on competitiveness if it can cause nominal currency appreciation or a surge in domestic inflation relative to foreign inflation. However, the link between environmental regulation and the nominal exchange rate is not straightforward—rather, it is indirect and ambiguous. A consideration of conventional exchange rate models tells us that environmental regulation may lead to nominal currency appreciation or depreciation, depending on how regulation is related to the explanatory variables of these models. For example, if environmental regulation has an adverse effect on economic growth then (according to the flexible-price monetary model) it will lead to currency depreciation. For a detailed discussion of the effect of environmental regulation on the nominal exchange rate, see Moosa and Ramiah (2014).

Environmental regulation has an adverse effect on inflation if it adversely affects productivity, production costs, and prices. Again, these effects are not clear. In considering any possible effects of environmental regulation on prices, Haveman and Christainsen (1981) make a distinction between a one-time increase in the general price level (also known as an inflationary burst) and a sustained increase in prices. They suggest that the relative prices of particular goods and services may rise on a one-time basis as a result of environmental regulation, but it cannot cause a sustained increase in the general price level or a rise in the inflation rate.

Esty and Porter (2002) present cross-sectional evidence on the link between environmental performance and competitiveness. They use three measures of environmental performance and the global competitiveness index and conclude that "on the tradeoff between green and competitive, we find no evidence that improving environmental quality compromises economic progress" and that "strong environmental performance appears to be positively correlated with competitiveness." They add that "the quality of a nation's environmental regulatory regime is strongly and positively correlated with its competitiveness." As an example, they suggest that Finland ranks at the top in terms of the environmental regulatory regime index and the competitiveness index. They interpret this result as supportive of the "soft version of the Porter hypothesis," suggesting that environmental progress can be achieved without sacrificing competitiveness.

Vassilopoulos (1999) summarizes the debate by suggesting that the impact of environmental regulation on competitiveness is essentially an empirical question. For example, those who argue that it does not have a significant effect do not deny the logic of the argument but merely claim that at present levels of environmental regulation, cross-country differences are not sufficiently large to affect the profitability of firms, trade, and investment patterns. He also argues that "one could summarize that there is no empirical evidence that high, or relatively high, environmental standards have a systematic negative impact on competitiveness at the macroeconomic or microeconomic level." He also contends that "most studies show insignificant relationships between stringent environmental regulations and competitiveness."

Moosa and Ramiah (2014) present stylized facts about the relation between competitiveness and environmental regulation. They show that high inflation was witnessed in the 1970s because of the increase in oil prices, while the deflation of 2009 was not caused by the dismantling of regulation but rather by the global financial crisis. This deflation actually coincided with the call for tighter regulation. Moosa and Ramiah (2014) provide cross-sectional evidence suggesting that a more conspicuous negative relation between inflation and environmental performance, when the association as envisaged by the opponents of regulation is positive. They also demonstrate a clear positive association between competitiveness and environmental performance (this relation, according to the proponents of environmental regulation, should be negative). Moosa and Ramiah conclude that "there is nothing in these stylized facts to show that environmental regulation boosts inflation and reduces competitiveness."

13.9 ECONOMIC GROWTH, ENVIRONMENTAL DEGRADATION, AND REGULATION

The rationale for believing that environmental regulation hurts economic growth is that higher production costs induced by regulation have an adverse effect on the production of commercial output. Furthermore, the diversion of resources to the production of items that are not included in GDP, such as clean air, reduces the level of commercial output even further. But it has been suggested that to envisage an adverse effect on growth (because of the diversion of resources away from the production of commercial output) is a reflection of the failure of national income accounting, which excludes national welfare from the measure of GDP (Office of Technology Assessment, 1994).

The relation between environmental regulation and economic growth stems from the perceived trade-off between economic growth and environmental degradation. Growing economic activity (production and consumption) requires progressively larger amounts of inputs of energy and materials while producing larger quantities of waste, hence causing environmental degradation. Daly (1991) argues that growing extraction of natural resources, accumulation of waste, and concentration of pollutants will "overwhelm the carrying capacity of the biosphere and result in the degradation of environmental quality and a decline in human welfare, despite rising incomes." Furthermore, it is arguable that degradation of the resource base will eventually put economic activity itself at risk. It follows that to save the environment, and even economic activity from itself, economic growth must slow down. If environmental regulation is intended to protect the environment, it must impede economic growth.

On the other hand, there are those who believe that the way to environmental improvement is along the path of economic growth as higher incomes boost the demand for goods and services that are less material intensive, as well as demand for improved environmental quality that leads to the adoption of environmental regulation. As Beckerman (1992) puts it, "strong correlation between incomes, and the extent to which environmental protection measures are adopted, demonstrates that in the longer run, the surest way to improve your environment is to become rich." This proposition is intuitively sound: poor people are more concerned about having the next meal than about whether or not the last breath they took has an acceptable level of sulfur dioxide. Esty and Porter (2002) argue that the promotion of economic growth is a key mechanism for improving environmental results.

Some economists believe that the relation between economic growth and environmental quality is time-varying along a country's development path (for example, Shafik and Bandyopadhyay, 1992; Panayotou, 1993; Grossman and Kruger, 1993; Selden and Song, 1994). It may change sign from positive to negative as a country reaches a level of income at which people can afford more efficient infrastructure and a cleaner environment. The implied inverted-U relation between environmental degradation and economic growth is called the "environmental Kuznets curve" (Kuznets, 1965, 1966). The important policy implication of the shape of this curve is that if environmental degradation rises with economic growth, this provides justification for strict environmental regulation and putting limits on economic growth. This is the policy-response explanation for the negative relation between environmental regulation and economic growth.

Empirical models of the environment and growth are typically reduced form single-equation specifications relating an environmental impact indicator to income per capita (which is a measure of economic development rather than economic growth). The dependent variable is typically emissions of a particular pollutant such as sulfur dioxide and carbon dioxide or it could be concentrations of various pollutants or some composite index of environmental degradation. Several functional specifications have been used, including quadratic, log quadratic, and cubic in income. The models are estimated using cross-sectional or panel data. Given the big ideological element surrounding this debate and various combinations that can be pursued to produce the results, the tendency would be to torture the data until results are obtained to confirm an ideologically motivated preconceived idea. This is what Moosa (2012) refers to as the use of "stir-fry regressions" to prove a point. As Panayotou (2000) puts it, "the ad hoc specifications and reduced form of these models turn them into a 'black box' that shrouds the underlying determinants of environmental quality and circumscribes their usefulness in policy formulation."

Lim (1997) explores the relation between economic growth and environmental quality in South Korea and finds an inverted U-shaped relation between environmental pressure and economic growth. The turning point of the curve is located around the early 1980s around the time when Korean environmental regulation and policies became more serious. The results show some evidence for the effectiveness of these policies.

Stern (2004) concludes that the empirical support for the environmental Kuznets curve is ambiguous. Zhang (2012) recommends caution in relying on the environmental Kuznets curve hypothesis to solve the environmental problems associated with economic growth, arguing that the inverted U-shaped curve may be "an artifact of restrictive functional forms in the sense that 'true' relationship could be N-shaped or an even more flexible shape." The policy implication of this strand of research is that we cannot guarantee that environmental quality will always improve once a certain level of income is reached. Moreover, causality between economic growth and the environment is not well established because of concerns about omitted variables and reverse causality. For example, while environmental regulation affects income and emissions simultaneously, it is typically omitted from the analysis. Zhang also argues that while environmental quality can be a productive input for economic growth, the feedback is ignored.

A number of studies deal with the impact of environmental regulation on GDP. Jorgenson and Wilcoxen (1990) analyze the impact of environmental regulation by simulating the long-term growth of the U.S. economy with and without regulation. The results reveal that over the period 1974–1985, the combined effect of these costs was to reduce the average growth rate of real gross national product (GNP) by about 0.2% points per year, with required investment having the biggest effect and operating costs the smallest. Meyer (1992) attempts to find out if strict environmental laws are associated with poor economic performance. He concludes that the pursuit of environmental quality does not hinder economic growth and development. Furthermore, there appears to be a moderate, yet consistent, positive association between environmentalism and economic growth. Jaffe et al. (1995) cast a shadow of doubt on the validity of Meyer's results, arguing that "his statistical analysis sheds very little light on a possible causal relationship between regulation and economic performance." His approach, they argue, does not control for factors other than the stringency of environmental laws that could affect economic performance.

While it is intuitive to think that environmental degradation, growth and environmental regulation are related, the underlying interaction is not straightforward. One scenario is that economic growth leads to environmental degradation, which makes it necessary to impose environmental regulation that exerts a negative effect on growth, thus reducing the pace of environmental degradation. Another scenario is that economic growth leads to environmental degradation, which makes it necessary to impose environmental regulation that exerts a positive effect on growth, thus accelerating the pace of environmental degradation. Yet another scenario is that economic growth leads to environmental improvement, which makes it unnecessary to impose environmental regulation, producing a positive effect on growth and further environmental improvement. The last scenario is that economic growth leads to environmental improvement, which makes it unnecessary to impose environmental regulation, producing a negative effect on the environment (at least in the sense that it will not improve). In this case environmental regulation is required to achieve environmental improvement. The stylized facts show that U.S. industrial production exhibits typical cyclical behavior that has nothing to do with environmental regulation. The last and biggest negative growth rate was caused by the global financial crisis whereas earlier ones were caused by the rise in the price of crude oil. The fact of the matter is that we cannot explain a typical cyclical behavior (of industrial production) in terms of a variable that does not move cyclically (environmental regulation).

13.10 AGGREGATE PRODUCTIVITY

A free market doctrinal explanation for the macroeconomic effects of environmental regulation on the level and growth of productivity is as follows. Environmental regulation, like any other form of regulation, represents an intervention in market processes, which causes production levels to deviate from those that can be attained without intervention. Holding output composition constant, this deviation means that additional inputs are required to reach any given level of output. Under these conditions, tighter regulation is associated with more significant deviations from the potential level of output and lower rates of growth of output per unit of input—that is, productivity. The channels whereby regulation affects either the output (numerator) or input (denominator) of productivity are complex. Jaffe et al. (1995) note that market-based regulation may have a very different effect on productivity from the traditional command-and-control type regulation. Because market-based controls provide incentives for firms to update and improve their pollution control methods on a regular basis, productivity may actually rise under this type of regulation. Berman and Bui (1999) suggest that the effects of environmental regulation need not be negative (and may be positive) and raise the question why there is no consensus on the effects of environmental regulation on productivity.

Several approaches have been followed in the literature to measure the productivity effects of environmental regulation. The three most common approaches are growth accounting, macroeconomic general equilibrium models, and single-equation models. A good example of the growth accounting methodology is provided by Denison (1979), who measures changes in total factor productivity and estimates the incremental cost brought about by environmental regulation in the post-1967 period. On the basis of his results, Denison concludes that environmental regulation was responsible for 13–20% of the productivity loss in that period. Haveman and Christainsen (1981) suggest that one of the striking aspects of Denison's study is the huge residual factor (which he labeled "advances in knowledge not elsewhere classified"). They argue that his study does not reveal the reasons for the sudden decline in the residual during the most recent period. The growth accounting approach does not consider the effects of the energy crisis, which involve the costs of adapting plants to be suitable for using substitute fuels and the increased obsolescence of some plant and equipment because of other factor substitutions. Another criticism of Denison's work is that it ignores the diversion of labor and capital to the redesign of products when energy prices induce a switch in the pattern of consumer demand (for example, from large to small fuel-efficient cars).

Jorgenson and Wilcoxen (1990) represent the U.S. economy by a general equilibrium macroeconomic model that includes a long-term growth component with and without environmental regulation. The model reveals that in the absence of environmental regulation, the capital stock would have been 3.79% higher and GNP would have been more than 2.50% higher. The macroeconomic study of Data Resources Incorporated indicates that environmental regulation reduces productivity as the induced pollution control investment crowds out alternative capital investments in plant and equipment (Data Resources Incorporated, 1979). In describing the results of the simulation analysis of labor productivity, it is stated that "the increased factor demands associated with the operating and maintenance and pollution abatement equipment resulted in a drop in labor productivity" because "any given firm would now require additional employees to produce the same level of output."

Single-equation models are used to study the pattern of changes in aggregate productivity over time in conjunction with the time pattern of other aggregate variables that might be expected to relate to, or explain, productivity. This approach, which is described by Haveman and Christainsen (1981) as "less ad hoc than the approach used by Denison," involves the estimation of regression equations in which productivity is the dependent variable and the explanatory variables are the hypothesized determinants of productivity (such as regulatory intensity and cyclical factors). The single-equation approach is used by Siegel (1979), who identifies structural breaks in the productivity series in 1967 and 1973. He explains these structural breaks in terms of the demographic composition of the labor force, which is found to be consistently important, and (from 1973 onward) changes in relative energy prices, which is the single most important explanatory variable. Kopcke (1980) follows a production function approach whereby current engineering knowledge determines the maximum, or potential, output per unit of material input that can be produced with a given stock of labor and capital. He reaches the conclusion that much of the slump in productivity growth resulted from a slower rate of capital accumulation. Specifically, he argues that one half of the decline in labor productivity for nonfarm nonresidential business is due to slower growth of the stock of plant and equipment. For manufacturing, slower capital accumulation may account for the entire drop in productivity growth.

The stylized facts presented by Moosa and Ramiah (2014) show that it does not make sense to blame the erratic behavior of the growth rate of labor productivity on environmental regulation. They also show that the pattern of productivity growth is similar across the G7 countries although the timing of the introduction of environmental regulation is not the same. Furthermore, they provide evidence indicating that labor productivity is positively related to either regulatory burden or environmental performance. If anything, they argue, "the association between environmental regulation and productivity seems to be positive."

13.11 CONCLUSION

The economic and financial effects of environmental regulation constitute a controversial issue. On balance, an examination of the views, stylized facts, and empirical evidence shows that the adverse effects of environmental regulation are exaggerated. Some evidence indicates that environmental regulation exerts positive effects on the indicators it is supposed to inflict damage on. According to standard economic analysis, regulation reduces productivity but the opposite view is that regulation can boost productivity by introducing cleaner, more efficient technologies in the workplace. Porter and van der Linde (1995b) reject the view that environmental regulation raises the costs of production and erodes productivity. They describe this view as "static" because everything except regulation is held constant. While they agree with the proposition that regulation raises the costs of production if technology, products, processes, and customer needs were all fixed, they cast doubt on its validity in "the real world of dynamic competition, not in the static world of much economic theory." Firms regularly find innovative solutions to pressures of all sorts arising from the activities and actions of competitors, customers, and regulators.

While the opponents of environmental regulation argue strongly that environmental regulation has adverse effects that may impinge on the decision-making process (for example, with respect to plant location and the level of production), the proponents of regulation suggest that the effect is either neutral or positive. While the empirical evidence is mixed, it mostly supports the proregulation view. From a macroeconomic perspective, it is bizarre to attribute the events of the 1970s to the advent of environmental regulation. Likewise, it does not make sense to explain sluggish growth in the aftermath of the global financial crisis. What happened in the 1970s reflected a malfunctioning of the economy. Blaming economic sluggishness in the aftermath of the global financial crisis on environmental regulation does not make sense because it is not that there was any intensification of environmental regulation as a result of the crisis. Most of the empirical evidence indicates that environmental regulation does not exert negative effects on employment, exports, and competitiveness.

One basic and overriding point is that the contribution of environmental regulation is not reflected in commercial output. Rather it is represented by improved health (implying less demand for medical care), longer lives, expanded outdoor recreation opportunities, greater enjoyment of existing recreation opportunities, and reduced demand for cleaning and other protective activities. If the standard productivity measures were effective indicators of economic welfare, these contributions should be included in the numerator of the measure (that is, output). Although they are difficult to quantify, numerous studies have indicated that environmental regulation produces social benefits. If this is the case, the effect of environmental regulation on productivity is bound to be less negative than the estimates produced by following various approaches. It could even be positive.

The business community tends to exaggerate the adverse effects of regulation in general, and environmental regulation in particular, on their businesses—the bottom line to be specific. But even if their fear of environmental regulation is justified because their bottom lines are adversely affected by environmental regulation, do we just not do anything and carry on destroying the planet in the name of free markets? Environmental regulation is a life-and-death matter. We cannot spend money on the military without any questions asked for the sake of national security and in the name of saving lives threatened by foreign enemies. The enemy within (environmental damage) is more conspicuous and saving lives should be about saving lives no matter what the threat is. Provoked by our actions, Mother Nature has already responded with a chemical attack on us. It is about time we get our priorities right.

REFERENCES

Babiker, M., 2009. The Impact of Environmental Regulations on Exports: A Case Study of Kuwait Chemical and Petrochemical Industry. Arab Planning Institute. Working Papers, No. 0209.

Barbera, A.J., McConnell, V.D., 1990. The impact of environmental regulations on industry productivity: direct and indirect effects. J. Environ. Econ. Manage. 18 (1), 50–65.

Bartik, T.J., 1988. The effects of environmental regulation on business location in the United States. Growth Change 19 (3), 22–44.

Beard, T.R., Ford, G.S., Kim, H., Spiwak, L.J., 2011. Regulatory Expenditures, Economic Growth and Jobs: An Empirical Study. Phoenix Center Policy Bulletin No. 28. Available at: http://www.phoenix-center.org/PolicyBulletin/PCPB28Final.pdf.

Becker, R., Henderson, V., 2000. Effects of air quality regulations on polluting industries. J. Polit. Econ. 108 (2), 379–421.

Becker, R., Henderson, V., 2001. Costs of air quality regulation. In: Carraro, C., Metcalf, G.E. (Eds.), Behavioral and Distributional Effects of Environmental Policy. University of Chicago Press, Chicago.

Beckerman, W., 1992. Economic growth and the environment: whose growth? Whose environment? World Dev. 20 (4), 481–496.

Berman, E., Bui, L.T.M., 1999. Environmental Regulation and Productivity: Evidence from Oil Refineries. Working Paper. Boston University, Boston.

Berman, E., Bui, L.T.M., 2001. Environmental regulation and labor demand: evidence from the south coast air basin. J. Public Econ. 79 (2), 265–295.

Bureau for Labor Statistics, 2011. Mass Layoff Statistics (MLS) Database. Available at: http://www.bls.gov/mls/#data.

Chen, K., Metcalf, R., 1980. The relationship between pollution control records and financial indicators revisited. Account. Rev. 55 (1), 168–180.

Crain, N.V., Crain, W.M., 2010. The Impact of Regulatory Costs on Small Firms. Small Business Administration, Washington, DC.

Crandall, R.W., 1980. Pollution controls and productivity growth in basic industries. In: Cowing, T., Stevenson, R. (Eds.), Productivity Measurements in Regulated Industries. Academic Press, New York.

Crandall, R.W., 1993. Manufacturing on the Move. The Brookings Institution, Washington, DC.

Daly, H., 1991. Steady-State Economics, second ed. Island Press, Washington, DC.

Data Resources Incorporated, 1979. The Macroeconomic Impact of Federal Pollution Control Programs: 1978 Assessment. Report Submitted to the Environmental Protection Agency and the Council on Environmental Quality, January.

Denison, E., 1979. Accounting for Slower Economic Growth: The U.S. in the 1970s. Brookings Institution, Washington, DC.

Dowell, G., Hart, S., Yeung, B., 2000. Do corporate global environmental standards create or destroy value? Manage. Sci. 46 (8), 1059–1074.

Duerksen, C.J., 1983. Environmental Regulation of Industrial Plant Siting. The Conservation Foundation, Washington, DC.

Eberly, J., 2011. Is Regulatory Policy a Major Impediment to Job Growth? Office of Economic Policy, Washington, DC.

Environmental Protection Agency, 1992. In: The Clean Air Market Place: New Business Opportunities Created by the Clean Air Act Amendments: Summary of Conference Proceedings. Office of Air and Radiation, Washington, DC.

Epping, M.G., 1986. Tradition in transition: the emergence of new categories in plant location. Arkansas Bus. Econ. Rev. 19 (3), 16–25.

Esty, D.C., Porter, M.E., 2002. Ranking national environmental regulation and performance: a leading indicator of future competitiveness?. In: World Economic Forum, Global Competitiveness Report 2001–2002. Oxford University Press, New York.

Feldman, S., Soyka, P., Ameer, P., 1996. Does Improving a Firm's Environmental Management System and Environmental Performance Result in a Higher Stock Price? ICF Kaiser, Washington, DC.

Friedman, J., Grlowski, D.A., Silberman, J., 1992. What attracts foreign multinational corporations? Evidence from branch plant location in the United States. J. Reg. Sci. 32 (4), 4113–4118.

Gardiner, D., 1994. Does environmental policy conflict with economic growth? Resources 115 (Spring), 20–21.

Goodstein, E.B., 1994. Jobs and the Environment: The Myth of a National Trade-Off. Economic Policy Institute, Washington, DC.

Graff, J.S., Neidel, Z.M., 2011. The Impact of Pollution on Worker Productivity. NBER. Working Papers, No. 17004.

Gray, W.B., Shadbegain, R.J., 1993. Environmental Regulation and Manufacturing Productivity at the Plant Level. NBER. Working Papers, No. 4321.

Greenstone, M., List, J.A., Syverson, C., September 2012. The Effects of Environmental Regulation on the Competitiveness of U.S. Manufacturing. Working Paper.

Grossman, G., Kreuger, A., 1993. Environmental Impacts of a North American Free Trade Agreement: The U.S.-Mexico Free Trade Agreement. MIT Press, Cambridge, MA.

Hamilton, J., 1995. Pollution as news: media and stock market reactions to the toxics release inventory data. J. Environ. Econ. Manage. 28 (1), 98–113.

Harris, M.N., Kónya, L., Mátyás, L., 2000. Modelling the Impact of Environmental Regulations on Bilateral Trade Flows: OECD, 1990–1996. Melbourne Institute. Working Papers, No. 11/00.

Hart, S.L., 1997. Beyond greening strategic for sustainable world. Harv. Bus. Rev. 75 (1), 66–76.

Haveman, R.H., Christiansen, G.B., 1981. Environmental regulations and productivity growth. In: Peskin, H.M., Portney, P.R., Kneese, A.V. (Eds.), Environmental Regulations and the US Economy. Resources for the Future, Washington, DC.

Jacobe, D., October 24, 2011. Gov't Regulations at Top of Small-business Owners' Problem List. Gallup Economy. Available at: http://wwwgallup.com/poll/150287/Gov-Regulations-Top-Small-Business-Owners-Problem-List.aspx.

Jaffe, A.B., Peterson, S.R., Portney, P.R., Stavins, R.N., 1995. Environmental regulation and the competitiveness of U.S. manufacturing: what does the evidence tell us? J. Econ. Lit. 33 (1), 132–163.

Jagg, B., Freedman, M., 1992. An examination of the impact of pollution performance on economic and market performance. J. Bus. Finance Account. 19 (5), 697–713.

Jorgenson, D.W., Wilcoxen, P.J., 1990. Environmental regulation and U.S. Economic growth. Rand J. Econ. 21 (2), 314–340.

Kalt, J.P., 1988. The impact of domestic environmental regulatory policies on U.S. International competitiveness. In: Spence, A.M., Hazard, H.A. (Eds.), International Competitiveness. Harper and Row, Cambridge, MA.

Klassen, R.D., McLaughlin, C.P., 1996. The impact of environmental management on firm performance. Manage. Sci. 42 (8), 1199–1214.

Kopcke, R.W., 1980. Capital accumulation and potential growth. Federal Reserve Bank of Boston. Conference Series No. 22.

Kuznets, S., 1965. Economic Growth and Structural Change. Norton, New York.

Kuznets, S., 1966. Modern Economic Growth. Yale University Press, New Haven.

Lanjouw, J., Mody, A., 1993. Stimulating Innovation and the International Diffusion of Environmentally Responsive Technology: The Role of Expenditures and Institutions. World Bank, Mimeo.

Leonard, H.J., 1988. Pollution and the Struggle for the World Product. Cambridge University Press, Cambridge.

Levinson, A., 1996. Environmental regulations and manufacturer's location choices: evidence from the census of manufacturers. J. Public Econ. 62 (1–2), 5–29.

Lim, J., 1997. Economic Growth and Environment: Some Empirical Evidences from South Korea. Working Paper. University of New South Wales, School of Economics.

Liu, H.C.K., June 18 2005. Scarcity Economics and Overcapacity. Asia Times Online.

Low, P., Yeats, A., 1992. Do Dirt Industries Migrate? World Bank. World Bank Discussion Paper Series, No. 159.

Lyne, J., October 1990. Service taxes, international site selection and the "green" moment dominate executives. Polit. Focus 5, 1134–1138.

Mahapatra, S., 1984. Investors reaction to corporate social accounting. J. Bus. Finance Account. 11 (1), 29–40.

Maitra, P., 2003. Environmental Regulation, International Trade, and Transboundary Regulation. Available at: http://www.eolss.net/Sample-Chapters/C13/E1-23-06-03.pdf.

McConnell, V.D., Schwab, R.M., 1990. The impact of environmental regulation on industry location decisions: the motor vehicle industry. Land Econ. 66 (1), 67–81.

McFadden, D., 1974. Conditional logit analysis of qualitative choice behavior. In: Zarembka, P. (Ed.), Frontiers in Econometrics. Academic Press, New York.

McGuire, J., Sundgren, A., Schneeweis, T., 1988. Corporate social responsibility and firm financial performance. Acad. Manage. J. 31 (4), 854–872.

McWilliams, A., Siegel, D., 2000. Corporate social responsibility and financial performance. Strategic Manage. J. 21 (5), 603–609.

Meyer, S.M., 1992. Environmentalism and Economic Prosperity: Testing the Environmental Impact Hypothesis. MIT, Mimeo.

Mill, G., 2006. The financial performance of and socially responsible investment over time and possible link with corporate social responsibility. J. Bus. Ethics 63 (2), 131–148.

Moosa, I.A., 2002. Foreign Direct Investment: Theory, Evidence and Practice. Palgrave, London.

Moosa, I.A., 2012. The failure of financial econometrics: "stir-fry" regressions as an illustration. J. Financ. Transform. 34 (1), 43–50.

Moosa, I.A., Cardak, B.A., 2006. The determinants of foreign direct investment: an extreme bounds analysis. J. Multinatl. Financ. Manage. 16 (2), 199–211.

Moosa, I.A., Ramiah, V., 2014. The Costs and Benefits of Environmental Regulation. Edward Elgar, Cheltenham.

Morgenstern, R.D., Pizer, W.A., Shih, J.S., 1998. The Cost of Environmental Protection. Resources for the Future. Discussion Paper No. 98-36.

Morgenstern, R.D., Pizer, W.A., Shih, J.S., 2000. Jobs versus the Environment: An Industry Level Perspective. Working Paper. Resources for the Future, Washington, DC.

Murray, A., Sinclair, D., Power, D., Gray, R., 2006. Do financial markets care about social and environmental disclosure? Auditing Account. J. 19 (2), 228–255.

Office of Management and Budget, 2011. Report to Congress on the Benefits and Costs of Federal Regulations and Unfunded Mandates on State, Local, and Tribal Entities. OMB, Washington, DC.

Office of Technology Assessment, 1994. Industry, Technology and the Environment: Competitive Challenges and Business Opportunities. U.S. Congress, Washington, DC.

Panayotou, T., 1993. Empirical Tests and Policy Analysis of Environmental Degradation at Different Stages of Economic Development. ILO Technology and Employment Programme. Working Papers, No. WP238.

Pannyotou, T., 2000. Economic Growth and the Environment. CID. Working Papers, No. 56.

Pearson, C.S. (Ed.), 1987. Multinational Corporations, Environment and the Third World. Duke University Press, Durham.

Pham, H.N.A., Ramiah, V., Moosa, I.A., 2014. Are European environmental regulations excessive? RMIT. Working Paper.

Porter, M.E., 1990. The Competitive Advantage of Nations. Free Press, New York.

Porter, M.E., 1991. America's green strategy. Sci. Am. 264 (4), 168.

Porter, M.E., van der Linde, C., 1995a. Towards a new conception of the environmental competitiveness relationship. J. Econ. Perspect. 9 (Fall), 97–118.

Porter, M.E., van der Linde, C., 1995b. Green and competitive: ending the stalemate. Harv. Bus. Rev. 73 (September–October), 120–134.

Ramiah, V., Martin, B., Moosa, I.A., 2013. How does the stock market react to the announcement of green policies? J. Banking Finance 37 (5), 1747–1758.

Ramiah, V., Pichelli, J., Moosa, I.A., 2015. Environmental regulation, the Obama effect and the stock market: some empirical results. Appl. Econ. 47 (7), 725–738.

Ramiah, V., Morris, T., Moosa, I.A., Gangemi, M., Puican, L. The effects of announcements of green policies on equity portfolios: evidence from the United Kingdom. Manage. Auditing J., forthcoming-a.

Ramiah, V., Pichelli, J., Moosa, I.A. The effects of environmental regulation on corporate performance: a Chinese perspective. Rev. Pac. Basin Financ. Mark. Policies, forthcoming-b.

Reinhardt, F., 1999. Market failure and the environmental policies of firm. J. Ind. Ecol. 3 (1), 9–21.

Repetto, R., 1990. Environmental productivity and why it is so important. Challenge 33 (September–October), 33–38.

Schmenner, R.W., 1982. Making Business Location Decisions. Prentice-Hall, Englewood Cliffs.

Schnietz, K., Epstien, M., 2005. Exploring the financial value of reputation for corporate social responsibility. Corp. Reput. Rev. 7 (4), 327–345.

Selden, T., Song, D., 1994. Environmental quality and development: is there a kuznets curve for air pollution emissions? J. Environ. Econ. Manage. 27 (2), 147–162.

Shafik, N., Bandyopadhyay, S., June 1992. Economic Growth and Environmental Quality: Time-Series and Cross-Country Evidence. World Bank Policy Research. Working Papers, No. 904.

Shapiro, I., Irons, J., April 2011. Regulation, Employment, and the Economy: Fears of Job Loss are Overblown. EPI Briefing Papers.

Siegel, R., 1979. Why has productivity slowed down? Data Resour. U.S. Rev. 8 (1), 59–65.

Sinclair, T., Vesey, K., 2012. Regulation, Jobs, and Economic Growth: an Empirical Analysis. George Washington University. Regulatory Studies Center. Working Paper.

Spicer, B., 1978. Investors' corporate social performance and information disclosure. Account. Rev. 53 (1), 94–111.

Stafford, H.A., 1985. Environmental Protection and industrial location. Ann. Assoc. Am. Geogr. 75 (2), 227–241.

Stern, D., 2004. The rise and fall of the environmental Kuznets curve. World Dev. 32 (8), 1419–1439.

Stewart, R., 1993. Environmental regulation and international competitiveness. Yale Law J. 102 (June), 2039–2106.

Tobey, J., 1990. The effects of domestic environmental policies on patterns of world trade: an empirical test. Kyklos 43 (2), 191–209.

Vassilopoulos, M., April 1999. Industrial Competitiveness and Environmental Regulation: Final Report. IPTS, Seville.

Waddock, S., Graves, S., 1997. The corporate social performance and financial performance link. Strategic Manage. J. 18 (4), 303–319.

Wagner, M., Vanphu, N., Azomahou, T., Wehrmeyer, W., 2002. The relationship between the environmental and economic performance for firms: an empirical analysis of European paper industry. Corp. Soc. Responsib. Environ. Manage. 9 (3), 113–146.

Wahba, H., 2008. Does the market value corporate environmental responsibility? An empirical examination. Corp. Soc. Responsib. Environ. Manage. 15 (2), 89–99.

Walter, I., 1982. Environmentally induced industrial relocation to developing countries. In: Rubin, S.J., Graham, T.R. (Eds.), Environment and Trade: The Relation of International Trade and Environmental Policy. Allanheld Osmun, Totowa.

Wheeler, D., Mody, A., 1992. International investment location decisions: the case of U.S. firms. J. Int. Econ. 33 (1–2), 57–76.

White, M., 1995. Does it Pay to Be Green? Corporate Environmental Responsibility and Shareholder Value. Working Paper. University of Virginia, Charlottesville.

Wintner, L., 1982. Urban Plant Siting. The Conference Board, New York.

Wu, S., 2013. The effect of environmental regulation of ASEAN on China's export trade. Commun. Inf. Sci. Manage. Eng. 3 (3), 145–153.

Yan, W., Carr, D., 2013. Federal environmental regulation impacts on local economic growth and stability. Econ. Dev. Q. 27 (3), 179–192.

Yang, J.L., November 13, 2011. Does Government Regulation Really Kill Jobs? Economists Say Overall Effect Minimal. Washington Post. http://www.washingtonpost.com/business/economy/does-government-regulation-really-kill-jobs-economists-say-overall-effect-minimal/2011/10/19/gIQALRF5IN_story.html.

Zhang, J., September 2012. Delivering Environmentally Sustainable Economic Growth: The Case of China. Working Paper. Asia Society.

CHAPTER 14

Environmental Challenges and Financial Market Opportunities

Colin Read

State University of New York College at Plattsburgh, Plattsburgh, NY, USA

Contents

14.1 INTRODUCTION

Traditional financial markets are efficient and adept in pricing accurately the vast majority of securities and commodities. However, it can only price the elements of commodities for which well-defined property rights exist. Until the Clean Air Act of 1990, though, there were no institutions that allowed the proper determination of environmental emissions like the sulfur dioxide that generates acid rain or the carbon dioxide that accelerates global warming. Since 1990, a combination of legislation and financial market innovation has created opportunities to trade emission rights. Already totaling tens of billions of US dollars per year, emissions markets are expected to grow dramatically over the century. I analyze the potential extent of these markets based on economic theory and published estimates and observations.

I begin in Section 14.2 with a review of the cap-and-trade system that was pioneered in the United States as part of the Clean Air Act of 1990. I also describe the economic theory that is embodied in the cap-and-trade system. I then describe the extension of this financial market innovation to other environmental emissions. In Section 14.3, I describe the Kyoto Protocol, its subsequent amendments, and the degree to which major nations have pledged to participate in future rounds designed to mitigate environmental damage. In Section 14.4, I describe various estimates of the extent of global emissions of greenhouse gases, and gage the degree to which financial markets must grow to accommodate and mitigate these changes. I then describe the necessary innovations in these financial markets and the degree to which such securitization of emissions will affect global capital flows in Section 14.5. I offer conclusions in Section 14.6.

Handbook of Environmental and Sustainable Finance
ISBN 978-0-12-803615-0
293

14.2 THEORY AND EXPERIENCES WITH CAP-AND-TRADE

One cannot trade what one does not own. In the 1960s, two economists honed our intuition with regard to the problem of incomplete markets. Decades earlier, the British economist Arthur Pigou put forward a theory that explores how the price system could accommodate pricing of a good or commodity for which the benefits or costs are also inflicted on third parties (Pigou, 1920). A positive externality confers a benefit upon a third party beyond the transaction, while the more commonly described negative externality imposes a cost on a third party who is not included in the financial transaction.

The classic example of such a negative externality is pollution or effluent discharge. If these by-products of production are not priced in proportion to the damage they inflict on other parties, the producer can be viewed as able to receive a necessary factor of production without any cost. In such a case, Pigou recommended the imposition of a *Pigouvian tax* that is equal to the cost of the damage inflicted. By doing so, the resulting product fully reflects and prices the factors commanded to produce the good. In the absence of such a cost, the producer would produce more than is optimal and impose damage to others to a greater degree.

Of course, such activity could also be regulated. However, the requisite monitoring costs and bureaucratic inefficiencies may out-swamp the damages themselves. Employment of the price system to determine these costs, and hence the optimal level of pollution is a more efficient way to remedy the emissions problem.

Intrinsic in this analysis, though, is the necessity to properly assign the appropriate property rights. In 1833, the British economist William Forster Lloyd produced a pamphlet that described the problems which arise when an unlimited set of shepherds are all allowed to graze their sheep on a common piece of public land. If the land belongs to everyone, it ultimately belongs to no one, Lloyd observed that overgrazing, and property destruction, often results. The ecologist Garrett Hardin developed this dilemma still further in his seminal 1968 article "The Tragedy of the Commons." (Hardin, 1968) Hardin is attributed to the generation of awareness among the environmental movement of the problem of unassigned property rights and nonsustainability.

Perhaps the most important contribution to our understanding of the nature of property and free markets came from Ronald Coase in his seminal article "The Problem of Social Cost" (Coase, 1960). In the most cited article in the history of law journals, Coase honed our intuition to the point that it could guide public and free market policy. Coase observed that private markets could properly price such factors as pollution if only pollution rights are assigned. His most intuitive observation was that it matters not how these rights are clearly assigned. Under certain circumstances, the market will produce the correct pricing results regardless of whether the polluter is given the right to pollute, the public is granted the right to clean air, or each is given the right up to some cap or floor.

Coase's intuition is instructive. If markets can efficiently mediate both sides of a transaction, then, should a polluter be granted the right to pollute, those who prefer no or less pollution can negotiate with the polluter to reduce pollution by paying the cost the polluter would incur to mitigate its discharge. The price these two sides of the market would arrive upon is one in which the marginal cost of a unit of pollution reduction is equal to the marginal benefit to the public of the pollution reduction. Beyond that point, further pollution reductions are presumed to become more expensive, while additional pollution abatement does not produce sufficiently large benefits to health or aesthetics. This mechanism is precisely how supply and demand interact to determine the price of commodities in more traditional markets.

Note that this same solution will also occur if the public is instead offered the right to clean air. Then, potential polluters could "purchase" the right to pollute from the public. If the market functions efficiently, it will do so until the costs saved by being permitted to pollute, on the margin, is equal to the compensation the polluter must offer to the public for the incremental right.

This intuition, which secured Coase the Nobel Memorial Prize in Economic Sciences in 1991, was the basis for a new financial market that was designed to alleviate the growing problem of acid rain in the Northeast United States in the 1980s. Sulfur dioxide and nitrogen oxide emissions as a by-product of coal-burning power plants and other industries were combining with moisture in the atmosphere and energy from the sun to form sulfuric and other acids, which then rained downwind of the plants in the lakes and streams of the Adirondacks and elsewhere. The resulting increasing concentration of acid in these waters progressively destroyed plant life and damage the ecosystem.

To mitigate these emissions and the acid rain damage, in 1990 President George Bush signed a revised Clean Air Act that incorporated the concept of cap-and-trade. Such a system sets a legal limit on the amount of an emission each producer could create. If a producer finds it too costly to stay below the rights assigned to it, this first producer must find a second producer who can reduce their emissions below their permitted target for a price less than the cost of the first producer. In essence, the first producer then buys excess discharge rights from the second producer.

The ethic of sanctioning pollution aside, this mechanism has some notable advantages over a command-and-control system. If each producer was rigidly required to maintain some sort of limit determined by a regulatory body, each producer will stay at the limit, even if it is very costly for one producer to do so, and even if another producer could easily reduce their discharges at only modest costs.

We can arrive at the same combined level of discharge under either cap-and-trade or command-and-control. But, only the cap-and-trade system induces cleaner technologies to innovate to become cleaner yet. In addition, the cap-and-trade system is self-policing because the market has the incentive to monitor the emissions of its participants. The private market, by mediating the exchange of pollution rights, can then also provide the ability to manage future discharge price risk. Finally, the government can monitor the overall price of emissions and can adjust the level of caps to ensure that the "price" of pollution corresponds to the social costs the discharges may impose. Rigid Pigouvian taxes do not provide such information.

The cap-and-trade system is not without some social controversy, though. By creating a market for pollution discharges, it sanctions pollution. However, any alternate system tolerates some level of discharge. The only issues are the overall level attained, the distributional effects of which sides of the market receives the market revenue and which incurs the costs, and whether the "price" of pollution properly reflects the social costs of its distribution. Should the price be inappropriate, the resulting distortion will produce either too much or too little pollution than is socially optimal.

The cap-and-trade system has been operating effectively through the Chicago Climate Futures Exchange through the Sulfur Dioxide Emission Allowance Trading Program since 1995. From a trading level of 5 million tons per year, trading has since risen to 11 million tons by 2003, and have generated trades totaling $4 billion US per year (The Chicago Climate Futures Exchange, 2004). This is a lower bound to the value of such trades, however. Much of the trading can occur internally within a multiplant corporation as it balances emissions between its operations (Figures 14.1 and 14.2).

By the early 2010s, improvements in changing technologies have substantially reduced the creation of sulfur dioxide to the extent that the market price for sulfur dioxide rights has become almost irrelevant. Instead, the focus has shifted on carbon dioxide emissions, based on their role in generating warmer average global

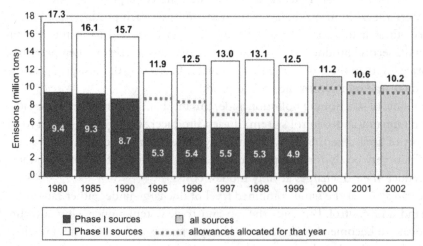

SO$_2$ Emissions under the Acid Rain Program

Figure 14.1 A reduction of 7 million tons of sulfur dioxide since 1980. *(Source: EPA (Capandtrade, 2015)).*

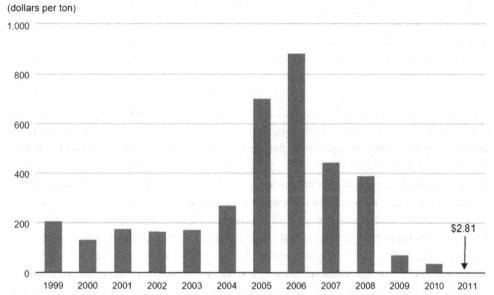

Weighted average price of spot sulfur dioxide for winning bids

(dollars per ton)

Figure 14.2 The evolution of spot sulfur dioxide prices since 1999. *(Source: U.S. Energy Information Administration (Todayinenergy, 2015)).*

temperature increases. While the nature of sulfur dioxide emissions tend to be more localized, carbon dioxide emissions anywhere contribute to greenhouse gas warming effects everywhere.

Cap-and-trade has been proposed to address this growing global problem. However, it is not the only potential solution. An alternative to the market-determined cap-and-trade price is a fixed price per unit of emissions. By imposing a fixed cost on the combustion of hydrocarbons in proportion to the amount of carbon dioxide released in its combustion, any emission of carbon dioxide is penalized in proportion to the amount of carbon dioxide released. Such a penalty will discourage carbon dioxide emissions and direct innovations and consumption toward more benign technologies.

Should the carbon tax be properly determined so that the market equilibrium coincides with the social optimum, such a carbon tax may have the same desired outcome as the cap-and-trade system. Of course, it is much broader because it impinges upon all producers of carbon dioxide. The primary difference, though, is the revenue created by the tax will flow inevitably to government agencies, which may or may not have the discipline to redirect these revenues toward the mitigation of greenhouse gas-created problems. Under such a regulatory regime, deep pocket polluters will have an increased incentive to lobby government to be exempted from the carbon tax.

In summary, a carbon cap-and-trade system can produce a limit on the quantity of emissions, while the market price varies to realize the desired quantity. On the other hand, a carbon tax fixes the price, but ultimately will allow variability in carbon emissions. While the latter solution is perhaps institutionally easier to administer by government, it may also lead to other challenges, including misplaced use of revenues, growth of government, and a greater incentive for lobbying and corruption to evade the regulatory costs.

Both of these mechanisms can potentially produce the same results, but in different ways. If the carbon tax is determined appropriately and optimally, so that it results in the optimal level of carbon discharge, then it has the same efficiency effect as a cap-and-trade system with an optimally determined emissions cap. Note, though, that the distributional and market effects differ substantially. In the cap-and-trade system, some participants profit, especially if they can become even cleaner than their discharge permits require. Transfers remain confined between participants in the private market. But, under a carbon tax, all dischargers pay, and all revenues flow to government coffers. This robs the private sector of new revenues that could be used by cleaner emitters to become even cleaner still. In effect, the cap-and-trade market system is a zero-sum exchange, while the carbon tax regime drains financial resources from all producers. It then remains a political decision whether these sums will be reinvested into the private sector, or will be used for other public finance purposes.

14.3 THE KYOTO PROTOCOL AND GLOBAL PARTICIPATION

While the concept is not without disagreement and controversy, the vast majority of scientists agree that the dramatic acceleration of the combustion of hydrocarbons since the onset of the Industrial Revolution has raised the level of carbon dioxide in the atmosphere from an average of less than 250 parts per million to over 400 parts per million (Figure 14.3).

Scientists generally agree that there is a strong positive correlation between the concentration of atmospheric carbon dioxide and the mean atmospheric temperature. Theory suggests that carbon dioxide, in addition to methane and other gases, helps to reflect back the emission of heat from the earth into deep space. In effect, the heat that is absorbed from the Sun is increasingly trapped within the atmosphere as the concentration grows of greenhouse gases. This greenhouse blanket induces a positive feedback loop that causes the additional release of naturally stored greenhouse gases within the soil, the ocean, and the melting tundra.

These scientific concerns have induced nations to come together to agree upon a series of measures that could mitigate, and eventually reverse, this buildup and positive feedback loop. Global action was galvanized through the 1992 United Nations Convention on Climate Change, and the result of the 1997 meeting of world leaders in Kyoto, Japan, which resulted in a Kyoto Protocol placed into effect in 2005. 192 nations and

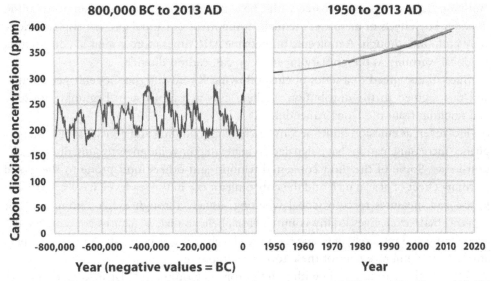

Figure 14.3 The trend in carbon dioxide concentration over the past 8000 centuries. *(Source: U.S. EPA (Climatechange, 2015)).*

groups are subscribers to the Protocol. The Kyoto Protocol required cutbacks of future growth of carbon emissions by nations, especially the world's most developed nations who generate the majority of emissions.

An ultimate agreement had to first balance the acceptable level of global warming with the appropriate level of emissions, mostly through mitigation of carbon dioxide discharges. The science of global warming is complicated and politically charged. If world leaders accept that carbon dioxide levels affect global warming, and if the proposition that human-produced emissions are the most easily influenced carbon dioxide source, among others, then the remaining issue is to decide the acceptable level of global warming and the requisite response through discharge limits.

A further round of scientific explorations and governmental negotiations in Copenhagen culminated in 2009 with the determination of an acceptable level of global warming by 2100.

The Intergovernmental Panel on Climate Change recommended that a reduction in carbon dioxide emissions must also be combined with technologies that further remove carbon dioxide from the atmosphere sometime between 2055 and 2070. This balance point would result in a net zero human carbon dioxide aggregate emission worldwide. If this point can be reached, scientists believe that global warming can reach a tipping point that can result in further reductions beyond 2070.

Similarly, the Panel proposed net zero methane and nitrous oxide emissions between 2080 and 2100. These greenhouse gases constitute 16% and 6% of greenhouse gas emission, respectively, as compared with a 65% greenhouse gas contribution from carbon dioxide. The effects of methane, the main component of natural gas, are disproportionately profound, though. A molecule of methane is 16 times more potent in contributing to global warming as the primary greenhouse gas, carbon dioxide.

Subscribers agreed to a series of cuts following Kyoto and the Copenhagen rounds, and then agreed to the distribution of these cuts in various steps beginning in 2012 and resulting from the Doha Amendment.

However, as emission limits become increasingly binding and penalties for noncompliance increasing, nations have signaled an increasing unwillingness to subscribe to additional cuts. Some of the most congested nations, and others most prone to the global warming effect of glacial melt and the concomitant rise in the sea level, have nonetheless introduced measures to curb global warming, usually through either carbon taxes or cap-and-trade exchanges. Many commentators believe that, as nations contemplate the accelerating and relatively dramatic rise in sea levels and ocean and atmospheric temperatures, future embracement of these tools will improve.

Meanwhile, nations mostly within the European Community, Japan, and some states and provinces in Australia, Canada, and the United States, have developed carbon reduction institutions. The most significant is the European Union Emissions Trading Scheme (EU ETS). The EU ETS has been integrated into the fabric of and embraced by the Kyoto Protocol by permitting trading of obligations among the Kyoto Protocol signees.

Within these groups, there are somewhat less-developed nations within Europe that were originally granted significant emission limits. While these higher limits were designed to allow these nations to develop their way to European gross domestic product (GDP) standards as they emerge from the former Soviet bloc economies, these nations often view their excess surpluses as an asset they can sell to mitigate the restructuring costs that are necessarily incurred as nations are forced to increasingly rely on more sustainable and benign energy sources. In effect, these rights act much like the transfer payments offered by more developed European Union members to its newer and less-developed members.

14.4 EXPECTED FUTURE EMISSIONS AND CAPS

The size of the resulting emission trading markets, and the size of intergovernmental transfers, all depends on the pattern of initial emission permits and the degree to which the caps will be reduced in the future. At first, the caps were designed to limit growth of emissions, but, as global warming accelerates, there is an increasing recognition that absolute emission levels must actually decrease, perhaps even substantially. Such emission gaps continue to evolve.

The United Nations Environmental Programme (UNEP) regularly publishes an analysis of the level of emissions and the gap between emissions and allowances. Their report from November of 2012 noted that the increasing trend for carbon dioxide emissions will well exceed the need for cutbacks to arrest and then reverse annual emissions consistent with the science that would allow the global average temperature to be restricted to a 2 °C increase (Figure 14.4).

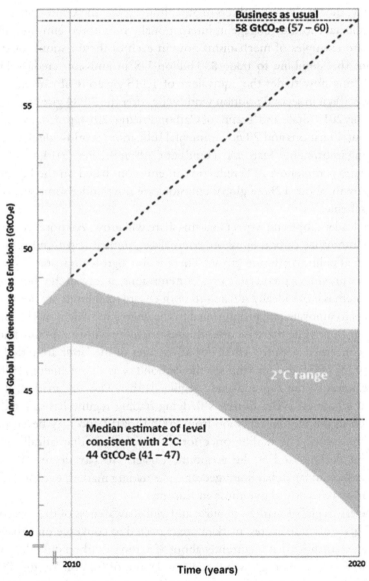

Figure 14.4 CO_2 emissions (gigatons/year) and temperature growth limitation range. *(Source: UNEP (Emissions Gap Report, 2012)).*

Clearly, there will be an increasing pressure for subscribers and other nations to reduce their allotments of emission permits if the global temperature rate of increase is to be limited to 2 °C. Reductions in the order of 25% over current trends will be necessary.

14.5 THE SIZE AND EXTENT OF ENVIRONMENTAL SECURITIZATION

Until recently, the two largest emitters of greenhouse gases, the United States and China, have been reluctant to fully engage in internationally monitored emission limits. However, there are examples of mechanisms now in each of these nations, or states within these nations that combine to trade $30 billion US in emission credits. The national market in China now trades the equivalent of 1.115 gigatons of carbon dioxide each year, which places it in second position worldwide, after the 2.084 gigatons traded within the EU ETS in 2013 (State and Trends of Carbon Pricing, 2014, p. 15). In addition, there are now about 40 nations and 20 governmental subnations worldwide that are imposing carbon pricing instruments (State and Trends of Carbon Pricing, 2014, p. 15). While scientific estimates commend a 25% reduction in emissions based on the baseline assumed emissions growth, about 12% of global emissions are now under some sort of trading or carbon tax scheme.

National leaders hope and expect that this share will grow. As more nations recognize the need to internalize carbon emission externalities, nonparticipation becomes increasingly difficult as political pressure grows. There is also a growing recognition that, while the carbon tax provides a predictable price for emissions, at least in the short run, the cap-and-trade system is more ideally situated to both ensure total limits are not exceeded and the incentives to innovate and proportional to the degree to which caps become increasingly binding. But, while there is currently more activity within cap-and-trade regimes, the nations concerned with the efficiency advantages of the latter are pitted against the subset of nations of the former who see the carbon tax as a convenient revenue source.

One weakness of the cap-and-trade regime is that, like any market, it suffers the vagaries of the business cycle. Europe's fledging trading regime has suffered as Europe went through its double-dip recession from 2008 to 2012 and may be on the verge of a triple-dip recession. The trading price for carbon declines dramatically as both economic output declines and as the economic capacity to pay decreases. At the same time, the recession in Japan has impinged upon its trading market, even if its equilibrium price remains significantly above those in Europe.

Nonetheless, as global markets stabilize and global awareness of the costs of unabated greenhouse gas growth increases, markets are expected to converge on a shadow carbon price for the externalities that is currently about $35/ton of carbon dioxide, and must rise from there to reduce carbon growth (State and Trends of Carbon Pricing, 2014, p. 18).

Such prices create opportunities for financial markets in two different ways. At a production rate of approximately 50 gigatons of carbon dioxide per year, the value of this shadow price represents $1.75 trillion dollars US per year, and growing. This sum will be divided between cap-and-trade markets, carbon taxes, and through private and public institutions that reward industries best positioned to reduce carbon dioxide production.

These calculations are consistent with other estimates. A recent review of current studies by Barker et al (Barker et al., 2007). While there are a number of potential baselines, based on various growth assumptions for national economies, and the interaction between potential future governmental policies, a range of costs arising from greenhouse gas discharges can be created. The authors estimate that the sum of net global GDP will decline between 0% and 3% by 2030, should greenhouse gas concentrations rise up to 710 parts per million. By 2050, the net effect in GDP ranges from a slight gain to a 5.5% loss.

Note that these estimates net out the negative externalities and positive externalities of greenhouse gas emissions. Typically, the coordination of policies has focused on the negative effects. However, benefits also accrue to some nations. For instance, the climates of Canada, Alaska, and Russia are projected to become more moderate. Growing and resource extraction seasons will expand, and domestic production will rise. These are the positive externalities that would normally command subsidies. It would be a political reach, though, to demand additional payments from those nations that benefit from the emissions of all, even if global leaders are becoming increasingly resolute at combining to deter carbon emissions to reduce the negative externality aspects of the discharges.

Public stimulation of the private sector devoted to energy sustainability and carbon emission reduction can follow a model that is increasingly being adopted in another huge private sector. The health industry in the United States is moving toward a managed care model in which providers are compensated to provide for the health of a population, not just for the procedures performed on the population. Such accountable care organizations are part of a financing approach called results-based financing (RBF). Under such a regime creates a pay-for-performance financial environment that aligns the interests of energy intensive industries or providers with public policy goals. Such new financial and industrial markets can be employed both in energy intensive industries and in industries organized to mitigate, reduce, or transition carbon intensive energy sources toward more sustainable energy production.

Some nations will join these protocols because they see the opportunity to create markets to price a commodity that has gone unpriced perhaps too long. Other nations will see value in the creation of either a trading regime or a carbon tax because they recognize the advantage of the stimulation of industry in sustainable energy production. Still others see an opportunity to engage in favorable intergovernmental transfers by negotiating their participation in protocols that afford them excess permits, which they can subsequently sell to other nations. Finally, others yet see the public finance advantage of a carbon tax that can yield significant revenues to governmental coffers.

14.6 CONCLUSIONS

While countries are increasingly relying on carbon taxes to discourage the use of green-house gas generating fuels, and while cap-and-trade schemes have been growing but with volatility induced by the Great Recession, the developed and rapidly developing nations have been steadily and increasingly expanding their subscription to institutions that will mitigate, and hopefully reverse, greenhouse gas emissions growth. There is increasing momentum behind these efforts as the effects of global warming become more obvious and the costs more pronounced. At current prices, the costs of carbon are approximately $1.75 trillion US, but are expected to grow steadily.

Financial markets can play a role in this transition. While the public finance of carbon taxes may represent an increasing source of government revenue, and substitute for borrowing on traditional financial markets, commodities future exchanges are best situated to facilitate carbon trading regimes. These cap-and-trade systems are perhaps the best mechanisms to limit carbon emission growth in a way that does not distort individual markets and best permits the confinement of emissions below predetermined limits. Finally, energy usage, as a share of GDP, in most nations rivals the size of the health sector. Increasingly, the health sector is relying on the expertise of insurance markets to manage risk and spur cost containing innovations. It is not unreasonable to imagine that clever financial risk management institutions can increasingly capitalize on the global need to better manage the risk of global warming. They can do so by offering to manage and contain the growth of greenhouse gas emissions.

Cap-and-trade regimes and RBF offer the best opportunity to adhere to carbon dioxide emission limits. These are tools that are well understood in free market econo-mies. Nations that have the most elaborate and efficient markets and the most sophisti-cated institutions for risk management will be in the best position to capitalize on what will necessarily be a growing share of the global economy.

Nations that prize innovation and have well-developed research and development sectors are also well positioned to capitalize on the growing need to contain energy costs in the wake of increasing shadow prices for carbon emissions. While the shadow price of carbon emissions, beyond the direct energy costs of industry and the overall economy, may only represent about 3% of global GDP today, this market for emissions trades or carbon taxes is expected to grow substantially over the next decades and increasingly be brought out of the shadows and into traditional public and private financial markets. The most innovative nations can capitalize twice—first in the development and maintenance of efficient and well-functioning markets, and second in the creation of energy-related industries that are well positioned to guide nations toward a more envi-ronmentally benign energy future. While there remain significant political challenges, the strong science of the dilemma implies that nations will be brought along. Those that have already arrived will be in a superior position when that time comes.

REFERENCES

Barker, T., Bashmakov, I., Alharthi, A., Amann, M., Cifuentes, L., Drexhage, J., Duan, M., Edenhofer, O., Flannery, B., Grubb, M., Hoogwijk, M., Ibitoye, F.I., Jepma, C.J., Pizer, W.A., Yamaji, K., 2007. Mitigation from a cross-sectoral perspective. In: Metz, B., Davidson, O.R., Bosch, P.R., Dave, R., Meyer, L.A. (Eds.), Climate Change 2007: Mitigation, Contribution of Working Group III to the Fourth Assessment Report of the Intergovernmental Panel on Climate Change. Cambridge University Press, Cambridge, United Kingdom and New York.

Coase, R.H., 1960. The problem of social cost. J. Law Econ. 3, 1–44.

http://www.epa.gov/capandtrade/documents/ctresults.pdf (accessed 13.01.15.).

http://www.epa.gov/climatechange/pdfs/print_ghg-concentrations-2014.pdf (accessed 13.01.15.).

Emissions Gap Report, 2012, p. 7.

Hardin, G., 1968. The tragedy of the commons. Science 16, 1243–1248.

Pigou, A.C., 1920. The Economics of Welfare. Macmillan, London.

State and Trends of Carbon Pricing, May 2014. World Bank Group on Climate Change, Washington.

The Chicago Climate Futures Exchange, 2004. The Sulfur Dioxide Emission Allowance Trading Program: Market Architecture, Market Dynamics, and Pricing. p. 8.

http://www.eia.gov/todayinenergy/detail.cfm?id=1330 (accessed 13.01.15.).

CHAPTER 15

Environmental Finance

John Anderson
University of New England, Armidale, NSW, Australia

Contents

15.1 INTRODUCTION

This chapter examined the rising prominence of what is commonly referred to as "Environmental Finance." While the notion of ethical investing that is seeking to avoid investments in industries perceived as providing socially negative outcomes such as tobacco companies, gambling and gaming companies etc., has long been a feature of the investment landscape, Environmental Finance has emerged as another investment class with a much greater concentration of projects with very defined environmental objectives.

A number of financial instruments have emerged to serve this growing market including Green Bonds, Green Bond Funds, REDD/REDD+, and Debt-for-Nature Swaps (DNS). While some instruments, such as Green Bonds, are used to finance environmentally centered projects, their popularity and acceptance in the capital markets have also risen due to investor demand for such investments and are available to wholesale and retail investors via Green Bond Funds.

Green Bonds are defined in many ways and provide considerable flexibility in terms of the scope of projects able to be funded under this banner to those raising funds. However, Environmental Finance can also encompass more specific objectives such as REDD/REDD+ projects where the objective is a more targeted funding model to counter "reducing emissions from deforestation and forest degradation." In a completely different approach, other innovations have emerged to tackle the financing of environmentally centered projects. One of these is the DNS where debt obligations are purchased at a discount by various bodies such as the World Wildlife Fund and the proceeds used to finance the conservation and management of large tracts of typically native forests.

15.2 GREEN BONDS

Green Bonds have proven to be a popular mechanism for funding environmentally focused projects. They have provided both users of funds and those seeking to invest in more environmentally targeted projects such as alternative energy, energy efficiency, pollution prevention and control, sustainable water, and green building. While other projects may also be considered for funding under the Green Bond model, these are the primary focus of this emerging asset class. Given the higher perceived social benefit of such projects, they have also provided a robust avenue for those fund managers aiming to promote environmentally friendly/ethical investment to a growing list of investors seeking to allocate funds to what may be seen as more socially positive projects.

15.2.1 Definition of Green Bonds

While there appears to be no universal definition for "Green Bonds," they have been defined by Barclays bank as:

> To be classified as a green bond, a security's use of proceeds must first fall within at least one of five eligible environmental categories: alternative energy, energy efficiency, pollution prevention and control, sustainable water, and green building.[1]

The World Bank also provides a definition for its Green Bond program highlighting support for the transition to low-carbon and climate resilient development and growth in client countries. This includes what it describes as mitigation of and adaptation to climate change. They also stress that these objectives must be met while observing the World Bank's safeguard policies for environmental and social issues.

The World Bank aims to promote funding methods for green technology and provide numerous examples to illustrate their objectives broken into mitigation projects and adaptation projects. Mitigation projects are targeted at reducing the levels of "Greenhouse Gas" (GHG) emissions and include projects such as:

- Solar and wind installations;
- Funding for new technologies that permit significant reductions in GHG emissions;

---

[1] Source: http://www.msci.com/resources/factsheets/Barclays_MSCI_Green_Bond_Index.pdf (accessed 12.01.15.).

- Rehabilitation of power plants and transmission facilities to reduce GHG emissions;
- Greater efficiency in transportation, including fuel switching and mass transport;
- Waste management (methane emissions) and construction of energy-efficient buildings; and
- Carbon reduction through reforestation and avoided deforestation.

Conversely, adaptation projects are targeted at adapting existing systems to improve areas including:

- Protection against flooding (including reforestation and watershed management);
- Food security improvement and implementing stress-resilient agricultural systems (which slow down deforestation); and
- Sustainable forest management and avoided deforestation.

In assessing whether the bond is classified as green or not Barclays Bank note the expanded definition for those provided by the World Bank for bonds to be included in the MSCI (Morgan Stanley Capital International) Green Bond Index. These activities are far broader than those proposed by the World Bank and would be more appealing to investment fund managers who may have part of their investment performance compared to the MSCI Green Bond Index. This expanded range of activities includes projects where 90% of the issuer's activities (as measured by revenues) fall within one or more of the eligible MSCI environmental categories. These include the following:

1. Alternative Energy: Products, services, or infrastructure projects supporting renewable energy and alternative fuels.

Wind	Biomass	Small hydro
Solar	Waste energy	Biogas
Geothermal	Wave tidal	Biofuels

2. Energy Efficiency: Products, services, infrastructure, or technologies addressing the demand for energy while minimizing environmental impact.

Demand-side management	Energy storage	Compact fluorescent
Battery	Superconductors	Insulation
Fuel cells/hydrogen systems	Natural gas combined heat and power	Hybrid/electric vehicles
Smart grid	LED lighting	Clean transportation infrastructure
Industrial automation	Environmental IT	IT optimization

3. Pollution Prevention and Control: Products, services, or projects that support pollution prevention, waste minimization, or recycling to alleviate unsustainable waste generation.

Environmental remediation   Waste reuse and recycling   Conventional pollution control
Waste treatment

4. **Sustainable Water:** Products, services, and projects that attempt to resolve water scarcity and water quality issues, including minimizing and monitoring current water use and demand increases, improving the quality of water supply, and improving the availability and reliability of water.

Water infrastructure   Smart metering   Drought-resistant seeds
  and distribution   devices
Rainwater harvesting   Desalinization   Wastewater treatment

5. **Green Building:** Design, construction, redevelopment, retrofitting, or acquisition of "green" certified properties, subject to local building criteria.

Green certified properties   Uncertified green property (top 15% on
  (LEED, BREEAM, etc.)   energy efficiency, etc.)

6. **Others:** Other environmental activities that do not fit into the categories above, including climate resilience projects (flood relief, mitigation) and sustainable forestry/afforestation.

As the previous tables show, this provides a far broader scope of activities to be declared as "green." Indeed, S&P provides a very open definition given as:

Green bonds are debt instruments issued to finance environmental projects. These projects are often focused on climate-change initiatives, but not always. When we refer to bonds as green, we typically mean that they have been labelled as such by their issuers at the time of their placement.[2]

No doubt these expanded definitions provide a far more appealing range of investment and issuers, especially those seeking to promote their green credentials either from stemming from genuine beliefs about the benefits of the projects through, to those aiming to use green credentials more cynically as a marketing tool.

15.2.2 Origins

The European Investment Bank (EIB) was the first to issue environmentally targeted bonds through their 5 year €600 million "Climate Awareness Bond" issuance in 2007

---

[2] Source: S&P Green Bond Index FAQ sheet at http://au.spindices.com/indices/fixed-income/sp-green-bond-index (accessed 13.01.15.).

Table 15.1 Key Features of First Green Bond Issued by EIB

Maturity Date:	28 June 2012 (5 year)
Coupon:	None
Redemption amount:	100% (capital protected)
Additional redemption amount:	Index linked
	Index: FTSE4Good Environmental Leaders Europe 40
CO2 option:	At maturity, investors have the option to use additional redemption amount in excess of 25% to purchase and cancel EU Allowances (EUAs) allocated and traded in accordance with the EU Emission Trading Scheme referred to under the EU Directive 2003/87/EC
Public offering:	In all 27 countries of the EU
Denomination:	EUR 100
Listing:	Luxembourg. In addition the Bonds have been included in the Regulated Unofficial Market (Freiverkehr) at the Stuttgart Stock Exchange. It is also possible that additional listings will take place on further EU stock exchanges[3]

by marketing a new EUR bond that combines innovative features focused on climate protection. The bonds had the following key features:

Table 15.1 provides a description of the key features of the first Green Bond, or "Climate Awareness Bond" issued by the RIB in 2007. As this was the first issuance of its type, it is a very conservatively packaged debt instrument with most features being very "plain vanilla" in style. It is a 5-year bond with a relatively short maturity minimizing some of the price risks associated with owning longer-dated maturity bonds, especially those with offering zero coupons. It is also 100% capital protected and index-linked bond with additional redemption amounts being tied to the performance of the FTSE4-Good Environmental Leaders Europe 40 index. To enhance retail investor appeal the bonds have also been publicly listed on the Luxembourg stock exchange with a face value of €100.

Since this first issuance by EIB in 2007, many other market participants have entered this space. The World Bank has individually issued 77 different bonds with a total value of US$77 billion with a range of coupons, maturities, and currencies including the Euro, US Dollar, Australian Dollar, British Pound, Russian Ruble, Mexican Peso, Japanese Yen, Swedish Kroner, Malaysian Ringgit, and Canadian Dollar. These bond issues have also been supported by a wide range of issuing financial institutions including Daiwa, JP Morgan, Bank of America Merrill Lynch, Goldman Sachs, Clariden Leu, Nomura, HSBC, Westpac, BNP Paribas, and Morgan Stanley.

---

[3] Source: http://www.eib.org/investor_relations/press/2007/2007-057-epos-ii—the-climate-awareness-bond-pan-eu-public-offering-extended.htm?lang=en (access date 13.01.15.).

Table 15.2 Largest Green Bond Index Eligible Issuers

Issuer Name	Sector	Market Value %
European Investment Bank	Supranational	19.7
KFW	Agency	9.3
GDF Suez	Utility	9.1
International Finance Corp	Supranational	6.9
Solar Star/Topaz Solar	Utility	6.3
World Bank (IBRD)	Supranational	6.2

In terms of the growth trajectory for Green Bonds, Bloomberg[4] reported that an estimated US$40 billion was issued in 2014 and this is expected to grow to US$100 billion in 2015. Table 15.2 shows the largest issuers of Green Bonds and the proportion of the market value for the largest issuers as reported by Barclays MSCI ESG Research as of November 28, 2014.

15.2.3 Green Bond Funds

While Green Bonds are able to be purchased and traded by retail investors, another avenue for investing in these bonds is via Green Bond Funds. The World Bank reports in their Green Bond Sixth Annual Investor Update 2014 a diverse range of investors in their bonds and these are shown in Table 15.3.

Table 15.3 reveals a substantial diversity of investors in Green Bonds ranging from insurance companies such as QBE and Zurich Insurance to asset managers including Aberdeen Asset Management and Trillium Asset Management, to retirement/superannuation funds including New York Common Retirement Fund and UniSuper, and other entities such as the Church of Sweden and Ikea Group.

15.2.4 Indexes

Three main Green Bond indexes have emerged to track the performance of this rapidly growing sector, these being the Barclays MSCI Green Bond Index, the S&P Green Bond Index, and the Solactive Green Bond Index. While many characteristics of these indexes are similar, comparison of these indexes with any Green Bond Fund manager's performance (or evaluation of Green Bonds as an asset class for potential investment) must be treated with some caution given the subtle differences that can emerge between them.

For example, the Barclays MSCI Green Bond Index includes only investment-grade bonds using the middle rating of Moody's, S&P, and Fitch, includes fixed-rate securities only. Furthermore, the flagship index does not have a 1-year minimum time to maturity

---

[4] Source: http://www.bloomberg.com/news/2014-06-26/green-bonds-show-path-to-1-trillion-market-for-climate.html (accessed 13.01.15.).

Table 15.3 Investors in World Bank-issued Green Bonds

Aberdeen Asset Management	California State Treasurer's Office	MISTRA	Sonen
ACTIAM (Formerly SNS AM)	CalSTRS	Natixis Asset Management	Standish Mellon Asset Management
Adlerbert Research Foundation	Calvert Investments	New York Common Retirement Fund	State Street Global Advisors
Aegon Asset Management	Church of Sweden	Nikko Asset Management	TIAA-CREF
AMP Capital	Colonial First State Global AM	Pax World Balanced Fund	Trillium Asset Management
AP2 and AP3 — Swedish National Pension Funds	Deutsche Asset & Wealth Management	Pictet	UN Joint Staff Pension Fund
Australia Local Government Super	Everence Financial	QBE Insurance Group Ltd	UniSuper
Australian Ethical Investment Ltd	FMO (Netherlands Development Finance)	Rathbone Greenbank	WWF–Sweden
BlackRock	Ikea Group	Sarasin	ZKB (Zürcher Kantonalbank)
Breckinridge Capital Advisors	LF Liv	SEB Ethos ratefund/SEB Fonden/SEB Trygg Liv	Zurich Insurance
Caisse Centrale de Reassurance	Mirova	Skandia Liv	Zwitserleven

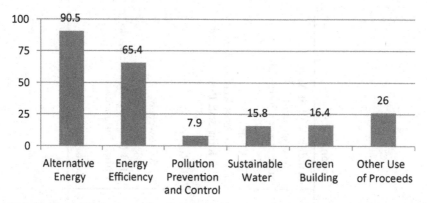

Figure 15.1 Barclays MSCI Green Bond Index—Percentage proportion of funding by use of proceeds.[5]

and the index holds bonds until final maturity. In contrast, the S&P Green Bond Index allows different types of bonds to be included in their index. Some of these differences include coupons that may be zero, fixed or fixed-to-float and allows bonds with numerous in-built option features such as bullets, callable, and puttable. Finally, the S&P Green Bond Index specifically excludes STRIPS, Bills, and Inflation-Linked bonds.

15.2.5 Projects Financed Using Green Bonds

The previous sections highlighted the rapidly growing use of Green Bonds to fund a variety of environmentally focused projects. Figure 15.1 shows the way funds are being allocated to projects included in the Barclays MSCI Green Bond Index.

Figure 15.1 clearly shows that the focus of the projects is heavily concentrated in the Alternative Energy sector with 90.5% of projects allocating funding to this sector. This is followed by Energy Efficiency projects at 65.4% of funding allocation to Green Bonds included in the Barclays MSCI Green Bond Index. It should be noted that many of the projects are not mutually exclusive and Figure 15.1 suggests that the Alternative Energy projects are being conducted in conjunction with other projects.

The following section presents nine brief case studies to illustrate examples of the types of projects currently being funded by Green Bonds issued by the World Bank. It also highlights the broad geographical dispersion and broad range of projects highlighting the potential applications of Green Bonds in funding environmentally focused activities.[6]

---

[5] Use of Proceeds classifications are not mutually exclusive and Green Bond may fund multiple projects.
[6] Source: http://treasury.worldbank.org/cmd/htm/MoreGreenProjects.html (accessed 15.01.15.).

Case Study 1	Brazil—Integrated Solid Waste and Carbon finance
Purpose	To improve the treatment and final disposal of urban solid waste
Time frame	2010—2015
Amount funded	US$50 million
Mitigation	Methane capture from waste facilities
Project details	In 2007, only 39% of cities in Brazil had adequate solid waste treatment and disposal facilities. The objective of this project is to improve the treatment and final disposal of municipal solid waste in Brazil including reducing methane emission.

This project supports:
- the closing of open dumps and the construction and operation of
- modern and environmentally safe landfills or alternatives to waste
- disposal improved municipal solid waste management practices
- reduction of poverty among waste pickers
- increased private sector participation in solid waste service provision
- capacity building of the federal executing agency Caixa Economica Federal to manage carbon finance projects.
- the capture of methane emissions in at least three landfills.

As for results so far, a new landfill was built resulting in two open dumps being closed and capture and flaring of methane emissions equivalent to 795,000 metric tons of CO_2.

Case Study 2	China—Beijing Rooftop Solar Photovoltaic Scale-Up (Sunshine Schools) Project
Purpose	To promote renewable energy in schools
Time frame	2014—2019
Amount funded	US$120 million
Mitigation	Renewable energy
Project details	The project aims to increase the share of clean energy in electricity consumption and to demonstrate the viability of the renewable energy service company model for scaling up the deployment of rooftop solar photovoltaic systems in schools and other educational institutions in Beijing Municipality. The "Sunshine Schools" program will install 100 mega-watts of solar power in about 1000 public schools and colleges in Beijing representing the biggest solar photovoltaic project in the public sector to date.

Case Study 3	China—Huai River Basin Flood Management and Drainage Improvement
Purpose	To improve flooding and drainage infrastructure
Time frame	2010—2016
Amount funded	US$200 million
Adaptation	Flood prevention
Project details	The Huai River Basin is the third largest river basin in China with a population of 165 million people in its watershed. Severe flooding and associated disasters occur every 3—5 years with extraordinary human and economic consequences. Climate change is estimated to increase average precipitation in the summer season by 5% over the next 50 years. The project supports improved flooding and drainage infrastructure (e.g., better dikes, drainage channels, maintenance) and institutional strengthening for disaster assessment and management. These kinds of infrastructure investments are needed to increase resilience of communities to the impacts of climate change, particularly floods. When the project is completed, about 9500 km^2 of rural and urban areas will be better protected from flooding, affecting about 6.6 million people.

Case Study 4	China—Urumqi District Heating Project
Purpose	To promote energy efficiency in district heating
Time frame	2011—2015
Amount funded	US$100 million
Mitigation	Energy efficiency
Project details	Urumqi, the capital of Xinjiang Uyghur Autonomous Region in northwest China, has been suffering from serious air pollution in winter, with home heating as a major culprit. The municipality of Urumqi has determined to improve the air quality for its 2.3 million residents. One of the major measures is to close down hundreds of dispersed boilers in urban areas and replace them with an integrated district heating network. The Urumqi District Heating Project is designed to support this effort. The project will finance construction of basic infrastructure to connect residents to district heating services with improved energy efficiency and environmental performance mainly in two districts, Shuimogou, Shayibake, and small part of Tianshan District in Urumqi.

Case Study 5	Dominican Republic—Emergency Recovery and Disaster Risk Management
Purpose	Provide infrastructure recovery and strengthen risk management capacity in tropical storm affected areas
Time frame	2008–2014
Amount funded	US$80 million
Adaptation	Risk management and protection against flooding
Project details	In 2007, the Dominican Republic was hit by two powerful tropical storms that not only left thousands of families homeless and damaged crops, but also destroyed the better part of country's infrastructure, including roads, bridges, electricity networks, and irrigation systems. With the help of World Bank-funded Emergency Recovery and Disaster Management Project, the national government is rebuilding the country's national electricity, irrigation, and water supply sectors. It is also strengthening its government agencies' capacity to manage water and electricity resources in order to mitigate potential effects of future emergencies. As these agencies update contingency plans and increase their risk management capacity, this project is illustrative of climate adaptation support in storm vulnerable tropical areas. The project is expected to rebuild irrigation for 11,577 ha damaged by the storms, restore 152 km transmission lines to "disaster-resistant standards, restore operation of the Santiago wastewater plant, restore 200 MW of damaged hydropower facilities, and improve dam safety standards".

Case Study 6	Mexico—Modernization of the National Meteorological Service for Improved Climate Adaptation
Purpose	To strengthen Mexico's meteorological service.
Time frame	2012–2017
Amount funded	US$250.6 million
Adaptation	Improved information to reduce vulnerability
Project details	The impacts of climate variability and global climate change in Mexico include an increase in average and extreme temperatures, a greater variability in rainfall extremes and thus changes in water flows. Given the population growth, the effects of extreme weather events can increase the threats to lives as well as to assets. Better capacity to monitor and forecast hydrometeorological events and climate variability is therefore essential to confronting the

challenges of climate change. The project aims to strengthen the human resource, institutional, and infrastructure capacity of Mexico's National Meteorological Service to meet this need of timely and accurate weather and climate information to better manage water resources and risks of disasters in the face of climate change and climate variability. By improving the observational infrastructure, data management and processing, and climate modeling and forecasting tools, the project will help the government to improve the quality and timeliness of weather forecasts and alerts. As Mexico is suffering from the impacts of longer droughts and severe cyclones and hurricanes, improved meteorological service is a key to adapt to climate change and help reduce impacts in vulnerable populations, such as small-scale farmers and marginal neighborhoods in cities.

Case Study 7	Mexico—Urban Transport Transformation Program
Purpose	To reduce carbon emissions and transform public transportation efficiency
Time frame	2010–2017
Amount funded	US$150 million
Mitigation	Lower carbon urban transport
Project details	Mexico has one of the most carbon-intensive transport sectors in Latin America, accounting for 18% of Mexico's total greenhouse gas (GHG) emissions. The rise in traffic, lack of street space, and old transport technologies (including inadequate fuel standards) account for some of the many reasons behind the inefficiency of urban transportation that contribute to overcrowding and high GHG emissions in Mexico's many cities. The Urban Transport Project will help transform urban transportation efficiency in Mexican cities. It will also reduce its transport sector carbon footprint by improving the quality of service provided by the urban transport systems in a cost-efficient manner, and by deploying equipment, infrastructure, and operational strategies that reduce CO_2 emissions, such as the modernization of the bus rapid transit system. As an example of the impacts, three bus lines in Mexico City reduce 100,000 metric tons of CO_2 and replace 126 cars, and induce more walking and bicycle use.

Case Study 8	Tunisia—Fourth Northwest Mountainous and Forested Areas Development
Purpose	To promote better protection and management of natural resources
Time frame	2010–2017
Amount funded	US$41.60 million
Adaptation	Watershed management
Project details	Tunisia's mountainous and forested areas of the northwest cover 1.2 million hectares and support watersheds supplying 75% of the water consumed in the country. It hosts half of the forests remaining in Tunisia. Land pressure from inadequate agriculture and livestock practices, naturally poor soils, and heavy winter precipitation combines to make soil erosion and forest degradation serious threats to these vital natural resources. Climate change is adding pressure as flash floods and drought exacerbates these problems. The Fourth Northwest Mountainous and Forested Areas Development Project is designed to improve the socioeconomic conditions of rural populations in the northwest region (about 318,000 people) through access to potable water, conservation of soil and water by improving agriculture and pasture practices, and better management of forest resources. In addition, climate change awareness-building activities and the dissemination of climate-appropriate practices reinforce livelihood and agro-ecosystem resilience.

Case Study 9	Ukraine—Power Transmission Project in Support of the Energy Sector Reform and Development Program
Purpose	Improve reliability and efficiency of energy supply to facilitate smooth operation of the energy market both domestically and internationally.
Time frame	2011–2016
Amount funded	US$200 million
Mitigation	Increased efficiency in electricity transmission
Project details	Despite significant surplus in installed generating capacity, some regions of Ukraine are experiencing difficulties covering their peak load due to inadequate power transmission capacity. This is due to congestion in the main transmission networks and poor reliability of the existing transmission infrastructure. Users resort to diesel-powered back-up generators, which cause additional CO_2 emissions. With demand for energy increasing, combined with

outdated equipment supporting the country's major power grids, UkrEnergo, Ukraine's transmission system operator, cannot supply energy efficiently to both Ukraine and the EU. The World Bank-funded Power Transmission Project helps mitigate GHG emissions by improving the transmission infrastructure and helping UkrEnergo reduce Ukraine's energy intensity by 10%. It also will help increase UkrEnergo's revenue through more electricity trade with the EU. The project's activities include:
- replacing high voltage equipment with cleaner energy systems in the eastern part of Ukraine;
- upgrading existing transmission lines and associated substations to an optimal voltage level; and
- establishing a Management Information System in Ukraine for the auditing and implementation of energy regulations across the country.

As the previous nine case studies reveal, Environmental Finance is highly diverse with respect to the nature of projects, amount of funding required, and geographical locations where projects are undertaken. While much of the funding is directed toward alternative energy production, the World Bank uses Green Bonds to fund other areas for environmental enhancement such as better waste management in Brazil, flood management in China, emergency recovery in the Dominican Republic, modernization of the meteorological service and urban transport improvements in Mexico, improved forestry management in Tunisia, and improvements to electricity distribution in Ukraine.

15.2.6 Environmental Investing

A number of so-called "Green Funds" have also emerged in recent years. However their history is far longer than we might expect with its origins reported to have been among religious communities, such as the Quakers, and their focus on what would now be described as ethical investing to benefit their communities. However, we must be certain we understand the differences between ethical investing, which is not investing in industries deemed to be unethical such as tobacco, gambling, and so on, and green investing in industries such as those identified in the earlier section on Green Bonds.

When considering Green Investments as an investable asset class, a vast array of firms and financial products are available to ethical investors. As an example, if green energy alone is considered it reveals a substantial number of funds, companies, and subsectors, and an example of these are shown in Table 15.4.

Table 15.4 reveals that just within the green energy space alone, a number of different sectors appear, such as solar, wind, and general clean energy funds, and this is only a very small sample of the environmentally focused exchange-traded funds available. It should

Table 15.4 Examples of Green Investment ETFs in the Alternative Energy Sector[7]

Solar Energy ETFs	Claymore/MAC Global Solar Energy Index ETF (NYSE: TAN)
	This Index follows companies within various business segments of the solar energy industry. These companies include OEMs, raw material suppliers, distribution, installation, and financing.
	Market Vectors Solar Energy ETF (NYSE: KWT)
	Market Vectors Solar Energy seeks to replicate the activity of the Ardour Solar Energy Index, SOLRX. This Index allows for an objective exposure of companies that derive at least 66% of their profits from solar energy.
Wind Energy ETFs	First Trust Global Wind Energy ETF (NYSE: FAN)
	This Index Fund is based upon the ISE Global Wind Energy Index. It is a modified market capitalization weighted index that indicates publicly traded companies active in the wind energy industry around the globe.
Clean Energy ETFs	First Trust NASDAQ Clean Edge US Liquid ETF (NASDAQ: QCLN)
	First Trust NASDAQ Clean Edge is an exchange-traded index fund that searches for the corresponding investment results to the price and yield of the equity index NASDAQ® Clean Edge® U.S. Liquid Series Index.
	PowerShares Cleantech ETF (NYSE: PZD)
	In order to categorize a company as a cleantech company for PZD, it must obtain 50% or more of its revenue from respective cleantech businesses.
	PowerShares WilderHill Progressive Energy ETF (NYSE: PUW)
	Based on the WilderHill Progressive Energy Index, the PowerShares WilderHill Progressive Energy Portfolio (Fund) indicates US listed companies that are notably concerned with transitional energy bridge technologies. The main concern of these companies revolves around improving the use of fossil fuels.
	PowerShares WilderHill Clean Energy ETF (NYSE: PBW)
	Similar to the aforementioned WilderHill Progressive Energy Index, the WilderHill Clean Energy Index recognizes companies that center around greener and often renewable sources of energy. They also focus on the technology that aids cleaner power production.
	Market Vectors Global Alternative Energy ETF (NYSE: GEX)
	This investment follows the Ardour Global Index (Extra Liquid). In proportion to their weightings, the fund invests up to 80% of total assets and 30% of assets in securities of non-US companies. The main focus of this index is companies that have engaged in generating power from environmentally friendly and nontraditional sources.

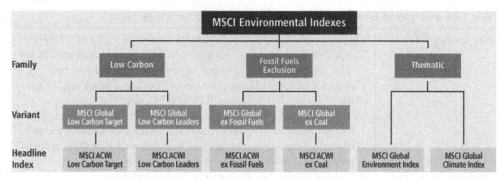

Figure 15.2 MSCI environmental indexes.

be noted that the wide range of indexes created to allow comparison of fund performance against index benchmarks. Table 15.4 shows that these include the Ardour Solar Energy Index (SOLRX), the ISE Global Wind Energy Index, and the Ardour Global Index (Extra Liquid).

MSCI has also created a range of benchmark environmental indexes shown in Figure 15.2, and the environmental indexes they have created fall into three main categories: low carbon, fossil fuels exclusion, and thematic. Under these three main index categories, they are then broken down into variants on these such as low carbon target and MSCI global ex-coal and outlined in Table 15.5. The thematic indexes are discussed in more detail in Table 15.6.[8]

Table 15.5 shows that these indexes allow fund managers to have their performance assessed against fairer benchmarks for those with funds denied access to some significantly large, though highly polluting, industries such as coal and fossil fuels. Should oil have had a year with substantial price increases, so benefiting fund managers who hold oil stocks, this would make the Green Fund manager appear to have a relatively insipid performance. However, when compared to the appropriate benchmark, such as the MSCI ACWI ex-Fossil Fuels Index, the performance may have been completely in line with the performance of those assets than can be held under the fund's mandate to not invest in fossil fuel firms.

Table 15.6 provides an overview of the Thematic indexes offered by MSCI. These are far broader indexes than those discussed in Table 15.5 and more focused on industry leaders drawn from small-, mid-, and large-cap companies providing goods and services in the renewable energy, future fuels, and clean technology and efficiency sectors.

To better understand the nature of the indexes and their construction, an in-depth look at an index is presented for the ISE Global Wind Energy Index from the

---

[7] Source: http://www.energyandcapital.com/report/green–energy–etfs/368 (accessed 20.01.15.).

[8] Source: http://www.msci.com/products/indexes/esg/environmental/ (accessed 20.01.15.).

Table 15.5 Overview of MSCI Environmental Indexes

Index Name	Broad Objectives
MSCI Global Low Carbon Target Indexes	Aims to reflect a lower carbon exposure than that of the broad market by overweighting companies with low carbon emissions (relative to sales) and those with low potential carbon emissions (per dollar of market capitalization)
MSCI Global Low Carbon Leaders Indexes	Aims to achieve at least 50% reduction in the carbon footprint by excluding companies with the highest carbon emissions intensity and the largest owners of carbon reserves (per dollar of market capitalization)
MSCI ACWI ex-Fossil Fuels Index	Aims to eliminate 100% of carbon reserves exposure by excluding companies that own oil, gas, and coal reserves
MSCI ACWI ex-Coal Index	Aims to significantly reduce carbon reserves exposure by excluding companies that own coal reserves

Table 15.6 Overview of MSCI Thematic Environmental Indexes

Index Name	Broad Objectives
MSCI Global Climate Index	An equal weighted index consisting of 100 companies that are leaders in mitigating the causes or the impact of climate change. Constituents are leaders in renewable energy, future fuels, and clean technology and efficiency, selected from a universe of small-, mid-, and large-cap companies in developed markets. Criteria for climate leadership include market share, strategic commitment, investment in R&D, intellectual property, and reputation.
MSCI Global Environment Index	Includes companies that derive 50% or more of their revenues from environmentally beneficial products and services. Constituents are drawn from a universe of large-, mid-, and small-cap companies in developed and emerging markets. The Index is an aggregation of the following: • MSCI Global Alternative Energy Index • MSCI Global Clean Technology Index • MSCI Global Sustainable Water Index • MSCI Global Green Building Index • MSCI Global Pollution Prevention Index

International Securities Exchange (ISE)[9]. In their description of the index, they define it as:

> *The ISE Global Wind Energy Index provides a benchmark for investors interested in tracking public companies that are active in the wind energy industry based on analysis of those companies' products and services. The Index uses a quintile-based modified capitalization weighted methodology for each group of companies. The methodology sets the weight of each quintile to a multiple of the weight of the lowest quintile, based on its market capitalization. The resulting linear weight distribution prevents a few large component stocks from dominating the index while allowing smaller companies to adequately influence index performance. Index components are reviewed semi-annually for eligibility, and the weights are re-set accordingly.*

The following steps were taken to select the initial components for the ISE Global Wind energy Index.:

1. Establish total population of companies that are involved in the Wind Energy industry.
2. If a component has multiple share classes, include the most liquid issue for that company (using average daily value traded during the prior 6-month period) and remove the remaining classes.
3. Companies having been identified as providing goods and services exclusively to the Industry are given an aggregate weight of 66.67% of the portfolio. Companies determined to be significant participants in the Industry despite not being exclusive to the Industry are given an aggregate weight of 33.33%.
4. Rank the stocks in descending order by market capitalization.
5. Divide each segment of the index into quintiles, assign a numerical score to each quintile based on its rank (e.g., first quintile gets a "5," second quintile gets a "4," etc.).
6. Adjust each stock's weighting to an equal weighted distribution of the aggregate weight of the stock's quintile.
7. If the total number of companies in the index segment is not divisible by 5 round down such that the remainder companies will reside in the last quintile. The adjustment described above results in a modified linear distribution where each component is weighted by a certain multiple of the smallest component.

Note that the index portfolio does not have a fixed number of stocks and attempts to include every stock in the industry that meets the eligibility requirements.

Therefore the index aims to capture the movements of prices for firms engaged in the production of wind power and associated industries with the aim of including all industry participants. For this reason the number of stocks in the index is not fixed as it is in indexes such as the S&P 500. As we would expect, the participants in this

---

[9] Source: http://www.ise.com/assets/files/Index/ETF_GWE_GlobalWind_0115f.pdf.

industry range from major energy companies such as Royal Dutch Shell, GE, and Siemens through to smaller niche participants such as Capstone Turbine Corp. The list of participants (as of 31 December 2014) included in the ISE Global Wind Energy Index is shown in Table 15.7.

Table 15.7 shows the component firms included in the ISE Global Wind Energy Index. It shows this to be a weighted index with companies weight in the index represented by the company's size, represented by its market capitalization, and a weighting depending on whether the firm is exclusively engaged in the provision of services to the Wind Energy industry or not. This means that firms that are purely engaged in Wind Energy constitute 66.67% of the index, and firms that have wind energy as only part of their corporate activities are given a total weighting of 33.33% of the index. While this method of index construction is not perfect, it does provide users with the ability to gain some form of snapshot of how the industry is performing financially. This also enables users to readily compare the performance of how this sector is performing compared to other sectors.

When using indexes such as these to compare performance, users of these indexes must also exercise caution about whether the index is a simple index or a total return index. Some indexes will simply aggregate the weighted prices of these firms and present the value as closing prices—with one key difference, it has not taken dividends into account. Conversely a total return index will provide a measure of performance of that sector by including not only changes in the underlying share prices but also include dividends. This difference is highlighted in Figure 15.3.

As Figure 15.3 shows, we can have a material difference in the performance of an index depending on whether dividends are taken into account and this disparity increases over time with the Total Return Index growing faster than indexes that do not consider the impact of dividends.

15.2.7 REDD/REDD+ Financing

Another initiative that has arisen to support environmentally engaged projects is REDD and REDD+ (REDD Plus) financing which aims to support projects with the goal of "reducing emissions from deforestation and forest degradation" in developing countries. It achieves this by providing financial incentives for developing countries to reduce carbon emissions through a reduction in conversion of forests to alternative land uses and recognizing the value of existing forests.

In the origins of REDD, the overall goals were to reduce emissions through incentives to diminish deforestation and forest degradation. It also aimed to more formally recognize to roles of sustainable management, conservation, and improving the carbon stocks of forests in developing countries. This definition was later expanded in Peskett et al. (2008) to REDD+ to include the reduction of emissions from deforestation by

Table 15.7 Component Companies in the ISE Global Wind Energy Index

Ticker	Name	Weight %
916 HK equity	China Longyuan Power Group Corporation Ltd—H Shares	7.38%
VWS DC equity	Vestas Wind Systems AS	7.35%
IBE SM equity	Iberdrola SA	7.33%
NDX1 GY equity	Nordex SE	6.07%
EDPR PL equity	EDP Renovaveis SA	5.88%
GAM SM equity	Gamesa Corp Tecnologica SA	5.60%
GES DC equity	Greentech Energy Systems	4.54%
PNE3 GY equity	PNE Wind AG	4.35%
182 HK equity	China WindPower Group Ltd	4.06%
EOLUB SS equity	Eolus Vind AB	3.10%
2766 JT equity	Japan Wind Development Co Ltd	2.96%
TEO FP equity	Theolia	2.94%
ARISE SS equity	Arise AB	2.30%
IFN AU equity	Infigen Energy	2.12%
RDS/A UN equity	Royal Dutch Shell PLC A ADR	2.00%
BP UN equity	BP ADR	1.95%
GE UN equity	General Electric Co	1.89%
DUK UN equity	Duke Energy Corp	1.88%
SIEGY US equity	Siemens AG ADR	1.84%
EOAN GY equity	E.ON SE	1.81%
NEE UN equity	NextEra Energy Inc	1.57%
ELE SM equity	Endesa SA	1.56%
8031 JT equity	Mitsui & Co	1.54%
SKFB SS equity	SKF AB B	1.52%
EDP PL equity	Electricidade de Portugal SA	1.45%
RWE GY equity	RWE AG	1.37%
LNT UN equity	Alliant Energy Corp	1.17%
TRN UN equity	Trinity Industries Inc	1.16%
AES UN equity	AES Corp	1.15%
NRG UN equity	NRG Energy	1.13%
EGPW IM equity	Enel Green Power SpA	1.09%
ALO FP equity	Alstom	1.06%
ATI UN equity	Allegheny Technologies Inc	0.85%
ELET3 BS equity	Eletrobras SA	0.80%
FDML UQ equity	Federal-Mogul Holdings Corp	0.79%
WWD UQ equity	Woodward Inc.	0.76%
ANA SM equity	Acciona SA	0.73%
BKW SW equity	BKW SA	0.72%
OTTR UQ equity	Otter Tail Corp	0.34%
CPST UQ equity	Capstone Turbine Corp	0.33%
1133 HK equity	Harbin Power Equipment Co Ltd—H Shares	0.33%
956 HK equity	China Suntien Green Energy Corp Ltd	0.32%
100130 KS equity	Dongkuk Structures & Construction Co Ltd	0.32%
658 HK equity	China High Speed Transmission Equipment Group Co Ltd	0.32%
TENERGY GA equity	Terna Energy S.A.	0.28%

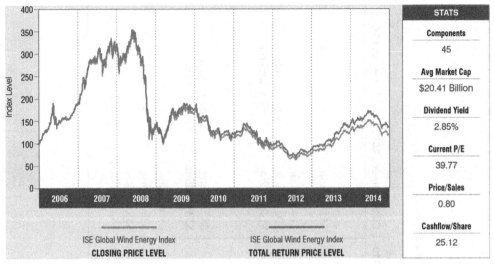

Figure 15.3 Historical index levels for the ISE Global Wind Energy Index.[10]

recognizing efforts to reduce forest degradation, conserve forest carbon stocks, more sustainable management of forests, and enhancing forest carbon stocks. Such activities would be likely to produce dual benefits of both positive environmental outcomes and also the capacity to provide projects that may help reduce poverty in rural areas of developing countries.

While REDD+ is still in its infancy, many countries have pledged funds for REDD-based activities and an example of this from the Voluntary REDD+ Database is shown in Table 15.8. Table 15.8 shows that despite numerous countries pledging funds in this area, many nations are still to make significant disbursements for the funds they planned to make available. This appears due to a number of reasons including issues related to the reporting of disbursements, however, others such as Canada have made no specific pledges and provides funding for projects such as $20 million to the Congo Basin Forest Fund and $4.5 million to the World Bank BioCarbon Plus Fund. Other nations such as Finland and Denmark have also not provided specific pledges though have funded projects that have met their criteria. It should also be noted that the EU has approached their REDD+ activities as providing 50% of funds for adaptation and 50% toward mitigation. The EU has included REDD+ projects as part of their mitigation budget and has not specified any particular amount for future REDD+ projects.

---

[10] http://www.ise.com/assets/files/Index/ETF_GWE_GlobalWind_0115f.pdf.

Table 15.8 Sample of Pledged Funds and Disbursements from the Voluntary REDD+ Database

Country	Fast–start Pledge	Fast–start Arrangements[2] 2010–2012	Disbursement 2010	Disbursement 2011	Disbursement 2012
Australia	$144 million	$14 million	$14 million		
Belgium	$14 million	$72 million		$41 million	$31.5 million
Canada		$46 million			
Denmark					
European Commission	See comment	$19.6 million	$5.6 million	$14 million	$12.5 million
Finland	None	$48 million	$7 million	$20 million	
France	$342 million	$373 million	$122 million	$148 million	$103 million
Germany	$503 million	$388.8 million			
Italy					
Japan	$500 million	$718.6 million	$302 million		
The Netherlands	$21 million		$5 million	$8 million	$8 million
Norway	$1000 million	$986 million	$461 million	$334 million	$51 million
Slovenia					
Spain	$81.9 million	$23.95 million	$18.57 million	$1.32 million	
Sweden		$71 million	$29 million	$28 million	$14 million
Switzerland		$34 million*	$215 million	$89 million	
United Kingdom	$493 million	$457.9 million	$191.5 million	$87.2 million	$32 million
United States	$1000 million	$249 million			

Source: http://www.fao.org/forestry/vrd/by/funders#fsf_report (accessed 19.02.15).

REDD+ financing is not without its challenges in financing this type of investment product through existing capital markets. O'Sullivan et al. (2010) prepared a report for the UK government and provide an overview of the key areas of the REDD+ market and the types of structures and challenges that may be present when using this type of financing to provide carbon credits to market participants. In 2009, developed nations pledged US$3.5 billion at the UN Climate conference in Copenhagen, although at this stage such funding appeared to be "in principle" as market structures were not yet present to provide a clear method of funding such activities. It was also unclear as to how risks, such as default risk, were going to spread among the market participants, making reliable pricing of these debt instruments virtually impossible. However some alternatives were identified in their report including:

- Guarantees: These may be provided by either sovereign nations, the World Bank, regional development banks, or a combination of these to enhance the attractiveness of such investments to private sector investors;
- Insurance: Insurance companies may be willing to participate in this area of risk transference to improve the risk profile to investors. However this may only be suitable for projects in relatively more developed markets where these risks are better understood, leaving projects in relatively lesser-developed economies unable to find insurance given a potentially higher risk profile;
- Securitization: This proposal would involve converting future REDD+ credits and emissions reductions into a security that could be traded on local and/or international exchanges.

Despite the socially and environmentally positive outcomes available to REDD+ financed projects, it would appear that the market is one still very much in its infancy with numerous challenges to overcome to reach its potential. O'Sullivan et al. (2010, p. 25) raise legitimate concerns about the realities of dealing in developing economies where they cite

> ...significant challenges of meeting the prerequisites include lack of investor protection; insufficient transparency in securities trading; problems with registration of shares; problems with enforcement of property rights; entrenched culture of "cronyism" with high levels of self-dealing (i.e. direct or indirect preferential treatment to certain insider groups); inadequate or non-existent rules and regulations; and serious problems with enforcement of rules and regulations.

While the challenges to such projects and financing models are present, the evidence of pledges and disbursements from governments and other bodies to fund such initiatives would indicate that the will is present to make such funding available in a responsible and practical manner. The key future challenge is to be able to provide products to conventional capital markets to provide access that will allow access to global investable capital flows.

15.2.8 Debt-for-Nature Swaps

The final topic considered in this chapter is DNS. Kessel (2006) defines these as "…an agreement between actors in a lending and borrowing country to reduce some of the borrowing country's debt in exchange for the support of a specific environmental project." Following the Latin debt crisis of the 1980s, it became apparent that many countries would simply be unable to pay debts they had amassed during the 1970s when relatively easy credit was extended to nations with little prospect for repayment. The DNS model provided a novel alternative to debt payment, with numerous environmental benefits available and a reduction in the real debt burden of these countries.

Given the dire financial position of these highly indebted countries, in 1984, Executive Vice President of the World Wildlife Fund (WWF), Thomas E. Lovejoy, proposed the first DNS model. It was analogous to the popular debt-for-equity swap in that the debt was able to be "swapped" for some other form of payment. In the case of DNS, it allowed debt-burdened governments to buy debt at a discount to its face value and swap these for environmental investments such as forests and so on.

The first DNS was in 1987 between the Bolivian government and Conservation International and involved an arrangement that allowed Bolivian bank debt with a face value of $650,000 to be purchased for $100,000. The debt was then traded to the Bolivian government and in exchange would establish an endowment fund to establish legal protection of rainforest land and manage the operating costs for a reserve of 2.7 million acres. It must be noted that in a DNS, the ownership of the land is not transferred under the swap with the aim being to secure robust protection/conservation objectives.

Since that original deal in Bolivia, DNS has provided a novel way for many indebted developing nations to be able to swap financial obligations for environmental obligations. Further examples are cited in Mawson's (n.d.) Ethics of Development in a Global Environment seminar where he discusses deals that followed including the WWF's DNS in Ecuador where $9 million in debt was purchased for $1 million with the goal of saving around 10 million acres of forest land. Similarly, Conservation International have engaged in numerous DNS deals with Costa Rico and Mexico whereby $16 million in debts has been purchased for $6.5 million to save and manage native forests.

The steps for creating a DNS are outlined in Kessel (2006) and these are shown in Table 15.9. It reveals how such a deal may be constructed and also highlights some of the uses of funds in a DNS arrangement including conservation, staff training, management of designated areas, and environmental education programs.

The concept of DNS remains popular with numerous projects that have been able to be funded through this branch of Environmental Finance. Numerous deals have been completed since the first project in Bolivia in 1987 and more recent

Table 15.9 Steps in Creating a Debt-for-Nature Swap

Step	Details
1. C.O.s want to protect the environment and see an added benefit of buying discounted debt on the secondary market.	The C.O. draws a plan and approaches governmental organizations to reach an agreement on the conservation program.
2. The C.O. verifies that the amount of debt relief will be sufficient for the envisioned projects.	
3. The debtor country's central bank and ministries of finance and environment must then approve of the deal.	The central bank converts the external debt to local currency.
4. Specific terms of the swap are established and the purchase price and redemption rate (percent of face value redeemed in local currency) of the debt comes from the secondary market price for debt.	The debt relief through the swaps is usually small in comparison to the overall external debt of a country. However, the funds for conservation may be much larger than what the host country had previously been spending in debt payments.
5. The debt process is carried out through the central bank of the indebted country. Specifically, the host country places the converted local currency bonds into an environmental trust fund where interest can be earned to fund conservation projects until the bond matures.	The fact that debt in secondary markets is bought at usually a small fraction of its face value indicates that commercial banks do not really expect to get their money back. The Debt-for-Nature deal is contingent upon the bank's willingness to sell the debts back at less than full value because they believe that the developing countries would otherwise not pay off their debts.
6. The conservation projects are implemented.	These could include: environmental conservation, natural resource management, designation and management of protected areas, park personnel training, and environmental education programs and activities. Debt-for-Nature swaps usually involve some form of land conservation, but conservation is not required as part of the agreement.

examples including the 2008 DNS between the governments of Madagascar and France providing €13 million to help preserve the biodiversity in Madagascar and the $28.5 million deal between the Nature Conservancy, WWF, and the governments of Indonesia and the US to help protect tropical rainforests in Kalimantan, Indonesia.

15.3 CONCLUSION

This chapter has revealed several alternative approaches to funding environmentally targeted projects. Green Bonds play a substantial role in this area as they provide a clear mechanism for both environmental bodies requiring funding and more green-centric investors to be able to provide an exchange of funds to meet their objectives. If underwriters and lead managers of these bonds can provide sufficient numbers and size of bonds being issued to global capital markets, this should provide sufficient liquidity and improved understanding of pricing and risk considerations to see this market remain buoyant in coming years. The more Green Bonds that are being issued and traded will certainly benefit the capacity for capital markets to better understand the risk characteristics of these instruments with more likelihood of Green Bonds becoming part of otherwise mainstream debt portfolios.

Of course, increased awareness of these instruments, and the indexes constructed for Green Bonds, should arguably lead to more widespread adoption of these instruments in global bond portfolios. While Green Bonds are still in their infancy, increased international focus on environmental matters will be likely to see the inclusion of these debt instruments in various portfolios, especially those seeking to use environmental credentials as part of their marketing strategies. The indexes developed by various organizations, such as MSCI, will also help to provide key performance benchmarks for this growing sector.

While it expected that Green Bonds will be the dominant form of finance in the foreseeable future, this chapter has shown that other alternatives do exist. More targeted approaches such as those seen in REDD/REDD+ and DNS also allow finance to play a key role in developing alternative financing mechanisms for environmentally targeted projects. DNS also provides the benefits of allowing those nations that are highly indebted, coupled with limited prospects of debt repayment, to utilize their natural resources in a way that is seen to be beneficial well beyond that nation's borders.

These types of creative approaches to problem solving in the environmental context show that many environmental problems can be addressed with creativity and foresight among the financial community. While there are a number of other approaches that have been developed, the objectives are typically similar to those highlighted in this chapter where modifications to existing financial tools are able to provide benefits in genuinely creative and far-reaching ways.

REFERENCES

Kessel, A., 2006. Debt-for-Nature Swaps: A Critical Approach, Comparative Environment and Development Studies: A Seminar in Cultural and Political Ecology. Macalaster College.
Mawson, S., n.d. Debt-for-Nature: Past and Future, Ethics of Development in a Global Environment Seminar, Stanford University.

O'Sullivan, R., Streck, C., Pearson, T., Brown, S., Gilbert, A., 2010. Engaging the private sector in the potential generation of carbon credits from REDD+; an analysis of issues. In: Report to the UK Department for International Development (DFID).

Peskett, L., Huberman, D., Bowen-Jones, E., Edwards, G., Brown, J., 2008. Making REDD Work for the Poor. A Poverty Environment Partnership (PEP) Report. Overseas Development Institute (ODI), London, UK.

CHAPTER 16

The Relationship between Screening Intensity and Performance of Socially Responsible Investment Funds

Philip Gharghori[1], Elizabeth Ooi[2]
[1]Department of Banking and Finance, Monash Business School, Monash University, Berwick, VIC, Australia;
[2]University of Western Australia Business School, University of Western Australia, Perth, WA, Australia

Contents

16.1 INTRODUCTION

The popularity of socially responsible investment (SRI) has gained considerable momentum in recent years. SRI is driven by the increasing awareness of social concerns by investors, companies, governments, activists, and the media. The Forum for Sustainable and Responsible Investment (2012)[1] estimates that SRI now covers $3.47 trillion out of the $33.3 trillion assets under management in the US market. Similar growth is observed in many other markets, particularly in Europe. As SRI is a relatively new area, academic literature is still inconclusive about the performance of SRI funds. This chapter is motivated by the lack of academic literature that examines the heterogeneity of SRI funds as well as the conflicting results on how SRI funds perform. We are also motivated by the value this type of research can provide to fund managers, regulators, and most importantly, investors.

The primary aim of this study is to examine how screening intensity affects financial performance. SRI funds apply social screens in order to choose companies aligned with

---

[1] Available at http://www.ussif.org/files/Publications/12_Trends_Exec_Summary.pdf (accessed 12.10.14.).

their social concerns. Thus, screening intensity is defined as the number of screens that an SRI fund utilizes when deciding on companies to invest in. The more screens applied, the higher the screening intensity. We also examine if there is a relationship between screening intensity and the types of companies SRI funds select. These issues of heterogeneity are important when considering the performance of SRI funds and to date, only a few other studies address these issues. Our findings show a negative curvilinear (∩-shaped) relationship between screening intensity and performance. This is in contrast to results of prior studies. For example, Barnett and Salomon (2006) find a positive curvilinear (U-shaped) relationship. Renneboog et al. (2008b) find a negative linear relationship using screens relating to social and corporate governance issues. Finally, Lee et al. (2010) find evidence of either an insignificant or negative linear relationship, depending on the measure of performance used.

We contend to both stakeholder theory and modern portfolio theory to explain the negative curvilinear relationship observed. Specifically, we argue that funds with low intensity do not benefit from the focus on socially responsible well-run firms whereas funds with high intensity shrink their potential investment pool to the extent that they are no longer sufficiently diversified. The consequence is that funds with both low and high screening intensity earn lower returns whereas funds with a mid-level of screening intensity gain the benefits (as argued by stakeholder theory) without the cost of being inadequately diversified and thus earn higher returns. The analysis of the types of companies that SRI funds invest in shows that as screening intensity increases, funds tend to invest more in growth companies. Our research sheds new light on the mixed results of previous literature and provides important insights on the relationship between social and financial performance.

The chapter proceeds as follows: Section 16.2 outlines the background, literature review, and research questions. Section 16.3 discusses the methods and data. Section 16.4 reports the results and finally Section 16.5 concludes.

16.2 BACKGROUND, LITERATURE REVIEW, AND RESEARCH QUESTIONS

Socially responsible investing is the practice of incorporating social goals into the investment decision-making process. These social goals come under a variety of concerns and include environmental, human rights, community involvement, and labor relations. They also include concerns regarding industries that are considered socially harmful such as alcohol, tobacco, gambling, and defense. Overall, the goal of SRI is to achieve financial returns in a way that does not harm society. The social goals are implemented into the investment decision by the process of screening. The three main methods of screening are "negative screening," which involves the exclusion of companies based on their involvement in what are commonly known as "sin" industries, "positive screening," which involves seeking out companies that enhance SRI practices, and a form of positive screening known as "best-in-class" screening, which considers companies that demonstrate best practices (despite the activities of the entire industry).

Screening intensity describes the degree to which stocks are filtered out of an SRI fund's investable universe. An SRI fund will be regarded as having a higher screening intensity if it adopts more screens across a wider variety of issues. For example, a fund that adopts an environmental screen, a social screen, and a governance screen would be considered more "intense" than a fund that employs just an environmental screen. The Forum for Sustainable and Responsible Investment counts 14 screens available for mutual funds to choose from. These 14 screens are grouped into four overall categories: environmental, social, governance, and products. Thus, a fund with low screening intensity would be one that employs one screen while a fund with high screening intensity would be one that employs the maximum 14 screens.[2] This is the approach that has been employed by two of the prior studies that have considered screening intensity (Barnett and Salomon, 2006; Lee et al., 2010).

Studies in the area of SRI are generally divided into two groups: those that argue for a negative relationship between social and financial performance and those that argue for a positive relationship. The negative relationship is justified by the cost-concerned hypothesis and modern portfolio theory while the positive relationship is justified by the value-creation approach and stakeholder theory (Hassel et al., 2005). The cost-concerned approach argues that the costs of social responsibility are not offset by financial returns. Hamilton et al. (1993) expand this and argue that social responsibility artificially inflates the demand for stocks by those who believe in social responsibility, which leads to overpricing and thus underperformance of socially responsible companies. With regard to modern portfolio theory, it is reasonable to assume that as screening intensity increases, the possible investment universe available to the fund decreases. Thus, because of employing SRI screening, the smaller universe will lead to the fund exhibiting a lower level of return for a given level of risk and therefore a lower possible risk-adjusted level of return.

Stakeholder theory argues that better long-term financial performance can be driven by how well a firm conducts its relationships with related parties (Freeman, 1984; Donaldson and Preston, 1995). In line with this theory, proponents of the value-creation approach argue that some of the financial benefits of socially responsible activities such as access to better employees (Greening and Turban, 2000), better corporate image (Fombrun, 1996), and better quality products or services (Moskowitz, 1972) will slowly be impounded in stock prices and subsequently lead to higher future returns.

---

[2] The 14 screens are climate/clean technology, pollution/toxics, environmental/other, community development, diversity & EEO, human rights, labor relations, board issues, executive pay, alcohol, animal testing, defense/weapons, gambling, and tobacco. Refer to the Forum for Sustainable and Responsible Investment's Web site (www.ussif.org) for more details.

Stakeholder theory implies that SRI fund managers select from a smaller but better-performing class of firms (Barnett and Salomon, 2006). Thus, firms which consider the interests of all stakeholders through SRI screening will achieve higher long-term financial returns.

Studies of SRI funds have been dominated by research that investigates the performance of SRI funds compared to non-SRI funds.[3] Despite the large amount of literature, no conclusive results have been reached. For example, some studies have found slightly superior returns for SRI funds (Mallin et al., 1995; Abramson and Chung, 2000) while a number of studies have found no statistical difference (Hamilton et al., 1993; Guerard, 1997; Sauer, 1997; Statman, 2000; Bauer et al., 2005; Bello, 2005). Others still have found slight underperformance of non-SRI funds (Goldreyer and Diltz, 1999; Girard et al., 2007). The major limitation of these studies is the dichotomous way in which they view mutual funds as either SRI or non-SRI. By grouping all SRI funds together and comparing them to their non-SRI counterparts, these studies assume homogeneity of SRI funds when in fact each SRI fund uses different types, numbers, and combinations of screens. That is, they are heterogeneous.

Only six published studies have examined the heterogeneity of SRI funds. The first, by Barnett and Salomon (2006), examines 67 U.S. mutual funds from 1972 to 2000. They do so by regressing risk-adjusted returns (RARs) against a screening intensity figure equal to the number of screens employed by the fund and find a positive curvilinear (U-shaped) relationship between social and financial performance. To explain this finding, the authors contend to both modern portfolio theory and stakeholder theory. They argue that according to modern portfolio theory, funds with weaker standards (lower intensity) will be able to choose from a larger investment universe, thereby increasing their diversification and possible returns. As a fund's screening intensity increases, it has fewer potential investments from which to select. However, according to stakeholder theory, funds are now able to choose from a smaller yet higher quality pool of companies providing above average financial returns. Thus, the curvilinear shape is formed, as SRI funds with both low and high intensity earn greater returns and funds that have a moderate level of screening bear all the costs without any of the benefits.

Renneboog et al. (2008b) analyze 440 international SRI funds from 1991 to 2003. In contrast to Barnett and Salomon (2006) who measure intensity as the number of screens a fund uses, Renneboog et al. (2008b) group screens into broad categories and use the number of screening categories a fund uses and the number of screens within a category as a measure of intensity. They find some evidence that the number of social screens has a negative effect on performance but no relationship between the number of sin, ethical, or environmental screens and performance.

---

[3] Renneboog et al. (2008a) provide a good review of this research.

Lee et al. (2010) examine 61 U.S. SRI mutual funds from 1989 to 2006. Following Barnett and Salomon (2006), they measure intensity as the number of screens a fund uses. The benefit of their approach is that they use a number of performance metrics. However, unlike Barnett and Salomon (2006), they find little evidence of a relationship between screening intensity and returns—most of their performance measures show no relationship but they do find a negative relationship between screening intensity and Carhart RARs. Lee et al. (2010) measure fund performance over a 3-year period, which is quite long for an asset pricing analysis and this could be affecting their findings. They also examine the relationship between screening intensity and the types of companies funds invest in and find that as intensity increases, funds tend toward value firms and recent winners. Further, Lee et al. (2010) find weak evidence that as intensity increases, funds target lower beta firms and larger firms.

In addition to the three key studies just discussed, three other papers have investigated SRI screening intensity. Humphrey and Lee (2011) analyze 27 Australian SRI funds for the period 1996—2008. They find no relationship between the degree of screening and performance. Capelle-Blanchard and Monjon (2014) examine 116 French SRI funds from 2004 to 2007. They document no relation between screening intensity and performance but they show that sectoral screens decrease financial performance. Lastly, Fernandez Sanchez and Luna Sotorrio (2014) study 184 European SRI funds spanning the period 1993—2012. They report a negative relationship between the amount of screening and fund performance.

Overall, these six papers rely on cross-sectional regressions in their analysis of screening intensity. Due to this methodological limitation, it is questionable whether their results would hold under broader testing procedures (i.e., in time-series analysis). Therefore, this study is motivated to contribute to the literature on SRI fund heterogeneity. The first research question will test if the linear and curvilinear relationships hold using different performance models and measures of screening intensity as follows:

Is there a relationship between screening intensity and financial performance of SRI mutual funds and if so what is the nature of the relationship?

We will also examine the types of companies in which SRI funds invest. Lee et al. (2010) find that funds with more screens tend to invest more heavily in large value stocks. Conversely, one could argue that SRI funds will target small growth companies because these companies would be more likely to adopt social concerns, as it allows them to compete with larger more established companies who may not see the need to adjust their policies to consider social issues. Empirically, there is no conclusive evidence one way or the other. Similarly, there is little evidence to suggest a skew toward SRI funds investing in high beta stocks or past winners as opposed to low beta stocks or past losers. Thus, the second research question is:

Is there a relationship between the types of companies SRI funds invent in and screening intensity?

16.3 DATA AND METHODS

16.3.1 Data

The sample comprises U.S. equity mutual funds over the period 1984–2009. There are 88 funds in the sample. Screening data are manually collected from the Social Investment Forum while monthly fund data are sourced from the CRSP mutual funds database. The Fama–French factors and the momentum factor are obtained from Kenneth French's Web site.

16.3.2 Time-Series Analysis

In order to test each of the research questions, a combination of time-series and cross-sectional asset pricing tests will be employed as outlined below. A combination of models and metrics will be applied to measure performance. Value-weighted portfolios are formed each month based on screening intensity. Intensity will be measured by the number of screens used by each fund. Table 16.1 explains the three portfolio classification methods for the time-series regressions. While there are 14 screens according to the Social Investment Forum, there are no funds that employ 2, 3, or 5 screens. Thus, in the first method, 11 portfolios will be created whereby portfolio 1 contains all SRI funds that employ one screen, portfolio 2 contains all SRI funds that employ four screens, and so on. The purpose of the second and third classification methods, which aggregate all funds into five and three portfolios, respectively, is to create portfolios that are comprised of a similar number of funds and to mitigate the noise associated with portfolios containing very few funds.

Table 16.1 Portfolio Classification Methods

Panel A: 11 Portfolios

Portfolio no.	1	2	3	4	5	6	7	8	9	10	11
No. of screens	1	4	6	7	8	9	10	11	12	13	14
No. of funds in portfolio	3	1	5	4	3	8	9	3	5	12	34

Panel B: 5 Portfolios

Portfolio no.	1	2	3	4	5
No. of screens	1,4,6	7,8,9	10,11	12,13	14
No. of funds in portfolio	9	15	12	17	34

Panel C: 3 Portfolios

Portfolio no.	1	2	3
No. of screens	1,4,6,7,8,9	10,11,12,13	14
No. of funds in portfolio	24	29	34

This table outlines each of the three portfolio classification methods applied to the 88 U.S. equity mutual funds in this study. Panel A outlines the first method whereby portfolio 1 contains all SRI funds that employ one screen, portfolio 2 contains all funds that employ four screens, and so on. Panels B and C outline the two additional classification methods. The last portfolio in each of the three classification methods contains the same 34 funds that use all 14 screens.

To test the effect of screening on financial performance, raw returns will first be used as a measure of performance followed by RARs (alpha). Monthly returns are employed throughout the analysis. The time-series regressions will be run using the CAPM, the Fama and French (1996) three-factor model, and the Carhart (1997) four-factor model. The RARs for all portfolios, as estimated by alpha in all three models, will then be reported to determine the nature of the relationship between returns and screening intensity.

For the analysis on the types of companies, the same time-series methodology outlined will be applied but with a focus on the Carhart regression coefficients. These will be scrutinized to examine if there is a relationship between any of the variables and screening intensity. For example, a coefficient greater than (less than) one on the market factor would indicate that funds invest more in high (low) beta stocks. A positive (negative) coefficient on SMB would indicate a tendency for funds to invest in small (big) companies, a positive (negative) coefficient on HML would indicate a tendency for funds to invest in value (growth) companies, and a positive (negative) coefficient on UMD would indicate a tendency for funds to invest in past winner (loser) stocks.

16.3.3 Cross-sectional Analysis

The cross-sectional analysis of performance involves regressions according to the following specification:

$$r_i = \gamma_0 + \gamma_1 \text{ Screening intensity }_i + \gamma_2 \text{ Screening intensity }_{i^2} + [\text{control variables}] + \varepsilon_i \tag{16.1}$$

where the dependent variable r_i will first measure raw performance using the 1-month ahead excess fund returns followed by RARs. RARs are calculated using the regression coefficients from CAPM, Fama—French, and Carhart regressions on fund returns over the past 3 years. In addition, the Sharpe ratio is used as a measure of RARs. The variable of interest, screening intensity, will be measured in the same way as in the time-series analysis. However, as cross-sectional regressions are based on individual funds, it does not matter that there are no funds that use 2, 3, or 5 screens. Thus, to reflect the true intensity of the fund, intensity will be measured up to a maximum score of 14. In a similar way to the time-series tests, we also consider two additional screening classification methods where intensity is measured to a maximum score of 5 and 3 as per Table 16.1. The regressions will also include a number of control variables, which prior research has shown are likely to have an effect on the financial performance of mutual funds. These variables are the size of the fund, the age of the fund, and a dummy variable indicating whether the fund is an institutional fund or not. Year dummies are also included in the regressions. In order to examine whether there is a linear relationship with performance, the screening intensity-squared variable will initially be excluded from the model. The screening intensity-squared variable is then included to investigate whether there is a nonlinear relationship between intensity and performance.

For the analysis on the types of companies, the Carhart coefficients from the time-series regressions using the fund's past 3-year returns will be used as the dependent variables, as per the cross-sectional regression specification below:

$$s_i = \gamma_0 + \gamma_1 \text{ Screening intensity}_i + \gamma_2 \text{ Screening intensity}_{i^2} \\ + [\text{control variables}] + \varepsilon_i \tag{16.2}$$

where s_i is the coefficient on SMB. Similar regressions will be run with the RMRF, HML, and UMD coefficients as the dependent variable in place of the SMB coefficients. The relationship between screening intensity and the Carhart factor loadings will be used to infer whether the types of companies funds invest in differs based on their screening intensity. For example, a positive (negative) coefficient on γ_1 in Eqn (16.2) would indicate that as screening intensity increases, funds tend to invest in smaller (larger) stocks.

16.4 RESULTS

16.4.1 Descriptive Statistics

The descriptive statistics in Table 16.2 display the characteristics of the sample of SRI mutual funds based on screening intensity. The number of funds column in Panel A shows that the majority of funds employ 14 screens, which is the highest degree of intensity. The highest average size ($828.73 million) and age of funds (145.40 months) occur when the number of screens used is 11 and 6, respectively. Panel B shows that the number of SRI funds has increased consistently over time in line with the growing popularity of SRI. Panel B also reports the average number of screens employed by the sample of SRI funds over the entire sample period. We see that the average number of screens has remained stable since 1997. Before 1996, there were less than 20 funds each year.

16.4.2 Performance Analysis

Table 16.3 reports the results of the time-series analysis. Specifically, the average monthly excess returns (raw return proxy), Sharpe ratios, and the time-series regression results for the CAPM, Fama–French, and Carhart asset pricing models are presented.[4] Panel A displays the results for the first portfolio classification method described in Table 16.1. Panels B and C present the regression results for the second and third portfolio classification

---

[4] The findings are reported on value-weighted portfolios. Equal weighted portfolios are also examined. The results are similar and are available on request.

Table 16.2 Descriptive Statistics

Panel A: Fund Characteristics					Panel B: Yearly Statistics				
	Average					**Average**			**Average**
No. of Screens	Size ($ million)	Age (months)	No. of Funds	Year	No. of Funds	No. of Screens	Year	No. of Funds	No. of Screens
1	24.07	197.33	3	1984	1	9.00	1997	21	8.50
2	–	–	–	1985	1	9.00	1998	28	8.73
3	–	–	–	1986	1	6.00	1999	33	8.71
4	9.10	108.00	1	1987	2	7.67	2000	35	9.02
5	–	–	–	1988	3	5.00	2001	45	8.85
6	727.90	145.40	5	1989	5	4.67	2002	47	8.71
7	2.85	30.00	4	1990	6	5.43	2003	49	8.74
8	35.47	110.33	3	1991	7	5.43	2004	50	8.74
9	137.16	76.13	8	1992	8	6.22	2005	51	8.70
10	36.69	68.78	9	1993	9	6.22	2006	61	8.56
11	828.73	139.33	3	1994	9	7.63	2007	68	8.21
12	120.00	115.20	5	1995	17	7.65	2008	78	8.35
13	135.59	118.31	13	1996	20	7.76	2009	82	8.29
14	75.75	117.32	34						
All funds	144.21	109.40	88						

This table reports descriptive statistics for the sample of 88 U.S. SRI mutual funds. Panel A reports the average size, age, and number of funds based on screening intensity. Size is measured as the total net assets for the fund in $US millions and age is measured as the number of months since inception. There are no funds that use two, three, or five screens and only one fund that uses four screens. Panel B reports the number of SRI funds and the average number of screens per year over the period 1984–2009.

methods. There is no discernible pattern between intensity and returns in Panel A. This could be due to the low number of funds in some of the portfolios as shown in Table 16.1. Consequently, the focus will be on Panels B and C. In Panel C, there is a negative relationship between intensity and raw returns, the Sharpe ratio and the CAPM alphas. For the Fama–French and Carhart alphas, however, a different result is observed. In Panels B and C, there is a negative curvilinear (∩-shaped) relationship between both the Fama–French and Carhart alphas and screening intensity. This is in direct contrast to the positive curvilinear association found by Barnett and Salomon (2006). Portfolio 1 in Panels B and C plays a key role in the relationships observed. It earns a relatively high raw return and thus contributes to the negative relationship seen in Panel C between intensity and raw returns, the Sharpe ratio, and the CAPM alpha. However, after risk adjustment with the Fama–French and Carhart models, the alpha of portfolio 1 is much lower. As a result, in Panels B and C, a negative curvilinear relationship is observed for both the Fama–French and Carhart RARs.

The findings of the cross-sectional analysis of intensity and performance are reported in Table 16.4. The dependent variable is initially raw returns and then CAPM, Fama–French, Carhart, and Sharpe RARs. The three intensity variables, Inty 3, Inty 5, and Inty 14 measure intensity using 3, 5, and 14 screening groupings, respectively. Panel A

Table 16.3 Time-Series Regressions for Screening Intensity and Performance

No. of Screens	Monthly Excess Return	Standard Deviation	Sharpe Ratio	Alpha	RMRF	Adjusted R^2
CAPM						
Panel A: Screening Intensity 1–11						
1	0.0044* (1.76)	0.0382	0.1161	−0.0002 (−0.25)	0.8301** (51.52)	0.933
2[a]	0.0003 (0.07)	0.0530	0.0066	−0.0001 (−0.05)	0.9591** (15.90)	0.777
3	0.0070** (2.13)	0.0496	0.1406	0.0021 (1.00)	0.8740** (10.58)	0.608
4	0.0039 (0.31)	0.0684	0.0564	0.0099** (1.97)	0.9640** (15.06)	0.855
5	0.0029 (0.78)	0.0484	0.0594	−0.0009 (−0.99)	0.9628** (46.62)	0.932
6	0.0046 (1.40)	0.0451	0.1014	0.0006 (0.45)	0.8893** (26.47)	0.849
7	0.0012 (0.35)	0.0442	0.0269	−0.0018 (−1.36)	0.8775** (27.14)	0.851
8	0.0053** (2.18)	0.0351	0.1516	0.0022* (1.82)	0.6855** (18.92)	0.772
9	0.0076* (1.87)	0.0619	0.1233	0.0022* (1.82)	0.6855** (18.92)	0.772
10	0.0040 (1.34)	0.0449	0.0885	−0.0014* (−1.89)	0.9755** (51.63)	0.929
11	0.0036 (1.26)	0.0438	0.0831	−0.0014 (−1.12)	0.9119** (17.39)	0.853
Panel B: Screening Intensity 1–5						
1	0.0067** (2.13)	0.0479	0.1409	0.0018 (0.97)	0.8918** (11.77)	0.679
2	0.0041 (1.25)	0.0449	0.0909	0.0002 (0.18)	0.9024** (31.03)	0.881
3	0.0048* (1.92)	0.0358	0.1337	0.0015 (1.35)	0.7203** (21.97)	0.820
4	0.0061* (1.84)	0.0502	0.1215	0.0003 (0.26)	1.0410** (44.30)	0.846
5	0.0036 (1.26)	0.0438	0.0831	−0.0014 (−1.12)	0.9119** (17.39)	0.853
Panel C: Screening Intensity 1–3						
1	0.0062** (2.07)	0.0453	0.1372	0.0015 (0.93)	0.8767** (14.09)	0.734
2	0.0061* (1.92)	0.0481	0.1271	0.0008 (0.63)	0.9960** (41.61)	0.842
3	0.0036 (1.26)	0.0438	0.0831	−0.0014 (−1.12)	0.9119** (17.39)	0.853

Table 16.3 Time-Series Regressions for Screening Intensity and Performance—cont'd
Fama–French

No. of Screens	Alpha	RMRF	SMB	HML	Adjusted R^2
Panel A: Screening Intensity 1–11					
1	−0.0005	0.8438**	0.0079	0.0686**	0.936
	(−0.81)	(60.23)	(0.32)	(2.71)	
2[a]	0.0002	0.9493**	0.0426	−0.1302	0.780
	(0.10)	(16.06)	(0.37)	(−1.28)	
3	−0.0012	0.9738**	0.1998**	0.6141**	0.753
	(−0.75)	(16.18)	(2.43)	(8.03)	
4	0.0089*	0.9403**	0.2876	−0.074	0.855
	(1.69)	(10.02)	(1.04)	(−0.68)	
5	−0.0007	0.9945**	−0.1645**	0.0045	0.949
	(−0.85)	(47.22)	(−5.71)	(0.14)	
6	−0.0002	0.9471**	−0.0759*	0.2064**	0.883
	(−0.16)	(30.76)	(−1.66)	(4.45)	
7	−0.0023*	0.9211**	−0.0971**	0.1293**	0.872
	(−1.79)	(35.31)	(−2.32)	(3.02)	
8	0.0015	0.7269**	−0.0414	0.1596**	0.799
	(1.27)	(22.45)	(−1.03)	(3.64)	
9	0.001	1.0625**	0.3252**	−0.0282	0.680
	(0.42)	(16.89)	(3.26)	(−0.33)	
10	−0.0009	0.9799**	−0.114**	−0.0702**	0.936
	(−1.22)	(51.38)	(−4.40)	(−3.22)	
11	−0.0014	0.9024**	0.0361	−0.0149	0.853
	(−1.08)	(17.45)	(0.87)	(−0.51)	
Panel B: Screening Intensity 1–5					
1	−0.0012	0.9754**	0.2081**	0.5471**	0.804
	(−0.89)	(17.70)	(3.32)	(7.85)	
2	−0.0004	0.9528**	−0.0821**	0.1658**	0.906
	(−0.36)	(35.17)	(−2.01)	(4.19)	
3	0.0008	0.7635**	−0.0518	0.1599**	0.848
	(0.77)	(27.23)	(−1.41)	(3.99)	
4	0.0003	1.0167**	0.1016*	−0.0303	0.851
	(0.25)	(40.20)	(1.67)	(−0.67)	
5	−0.0014	0.9024**	0.0361	−0.0149	0.853
	(−1.08)	(17.45)	(0.87)	(−0.51)	
Panel C: Screening Intensity 1–3					
1	−0.0011	0.9489**	0.162**	0.4603**	0.831
	(−0.85)	(21.06)	(2.84)	(8.02)	
2	0.0009	0.9717**	0.083	−0.0456	0.846
	(0.73)	(38.33)	(1.37)	(−1.01)	
3	−0.0014	0.9024**	0.0361	−0.0149	0.853
	(−1.08)	(17.45)	(0.87)	(−0.51)	

Continued

Table 16.3 Time-Series Regressions for Screening Intensity and Performance—cont'd

Carhart

No. of Screens	Alpha	RMRF	SMB	HML	UMD	Adjusted R^2
Panel A: Screening Intensity 1–11						
1	−0.0004	0.8383**	0.0090	0.0639**	−0.0124	0.936
	(−0.56)	(48.27)	(0.36)	(2.15)	(−0.49)	
2[a]	0.0003	0.9561**	0.0439	−0.1319	0.0087	0.778
	(0.11)	(10.07)	(0.41)	(−1.30)	(0.09)	
3	−0.0000	0.9143**	0.2114**	0.5635**	−0.1338**	0.770
	(−0.01)	(17.06)	(2.85)	(8.73)	(−2.37)	
4	0.0071	0.8706**	0.2604	−0.1776*	−0.1331	0.868
	(1.47)	(11.75)	(0.86)	(−1.90)	(−1.63)	
5	−0.0004	0.9757**	−0.1603**	−0.0103	−0.0358	0.950
	(−0.49)	(46.30)	(−5.61)	(−0.36)	(−1.70)	
6	0.0002	0.9214**	−0.0702*	0.1850**	−0.0514**	0.886
	(0.20)	(26.25)	(−1.69)	(4.01)	(−2.18)	
7	−0.0019	0.8983**	−0.0922**	0.1121**	−0.0434*	0.874
	(−1.53)	(29.53)	(−2.22)	(2.63)	(−1.77)	
8	0.0017	0.7152**	−0.0383	0.1512**	−0.0242	0.800
	(1.39)	(21.30)	(−0.98)	(3.27)	(−1.00)	
9	0.0027	0.9775**	0.3418**	−0.1005	−0.1911**	0.703
	(1.16)	(16.19)	(3.63)	(−1.12)	(−3.19)	
10	−0.0009	0.9832**	−0.1146**	−0.0674**	0.0075	0.936
	(−1.26)	(51.45)	(−4.40)	(−3.29)	(0.43)	
11	−0.001	0.8830**	0.0398	−0.0314	−0.0435**	0.854
	(−0.74)	(15.83)	(0.96)	(−1.11)	(−2.22)	
Panel B: Screening Intensity 1–5						
1	−0.0002	0.9229**	0.2184**	0.5024**	−0.1181**	0.819
	(−0.11)	(19.58)	(3.86)	(8.63)	(−2.24)	
2	0.0000	0.9270**	−0.0762**	0.1447**	−0.0511**	0.909
	(0.02)	(30.53)	(−2.08)	(3.76)	(−2.68)	
3	0.0010	0.7509**	−0.0483	0.1506**	−0.0263	0.849
	(0.95)	(26.04)	(−1.37)	(3.57)	(−1.24)	
4	0.0009	0.9879**	0.1073*	−0.0548	−0.0648**	0.854
	(0.71)	(35.59)	(1.80)	(−1.20)	(−2.48)	
5	−0.001	0.8830**	0.0398	−0.0314	−0.0435**	0.854
	(−0.74)	(15.83)	(0.96)	(−1.11)	(−2.22)	
Panel C: Screening Intensity 1–3						
1	−0.0002	0.9054**	0.1709**	0.4237**	−0.0975**	0.843
	(−0.15)	(22.28)	(3.35)	(8.37)	(−2.49)	
2	0.0014	0.9469**	0.0877	−0.0666	−0.0558**	0.849
	(1.11)	(34.1)	(1.48)	(−1.47)	(−2.22)	
3	−0.001	0.8830**	0.0398	−0.0314	−0.0435**	0.854
	(−0.74)	(15.83)	(0.96)	(−1.11)	(−2.22)	

This table reports average monthly excess returns, Sharpe ratios, and the results from time-series regressions for value-weighted portfolios of SRI funds. Regressions for each portfolio are run using the CAPM, Fama–French, and Carhart models for each of the three portfolio classification methods outlined in Panels A, B, and C. T-statistics are reported in parentheses. Significant at **5% and *10% levels.
[a]Note, there is only one fund in portfolio 2 of Panel A.

Table 16.4 Cross-sectional Regressions for Screening Intensity and Performance

Panel A: Intensity

Variable	Raw			CAPM RAR			Fama–French RAR			Carhart RAR			Sharpe RAR		
Constant	0.0581 (1.19)	−0.0026 (−0.09)	−0.0043 (−0.15)	−0.0268 (−1.22)	−0.0267 (−1.22)	−0.0268 (−1.22)	−0.0231 (−1.17)	−0.0231 (−1.17)	−0.0234 (−1.19)	−0.0144 (−0.73)	−0.0144 (−0.73)	−0.0147 (−0.75)	0.6918 (0.60)	0.6916 (0.60)	0.6901 (0.60)
Inty 3	−0.0006 (−0.79)			−0.0006 (−1.61)			−0.0005* (−1.68)			−0.0004 (−1.29)			−0.0129 (−0.70)		
Inty 5		−0.0007** (−2.27)			−0.0003 (−1.59)			−0.0002 (−1.41)			−0.0002 (−1.06)			−0.0064 (−0.63)	
Inty 14			−0.0002 (−1.31)			−0.0001 (−1.18)			−0.0001 (−0.81)			−0.00002 (−0.55)			−0.0020 (−0.48)
Size	1.3000 (0.92)	0.9950 (0.72)	1.1200 (0.81)	−0.0653 (−0.1)	−0.1100 (−0.17)	0.0009 (0.00)	−0.1720 (−0.30)	−0.1870 (−0.33)	0.0836 (−0.15)	−0.5270 (−0.93)	−0.5360 (−0.94)	−0.4550 (−0.81)	27.5000 (0.83)	26.9000 (0.80)	29.2000 (0.88)
Age	0.0006 (0.05)	0.0034 (0.33)	0.0035 (0.34)	0.0073 (1.20)	0.0073 (1.20)	0.0070 (1.14)	0.0029 (0.52)	0.0030 (0.54)	0.0030 (0.54)	0.0130** (2.38)	0.0131** (2.40)	0.0131** (2.39)	−0.0927 (−0.29)	−0.0913 (−0.29)	−0.0981 (−0.30)
Institutional	0.4750 (0.29)	0.2990 (0.22)	0.0203 (0.01)	0.3670 (0.5)	0.3030 (0.41)	0.1900 (0.25)	0.8600 (1.31)	0.8200 (1.24)	0.7750 (1.13)	0.2220 (0.34)	0.1920 (0.29)	0.1710 (0.25)	25.7210 (0.67)	24.5420 (0.63)	21.9970 (0.55)
Adjusted R^2	0.1224	0.1119	0.1116	0.0181	0.0180	0.0179	0.0062	0.0061	0.0059	0.0162	0.0162	0.0160	0.1338	0.1338	0.1338

Panel B: Intensity Squared

Variable	Raw			CAPM RAR			Fama–French RAR			Carhart RAR			Sharpe RAR		
Constant	0.0517 (1.05)	−0.0050 (−0.17)	−0.0073 (−0.25)	−0.0334 (−1.51)	−0.0289 (−1.31)	−0.0290 (−1.32)	−0.0297 (−1.49)	−0.0264 (−1.33)	−0.0264 (−1.33)	−0.0203 (−1.03)	−0.0176 (−0.89)	−0.0174 (−0.89)	0.6521 (0.56)	0.6254 (0.54)	0.5364 (0.46)
Inty 3	0.0056 (0.89)			0.0059** (2.09)			0.0059** (2.34)			0.0055** (2.16)			0.1397 (0.94)		
Inty 3 squared	−0.0015 (−1.00)			−0.0016** (−2.30)			−0.0016** (−2.56)			−0.0014** (−2.34)			−0.0373 (−1.04)		
Inty 5		0.0001 (0.05)			0.0012 (0.91)			0.0019* (1.68)			0.0020* (1.73)			0.0382 (0.57)	
Inty 5 squared		−0.0001 (−0.35)			−0.0002 (−1.17)			−0.0003 (−1.92)			−0.0003* (−1.91)			−0.0070 (−0.67)	
Inty 14			0.0010 (1.39)			0.0005 (1.27)			0.0007** (2.04)			0.0007* (1.95)			0.0081 (0.39)
Inty 14 squared			−0.0001* (−1.70)			−0.00004 (−1.54)			−0.00005** (−2.25)			−0.00004** (−2.10)			−0.0006 (−0.50)
Size	1.4700 (1.03)	1.0000 (0.73)	0.6390 (0.45)	0.1160 (0.18)	0.0224 (0.03)	−0.2010 (−0.31)	0.0098 (0.02)	0.0093 (0.02)	−0.3500 (−0.61)	−0.3620 (−0.64)	−0.3420 (−0.59)	−0.7020 (−1.22)	31.8000 (0.95)	30.9000 (0.91)	25.7000 (0.76)
Age	0.0002 (0.01)	0.0040 (0.39)	0.0081 (0.76)	0.0068 (1.12)	0.0076 (1.24)	0.0085 (1.36)	0.0024 (0.44)	0.0034 (0.61)	0.0049 (0.88)	0.0126** (2.30)	0.0135** (2.47)	0.0150** (2.68)	−0.1040 (−0.32)	−0.0832 (−0.26)	−0.0726 (−0.22)
Institutional	0.7970 (0.48)	0.3460 (0.26)	0.8920 (0.61)	0.7000 (0.94)	0.5600 (0.73)	0.8170 (0.94)	1.1940* (1.78)	1.2000* (1.74)	1.6030** (2.06)	0.5240 (0.79)	0.5690 (0.83)	0.9380 (1.21)	33.5800 (0.86)	32.3420 (0.80)	32.7090 (0.72)
Adjusted R^2	0.1224	0.1118	0.1117	0.0188	0.0181	0.0181	0.0071	0.0065	0.0065	0.0170	0.0166	0.0166	0.1338	0.1337	0.1337

This table presents the cross-sectional regression results. The dependent variable, fund performance is first measured using raw returns and then risk–adjusted returns (RARs). RARs are calculated using regression coefficients estimated from time–series regressions of the past 3-year fund returns against the CAPM, Fama–French, and Carhart models. The Sharpe ratio is also used as a measure of RARs. Screening intensity is measured by the number of screens a fund uses in each of the three screening classification methods. Hence, Inty 3, Inty 5, and Inty 14 represent the 3, 5, and 14 screening groupings, respectively. The control variables include size, which measures each fund's total assets in $US millions, age, measured by the number of months since the fund's inception, and institutional, a dummy variable equal to 1 if the fund is an institutional fund and 0 otherwise. The size coefficient is multiplied by 1,000,000 while the age and institutional coefficients are multiplied by 1000. Year dummies are also included in the regression but are not reported. Panel A reports regressions that do not include a screening intensity–squared variable whereas this variable is included in Panel B. T-statistics are reported in parentheses. Significant at **5% and *10% levels.

presents the results where the screening intensity-squared variable is suppressed from the regression and is thus an examination of the linear relationship between intensity and performance. The coefficients on each intensity variable, Inty 3, Inty 5, and Inty 14 and for each measure of performance are negative. However, only the Inty 5 coefficient for the raw returns analysis is significant at the 5% level. For the Fama—French RARs, the Inty 3 coefficient is significant at 10%. In all other cases, the intensity coefficients are not significant. On balance, the results in Panel A suggest that the relationship between intensity and performance is not linear.

Panel B of Table 16.4 reports the results where the screening intensity-squared variable is included in the regression and is thus an analysis of the nonlinear relationship between intensity and performance. Of particular interest in these regressions is the coefficient on the intensity-squared variable. Generally, a significantly positive (negative) coefficient on the intensity-squared variable indicates a positive (negative) curvilinear relationship with performance. However, the coefficient on the regular intensity variable can also influence the nature of the relationship. Therefore, in addition to reporting the regression results in Panel B, plots are produced using the coefficients on the intensity and the intensity-squared variable to display the estimated impact of screening intensity on returns pictorially. These plots are presented in Figure 16.1. From Panel B, we see that all of the coefficients on the intensity-squared variable are negative. Further, these coefficients are significant at either 5% or 10% for Inty 14 squared and raw returns, Inty 3 squared and CAPM RARs, and for all three intensity classifications for both the Fama—French and Carhart RARs. Therefore, the regressions indicate that there is a negative curvilinear relationship between screening intensity and performance. Figure 16.1 reinforces these results, as a negative curvilinear relationship is observed in almost all cases—possible exceptions are the raw and CAPM RARs in Chart B.

Taken together, the time-series and cross-sectional analyses suggest that there is a negative curvilinear relationship between screening intensity and performance. This is particularly so when Fama—French and Carhart RARs are used to measure performance. However, in the time-series analysis, there is some evidence of a negative relationship between intensity and both raw returns and CAPM RARs. On balance though, the findings suggest a negative curvilinear relationship. This is in direct contrast to Barnett and Salomon (2006) who find a positive curvilinear relationship and also conflicts with Renneboog et al. (2008b) and Lee et al. (2010) who generally find no relationship. We believe that the divergent findings observed in the current study compared to results of prior studies are a function of the more robust and wide-ranging analysis we conduct.

In terms of a theoretical justification for the negative curvilinear association observed, in a similar vein to Barnett and Salomon (2006) but in the opposite direction, one can contend to stakeholder theory and modern portfolio theory. At low levels of screening intensity, the financial benefits of focusing on companies that are socially responsible and well-run are limited. It is only at medium to high levels of intensity that the financial

Chart A: 3 screening groupings

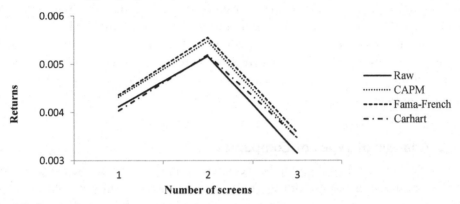

Chart B: 5 screening groupings

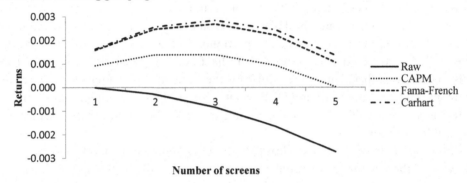

Chart C: 14 screening groupings

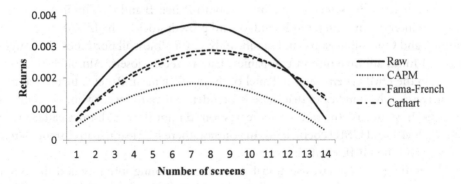

Figure 16.1 *Impact of screening intensity on performance.* These plots show the estimated impact of screening intensity on returns after controlling for variables known to explain fund returns (size, age, and the institutional dummy). They are derived using the coefficients on the intensity and the intensity-squared variable in Panel B of Table 16.4. Plots are produced for the raw return regressions as well as for the risk-adjusted return regressions for the CAPM, Fama–French, and Carhart models. Chart A presents the plot for the 3 screening groupings, Chart B presents the plot for the 5 screening groupings, and Chart C presents the plot for the 14 screening groupings.

benefits are observed. However, when screening intensity becomes too high and as a result, the investible universe becomes too small, the lack of diversification will cause RARs to fall. Therefore, a negative curvilinear relationship is formed where low, high intensity funds earn lower returns, and medium intensity funds earn higher returns. In particular, medium intensity funds benefit from both the focus on socially responsible well-run firms and having an investible universe large enough to maximize risk-adjusted performance.

16.4.3 Analysis of Types of Companies

The examination of the relationship between screening intensity and the types of companies funds invest in begins with the time-series regression results in Table 16.3. In contrast to the performance analysis, which concentrates on the model alphas, we are now interested in the coefficients on the factors from the asset pricing models. The focus shall again be on the second and third portfolio classification methods, which are reported in Panels B and C, as a number of the portfolios in Panel A have very few funds. Table 16.3 reports coefficient estimates for the factors from all three models. The RMRF coefficients (betas) are quite similar for the CAPM, Fama–French, and Carhart regressions. Likewise, the SMB and HML coefficients in the Fama–French and Carhart regressions are comparable. As such, the ensuing discussion of patterns in the coefficients is applicable to all three models.

With regard to beta, there is no clear pattern with intensity in Panel B whereas Panel C shows that there is a negative curvilinear association. For the SMB coefficients, there is no trend in Panel B but a negative relationship with intensity in Panel C. The HML coefficients show the strongest pattern—there is a clear negative relationship between the HML coefficients and screening intensity in both Panels B and C. This indicates that as screening intensity increases, funds tend toward growth stocks. The UMD coefficients in Panels B and C are all negative and significantly so for almost all portfolios. This suggests that SRI funds are investing in companies that are recent losers. Although there is no pattern in the UMD coefficients in Panel B, the coefficients in Panel C increase as intensity increases, which indicates that there is a tendency toward recent winners as intensity increases. In summary, the time-series regressions do not show a distinct pattern for the RMRF, SMB, and UMD coefficients. In contrast, there is a clear negative trend between intensity and the HML coefficients.

The findings for the cross-sectional analysis of screening intensity and the types of companies funds invest in are presented in Table 16.5. Similar to Table 16.4, Panel A reports results with the intensity-squared variable excluded from the regressions while in Panel B this variable is included. Table 16.5 Panel A shows that there is a significant linear relationship between each of the four coefficients and intensity for each of the three measures of intensity (Inty 3, Inty 5, and Inty 14). There is a positive relationship between

Table 16.5 Cross-sectional Regressions for Screening Intensity and Types of Companies

Variable	Beta			SMB		
Panel A: Intensity						
Constant	1.1866** (8.66)	1.1686** (8.58)	1.1504** (8.45)	0.4422 (1.53)	0.4418 (1.53)	0.4414 (1.53)
Inty 3	0.0283** (12.93)			0.0193** (4.19)		
Inty 5		0.0186** (15.45)			0.0098** (3.84)	
Inty 14			0.0076** (15.50)			0.0033** (3.15)
Size	−16.4000** (−4.16)	−12.2000** (−3.09)	−17.2000** (−4.39)	11.3000 (1.36)	12.3000 (1.47)	9.1100 (1.10)
Age	−0.4810** (−12.66)	−0.4710** (−12.45)	−0.4270** (−11.19)	0.0024 (0.03)	0.0009 (0.01)	0.0141 (0.17)
Institutional	−47.1390** (−10.31)	−42.4700** (−9.30)	−30.1590** (−6.38)	−156.9210** (−16.32)	−155.0740** (−16.04)	−150.6090** (−15.04)
Adjusted R^2	0.1384	0.1482	0.1485	0.1082	0.1078	0.1071
Panel B: Intensity Squared						
Constant	1.3714** (10.02)	1.2881** (9.51)	1.1793** (8.65)	0.5740** (1.98)	0.8536** (3.02)	0.5478* (1.90)
Inty 3	−0.1531** (−8.76)			−0.1101** (−2.97)		
Inty 3 squared	0.0444** (10.46)			0.0317** (3.51)		
Inty 5		−0.0619** (−7.91)			−0.2676** (−16.39)	
Inty 5 squared		0.0127** (10.40)			0.0437** (17.18)	
Inty 14			−0.0001 (−0.03)			−0.0249** (−4.85)

Continued

Table 16.5 Cross-sectional Regressions for Screening Intensity and Types of Companies—cont'd

Variable	Beta			SMB		
Inty 14 squared			0.0004**			0.0016**
			(3.21)			(5.60)
Size	−21.500**	−19.5000**	−14.5000**	7.6600	−12.7000	18.8000**
	(−5.46)	(−4.90)	(−3.64)	(0.92)	(−1.53)	(2.22)
Age	−0.4680**	−0.4850**	−0.4470**	0.0121	0.0494	−0.0571
	(−12.41)	(−12.94)	(−11.56)	(0.15)	(−0.63)	(−0.70)
Institutional	−56.4840**	−56.5480**	−38.2940**	−163.5820**	−203.5810**	−180.5640**
	(−12.23)	(−11.97)	(−7.14)	(−16.71)	(−20.66)	(−15.93)
Adjusted R^2	0.1537	0.1632	0.1498	0.1099	0.1495	0.1115

Panel A: Intensity

Variable	HML			UMD		
Constant	0.3345	0.3919*	0.4297*	−0.1793	−0.1739	−0.1821
	(1.41)	(1.68)	(1.82)	(−1.11)	(−1.07)	(−1.12)
Inty 3	−0.0992**			0.0233**		
	(−26.10)			(9.00)		
Inty 5		−0.0636**			0.0100**	
		(−30.96)			(7.23)	
Inty 14			−0.0240**			0.0041**
			(−28.42)			(6.99)
Size	53.3000**	39.4000**	58.0000**	−9.2900**	−8.9100*	−11.8000**
	(7.77)	(5.82)	(8.59)	(−1.99)	(−1.89)	(−2.53)
Age	−0.2080**	−0.2410**	−0.3640**	−0.5830**	−0.5890**	−0.5670**
	(−3.15)	(−3.73)	(−5.52)	(−12.97)	(−13.08)	(−12.46)
Institutional	−19.3720**	−35.0810**	−71.8890**	23.9380**	25.5030**	31.9370**
	(−2.44)	(−4.49)	(−8.79)	(4.44)	(4.69)	(5.67)
Adjusted R^2	0.2604	0.2899	0.2742	0.1329	0.1288	0.1283

Variable	HML		UMD		
Panel B: Intensity Squared					
Constant	0.3672 (1.53)	0.3268 (1.40)	−0.1069 (−0.66)	−0.0854 (−0.53)	−0.0781 (−0.48)
Inty 3	−0.1313** (−4.29)		−0.0478** (−2.29)		
Inty 3 squared	0.0079 (1.06)		0.0174** (3.44)		
Inty 5	−0.0198 (−1.46)			−0.0492** (−5.25)	
Inty 5 squared	−0.0069** (−3.29)			0.0094** (6.43)	
Inty 14		0.0263** (6.34)			−0.0234** (−8.16)
Inty 14 squared		−0.0029** (−12.38)			0.0016** (9.77)
Size	52.4000** (7.58)	43.3000** (6.31)	−11.3000** (−2.40)	14.3000** (−3.00)	−2.3600 (−0.50)
Age	−0.2060** (−3.12)	−0.2330** (−3.61)	−0.5770** (−12.86)	−0.6000** (−13.35)	−0.6360** (−13.92)
Institutional	−21.0260** (−2.60)	−27.4140** (−3.37)	20.2800** (3.69)	15.0820** (2.67)	2.6690 (0.42)
Adjusted R^2	0.2604	0.2910	0.1344	0.1346	0.1418

This table presents the cross-sectional regression results for the analysis of screening intensity and the types of companies funds invest in. The Carhart coefficients from the time-series regressions using the fund's past 3-year returns are the dependent variables in these regressions. Thus, in turn, regression results are reported where the dependent variables are the RMRF, SMB, HML, and UMD coefficients. Screening intensity is measured by the number of screens a fund uses in each of the three screening classification methods. The control variables include size, which measures each fund's total assets in $US millions, age, measured by the number of months since the fund's inception, and institutional, a dummy variable equal to 1 if the fund is an institutional fund and 0 otherwise. The size coefficient is multiplied by 1,000,000 while the age and institutional coefficients are multiplied by 1000. Year dummies are also included in the regression but are not reported. Panel A reports regressions that do not include a screening intensity-squared variable whereas this variable is in included in Panel B. T-statistics are reported in parentheses. Significant at **5% and *10% levels.

the RMRF, SMB, and UMD coefficients and intensity and a negative relationship between the HML coefficient and intensity. However, with the exception of the HML coefficient, these findings are not corroborated in Panel B.

For beta, the intensity-squared term is significantly positive for all three intensity measures in Panel B suggesting a positive curvilinear relationship. However, the negative significance of the regular intensity variable for Inty 3 and Inty 5 makes it difficult to infer the exact nature of the relationship. This issue also arises with the SMB, HML, and UMD coefficients. As such, in a similar fashion to the performance analysis with Figures 16.1 and 16.2 is a pictorial representation of the impact of screening intensity on the Carhart regression coefficients. The plots are constructed from the coefficients on the intensity and the intensity-squared variable in Table 16.5 Panel B. The three beta plots show a slightly positive relationship with intensity, which is generally consistent with the linear results in Panel A. With SMB, there is more of a positive curvilinear relationship, particularly in Charts B and C for Inty 5 and 14—this is inconsistent with the linear results in Panel A. For UMD, the plot in Chart A shows a positive association with intensity whereas in Charts B and C the relationship is positive curvilinear. As indicated earlier, it is only with the HML coefficient results that there is clear consistency between the linear and nonlinear cross-sectional analysis. There is a negative relationship between intensity and the HML coefficient in Figure 16.2, which means that as intensity increases, funds tend to invest more in growth companies. This is particularly so in Charts A and B whereas in Chart C, the relationship is flat for intensity levels from 1 to 6 and downward sloping thereafter. In addition, the strength of the association between intensity and the HML coefficient is much stronger than for the other coefficients. Only the positive curvilinear relationship between intensity and SMB in Chart B is of a similar magnitude to the relationship observed in the plots between intensity and HML.

Overall, the examination of screening intensity and the types of companies funds invest in is inconclusive with the exception of the findings on HML. In both the time-series and cross-sectional analysis and for both the linear and nonlinear cross-sectional regressions, there is a negative relationship observed between screening intensity and the HML coefficients. There are patterns observed between the RMRF, SMB, and UMD coefficients with intensity but the patterns are not consistent across the different tests. Thus, we conclude that there is no clear relationship between screening intensity and the beta, size, or momentum of the companies in which SRI funds invest. Conversely, we find that as intensity increases, funds tend to target growth companies. Our findings contrast with those of Lee et al. (2010) who show that as intensity increases, funds invest more in value companies and recent winners. We believe that our wide-ranging tests are more robust, encompassing both time-series and cross-sectional regressions. Further, our cross-sectional regressions examine both linear and nonlinear associations.

Chart A: 3 screening groupings

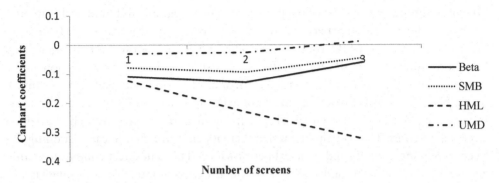

Chart B: 5 screening groupings

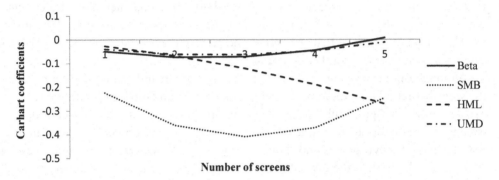

Chart C: 14 screening groupings

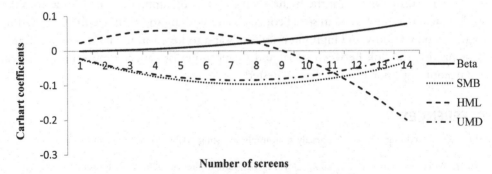

Figure 16.2 *Impact of screening intensity on the types of companies.* These plots show the estimated impact of screening intensity on the types of companies funds invest in. They are derived using the coefficients on the intensity and the intensity-squared variable in Panel B of Table 16.5. Plots are produced for beta and for the coefficients on the SMB, HML, and UMD factors. Chart A presents the plot for the 3 screening groupings, Chart B presents the plot for the 5 screening groupings, and Chart C presents the plot for the 14 screening groupings.

16.5 CONCLUSION

One of the limitations of most of the prior studies that examine SRI funds is that in order to analyze the relationship between social and financial performance, SRI funds are compared to their non-SRI counterparts. This approach ignores the heterogeneity of SRI funds. In particular, SRI funds differ on how strict or intense their selection criteria may be in deciding whether the social performance of a firm is sufficient to allow investment. This is one of the first studies to examine heterogeneity in SRI funds and we do so by analyzing the link between social and financial performance through the lens of screening intensity. Our key finding is that there is a negative curvilinear relationship between screening intensity and financial performance. This is in direct contrast to Barnett and Salomon (2006) who find a positive curvilinear relationship and it also conflicts with the findings of Renneboog et al. (2008b) and Lee et al. (2010). We also examine the relationship between screening intensity and the types of companies that funds invest in and find that as intensity increases, funds tend toward growth companies. This also conflicts with Lee et al. (2010) who find that as intensity increases, funds are more likely to invest in value firms. We believe that the robust and wide ranging analysis conducted in this study is the cause of the conflicting results.

Although our key finding is in direct contrast to Barnett and Salomon (2006), we use the same theoretical underpinnings of stakeholder theory and modern portfolio theory to explain our result. Specifically, at low levels of screening intensity, funds do not benefit from the focus on socially responsible well-run firms. It is only at higher levels of intensity that the benefits envisaged by stakeholder theory arise. However, if screening intensity becomes too high, then the investible universe available to the fund becomes too small and then according to modern portfolio theory, RARs fall. Therefore, a negative curvilinear relationship is formed where low and high intensity funds earn lower returns, and medium intensity funds earn the highest returns. The recommendation arising from this study to investors interested in social concerns and wishing to maximize financial performance is to avoid low and high intensity funds and to invest in medium intensity funds. It seems that the link between social and financial performance is more nuanced than the majority of prior research has suggested.

REFERENCES

Abramson, L., Chung, D., 2000. Socially responsible investing: viable for value investors? J. Invest. 9 (3), 73–80.

Barnett, M.L., Salomon, R.M., 2006. Beyond dichotomy: the curvilinear relationship between social responsibility and financial performance. Strategic Manag. J. 27 (11), 1101–1122.

Bauer, R., Koedijk, K., Otten, R., 2005. International evidence on ethical mutual fund performance and investment style. J. Bank. Finance 29 (7), 1751–1767.

Bello, Z.Y., 2005. Socially responsible investing and portfolio diversification. J. Financial Res. 28 (1), 41–57.

Capelle-Blanchard, G., Monjon, S., 2014. The performance of socially responsible funds: does the screening process matter? Eur. Financ. Manag. 20 (3), 494–520.

Carhart, M.M., 1997. On persistence in mutual fund performance. J. Finance 52 (1), 57—82.

Donaldson, T., Preston, L.E., 1995. The stakeholder theory of the corporation: concepts, evidence and implications. Acad. Manag. Rev. 20 (1), 65—91.

Fernandez Sanchez, J.L., Luna Sotorrio, L., 2014. Effect of social screening on funds' performance: empirical evidence of European equity funds. Span. J. Finance Account. 43 (1), 91—109.

Freeman, R., 1984. Strategic Management: A Stakeholder Approach. Pitman, Boston, MA.

Fama, E.F., French, K.R., 1996. Multifactor explanations of asset pricing anomalies. J. Finance 51 (1), 55—84.

Fombrun, C.J., 1996. Reputation: Realizing Value from the Corporate Image. Harvard Business School Press, Boston, MA.

Girard, E., Rahman, H., Stone, B., 2007. Socially responsible investments: goody-two-shoes or bad to the bone? J. Invest. 16 (1), 96—110.

Goldreyer, E.F., Diltz, J.D., 1999. The performance of socially responsible mutual funds: incorporating socio-political information in portfolio selection. Manag. Finance 25 (1), 23—36.

Greening, D.W., Turban, D.B., 2000. Corporate social performance as a competitive advantage in attracting a quality workforce. Bus. Soc. 39 (3), 254—280.

Guerard, J.B., 1997. Is there a cost to being socially responsible in investing? J. Forecast. 16 (7), 475—490.

Hamilton, S., Jo, H., Statman, M., 1993. Doing well while doing good? The investment performance of socially responsible mutual funds. Financial Analysts J. 49 (6), 62—66.

Hassel, L., Nilsson, H., Nyquist, S., 2005. The value relevance of environmental performance. Eur. Account. Rev. 14 (1), 41—61.

Humphrey, J.E., Lee, D.D., 2011. Australian socially responsible funds: performance, risk and screening intensity. J. Bus. Ethics 102 (4), 519—535.

Lee, D.D., Humphrey, J.E., Benson, K.L., Ahn, J.Y., 2010. Socially responsible investment fund performance: the impact of screening intensity. Account. Finance 50 (2), 351—370.

Mallin, C.A., Saadouni, B., Briston, R.J., 1995. The financial performance of ethical investment funds. J. Bus. Finance Account. 22 (4), 483—496.

Moskowitz, M.R., 1972. Choosing socially responsible stocks. Bus. Soc. Rev. 1 (1), 71—75.

Renneboog, L., Ter Horst, J., Zhang, C., 2008a. Socially responsible investments: institutional aspects, performance, and investor behavior. J. Bank. Finance 32 (9), 1723—1742.

Renneboog, L., Ter Horst, J., Zhang, C., 2008b. The price of ethics and stakeholder governance: the performance of socially responsible mutual funds. J. Corp. Finance 14 (3), 302—322.

Sauer, D.A., 1997. The impact of social-responsibility screens on investment performance: evidence from the Domini 400 social index and Domini equity mutual fund. Rev. Financial Econ. 6 (2), 137—149.

Statman, M., 2000. Socially responsible mutual funds. Financial Analysts J. 56 (3), 30—39.

CHAPTER 17

Using CO$_2$ Emission Allowances in Equity Portfolios

Michael Graham, Anton Hasselgren, Jarkko Peltomäki
Stockholm Business School, Stockholm University, Stockholm, Sweden

Contents

17.1 INTRODUCTION

Environment-related investments are popular choices among investors as evidenced by the increasing investor use of corporate sustainability reporting to inform investment strategies. For example a class of investors, referred to as *carbon investors*, integrate sustainable and environmentally friendly practices into their investment choices to reduce risk and enhance financial returns. This stems from the near global acceptance that greenhouse gases (GHG) in the atmosphere are of significant concern to the economy (see Stern, 2007). Such investments, however, are rarely considered as *alternative investments* and portfolio diversifiers. In this chapter, we explore the carbon futures allowance market, which is a relatively immature derivatives market, as a potential source of portfolio diversification and yield to equity portfolios. The carbon emission allowance futures market is a market for carbon emissions trading, with the underlying instrument being the emission allowances.

Two interesting characteristics of investments in carbon emission allowances make it interesting companions for equity portfolios. First, as shown by Trueck et al. (2014), futures contracts in the carbon emission allowance market tend to trade with significant convenience yields which raise the possibility of earning an additional premium from continuously rolling over positions in these contracts. Second, in the carbon emissions market under the emissions trading scheme (ETS), firms that emit more than the allowances allocated to them (e.g., through higher production) would wish to purchase allowances from firms with spare allowances (e.g., through reduction in production or technological change) to avoid paying penalties. Consequently, industrial firms have motivation to take long positions in futures contracts as hedge against higher emission allowance prices in futures. Increasing emission allowance prices in futures may come about as a result of stronger industrial activity, which is a likely result of economic growth. This implies that increases in industrial production, which has been shown to increase stock prices (see e.g., Nasseh and Strauss, 2000), may also increase the prices of carbon emission allowance futures prices. It, therefore, follows that the prices of carbon emission allowance futures may also increase with higher industrial production and economic activity. Consequently an investment strategy of taking short positions in carbon emission allowance futures contracts offers a logical counterpart for equity portfolios as the prices of stocks and carbon emission allowance futures should relate mutually. This strategy can also offer an opportunity to earn positive returns from the short positions in the futures contracts given that carbon emission allowance futures trade with significant and negative convenience yields (Trueck et al., 2014). In an earlier study, Kosobud et al. (2005) provide a generalized economic rationale for tradable pollution rights in a well-diversified portfolio.

Motivated by the hedging and yield characteristics of the carbon emission allowance market, this chapter investigates the exposures of carbon emission returns in different states of stock market distress. We expect the taking of short positions in carbon emission futures to be a valid investment alternative. This emanates from two factors: carbon emission futures being potential hedgers due to their link to industrial production and the market being shown to be in contango (Trueck et al., 2014). However, the dependence of the exposure in different market states is yet to be explored, which we do in this chapter. This examination contributes to the financial stability literature as it provides valuable insights into efficient management of financial crises.

Our analysis shows three main results: first, the returns of a trading strategy that rolls over short positions in carbon emission allowance futures contracts has a low correlation with global stock returns but the correlation increases with the level of stock market stress. Second, present evidence for negative convenience yields in carbon emission allowance futures and short positions in them can generate alpha in times of stock market stress. Third, the exposure of the returns on carbon futures returns to equity returns depends on the state of the stock market distress. The findings of our study imply that the carbon emission allowance market is an emerging asset class that can offer attractive investment opportunities for global portfolio managers.

The remainder of our chapter is organized as follows: Section 17.2 discusses carbon allowances and briefly presents the extant literature. Section 17.3 presents the data and methodology. The empirical results are discussed in Section 17.4, and Section 17.5 concludes.

17.2 CARBON ALLOWANCES AND LITERATURE REVIEW

The pervasive impact of the increasing concentration of GHG on human activities has been a significant concern for society. In response to this, the Kyoto Protocol, a binding treaty to decrease emissions of GHG, was agreed upon by the most industrialized countries and has currently been ratified by over 170 countries. A mechanism of GHG reduction included in the Kyoto Protocol is the trading of carbon dioxide (CO$_2$) emission allowances or permits in organized financial markets.[1] The emission targets vary among the signatories to the deal and the emission allowances trading enables countries to purchase and dispose-off their agreed allocation of GHG emissions. A country's agreed allocation is given out as emission permits to institutions representing emission sources within the respective country, which enables the institutions to participate in a trading system. If an institution does not need all of its allocated emission permits, they can trade that extra allowance to another institution to cover their emission needs and vice versa. This trading scheme rewards overachievement by providing financial compensation for reduction of pollution below required levels (e.g., through reduction in production or technological change). The market price determined through emissions trading encompasses the cost of pollution reduction and facilitates the decision as to whether it is more cost-effective to reduce carbon emissions or to buy allowance permits (see Hasselknippe and Høibye, 2001).[2]

Several regional emission markets that have been established, with the European Union Emissions Trading Scheme (EU ETS) being the world's largest single market for CO$_2$ emission allowances.[3] The EU ETS is also the most liquid of the carbon markets and accounts for approximately 90% of the global transactions (see Daskalakis et al., 2009; Linacre et al., 2011) and 45% of the total GHG emissions.[4] Currently, there are ongoing efforts to link the EU ETS to other emission trading markets. EU emissions allowance futures in the EU ETS are mainly traded in the European Climate Exchange (ECX) with varying degrees of dates.[5] They are deliverable with allowances delivery typically occurring by mid-month of the expiration month. Any institution with an open position at end of trading for a contract month is obligated to either make or take delivery of emission allowances to or from national registries. The unit of trading

---

[1] For theoretical arguments for and against such schemes (e.g., Cooper, 1996; Klepper et al., 2003; McKibbin et al., 1999; McKibbin and Wilcoxen, 2006).

[2] See Springer (2003) for an overview of factors influencing CO$_2$ allowance price levels.

[3] The existing emissions trading schemes include the Switzerland Emissions Trading Scheme; California Cap-and-Trade Program (US); Regional Greenhouse Gas Initiative (RGGI); Alberta Greenhouse Gas Reduction Program (Canada); Quebec Cap-and-Trade System (Canada); Kazakhstan Emissions Trading Scheme; New Zealand Emissions Trading Scheme (NZ ETS); Japan (various schemes); and China Emissions Trading Schemes (Beijing, Guangdong, Hubei, Shanghai, Shenzhen, and Tianjin).

[4] See http://ec.europa.eu/clima/policies/ets/index_en.htm.

[5] See Mizrach and Otsubo (2014) for an analysis on the microstructure of this market.

is 1 lot of 1000 emission allowances and an emission allowance entitles the holder to emit 1 ton of CO_2 equivalent gas (see e.g., Ibikunle et al., 2014).

The pricing of carbon futures contracts is characterized by the fact that carbon emissions are not storable, although it may be possible to store the emission rights. This makes the pricing of carbon emission futures slightly different from the usual storable commodities such as oil and metals. In any case, the existence of convenience yield in carbon emission futures contracts would imply that market participants could earn extra yield from investing in carbon futures contracts. The so-called *carbon investors* represent an important group of participants in the emission allowance markets worldwide as evidenced by the availability of numerous carbon funds. Fusaro (2007) reports scores of specialized carbon funds in operation and the World Bank estimates the worth of the world's ETSs to be about US$30 billion.[6]

The existing literature has examined several dimensions of the carbon emissions market. We briefly discuss this literature here. Kanamura (2012) studies the characteristics of carbon emission allowance futures and suggests that the pricing of carbon futures rather resembles security futures than energy commodity returns. The study also documents a negative relation between the convenience yield of carbon emission future prices and their price returns. Uhrig-Homburg and Wagner (2009), using data on the trial period of trading of EUA futures to investigate the future price dynamics of carbon emission allowances, show that their convenience yield is not consistent over time. Daskalakis et al. (2009) further find evidence of positive convenience yields.

El Hédi Arouri et al. (2012) consider nonlinearity between carbon future and spot prices and find that the relationship between them is nonlinear, showing that the nature of convenience yield in the carbon emission futures market is more complicated. In recent studies, Charles et al. (2013) and Gorenflo (2013) examine the cost-of-carry relationship between futures prices and spot prices in the carbon emission market. As the two studies consider cost-of-carry models with zero convenience yields and find that the cost-of-carry hypothesis does not hold, they show undeviating evidence for the existence of the convenience yield of carbon futures emission prices. Trueck et al. (2014) show evidence that the carbon emission futures market has changed from backwardation to contango, implying that the convenience yield in carbon emission futures is negative and that investors investing in short positions in carbon emission contracts earn an additional yield.

The evidence on market efficiency in the European carbon market is inconclusive. The evidence of Charles et al. (2013) using the cost-of-carry model suggests that the carbon emission market is inefficient. In other studies, Tang et al. (2013) present evidence that the EU ETS market can accurately predict the spot price 1 month ahead, suggesting

---

[6] State and trend of carbon pricing 2014. World Bank Group (2014) (http://www.ecofys.com/files/files/world-bank-ecofys-2014-state-trends-carbon-pricing.pdf).

that the European carbon futures market is efficient. Daskalakis (2013) also examine the efficiency of the European carbon market and suggests that the market has been becoming more efficient after 2009.

The pricing of carbon emission prices has also been shown to be linked to other assets. Byun and Cho (2013) examine whether using energy volatilities predict the volatility of carbon future prices using data for the ECX futures contracts. Their results suggest that Brent Oil, coal, and electricity forecast carbon futures volatility with 1 day lag, implying that the volatility of carbon futures prices is linked to the volatility of energy markets. Kanamura (2013) further investigates the linkages between carbon allowance prices and stock indexes and finds that the correlation between carbon allowance price changes and stock returns increase with poor stock market performance. These studies show that the pricing of carbon emission futures is linked to other assets and that the dependence between the carbon emission futures market returns and the stock returns is stronger in market downturns.

17.3 DATA AND METHODOLOGY

17.3.1 Data

This chapter investigates the exposures of carbon emission returns in different states of stock market distress. To do this, we utilize two carbon emission future returns net-asset-value (NAV) and market returns of iPath Global Carbon Exchange Traded Notes obtained from the iPath Web site (www.ipathetn.com) and Finance Yahoo (www.finance.yahoo.com), respectively. We employ both of these returns in our empirical analyses to reduce the impact of nonsynchronous trading on the empirical output as the Exchange Traded Note has poor liquidity and is traded infrequently. For equity returns, we include returns of different exchange-traded products (ETPs) including iShares MSCI Emerging Markets Exchange Traded Fund (EMETF) and Vanguard Total World Stock Exchange Traded Fund (WETF). We include both of these exchange traded funds since emerging market equity returns might have a higher exposure to carbon emission returns. Data on these ETPs are sourced from Finance Yahoo (finance.yahoo.com). The sample period for our empirical investigation is from 27th June 2008 to 29th September 2014. All returns are denominated in US dollars and adjusted for splits and dividends.

Table 17.1 presents the descriptive statistics for the carbon emission and equity indexes included in our data set. We detect negative average values of −0.05% and −0.01% for the NAV and market returns of iPath Global Carbon Exchange Traded Note, respectively. Table 17.1 also shows positive average return for iShares MSCI EMETF and Vanguard Total World Stock Exchange Traded Fund (WETF). Additionally the NAV of iPath Global Carbon Exchange Traded Note and the Total WETF index series show negative skewness whiles the market returns of iPath

Table 17.1 Descriptive Statistics

	CEFR		EMETF	WETF	CFSI
	NAV	**Mkt Returns**			
Mean	−0.05%	−0.01%	0.04%	0.04%	13.23
Median	0.00%	0.00%	0.08%	0.10%	14.83
Maximum	29.85%	24.42%	22.75%	13.46%	23.89
Minimum	−39.46%	−30.73%	−16.18%	−12.12%	1.09
Standard deviation	3.74%	4.66%	2.38%	1.66%	6.22
Skewness	−0.24	0.11	0.90	−0.08	−0.44
Kurtosis	17.56	8.76	18.71	11.52	1.84

This table presents the descriptive statistics on the indexes contained in our data set. CEFR is the carbon emission returns; NAV and Mkt Returns are the net-asset-value (NAV) and market returns of iPath Global Carbon Exchange Traded Note, respectively; EMETF is the returns on iShares MSCI Emerging Markets Exchange Traded Fund; WETF is the returns on Vanguard Total World Stock Exchange Traded Fund; and CFSI is the Cleveland Financial Stress Index. The data period is between 27th June 2008 and 29th September 2014 comprising 1511 observations.

Global Carbon Exchange Traded Note, and the iShares MSCI EMETF index series show positive skewness.

We employ the indicator of contributions of stock market crashes to Cleveland Financial Stress Index (CFSI), obtained from the Web site of Federal Reserve Bank of St. Louis (http://research.stlouisfed.org/fred2/) to delineate the states of stock market distress. The indicator is a measure of units of stock market stress measured as the ratio of the current value of the S&P 500 Financial Index relative to its maximum value over the 365 days. Financial stress indicators provide information on current levels of the financial variables which can influence the future economic conditions (Hatzius et al., 2010). Thus they can be used to forecast the future financial condition and provide focused and accurate signals on financial stress. To differentiate between the different economic states of the CFSI we use 10, 15, and 20 units of stress as naive but practical thresholds. The CFSI stock market stress indicator is shown to have a mean of 13.23 and a negative skewness for the data period. Figure 17.1 plots the changes in the CFSI over the stated data period. As it can be observed from Figure 17.1, the level of the CFSI stock market stress indicator ranges between 20 and 25 demonstrating stock market stress during the period. Thus the naïve threshold of 20 is an appropriate threshold for characterizing extreme financial stress.

Additionally, we include data on the risk-free rate of return and the value (HML) and size (SMB) factors, obtained from Kenneth French's Web site, as additional variables in our empirical model.

Table 17.2 depicts the Pearson pairwise correlations between the returns on the equity indexes and carbon emission returns. This allows for an initial understanding of the relationship between the two assets. Panel A shows the correlations for the whole sample.

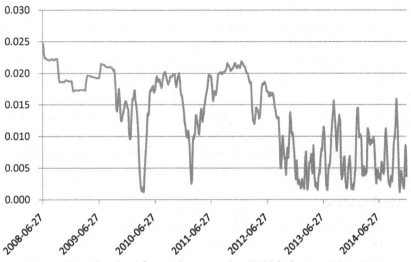

Figure 17.1 Cleveland financial stress index (CFSI) behavior: 2008–2014.

Table 17.2 Correlation Statistics

	CEFR			
	NAV	**Mkt Returns**	**EMETF**	**WETF**
Panel A				
NAV	1.00	0.37	0.13	0.16
Mkt returns		1.00	0.13	0.14
WETF				1.00
Panel B				
CFSI < 10 (N = 478)				
NAV	1.00	0.32	0.00	−0.01
Mkt returns		1.00	0.06	0.07
WETF				1.00
Panel C				
CFSI > 10 (N = 1000)				
NAV	1.00	0.42	0.21	0.25
Mkt returns		1.00	0.17	0.18
WETF				1.00
Panel D				
CFSI > 15 (N = 732)				
NAV	1.00	0.45	0.23	0.28
Mkt returns		1.00	0.20	0.21
WETF				1.00
Panel E				
CFSI > 20 (N = 190)				
NAV	1.00	0.28	0.22	0.27
Mkt returns		1.00	0.18	0.13
WETF				1.00

This table presents the Pearson correlations of returns statistics for pairs of equity and carbon emission future returns in our data set. CEFR is the carbon emission returns; NAV and Mkt Returns are the net-asset-value (NAV) and market returns of iPath Global Carbon Exchange Traded Note, respectively; EMETF is return on the iShares MSCI Emerging Markets Exchange Traded Fund; and WETF is return on the Vanguard Total World Stock Exchange Traded Fund.

Correlations between the different pair of indexes are all below 0.40. The practical implication of this result is that carbon emission futures may offer diversification benefits for equity investors. As indicated above, we follow a practical approach in defining the market states by using the CFSI delineation. The results of this exercise, presented in Panels B–E, confirm the low correlations between the equity indexes and carbon-equity portfolios. The low correlations to demonstrate their potential diversification benefit. It can be also noted from Table 17.2 that the correlation between carbon emission returns and equity returns is higher in stock market crash states. When emerging market equity index returns are compared with global index returns, it appears that emerging market index returns have slightly lower correlations with carbon emission returns.

17.3.2 Methodology

We commence the empirical analyses by estimating the following OLS model to characterize the relation between carbon future returns and equity returns:

$$
\begin{aligned}
\mathrm{CEFR}_t = {} & \alpha + \beta_1 \mathrm{WETF}_t + \beta_2 \mathrm{S\&P500ETF}_t + \beta_3 \mathrm{EMETF}_t \\
& + \beta_4 \mathrm{HML} + \beta_5 \mathrm{SMB} + \varepsilon_t,
\end{aligned}
\tag{17.1}
$$

where CEFR is the carbon emission future returns measured by the NAV and the returns on iPath Global Carbon Exchange Traded Note; WETF is the returns on Vanguard Total World Stock Exchange Traded Fund; EMETF is the returns on iShares MSCI Emerging Markets Exchange Traded Fund; HML and SMB are the value and small firm premiums, respectively.

We subsequently break the data into periods encompassing different stock market stress and reestimate Eqn (17.1).

17.4 RESULTS

Table 17.3 presents the results from multivariate analysis of carbon emission future returns to stock market returns for the whole sample. The results show a positive relation between the returns on Vanguard Total World Stock Exchange Traded Fund and carbon emission futures returns. The coefficient for the MSCI EMETF is not statistically significant. The positive sign and statistical significance for Vanguard Total World Stock Exchange Traded Fund does not change when we reestimate the model without the statistically insignificant Exchange Traded Fund coefficients. As the coefficients for MSCI EMETF are not statistically significant, these results suggest in line with the correlation statistics in Table 17.2 that carbon emission returns are not particularly important with respect to emerging market equity index returns.

The value premium is shown to relate negatively to carbon emission future returns, indicating that short positions in carbon emission allowance futures would associate with

Table 17.3 Multivariable Analysis of Carbon Emission Allowance Returns

Variable	CEFR: NAV		CEFR: Mkt Returns	
	Coefficient	t-Statistic	Coefficient	t-Statistic
C	−0.066	−0.72	−0.026	−0.31
WETF	0.474***	5.54	0.349***	2.59
EMETF	−0.057	−0.96	0.060	0.79
HML	−0.245*	−1.83	−0.251	−1.35
SMB	0.067	0.49	0.045	0.27
Adj. R^2	0.03		0.02	
F-statistic	10.70		7.59	
Probability	0.000		0.000	
N	1511		1511	

This table reports ordinary least squares model results for the relation between carbon allowance futures returns and equity returns. The returns are in excess of the 1-month T-bill rate. The standard errors are both heteroskedasticity and autocorrelation robust. CEFR is the carbon emission returns; NAV and Mkt Returns are the net-asset-value (NAV) and market returns of iPath Global Carbon Exchange Traded Note, respectively; EMETF is return on the iShares MSCI Emerging Markets Exchange Traded Funds; and WETF is return on the Vanguard Total World Stock Exchange Traded Funds. * refers to statistical significance at the 0.1 level; ** refers to statistical significance at the 0.05 level; *** refers to statistical significance at the 0.01 level. HML and SMB are the value and small firm premiums, respectively.

the value investment strategy. From another point of view, firms that take long positions in carbon emission allowance futures have a negative exposure to the value premium. Additionally the finding means that corporate policies that reduce any need to hedge against rising carbon emission prices can reduce costs which arise from hedging against higher carbon emission prices. Thus environmentally responsible management policies that aim at reducing carbon emissions could return reduced hedging costs related to the value premium. This association between the value premium and carbon emission returns is a new finding and has not been discussed in previous studies such as Byun and Cho (2013) and Kanamura (2013), and lends itself to new lines of empirical investigations.

In further analyses, we investigate how carbon emission future returns relate to stock market returns at different levels of stock market stress (see Table 17.4). We show the positive relation between higher thresholds of stock market stress. The coefficient for the MSCI EMETF is not statistically significant except for the highest threshold (CFSI > 20).

Thus the exposures of carbon futures returns to equity return increases with stock market stress. This finding is similar to the results of Kanamura (2013) which suggest that the correlation between financial and carbon asset increases in financial turmoil.

Table 17.4 Carbon emission allowances returns and stock Market stress

Variable	CFSI<10		CFSI>10		CFSI>15		CFSI>20	
	Coefficient	t-Statistic	Coefficient	t-Statistic	Coefficient	t-Statistic	Coefficient	t-Statistic
CEFR: NAV								
C	0.030	0.15	−0.126	−1.42	−0.215**	−2.16	−0.204	−0.87
WORLDM	−0.306	−0.58	0.560****	6.50	0.587****	6.90	0.684**	2.42
EMR	0.053	0.18	−0.068	−1.38	−0.083*	−1.83	0.134	1.11
HML	0.532	1.04	−0.346****	−2.66	−0.359***	−2.55	−1.048***	−3.11
SMB	0.990*	1.79	−0.057	−0.45	−0.013	−0.09	0.68***7	2.36
Adj. R²	0.00		0.06		0.08		0.14	
F-statistic	0.79		18.07		17.48		8.97	
Probability	0.534		0.000		0.000		0.000	
N	478		1000		732		190	
CEFR: Mkt Returns								
C	0.181	0.95	−0.117	−1.31	−0.206**	−2.15	−0.257	−1.08
WORLDM	0.395	0.61	0.342**	2.39	0.367**	2.48	−0.443	−0.98
EMR	0.100	0.26	0.065	0.86	0.072	0.88	0.632**	2.43
HML	−0.862	−1.22	−0.153	−0.74	−0.216	−0.99	−0.280	−0.59
SMB	0.627	0.85	−0.080	−0.47	−0.070	−0.37	0.411	1.10
Adj. R²	0.00		0.03		0.04		0.02	
F-statistic	1.32		8.65		9.21		2.19	
Probability	0.260		0.000		0.000		0.072	
N	478		1000		732		190	

This table reports ordinary least squares model results for the relation between carbon allowance futures returns and equity returns in different states of stock market stress. The returns are in excess of the 1-month T-bill rate. The standard errors are both heteroskedasticity and autocorrelation robust. CEFR is the carbon emission returns; NAV and Mkt Returns are the net-asset-value (NAV) and market returns of iPath Global Carbon Exchange Traded Note, respectively; EMETF is return on the iShares MSCI Emerging Markets Exchange Traded Funds; and WETF is return on the Vanguard Total World Stock Exchange Traded Funds. * refers to statistical significance at the 0.1 level; ** refers to statistical significance at the 0.05 level; *** refers to statistical significance at the 0.01 level. HML and SMB are the value and small firm premiums, respectively.

17.5 CONCLUSION

In this chapter, we examine exposures of carbon emission returns in different states of stock market distress. Our findings also show that the exposure of carbon emission futures returns to global equity returns increases with the level of stock market stress. This implies, as expected, that an investment strategy of taking short positions in carbon emission allowance futures contacts offers a natural counterpart for equity portfolios, yet it becomes apparent only during times of stock market stress. The findings reported are in line with Trueck et al. (2014) that the carbon emission futures market has changed from backwardation to contango as we show that possible to earn an additional premium from investing in short positions in these contracts. We also add to this evidence by showing that this investment strategy does not only offer significant yield opportunities but it also has low correlations with global and emerging market equity indexes. These findings show that carbon emission allowances and environmental investments of its kind can be important portfolio diversifiers.

REFERENCES

Byun, A.-J., Cho, H., November 2013. Forecasting carbon futures volatility using CARCH models with energy volatilities. Energ. Econ. 40, 207–221.

Charles, A., Darn, O., Fouilloux, J., September 2013. Market efficiency in the European carbon markets. Energy Policy 60, 785–792.

Cooper, R., 1996. A Treaty on Global Climate Change: Problem and Prospects. In: Working Paper Series 97–7. Weatherhead Center for International Affairs. Harvard University, U.S.A.

Daskalakis, G., March 2013. On the efficiency of the European carbon market: new evidence from phase II. Energ. Policy 54, 369–375.

Daskalakis, G., Psychoyios, D., Markellos, R.N., 2009. Modelling CO$_2$ emission allowance prices and derivatives: evidence from the European trading scheme. J. Bank. Financ. 33 (7), 1230–1241.

El Hédi Arouri, M., Jawadi, F., Nguyen, D.C., 2012. Nonlinearities in carbon spot-futures price relationships during phase II of the EU ETS. Econ. Model. 29 (3), 884–892.

Fusaro, P.C., September 2007. Energy and environmental hedge funds. Commod. Now. 1–2.

Gorenflo, M., 2013. Futures Price dynamics of CO$_2$ emission allowances. Empir. Econ. 45 (3), 1025–1047.

Hasselknippe, H., Høibye, G., 2001. Meeting the Kyoto Protocol Commitments: Summary: Domestic Emissions Trading Schemes. Energy and Environment Confederation of Norwegian Business and Industry, Norway.

Hatzius, J., Hooper, P., Mishkin, F.S., Schoenholtz, K.L., Watson, M., 2010. Financial Conditions Indexes: A Fresh Look after the Financial Crisis. NBER. Working Paper No. 16150.

Ibikunle, G., Gregoriou, A., Pandit, N.R., 2014. Price impact of block trades: the curious case of downstairs trading in the EU emissions futures market. Eur. J. Financ. Forthcoming. Available at: http://dx.doi.org/10.1080/1351847X.2014.935871.

Kanamura, T., 2012. Comparison of futures pricing models for carbon assets and traditional energy commodities. J. Altern. Invest. 14 (3), 42–54.

Kanamura, T., 2013. Financial turmoil in carbon markets. J. Altern. Invest. 15 (3), 92–113.

Klepper, G., Peterson, S., Springer, K., 2003. DART97: A Description of the Multiregional, Multisectoral Trade Model for the Analysis of Climate Policies. Kiel Working Papers 1149. Kiel Institute for World Economics, Kiel.

Kosobud, R.F., Stokes, H.H., Tallarico, C.D., Scott, B.L., 2005. Valuing tradable private rights to pollute the public's air. Rev. Account. Financ. 4, 50–71.

Linacre, N., Kossoy, A., Ambrosi, P., 2011. State and Trends of the Carbon Market 2011. The World Bank Report, Washington, DC.

McKibbin, W.J., Ross, M.T., Shackleton, R., Wilcoxen, P.J., 1999. Emissions trading, capital flows and the kyoto protocol. Energy J. 20, 287—333. Special issue: The costs of the Kyoto protocol: a multi-model evaluation.

McKibbin, W., Wilcoxen, P.J., 2006. A Credible Foundation for Long Term International Cooperation on Climate Change. Lowy Institute for International Policy. Working Paper No. 1.06.

Mizrach, B., Otsubo, Y., 2014. The market microstructure of the European climate change. J. Bank. Financ. 32 (2), 2022—2032.

Nasseh, A., Strauss, J., 2000. Stock prices and domestic and international macroeconomic activity: a cointegration approach. Q. Rev. Econ. Finance 40 (2), 229—245.

Springer, U., 2003. The market for tradable GHG permits under the Kyoto Protocol: a survey of model studies. Energ. Econ. 25 (5), 527—551.

Stern, N., 2007. The Economics of Climate Change: The Stern Review. Cambridge University Press, Cambridge, MA.

Tang, B.-J., Shen, C., Gao, C., 2013. The efficiency analysis of the European CO_2 futures market. Appl. Energ. 112, 1544—1547.

Trueck, S., Hardle, W., Weron, R., 2014. The relation between spot and futures CO_2 emission allowance prices in the EU-ETS. In: Gronwald, M., Hintermann, B. (Eds.), Emission Trading Systems as a Climate Policy Instrument—Evaluation and Prospects. MIT Press.

Uhrig-Homburg, M., Wagner, M., 2009. Futures price dynamics of CO_2 emission allowances: an empirical analysis of the trial period. J. Deriv. 17 (2), 73—88.

CHAPTER 18

The Returns from Investing in Water Markets in Australia

Sarah Ann Wheeler[1,2], Peter Rossini[1], Henning Bjornlund[1], Belinda Spagnoletti[3]

[1]School of Commerce, University of South Australia, Adelaide, SA, Australia; [2]Global Food Studies, University of Adelaide, Adelaide, SA, Australia; [3]Nossal Institute for Global Health, Melbourne School of Population and Global Health. University of Melbourne, Melbourne, VIC, Australia

Contents

18.1 INTRODUCTION

In the previous chapter, Settre and Wheeler (2016) provided a historical perspective about the shift in Australian water policy from regulation and engineering to economic-based instruments. This chapter provides a more in-depth study of Australian water markets, focusing specifically on Australia's largest water market, the Goulburn—Murray Irrigation District (GMID) in the state of Victoria. Australia's institutions for managing water have received much international attention as they represent radical and farsighted responses to extreme water scarcity, water overallocation, and associated environmental impacts. Consequently, any study of institutions and instruments in Australia may provide important insights for other countries facing water scarcity.

Water scarcity around the world is predicted to increase, with water shocks anticipated to be one of the greatest risks facing society (World Economic Forum, 2015). Reflecting this, current and future scarcities are starting to impact stock markets. For example, in the past 5 years, investments in water-related stocks (for example, the S&P Global Water Index, which measures the performance of 50 global companies engaged in water-related activities) have outperformed gold, oil, and the wider stock market, with this trend predicted to continue (Dyson, 2014).

Handbook of Environmental and Sustainable Finance
ISBN 978-0-12-803615-0

Although research into water-related investments is on the rise, very little work has been done that investigates the viability of investing in various water rights. There are two main reasons for this. One is social concern and opposition to treating water as an economic good to be traded at a profit, as water is argued to be a social good and necessary to sustain human life and, as such, should not be subject to mere market forces. The second reason is the lack of liquid water markets around the world; although a number of water markets do exist. Some of the countries that have developed water trading include the United States (mainly Western states), Chile (Limarí River Valley), Australia (Murray–Darling Basin, MDB), India, Spain, Canada (Alberta), South Africa, China, Brazil, Mexico, Oman, Pakistan, and Tanzania.

However, in many emerging economies water markets remain in their infancy (Grafton and Wheeler, 2015). They have been slow in developing into liquid markets that can shift substantial volumes of water on a regular basis between a large number of buyers and sellers. There are also legal issues associated with the specification, ownership, and control of water rights, as well as the absence of secure registers and markets for these rights.

Grafton et al. (2011) compared water markets in five countries and found Australia has one of the most developed and longest running water markets in the world, while Bjornlund et al. (2013) found that the level of market participation and market maturity within some Australian water markets has reached a level comparative to other asset classes. Because of this historical longevity and the increased level of activity in these markets, sufficient time series data are now available to conduct financial analyses on water investments and returns in Australia.

18.2 BACKGROUND OF WATER MARKETS IN AUSTRALIA

As discussed in the previous chapter, the emergence of a mature water market in Australia was driven by severe water scarcity, overallocation of water rights to consumptive uses, and legal separation of property rights and water rights. This study focuses on water market data from the MDB in Australia. In the MDB there are two main water markets, namely (1) the entitlement market (also known as the permanent water market, where the long-term rights to receive water allocations are traded) and (2) the allocation market (also known as the seasonal or temporary water market, where the water allocations yielded by water entitlements are traded).

Since the late 1980s, water trade has been increasingly adopted by irrigators. The adoption of water trade has grown exponentially (Figure 18.1). Figure 18.1 illustrates the cumulative adoption of water markets in the southern MDB and shows that by 2011, 68% of all irrigators in the southern MDB had conducted at least one water allocation trade, while 28% had conducted at least one water entitlement trade. This growing adoption and use of the water market is very similar to the adoption of other agricultural innovations and practices (Wheeler et al., 2009).

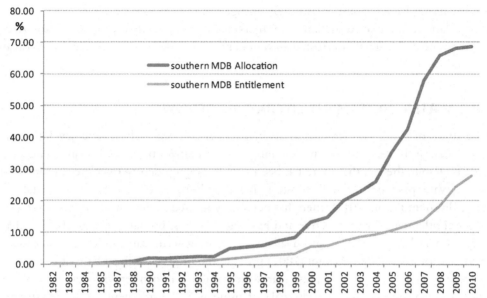

Figure 18.1 Cumulative adoption of water allocation and entitlement trade in the southern MDB from 1983–1984 to 2010–2011. *(Source: UniSA data.)*

In Australia there are both water rights linked to regulated and unregulated water sources. Water rights linked to regulated sources have different levels of reliability (namely high, general, and low security). High security licenses generally have an estimated reliability of 90–95% (e.g., 100% allocations are expected at least 90 years out of 100), while general security licenses have an expected reliability of approximately 70% and low security have an expected reliability of up to 30%. During the Millennium Drought of the 2000s in Australia, many license holders received severe restrictions to their allocations, when water storages fell to a low level of only 15% (Wheeler et al., 2014). At the beginning of each growing season (July to August), in each water district, the local water authority announces an initial opening allocation as a percentage of an irrigator's full entitlement, and is based on water availability in the storages plus minimum expected inflows. Allocations are then revised every 2 weeks during the season as, or if, additional water becomes available until final allocations are known (which normally occurs by February, though in wet years it occurs much earlier).

Irrigators have traditionally been the foremost market participants, purchasing entitlements, and allocations to manage risk associated with current levels of production, as well as to facilitate expansion. However, since unbundling occurred (see Chapter 17), nonlandowners can engage in the water market. This includes foreign owners, corporate companies, and superannuation investors. In particular, in recent years environmental water purchasers, mainly from government but also from some nongovernmental organizations, have become increasingly engaged in the MDB water market. However, foreign involvement in Australian water markets still seems relatively limited, as the ABS (2011) estimated that 91% of Australian water entitlements were domestically owned as at the end of 2010.

18.3 WATER INVESTMENT LITERATURE REVIEW

The study of water entitlements as an investment is a relatively recent development, given that it is very much an emerging issue internationally (Roca and Tularam, 2012). Despite the fact that there has been a wide-ranging debate over the appropriateness of water passing from rural (consumptive irrigation use) to urban/environmental use (e.g., Bjornlund et al., 2011), there has been only limited investigation into investing in water entitlements for prospective and existing investors, primarily due to the fact that there is a lack of access to, and availability of, long-term data on water markets in Australia. Bjornlund and Rossini (2007) provide the only research so far that compared the return on investment in an Australian water entitlement market with that of the share market. They investigated the rationale of using water entitlements as an investment vehicle, comparing the returns of such investment under different investment, and cash flow scenarios over the 13-year period from 1998 to 2005 (Bjornlund and Rossini, 2007). These analyses suggest that investments in water entitlements are subject to a substantial risk premium. They found that during the first half of this period, mean returns on water investments would have been 22.44% pa ($s = 5.09$) compared to 12.00% pa ($s = 1.67$) for investments in the S&P ASX accumulation index; and during the second half it was 16.83% pa (standard deviation 4.16) for water compared to 8.20% pa (standard deviation 2.05) for the ASX. Thus, investment in water entitlements shows substantially higher returns, but with corresponding higher uncertainty (risk).

Given that water markets have now matured considerably in Australia (e.g., Wheeler et al., 2009, 2010) and are showing strong characteristics required of an efficient market (for example, an increasing proportion of water used each year is traded in the market and an increasing proportion of water entitlements are traded each year (Bjornlund et al., 2013)), it may be an opportune time to include water entitlements as an asset class into broader investment portfolios to benefit from further diversification.

18.4 DATA

18.4.1 Case Study Information

We used prices from the markets for water entitlements and water allocations over a 22-year period in the GMID in northern Victoria. Irrigation within the district is mainly supplied by two major sources: the Goulburn and the Murray rivers. It is a fitting area in which to study water market investments because by 2010—2011, 77% of all irrigators in

Victoria had engaged in at least one water market trade (Wheeler et al., 2014). Dairy, fruit, and grape producers have been the most significant buyers in the market, whereas cereal, grazing, and mixed farmers constitute the majority of water sellers (Wheeler et al., 2009). As mentioned previously, nonlandowner investors have only had the ability to own water entitlements since the mid-2000s. Based on available information, it is difficult to establish the exact amount of corporatization or nonlandowner ownership of water, but what is known is that the volume of water entitlements owned by businesses with some foreign ownership increased from 1.2 million megaliters in 2010 to 1.8 million megaliters in 2013. Foreign ownership represented 14% of all reported water entitlements in Australia in 2013, up from 9% in 2010 (ABS, 2014).

In most years, irrigators within the Goulburn system have received their full allocation, with the exception of the severe drought years of 2002–2003 and 2006–2009 (Table 18.1). In many years, opening allocations have started at 0%; for example, within the Goulburn system, this was the case for 14 successive years from 1998 to 2012 (Bjornlund et al., 2013). This exposes irrigators to significant risk when making planting decisions early in the season. In other words, irrigators have little idea how much water allocation their entitlements will yield by the end of the season. If they use more water allocations than their entitlements yields and fail to balance that with purchases in the market within a certain period of time, they will be subjected to large penalty payments. To manage that risk and enable them to use water beyond their allocation, irrigators can purchase water allocations from other entitlement holders.

The water market prices for GMID entitlements and allocations in this chapter have been collected by the authors for two decades using a number of sources: (1) irrigator surveys; (2) water brokers; (3) online water exchanges; and (4) Goulburn–Murray Water, the local authority managing the GMID. From the data available, mean monthly prices were computed both for entitlements and allocations. Price information on both high security entitlement and water allocations is available from January 1993 onward. Table 18.2 provides descriptive statistics of the data used.

18.4.2 Basic GMID Water Market Statistics

To profile fluctuations and trend in water entitlement and allocation prices over the more than two decades of our study, we first use a simple ratio-to-moving average approach to describe the data in terms of long-term trend, seasonality, and cycles.

Over the period July 1993 to July 2014 there has been a strong upward trend in the prices of water entitlements (Figure 18.2). The trend shows an average annual compound growth of 11.8%. Prices have little or no seasonal component but very significant cyclical fluctuations (see Figure 18.4) which are influenced by high demand from dairy and wine grape growers, the government entering and exiting the market to buy water for the environment, periods of drought and the end of drought periods. For example, the

Table 18.1 Final Seasonal Water Allocations in the Southern Murray—Darling Basin and Long-Term Average Annual Yield (LTAAY) Entitlement

Year	High Reliability Entitlements					Lower Reliability Entitlements			
	Vic Goulburn	Vic Murray	NSW Murray	NSW Murrumbidgee	SA Murray	Vic Goulburn (Low)	Vic Murray (Low)	NSW Murray (General)	NSW Murrumbidgee (General)
1998–1999	100%	100%	100%	100%	100%	0%	100%	93%	85%
1999–2000	100%	100%	100%	100%	100%	0%	90%	35%	78%
2000–2001	100%	100%	100%	100%	100%	0%	100%	95%	90%
2001–2002	100%	100%	100%	100%	100%	0%	100%	105%	72%
2002–2003	57%	100%	100%	100%	100%	0%	29%	10%	38%
2003–2004	100%	100%	97%	95%	95%	0%	0%	55%	41%
2004–2005	100%	100%	97%	95%	95%	0%	0%	49%	40%
2005–2006	100%	100%	69%	95%	100%	0%	0%	63%	54%
2006–2007	29%	95%	69%	90%	60%	0%	0%	0%	10%
2007–2008	57%	43%	50%	90%	32%	0%	0%	0%	13%
2008–2009	33%	35%	95%	95%	18%	0%	0%	9%	21%
2009–2010	71%	100%	97%	95%	62%	0%	0%	27%	27%
2010–2011	100%	100%	100%	100%	67%	0%	0%	100%	100%
2011–2012	100%	100%	100%	100%	100%	0%	0%	100%	100%
2012–2013	100%	100%	100%	100%	100%	0%	0%	100%	100%
LTAAY[1]	**95%**	**95%**	**95%**	**95%**	**9%**	**35%**	**24%**	**81%**	**64%**

[1]Percentage of total LTAAY accounted for by that type of water security in that region. LTAAY represents the average yield expected over 100-year time period.
Source: Adapted and updated from NWC (2011).

Table 18.2 Descriptive Statistics

	Water Entitlement Prices ($AUD)	Water Allocation Prices ($AUD)
Mean	1139.11	143.39
Standard error	39.58	13.26
Minimum	220.00	5.65
Maximum	2400.00	989.75

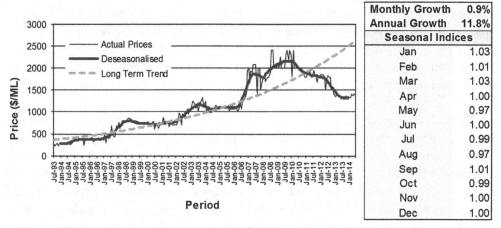

Figure 18.2 Ratio-to-moving average model of water entitlements 1993–1994 to 2013–2014.

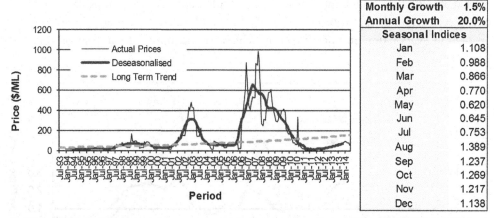

Figure 18.3 Ratio-to-moving average model of water allocations 1993–1994 to 2013–2014.

rain and flooding that broke the Millennium Drought in 2010–2011 signaled a change in the upward projection of water entitlement prices.

Water allocation prices have been trending upward at a strong average annual growth of 20% (Figure 18.3). Prices respond quickly and dramatically to drought with high peaks during 2002/2003 and 2006—2010, periods with below 100% allocations (Table 18.1). Unlike entitlement prices there is a strong seasonal component in the water allocation prices, with a low point in May with prices on average being 62% of the annual average and a high point in August at 1.38 times the yearly average. These seasonal differences are clearly evident in Figure 18.3 and are driven by irrigators' risk management decisions early in the season in response to low opening allocations and disposal of excess water at the end of the season. It can also be noted that these fluctuations seem to even out after 2008 and especially after 2011 as the capability to carry unused water over from one season to the next became permanent and increasingly flexible, allowing irrigators to manage risk by carrying over unused water from one season to the next or by buying water late in the season and carrying it over into the next season (Bjornlund et al., 2013). Both strategies evened the price out across years and seasons. This is also evident in the flattening out of the price cycle in Figure 18.4. Both markets show similar cyclical behavior (Figure 18.4).

An analysis of price cycles (Figure 18.4) shows that until 1999 prices in the entitlement markets lead the allocation market, and from then until 2001 they moved almost in unison. However, after 2001 prices in the water allocation market clearly lead the entitlement market. This is reflective of what happens in other capital markets such as shares and property, where changes in annual dividend and rental income respectively lead changes in the value of the underlying assets. This further suggests that trading in water assets is maturing and it might be time to include these assets in wider investment portfolios. This distinction is clear in our previous papers. The cycle ratios in the allocation market also fluctuate much more than in the entitlement market with peaks in the allocation cycle more than four times higher at times. This reflects the market's sensitivity to changes in evaporation and precipitation, which has an immediate impact on demand.

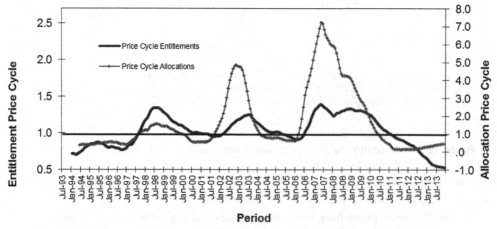

Figure 18.4 Water entitlements and allocations price cycles 1993—1994 to 2013—2014.

The next part of the analysis estimates the return from investing in water in the GMID over the study period.

18.5 METHODOLOGY

18.5.1 Discounted Cash Flow Analysis

To estimate the return from investing in water entitlements, a discounted cash flow approach was used to compute the internal rate of return (IRR) over a 5-year investment period. The IRR is calculated based on selling the water allocation during different periods of the year. The maximum IRR is based on selling as much water as possible during the periods when allocation prices are highest for that water year (1st ranked price). By comparison the minimum IRR assumes water is sold during periods when prices are at their lowest for that water year (12th ranked price) while the medium IRR assumes water is sold during the period when prices are in the middle (6th ranked price for the year). This covers the full range of outcomes that could have been achieved by the seller in that water year with the median IRR being the most likely outcome. Sales are based only on the level of water allocations available at the time. Additional allocations made available due to further increases in seasonal water are sold at the next available opportunity based on the same scenario. For example, if the highest ranked price is in October but only 70% of water is allocated, then 70% is assumed to be sold at that point. If the next highest price is in December and 100% has been allocated at that point, then the remaining 30% is sold at that time.

The monthly cash flows are estimated over a 5-year holding period. The purchaser is assumed to buy the investment at the mean entitlement price at the time of purchase, hold the investment for 5 years, while selling the annual allocations as mentioned, and then sell the investment after 5 years at the mean entitlement price at that time.

The following additional assumptions are made: (1) commission at a rate of 2% is deducted from the proceeds of the sale of allocations and entitlements; (2) the cost associated with holding the water entitlement is paid in December (the middle of the water season); and (3) as the IRR is calculated as a monthly rate, it is converted to an effective annual rate.

The IRR for each 5-year (60-month) cash flow is established for each time period by solving for the rate (r) in the following equation:

$$\sum_{t=1}^{60} \frac{(PA_t * A_t - C_t)}{(1 + r)^t} + \frac{(PE_{60} * (1 - Psellcom_{60}))}{(1 + r)^{60}} - (PE_0 * (1 - Pbuycom_0)) = 0$$

where,

PE_t = price of water on the entitlement market at time t

$Pbuycom_0$ = commission paid on purchase of an entitlement at time 0

$Psellcom_{60}$ = commission paid on sale of an entitlement at time 60

PA_t = price of water on the allocation market at time t

A_t = volume of allocation sold at time t

C_t = cost paid for water at time t

r = IRR or equated yield

For convenience the IRR is shown on the chart at the end of the 5-year holding period (the IRR for July 1998 is for a 5-year hold starting in July 1993). The most likely IRR is shown as a thicker line and the possible IRR range indicated by the band, which is based on the minimum and maximum scenarios as previously explained.

Figure 18.5 shows the analyzed IRR (return) over the study period. The band of possible outcomes becomes narrower over the time period as the market matures and investors become more familiar with the best times to sell and the variation of prices decreases from month-to-month.

In the holding periods that ended from 1998 through 2011 the median returns were generally around or above 15% per annum. During the early years, market returns showed a high level of variability as they were sensitive to the time of buying and selling entitlements and allocations, but as markets became more established returns stabilized for holding periods ending between 2003 and 2006 and then started fluctuating again for holding periods ending between mid-2006 and early 2009. Returns then stabilized at a higher level for holding periods ending from mid-2009 to mid-2011 with investors benefitting from high entitlement prices generated by the government entering the market to secure water for the environment. Investment returns during holding periods ending late 2011 then plummeted and negative returns emerged for holding periods ending from early 2012 and onward. During this period, water allocation prices were low due to successive years of full allocation and the price of the entitlement at the

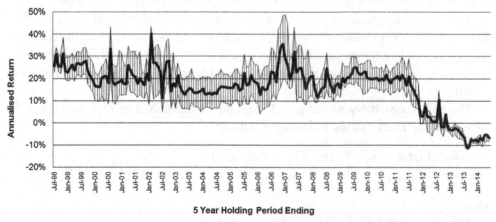

Figure 18.5 Internal rate of return of the Goulburn—Murray Irrigation District water market 1993—1994 to 2013—2014 when selling the allocation at the minimum, median, and maximum monthly prices.

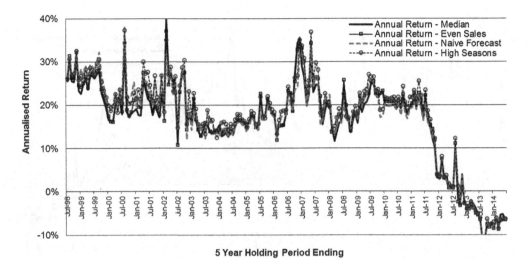

Figure 18.6 Internal rate of return of the Goulburn–Murray Irrigation District water market 1993–1994 to 2013–2014 when selling allocations evenly over the water years and using a naïve forecast.

end of the holding period was low in response to the government exiting the market. Hence, the annual costs and commissions were greater than the returns from selling the allocations and capital gains were either very low or turned negative as well (see Figures 18.2 and 18.3, as well as Table 18.1). Because investments that end after 2011 have relatively lower water allocation prices the variation between the maximum and minimum returns is lower.

Figure 18.6 provides the IRR of investing in water markets as well, this time using some different assumptions such as selling allocations evenly over the water years and using a naïve forecast. It is obvious from these figures that as the water market matured, and the drought ended, the differences in returns achieved from the different strategies have become insignificant.

18.5.2 Comparative Advantage of Water Market Investments

In the early market (for example, for the water market pre-2003), the majority of the returns from a 5-year hold in water entitlements were derived from capital growth. From 2003 to 2011, total return became more balanced between annual return from selling seasonal allocations and capital growth. This is similar to the situation with most equities such as property or shares and reflects a maturing market with assets more suitable for inclusion in broader asset portfolios. Figure 18.7 shows the breakdown of the total return from a 5-year holding period of water entitlements between capital growth and annual return and also provides a comparison to a 5-year holding period in the Australian share market (based on the S&P/ASX 200 accumulation index).

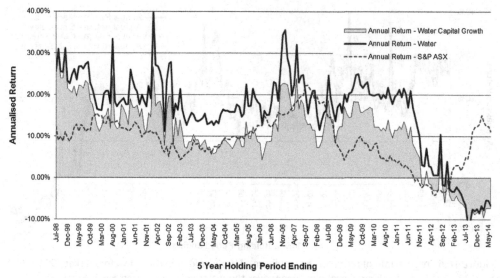

Figure 18.7 Total water market return, capital growth, and the S&P ASX accumulation index.

A breakdown of the water investment returns shows that capital growth has been the major source of return until around 2008 when the capital growth stabilized and then started to drop after 2009. The result is that a 5-year hold purchased after 2007 (ending late 2011 and onward) showed largely negative capital gain and hence overall losses with the return from allocations not offsetting the losses after 2012.

Until late 2012 investments in the water market outperformed investments in the share market with quite a significant margin, except for some volatile months during 2007/08. For holding periods ending between 1998 and 2005, returns in the water markets were close to double that in the share market. The same was the case from mid-2008 to late 2010 when returns in the share market began to dip with the global financial crisis and reached their lowest point with investments that ended in mid-2012. Since late 2012 the share market has outperformed the water market with a substantial margin but seems to be narrowing again at the end of the study period. The share market and water market are to some extent countercyclical, which indicates that investing in both may be useful in a portfolio situation. Further work on this is warranted.

18.6 DISCUSSION AND CONCLUSIONS

This study has analyzed market activity and actual prices paid and accepted for water allocations and water entitlements in the water market within the GMID, Australia's largest irrigation district from 1993—1994 to 2013—2014. The purpose was to evaluate whether the fundamentals are emerging to support the inclusion of water assets into wider investment portfolios.

Historically, there have been a number of impediments to the efficient operation of water markets as a result of spatial and other restrictions on trade, water being attached to land, lack of secure registers for water entitlements, and unresolved issues related to environmental flows. This is no longer the case, as the water market and institutional structures in Australia have evolved correspondingly, with the unbundling of the property rights in water, the permanent introduction of increasingly flexible carryover abilities, and the implementation of the MDB Plan. Water trading has increasingly been adopted by irrigators as a response to institutional change and increasing water scarcity, and price behavior has become more reflective of efficient markets.

The financial analysis in this paper suggests that while entitlement and allocation prices still follow the same cyclical pattern, returns to water markets have been less in the past 5 years. Although returns from investing in water entitlements over a 5-year holding period did consistently outperform the Australian share market, this has not been the case in recent years, but there is evidence of countercyclical returns between water and share markets which may indicate the importance of water investments for a diversified portfolio.

As discussed in the previous chapter, Australia has come a long way in its water policies and institutions. The foundation should therefore be established for treating water assets (associated with the water available for consumptive use) as any other financial assets and included in wider asset portfolios held by superannuation funds and other asset managers. In reality, many businesses, including farmers, are exceedingly offloading expensive assets from their balance sheets to focus their available capital on the activities where they can obtain the best return. Today many farmers rely on leasing land and equipment; there is no logical reason as to why water assets should not be treated the same way.

ACKNOWLEDGMENT

This research is part of a larger project funded by the Australian Research Council, DP140103946.

REFERENCES

ABS (Australian Bureau of Statistics), 2011. Agricultural Land and Water Ownership, December 2010. Publication 7127.0, Canberra.

ABS (Australian Bureau of Statistics), 2014. Agricultural Land and Water Ownership, June 2013. Publication 7127.0, Canberra.

Bjornlund, H., Rossini, P., 2007. An analysis of the returns from an investment in water entitlements in Australia. Pac. Rim Prop. Res. J. 13 (3), 334–360.

Bjornlund, H., Wheeler, S., Cheesman, J., 2011. Irrigators, water trading, the environment, and debt: perspectives and realities of buying water entitlements for the environment. In: Grafton, Q., Connell, D. (Eds.), Murray-Darling Basin Choices: Planning for a Better Future. Australia National University Press, Canberra.

Bjornlund, H., Wheeler, S., Rossini, P., 2013. Water markets and their environmental, social and economic impact in Australia. In: Maestu, J. (Ed.), Water Trading and Global Water Scarcity: International Perspectives. Francis Taylor, UK, pp. 68–93.

Dyson, R., September 8, 2014. Forget Gold — Invest in Water. The Telegraph. Available at: http://www. telegraph.co.uk/finance/personalfinance/investing/11081169/Forget-gold-investing-in-water-could-generate-far-greater-returns-over-time.html.

Grafton, R.Q., Libecap, G.D., McGlennon, S., Landry, C., O'Brien, R., 2011. An integrated assessment of water markets: a cross-country comparison. Rev. Environ. Econ. Policy 5 (2), 219–239.

Grafton, R.Q., Wheeler, S., 2015. Water economics. In: Halvorsen, R., Layton, D. (Eds.), Handbook on the Economics of Natural Resources. Edward Elgar Publishing.

NWC (National Water Commission), 2011. Australian Water Markets: Trends and Drivers, 2007–08 to 2009–11. National Water Commission, Canberra.

Roca, E., Tularam, G.A., 2012. Which way does water flow? An econometric analysis of the global price integration of water stocks. Appl. Econ. 4 (23), 2935–2944.

Settre, C., Wheeler, S., 2016. Environmental water governance in Australia: the movement from regulation and engineering to economic-based instruments. In: Gregoriou, G., Ramiah, V. (Eds.), Handbook of Environmental and Sustainable Finance. Elsevier.

Wheeler, S., Bjornlund, H., Shanahan, M., Zuo, A., 2009. Who trades water allocations? Evidence of the characteristics of early adopters in the Goulburn-Murray Irrigation District 1998–99. Agric. Econ. 40 (6), 631–643.

Wheeler, S., Bjornlund, H., Zuo, A., Shanahan, M.A., 2010. The changing profile of water traders in the Goulburn-Murray Irrigation District, Australia. Agric. Water Manage. 97 (9), 1333–1343.

Wheeler, S., Loch, A., Zuo, A., Bjornlund, H., 2014. Reviewing the adoption and impact of water markets in the Murray-Darling Basin, Australia. J. Hydrol. 518, 28–41.

World Economic Forum, 2015. Global Risks 2015, tenth ed. WEF, Switzerland.

CHAPTER 19

Product Market Competition and Corporate Environmental Performance

Kartick Gupta[1], Chandrasekhar Krishnamurti[2]

[1]Centre for Applied Financial Studies (CAFS), School of Commerce, UniSA Business School, University of South Australia, Adelaide, SA, Australia; [2]Faculty of Business, Education, Law and Arts, University of Southern Queensland, Toowoomba, QLD, Australia

Contents

19.1 INTRODUCTION

Corporate social responsibility (CSR) has been characterized as a crucial element of the social contract between business and society. However, despite decades of research, the purpose of this contract is the subject of debate. Two opposing views have been posited by analysts—the altruistic view and the strategic view. The altruistic view of CSR contends that firms are willing to give up profits for the sake of social interest (Elahuge, 2005). The alternate strategic view maintains that firms engage in profit maximizing CSR (Baron, 2001; McWilliams and Siegel, 2001) and that companies "do well by doing good." At present there is no consensus and both views prevail.

A significant stream of research on CSR examines the association between CSR and financial performance with mixed results. If the strategic view of CSR holds, then it would appear that there would be a positive relationship between CSR and financial performance. An alternate approach has emerged to distinguish between the two views of CSR which

Handbook of Environmental and Sustainable Finance
ISBN 978-0-12-803615-0

entails examining the impact of product market competition on CSR. Consistent with the strategic view of CSR, Fernandez-Kranz and Santalo (2010) find strong evidence that firms operating in more competitive industries are more socially responsible. Utilizing a quasi-natural experiment, Flammer (2014) finds that US domestic companies respond to import tariff reduction by increasing their CSR engagement to maintain their competitiveness. Using a dynamic simulation exercise, Hawn and Kang (2013) show that product market competition reduces CSR engagement in the US. This mixed evidence serves as a motivation for further work especially in a cross-country setting.

We contribute to the literature on CSR-Competition link in three ways. First, while prior work has predominantly concentrated on the US market, our paper extends the research to a cross-country setting. Our second contribution is to use a data set other than that of the Kinder, Lydenberg, and Domini Research and Analytics database (KLD). While KLD data set is used extensively by CSR researchers, Chatterji et al. (2009) point out that KLD is not optimally aggregating historical data resulting in lower predictive power of future emissions and penalties compared to simpler alternatives. Finally, we use more recent data, increasing the relevance of our research.

Our empirical results which are based on a comprehensive cross-country sample covering 69 countries support the altruistic view of CSR since there is a negative relationship between industry level competition and CSR. Our results are robust to alternate measures of competition. These findings are different from the conclusions based on the US studies, therefore our research highlights the importance of country context while examining the role of competition on CSR.

The rest of this chapter is organized as follows: In Section 19.2, we provide a brief review of the literature. In Section 19.3, we describe the data used and the methodology employed in our empirical tests. In Section 19.4, we present our empirical results. Finally in Section 19.5, we discuss our findings and offer our concluding remarks.

19.2 LITERATURE REVIEW

The management literature contends that companies can sustain their competitive advantage by leveraging the resources and capabilities in which they have a comparative advantage (e.g., Helfat and Peteraf, 2003; Hooley et al., 2006; Peteraf, 1993; Wernerfelt, 1984, 1995). In the context of global trade, domestic companies have a comparative advantage over foreign companies in their relationship with local stakeholders even if they are less competitive on a cost basis. Therefore, some researchers (Flammer, 2014 among others) argue that domestic companies may respond to increased foreign competition by strengthening their relations with local stakeholders such as consumers, employees, and the general public. By stepping up their social and environmental initiatives, companies can differentiate themselves and erect a "soft" trade barrier to disadvantage their foreign competitors.

The CSR literature has recognized the value of strengthening firms' relations with their stakeholders. For instance, Freeman's (1984) stakeholder theory advocates that companies should consider the interests of a broader group of stakeholders. Further extensions of stakeholder theory such as the instrumental stakeholder theory (Jones, 1995), which holds that CSR efforts can be instrumental in obtaining necessary resources or stakeholder support, have been proposed (see Agle et al., 2008). Likewise, companies may engage in CSR in order to improve their efficiency and enhance their reputation, brand, and trust (Barney, 1991; Hart, 1995; Porter, 1991; Russo and Fouts, 1997). This view is related to Porter and Kramer (2006, 2011) who highlight the strategic importance of considering a broader business environment and creating "shared value" for both society and the company. The creation of shared value is fundamental to a company's maximization of long-term shareholder value and its competitiveness in the global market place. Summing up, this line of thinking which is labeled as the "strategic view" of CSR holds that firms increase their CSR engagement in the wake of competition.

In contrast, the altruistic theories of CSR (Aupperle et al., 1985; Baron, 2006) posit that CSR policies entail an inevitable sacrifice of profits of the firm for the sake of social interest (Elahuge, 2005). Therefore, firms that perform responsibly suffer a competitive disadvantage because they incur additional costs that should be avoided entirely or borne by third parties (Waddock and Graves, 1997). Moreover, altruistically driven firms may deliberately abstain from engaging in certain lucrative activities since they do not wish to profit from unethical or odious actions (Rosen et al., 1991). In the modern firm, the source of this corporate philanthropy may originate either from altruistic managers or from the benevolence of owner—shareholders. If corporate owners are altruistic, they will empower managers to run the firm in a philanthropic manner. Alternately, shareholders may be selfish, and thus may be unwilling to accept lower returns in exchange for better firm social performance, but managers' preferences for CSR diverge from those of their principals. According to this view, CSR is simply an additional managerial perquisite and another manifestation of the agency problem between managers and shareholders (Friedman, 1970).

There is limited empirical evidence on whether the altruistic view holds or the strategic view holds especially in the context of competition. Flammer (2014) using the US data in the context finds support for the strategic view. Hawn and Kang (2013), who also use the US data, find mixed evidence. Their simulation results show that firms reduce their CSR when faced with increased competition. The absence of international evidence on this important issue motivates our research.

19.3 DATA AND METHODOLOGY

19.3.1 Environment Sustainability Index

We collect data from a number of sources. The environmental performance data come from a unique data set Asset4, owned by Thomson Reuters. This database is extensively used by investment professionals around the globe to determine various aspects of the environmental, social, and governance performance of a firm. Founded in Switzerland, trained research analysts collect over 900 evaluation points per company where all the primary data used are objective and publicly available (Chatterji et al., 2014). Subsequently, these evaluations are categorized into 250 key performance indicators and further into 4 major categories and 18 subcategories. Research analysts do not solely rely on the feedback of the company as multiple sources, e.g., stock exchanges filings, annual reports, company Web sites, and various other media outlets are used to verify the accuracy and quality of information. In this study, we only focus on environment.

The performance of each company is standardized into z-scores measuring the value of units of standard deviation of that value from the mean value of all companies. By construction, z-scores are normalized to position the score between 0% and 100%. However, we do not use this measure as environment z-scores are not provided. To construct our own environmental performance score, we retrieve information on each environment-related indicator provided by Asset4. For each indicator, Asset4 provides us with binary information on whether or not the firm meets with the particular indicator. We score 1 if a company fulfills the indicator or else 0. Next, we add all positive indicators and divide by the total number of indicators to arrive at the overall environmental performance index.

We carefully scrutinize the data and check whether the questions are conflicting.[1] We also find that in some instances two questions and responses are joined together and reported as a single indicator. For example, we split a single question on environment-emission reduction-implementation to form the following two questions:

1. Does the company describe the implementation of its emission reduction policy through a public commitment from a senior management or board member?
2. Does the company describe the implementation of its emission reduction policy through the processes in place?

After removing indicators with missing or incomplete information, our index is constructed based on 38 individual indicators. By the measure of construction, the maximum score a firm can achieve is 1, if it satisfies all 38 environment indicators. This is in line with the previous literature where an additive index is constructed from a set of individual indicators (Aggarwal et al., 2010; Bebchuk and Cohen, 2005; Drobetz et al., 2004; Gompers et al., 2003). We term our measure as the environment (ENV) index. The details of each of the variables are provided in Table 19.1. We rank each indicator based on the number of firms as a percentage of all firms that comply with it. We use 11 years of data from 2002 to 2012 since the starting year of Asset4 coverage is 2002 but our main regression results are from 2004 to 2012 as private sales data prior to 2004 are not available.

---

[1] This is necessary as by the measure of construction, the overall index is additive.

Table 19.1 Asset4 Environmental Indicators and Percentage of Firm's Complying

	Description	Percentage
1	Does the company describe the implementation of its resource efficiency policy through the processes in place?	54
2	Does the company have a policy for reducing the use of natural resources?	54
3	Does the company have a policy for reducing environmental emissions or its impacts on biodiversity?	43
4	Does the company report on initiatives to increase its energy efficiency overall?	39
5	Does the company describe the implementation of its emission reduction policy through the processes in place?	37
6	Does the company report on initiatives to recycle, reduce, reuse, substitute, treat or phase out total waste, hazardous waste, or wastewater?	36
7	Does the company describe the implementation of its resource efficiency policy through a public commitment from a senior management or board member?	34
8	Does the company report on partnerships or initiatives with specialized NGOs, industry organizations, governmental or supragovernmental organizations that focus on improving environmental issues?	32
9	Does the company have a policy for maintaining an environmental management system?	31
10	Does the company have a policy to lessen the environmental impact of its supply chain?	31
11	Does the company have an environmental product innovation policy (ecodesign, life cycle assessment, dematerialization)?	29
12	Does the company monitor its resource efficiency performance?	29
13	Does the company report about take-back procedures and recycling programs to reduce the potential risks of products entering the environment? Or does the company report about product features and applications or services that will promote responsible, efficient, cost-effective, and environmentally preferable use?	28
14	Is the company aware that climate change can represent commercial risks and/or opportunities?	27
15	Does the company report on initiatives to reduce the environmental impact of transportation of its products or its staff?	24
16	Does the company use environmental criteria (ISO 14000, energy consumption, etc.) in the selection process of its suppliers or sourcing partners?	23
17	Does the company report on initiatives to use renewable energy sources?	22
18	Does the company have environmentally friendly or green sites or offices?	21
19	Does the company set specific objectives to be achieved on emission reduction?	20
20	Does the company describe the implementation of its environmental product innovation policy?	19

Continued

Table 19.1 Asset4 Environmental Indicators and Percentage of Firm's Complying—cont'd

	Description	Percentage
21	Does the company describe, claim to have or mention the processes it uses to accomplish environmental product innovation?	19
22	Does the company report on at least one product line or service that is designed to have positive effects on the environment or which is environmentally labeled and marketed?	19
23	Does the company report on initiatives to reuse or recycle water? Or does the company report on initiatives to reduce the amount of water used?	18
24	Does the company report on its environmental expenditures or does the company report to make proactive environmental investments to reduce future risks or increase future opportunities?	18
25	Does the company report on initiatives to protect, restore, or reduce its impact on native ecosystems and species, biodiversity, protected and sensitive areas?	18
26	Does the company set specific objectives to be achieved on resource efficiency?	15
27	Does the company report or provide information on company-generated initiatives to restore the environment?	15
28	Does the company monitor its emission reduction performance?	15
29	Does the company describe the implementation of its emission reduction policy through a public commitment from a senior management or board member?	15
30	Does the company show an initiative to reduce, reuse, recycle, substitute, phased out or compensate CO_2 equivalents in the production process?	14
31	Does the company report on the concentration of production locations in order to limit the environmental impact during the production process? Or does the company report on its participation in any emissions trading initiative? Or does the company report on new production techniques to improve the global environmental impact (all emissions) during the production process?	13
32	Has the company received product awards with respect to environmental responsibility? Or does the company use product labels (e.g., FSC, Energy Star, MSC) indicating the environmental responsibility of its products?	12
33	Does the company report on specific products which are designed for reuse, recycling, or the reduction of environmental impacts?	9
34	Does the company report or show to be ready to end a partnership with a sourcing partner, if environmental criteria are not met?	8
35	Does the company develop products or technologies for use in the clean, renewable energy (such as wind, solar, hydro and geothermal, and biomass power)?	8
36	Does the company report on initiatives to reduce, substitute, or phase out ozone-depleting (CFC-11 equivalents, chlorofluorocarbon) substances?	7
37	Does the company set specific objectives to be achieved on environmental product innovation?	5
38	Does the company comment on the results of previously set objectives?	2

19.3.2 Product Market Competition

19.3.2.1 Sales HHI

We construct Herfindahl–Hirschman index (HHI) as a proxy of product market competition. This measure is widely used in the industrial organization literature and extensively used in the extant literature.[2] It is calculated by squaring the market share of each firm in the industry and then adding the resulting numbers. It can be represented in formula as following:

$$\text{Sales HHI}_{j,t} = \sum_{i=1}^{N_j} s_{i,j,t} \tag{19.1}$$

where $s_{i,j,t}$ is the sales market share of firm i in industry j in year t. The range of this measure lies between 0 and 1, with a high HHI index suggesting monopoly in the given industry. We use multiple sources to calculate this measure. We start with Worldscope and obtain sales of publicly listed companies of all companies in the database. Next, we exclude those observations with missing or negative sales. We, then, convert the local sales to the US dollar equivalent sales figure at the exchange rate on the corresponding date. However, relying just on the public company sales data may result in poor proxy of product market competition as pointed out by Ali et al. (2009). They further add that continuing to use industry concentration measure without including data on private firms may result in incorrect inferences. A similar concern is expressed in a paper by Hay and Morris (1991).

To address the concerns listed above, we include both public and private data while calculating industry concentration. Private data are retrieved from Bureau van Dijk's Orbis database that provides an extensive coverage of private firms across the globe. We further restrict the sample to only those firms that have at least US$100,000 in sales. This filter is to improve the quality of the data retrieved from Orbis. Borell et al. (2010) suggest imposing some filters on basic information reported by companies, as reported in Orbis' data set in order to limit the impact of "phantom" companies that are either recorded incorrectly or the values reported by the firms are questionable.

Additionally, we remove companies that do not report sales figure or have negative sales. Duplicate companies are removed if their unique identifiers are found to be the same for multiple companies in the same year. Next, we extract public companies' data from Worldscope for all companies included in the database. Finally, we match the databases of Orbis and Worldscope by the using the first two digits of the International Securities Identification Number (ISIN) code. We calculate Sales HHI if at least

---

[2] For a detailed discussion, please see Tirole (1988).

four companies exist in the industry. The maximum number of companies that we use in an industry is 50. This is consistent with the US Census that uses the 50 largest companies within each industry or the largest four-firm concentration ratio.

In this study, we also consider asset concentration, number of employees, and number of companies in an industry as other proxies of product market competition.

19.3.2.2 Asset Concentration

In this measure, we substitute sales with total assets and construct industry-wide asset concentration ratio as in Graham (2000). The intuition is that large firms in an industry may create entry barriers for new companies or arguably sustain losses in the short term to dissuade competing firms from entering the industry. Also, due to a large asset base, companies may have economies of scale and create entry barriers. The measure is constructed on the same lines of Sales HHI and both public and private companies' data are used to increase the reliability of this measure.

$$\text{Asset HHI}_{j,t} = \sum_{i=1}^{N_j} \left(\frac{a_{i,j,t}}{A_{j,t}} \right)^2 \tag{19.2}$$

Where $a_{i,j,t}$ is the assets of firm i in industry j in year t, and A is the sum of all assets in industry j in year t. The range of this measure lies between 0 and 1, with a high HHI index suggesting less competition in the given industry. This measure is also based on both private and public companies.

19.3.2.3 Number of Employees

We define Employee HHI as the concentration of employees in an industry. The notion here is that companies enjoying less competition have a large workforce, as compared to their competitors. This suggests a high concentration in an industry when the Employee HHI is high, indicating that majority of the workforce is employed in a few companies of the industry. We, however, would like to point out potentially confounding results, as it may also be the case that other companies with fewer employees operate at higher efficiency.

$$\text{Employee HHI}_{j,t} = \sum_{i=1}^{N_j} \left(\frac{e_{i,j,t}}{E_{j,t}} \right)^2 \tag{19.3}$$

Where $e_{i,j,t}$ is the number of employees working for firm i in industry j in year t, and E is the sum of all employees in industry j in year t. The range of this measure lies between 0 and 1, with a high HHI index suggesting less competition in the given industry. Similar to our previous measures, we use both private and public companies' data to construct this measure.

19.3.2.4 Number of Companies

In our final measure of market competition, we follow the idea proposed by Porter (1980), where the author argues that the number of firms operating in an industry indicates rivalry among the firms. Balakrishnan and Cohen (2011) further add that too many firms in an industry may pose difficulty in raising capital as the companies have to compete against a limited pool of funding. Following Datta et al. (2013), we calculate this measure as the log of number of companies operating in the industry. A large number of companies in an industry suggest intense competition and vice versa.

19.3.3 Industry Classification

We primarily rely on the industry classification provided by Fama and French's 48 industries' classification (Fama and French, 1997).

19.3.4 Control Variables

In order to control the effect of other factors known to affect firm environment performance, we use illiquidity, standard deviation, log of market value, and blockholding. We control for firm size by using logarithm of total assets. We use closely held shares (% shares closely held) provided by the Worldscope to control for this factor and include an illiquidity measure proposed by Amihud and Mendelson (1986). In this measure, the illiquidity is calculated as the daily ratio of absolute stock return to dollar volume. Next, we use the daily standard deviation of stock returns over the last 1 year as a measure of risk. The control variables are winsorized at 1 and 99 percentile to minimize the effect of outliers.

We estimate the effect of product market competition on environment sustainability by performing the following regression:

$$ENV_{i,j,t} = \alpha_i + \alpha_t + \beta'\left(HHI_{i,j,t}\right) + \sum \gamma' Controls + \varepsilon_{i,j,t} \qquad (19.4)$$

where $ENV_{i,j,t}$ is the environment sustainability index of firm i in country j in year t. α_i and α_t are firm and year fixed effects to control for unobservable heterogeneity and omitted factors related to both HHI and CSR index. The control variables include log of asset, blockholding, illiquidity, and standard deviation. The results are based on robust standard errors but remain qualitatively the same if standard errors are clustered at the firm, industry, or country level.

19.4 EMPIRICAL RESULTS

19.4.1 Sample Characteristics and Distribution

In Table 19.2, we provide sample distribution across countries showing the number of firm-year observations per country. The final sample contains 18,854 firm-year observations from 69 countries. The US is excluded to preclude the impact of its domination

Table 19.2 Sample Distribution

No.	Country	Observations
1	Australia	1598
2	Austria	147
3	Belgium	202
4	Bermuda	379
5	Brazil	173
6	British Virgin Islands	8
7	Canada	1586
8	Cayman Islands	211
9	Chile	78
10	China	273
11	Colombia	27
12	Cyprus	10
13	The Czech Republic	14
14	Denmark	200
15	Egypt	19
16	Finland	231
17	France	742
18	Germany	524
19	Gibraltar	7
20	Greece	107
21	Guernsey	4
22	Hong Kong	377
23	Hungary	15
24	Iceland	6
25	India	326
26	Indonesia	27
27	Ireland	239
28	Isle of Man	3
29	Israel	51
30	Italy	398
31	Japan	3520
32	Jersey	80
33	Jordan	3
34	Kazakhstan	4
35	Kuwait	6
36	Liberia	8
37	Luxembourg	51
38	Malaysia	147
39	Marshall Islands	4
40	Mauritius	5
41	Mexico	54
42	Morocco	13
43	The Netherlands	285

Table 19.2 Sample Distribution—cont'd

No.	Country	Observations
44	New Zealand	92
45	Nigeria	4
46	Norway	165
47	Oman	4
48	Panama	20
49	Papua New Guinea	9
50	Peru	7
51	The Philippines	58
52	Poland	84
53	Portugal	101
54	Puerto Rico	5
55	Qatar	7
56	Russia	125
57	Saudi Arabia	23
58	Singapore	346
59	South Africa	351
60	South Korea	236
61	Spain	381
62	Sri Lanka	3
63	Sweden	455
64	Switzerland	563
65	Taiwan	430
66	Thailand	84
67	Turkey	92
68	The United Arab Emirates	3
69	The United Kingdom	2693

in the sample.[3] The country coverage is comprehensive as it includes developed, developing, and transitional economies. The sample also includes offshore financial centers such as Bermuda, Cayman Islands, Jersey etc., but they do not dominate the result as excluding them does not qualitatively change our results. Australia, Japan, the United Kingdom, and Canada dominate the sample with a large number of firm-year observations. The sample distribution, however, from many other countries is not high. A number of reasons may be responsible for this. For example, it may be the case that Asset4 only covers the major firms in each country instead of all firms in the country.

In Table 19.3, we show the mean Sales HHI scores across different industries based on the Fama—French methodology (Fama and French, 1997). The results from Table 19.3

---

[3] The results remain qualitatively same even if the US is included in the sample.

Table 19.3 Product Market Competition in Fama and French 48 Industry Classification

FF48	Description	Mean	Standard Deviation	Median	Observations
1	Agriculture	0.32	0.161	0.304	86
2	Aircraft	0.489	0.29	0.405	105
3	Almost nothing	0.387	0.226	0.346	70
4	Apparel	0.23	0.163	0.195	96
5	Automobiles and trucks	0.192	0.14	0.122	436
6	Banking	0.183	0.111	0.153	1157
7	Beer and liquor	0.371	0.235	0.361	106
8	Business services	0.095	0.116	0.057	1134
9	Business supplies	0.243	0.187	0.228	127
10	Candy and soda	0.403	0.294	0.341	95
11	Chemicals	0.152	0.123	0.112	559
12	Coal	0.225	0.154	0.155	120
13	Communication	0.339	0.171	0.316	776
14	Computers	0.227	0.181	0.199	286
15	Construction	0.101	0.093	0.065	727
16	Construction materials	0.18	0.18	0.082	402
17	Consumer goods	0.233	0.205	0.107	289
18	Defense	0.999	0	0.999	2
19	Electrical equipment	0.241	0.255	0.127	191
20	Electronic equipment	0.18	0.156	0.145	489
21	Entertainment	0.199	0.186	0.159	125
22	Fabricated products	0.428	0.337	0.388	13
23	Food products	0.19	0.186	0.088	339
24	Health care	0.2	0.092	0.18	87
25	Insurance	0.257	0.184	0.2	647
26	Machinery	0.154	0.138	0.099	550
27	Measuring and control equipment	0.199	0.169	0.094	109
28	Medical equipment	0.341	0.219	0.265	159
29	Nonmetallic and industrial metal mining	0.277	0.156	0.314	477
30	Personal services	0.113	0.134	0.069	62
31	Petroleum and natural gas	0.24	0.219	0.2	912
32	Pharmaceutical products	0.27	0.228	0.197	388
33	Precious metals	0.186	0.065	0.162	373
34	Printing and publishing	0.216	0.164	0.189	187
35	Real estate	0.111	0.076	0.084	853
36	Recreation	0.281	0.151	0.198	117
37	Restaurants, hotels, motels	0.188	0.109	0.143	229
38	Retail	0.125	0.099	0.077	812
39	Rubber and plastic products	0.138	0.109	0.086	51
40	Shipbuilding, railroad equipment	0.202	0.059	0.187	60
41	Shipping containers	0.348	0.245	0.182	55

Table 19.3 Product Market Competition in Fama and French 48 Industry Classification—cont'd

FF48	Description	Mean	Standard Deviation	Median	Observations
42	Steel works, etc.	0.202	0.162	0.137	421
43	Textiles	0.181	0.07	0.214	36
44	Tobacco products	0.439	0.202	0.292	37
45	Trading	0.181	0.161	0.142	558
46	Transportation	0.145	0.145	0.09	875
47	Utilities	0.192	0.145	0.15	746
48	Wholesale	0.12	0.112	0.089	482

suggest that there is substantial variation in concentration across industries. The business services industry is the most competitive with a mean HHI of 0.095. The most concentrated industry is defense with mean HHI of 0.999. The second most concentrated industry is the aircraft industry with mean HHI of 0.489.

Industry-wise breakup of environment performance is reported in Table 19.4. The industry classification is based on the Fama—French methodology outlined in Fama and French (1997). The results from Table 19.4 suggest that there is substantial variation in environmental performance across industries. The lowest score of 0.115 is recorded for the personal services industry, while the highest score of 0.523 occurs in the automobile and trucks industry.

We report a year-wise variation in the environmental performance index in Table 19.5. The mean score of environment performance has increased overall from 2002 to 2012. The mean score in 2002 was 0.198 and it increased to 0.387 by the end of 2012. The highest jump is seen in the year 2007 where it increased from 0.246 to 0.346.

In Table 19.6, we report the distribution of main variables. The daily average standard deviation of stock returns is 2.16%. The Sales HHI of industry concentration averages 0.193 indicating that the average firm is in a competitive industry. The 5th percentile for HHI is 0.041 and the 95th percentile value is 0.549 indicating that there is sufficient diversity in values for conducting robust statistical tests. The ENV index averages 0.320, with the 5th percentile and the 95th percentile values at 0.000 and 0.763, once again indicating a wide variation in this variable. Blockholding average at 0.305 indicates that the average firm has concentrated shareholding. The values of other variables such as log of assets, illiquidity, and standard deviation of returns are within the expected range of values.

In Table 19.7, we report correlations between key variables. There is a statistically significant positive correlation between ENV index and HHI indicating support for the view that firms in concentrated industries have greater CSR engagement. HHI is also positively correlated with blockholding and illiquidity indicating that firms in concentrated industries tend to be less liquid and have concentrated ownership. ENV index is positively correlated

Table 19.4 Environment Index in Fama and French 48 Industry Classification

FF48	Description	Mean	Standard Deviation	Median	Observations
1	Agriculture	0.294	0.202	0.316	91
2	Aircraft	0.404	0.237	0.395	122
3	Almost nothing	0.357	0.255	0.368	75
4	Apparel	0.342	0.256	0.289	101
5	Automobiles and trucks	0.523	0.27	0.553	482
6	Banking	0.209	0.2	0.158	1234
7	Beer and liquor	0.409	0.258	0.408	138
8	Business services	0.246	0.226	0.184	1217
9	Business supplies	0.487	0.249	0.5	139
10	Candy and soda	0.369	0.252	0.368	104
11	Chemicals	0.452	0.24	0.447	608
12	Coal	0.245	0.183	0.211	127
13	Communication	0.286	0.251	0.237	857
14	Computers	0.368	0.315	0.316	295
15	Construction	0.329	0.254	0.289	779
16	Construction materials	0.451	0.238	0.421	423
17	Consumer goods	0.483	0.267	0.5	315
18	Defense	0.4	0.286	0.303	10
19	Electrical equipment	0.504	0.267	0.5	205
20	Electronic equipment	0.422	0.251	0.421	512
21	Entertainment	0.131	0.129	0.105	140
22	Fabricated products	0.275	0.235	0.237	33
23	Food products	0.304	0.24	0.263	359
24	Health care	0.197	0.181	0.184	94
25	Insurance	0.208	0.203	0.158	705
26	Machinery	0.416	0.251	0.395	615
27	Measuring and control equipment	0.398	0.282	0.368	121
28	Medical equipment	0.312	0.238	0.263	188
29	Nonmetallic and industrial metal mining	0.315	0.241	0.289	502
30	Personal services	0.115	0.111	0.079	66
31	Petroleum and natural gas	0.281	0.246	0.211	1003
32	Pharmaceutical products	0.324	0.246	0.289	439
33	Precious metals	0.207	0.186	0.158	385
34	Printing and publishing	0.26	0.224	0.211	212
35	Real estate	0.235	0.229	0.184	904
36	Recreation	0.498	0.321	0.526	137
37	Restaurants, hotels, motels	0.282	0.217	0.237	252
38	Retail	0.271	0.249	0.211	869
39	Rubber and plastic products	0.475	0.184	0.474	61
40	Shipbuilding, railroad equipment	0.35	0.224	0.289	68

Table 19.4 Environment Index in Fama and French 48 Industry Classification—cont'd

FF48	Description	Mean	Standard Deviation	Median	Observations
41	Shipping containers	0.492	0.221	0.474	72
42	Steel works, etc.	0.407	0.238	0.368	453
43	Textiles	0.457	0.3	0.461	42
44	Tobacco products	0.514	0.191	0.526	63
45	Trading	0.144	0.167	0.079	598
46	Transportation	0.328	0.221	0.316	974
47	Utilities	0.459	0.221	0.474	793
48	Wholesale	0.287	0.234	0.237	521

Table 19.5 Yearly Distribution of Environment Index

Year	Mean	Standard Deviation	Median	Observations
2002	0.198	0.173	0.184	431
2003	0.189	0.170	0.158	730
2004	0.173	0.170	0.132	1264
2005	0.191	0.175	0.158	1386
2006	0.246	0.209	0.211	1421
2007	0.346	0.241	0.316	1542
2008	0.377	0.264	0.368	1889
2009	0.355	0.264	0.316	2231
2010	0.354	0.256	0.342	2627
2011	0.373	0.267	0.342	2835
2012	0.387	0.273	0.368	2147

with log of total assets, suggesting that large firms are more socially engaged ostensibly to avoid political costs. ENV index is negatively correlated with blockholding and standard deviation of returns, implying that blockholders perhaps do not support social engagement and firms with volatile returns (risky firms) do less CSR. Illiquidity is strongly positively correlated with standard deviation of returns (0.901) suggesting potential multicollinearity problems. With the exception of this correlation other correlations are relatively lower and do not indicate any issues of multicollinearity.

19.4.2 Regression Results

The baseline regression results are reported in Table 19.8 in column 1. The main variable of interest is the proxy for product market competition measured by Sales HHI. Under the altruistic view, the coefficient of Sales HHI is expected to be positive, while under the alternate strategic view, coefficient of Sales HHI should be negative. We find that Sales HHI is strongly positively related to ENV index suggesting that firms in concentrated industries engage more in CSR. Therefore, in a cross-country setting we find empirical

Table 19.6 Summary Statistics of Main Variables

	Mean	Standard Deviation	Distribution				
			5th	25th	50th	75th	95th
HHI	0.193	0.171	0.041	0.074	0.137	0.248	0.549
ENV index	0.320	0.254	0.000	0.105	0.263	0.526	0.763
Log of asset	15.503	1.646	12.886	14.418	15.416	16.550	18.394
Blockholding	0.305	0.239	0.001	0.105	0.261	0.500	0.734
Illiquidity	4.0882E-08	4.0368E-07	2.7047E-12	2.376E-11	2.622E-10	2.1632E-09	6.9331E-08
Standard deviation	0.023	0.010	0.011	0.015	0.020	0.027	0.043

Table 19.7 Correlation Matrix

	HHI	ENV index	Log of Asset	Blockholding	Illiquidity	Standard Deviation
HHI	1					
ENV index	0.103***	1				
Log of asset	−0.0168**	0.338***	1			
Blockholding	0.0941***	−0.0606***	−0.0741***	1		
Illiquidity	0.0145**	0.00115	−0.0485***	0.00686	1	
Standard deviation	0.00615	−0.0407***	−0.263***	0.0313***	0.901***	1

Note: *p < 0.1, **p < 0.05, ***p < 0.01.

Table 19.8 Baseline Regressions

	(1) ENV Index	(2) ENV Index	(3) ENV Index	(4) ENV Index
Sales	0.143***			
	(12.61)			
Asset		0.0996***		
		(10.34)		
Employee			0.0992***	
			(10.09)	
Log companies				−0.00571***
				(−5.37)
Log of asset	0.0531***	0.0528***	0.0530***	0.0538***
	(47.84)	(50.01)	(49.82)	(48.02)
Blockholding	−0.0809***	−0.0767***	−0.0785***	−0.0801***
	(−10.38)	(−10.20)	(−10.28)	(−10.24)
Illiquidity	2743.2	2292.0	2670.2	2156.6
	(0.62)	(0.57)	(0.62)	(0.48)
Standard	−0.514**	−0.596***	−0.650***	−0.432**
deviation	(−2.54)	(−3.12)	(−3.36)	(−2.13)
Country fixed effects	Yes	Yes	Yes	Yes
Year fixed efffects	Yes	Yes	Yes	Yes
N	17,013	18,018	17,537	17,013
Adjusted R^2	0.360	0.364	0.365	0.355

Note: *$p < 0.1$, **$p < 0.05$, ***$p < 0.01$.

support for the altruistic view. Three of the control variables show statistically significant coefficients. The log of total assets has a significantly positive effect on ENV index implying that large firms are more involved in CSR activities. Blockholding has a significantly negative effect on ENV index. This result suggests that blockholders view CSR as charity without associated benefits to the firm where the larger the shareholding of blockholders the lower the level of CSR. The standard deviation of stock returns proxies for riskiness of the firm which has a significantly negative impact on ENV index implying that risky firms choose not to deploy their relatively scarce capital on CSR. Instead they may conserve it for a rainy day. Overall, our results are inconsistent with the strategic view.

We conduct several robustness checks. First, we replace Sales HHI by other proxies for industry concentration such as Asset HHI, Employee HHI, and log Companies HHI. Large firms in an industry may preclude healthy competition by creating entry barriers. Thus relative size based on asset size constitutes an alternate concentration measure. We compute Asset HHI as shown in the methodology section and use this instead of Sales HHI in column 2. Asset HHI is strongly positively associated with ENV index, once again supporting the altruistic view. The second alternate proxy for industry competition is Employee HHI which measures the concentration of employees in an industry. Companies employing a large workforce may arguably be deemed to be a dominant firm and the industry highly concentrated. The technique for computing Employee HHI is shown in the methodology section. Using this alternate concentration measure in column 3, we once again find a statistically significant positive coefficient. The final measure of concentration is based on number of companies in an industry. The log of number of companies proxies for competitiveness in the industry. In column 4, we use the log of number of companies as the measure of competitiveness and find a statistically significant negative coefficient suggesting that firms in competitive industries do less CSR.

Overall, our results support the altruistic view. Our findings are at inconsistent with the US studies such as Flammer (2014). Thus it appears that country context is relevant for the CSR-Competition relation.

19.5 CONCLUSION

In this study, we extend the prior research on competition-CSR relationship using a comprehensive cross-country sample. Our empirical results using 18,854 firm-year observations from 69 countries show that the industry competitiveness is negatively related to CSR. Our results are consistent with the "altruistic view" which posits that firms may be willing to give up profits for the sake of society. Our results are consistent with the view that firms are altruistic when they belong to a concentrated industry. Firms in competitive industries tend to do less CSR, ostensibly, due to the resource constraints they face while operating in a competitive industry. Our results are opposite to the findings of studies based on the US firms. A possible reason for this divergence may be that an informationally efficient market that aggregates environmental information is a prerequisite for obtaining competitive advantages through CSR.

REFERENCES

Aggarwal, R., Erel, I., Stulz, R., Williamson, R., 2010. Differences in governance practices between US and foreign firms: measurement, causes, and consequences. Rev. Financ. Stud. 23 (3), 3131–3169.

Agle, B.R., Donaldson, T., Freeman, R.E., Jensen, M.C., Mitchell, R.K., Wood, D.J., 2008. Dialogue: towards superior stakeholder theory. Bus. Ethics Q. 18 (2), 153–190.

Ali, A., Klasa, S., Yeung, E., 2009. The limitations of industry concentration measures constructed with compustat data: implications for finance research. Rev. Financ. Stud. 22 (10), 3839–3871.

Amihud, Y., Mendelson, H., 1986. Asset pricing and the bid-ask spread. J. Financ. Econ. 17 (2), 223–249.

Aupperle, K.E., Carroll, A.B., Hatfield, J.D., 1985. An empirical examination of the relationship between corporate social responsibility and profitability. Acad. Manage. J. 28 (2), 446–463.

Balakrishnan, K., Cohen, D.A., 2011. Product Market Competition and Financial Accounting Misreporting. Working paper. University of Pennsylvania, USA.

Barney, J., 1991. Firm resources and sustained competitive advantage. J. Manage. 17 (1), 99–120.

Baron, D.P., 2001. Private politics, corporate social responsibility, and integrated strategy. J. Econ. Manage. Strategy 10 (1), 7–45.

Baron, D.P., 2006. A Positive Theory of Moral Management, Social Pressure, and Corporate Social Performance. Working Paper. Stanford University, Graduate School of Business, USA.

Bebchuk, L.A., Cohen, A., 2005. The costs of entrenched boards. J. Financ. Econ. 78 (2), 409–433.

Borell, M., Tykvová, T., Schmitt, C., 2010. How Do Industry Characteristics and Persistence Shape Private Equity Investments? Working Paper Centre for European Economic Research (ZEW), Mannheim, Germany.

Chatterji, A., Levine, D.I., Toffel, M.W., 2009. How well do social ratings actually measure corporate social responsibility. J. Econ. Manage. Strategy 18 (1), 125–169.

Chatterji, A., Durand, R., Levine, D., Touboul, S., 2014. Do Ratings of Firms Converge? Implications for Strategy Research. IRLE. Working Paper No. 107–114.

Datta, S., Iskandar-Datta, M., Singh, V., 2013. Product market power, industry structure, and corporate earnings management. J. Bank. Finance 37 (8), 3273–3285.

Drobetz, W., Schillhofer, A., Zimmermann, H., 2004. Corporate governance and expected stock returns: evidence from Germany. Eur. Financ. Manage. 10 (2), 267–293.

Elahuge, E., 2005. Corporate managers' operational discretion to sacrifice profits in the public interest. In: Hay, B., Stavins, R., Vietor, R. (Eds.), Environmental Protection and the Social Responsibility of Firms. Washington, DC.

Fama, E., French, K., 1997. Industry costs of equity. J. Financ. Econ. 43 (2), 153–193.

Fernandez-Kranz, D., Santalo, J., 2010. When necessity becomes a virtue: the effect of product market competition on corporate social responsibility. J. Econ. Manage. Strategy 19 (2), 453–487.

Flammer, C., 2014. Does product market competition foster corporate social responsibility? Evidence from trade liberalization. Strategic Manage J 36 (10), 1469–1485.

Freeman, R.E., 1984. Strategic Management: A Stakeholder Approach. Pitman, Boston.

Friedman, M., September 1970. The social responsibility of business is to increase profits. N.Y. Times Mag. 13, 32–33.

Gompers, P.A., Ishii, J.L., Metrick, A., 2003. Corporate governance and equity prices. Q. J. Econ. 118 (1), 107–155.

Graham, J.R., 2000. How big are the tax benefits of debt? J. Finance 55 (5), 1901–1941.

Hart, S.L., 1995. A natural resource-based view of the firm. Acad. Manage. Rev. 20 (4), 986–1014.

Hawn, O., Kang, H.G., 2013. The Market Competition for Corporate Social Responsibility (CSR): How Industry Structure Determines CSR. Working Paper. Boston University, USA.

Hay, D.A., Morris, D.J., 1991. Industrial Economics and Organization: Theory and Evidence. Oxford University Press, Oxford.

Helfat, C.E., Peteraf, M.A., 2003. The dynamic resource-based view: capability lifecycles. Strategic Manage. J. 24 (10), 997–1010.

Hooley, G., Broderick, A., Moeller, K., 2006. Competitive positioning and the resource-based view of the Firm. J. Strategic Mark. 6 (2), 97–113.

Jones, T.M., 1995. Instrumental stakeholder theory: a synthesis of ethics and economics. Acad. Manage. Rev. 20 (2), 404–437.

McWilliams, A., Siegel, D., 2001. Corporate social responsibility: a theory of the firm perspective. Acad. Manage. Rev. 26 (1), 117–127.

Peteraf, M.A., 1993. The cornerstones of competitive advantage: a resource-based view. Strategic Manage. J. 14 (3), 179–191.

Porter, M.E., 1980. Competitive Strategies: Techniques for Analyzing Industries and Competitors. Free Press, New York.

Porter, M.E., 1991. America's green strategy. Sci. Am. 264 (4), 33–35.

Porter, M.E., Kramer, M.R., 2006. Strategy and society: the link between competitive advantage and corporate social responsibility. Harv. Bus. Rev. 84 (12), 78–92.

Porter, M.E., Kramer, M.R., 2011. The big idea: creating shared value. Harv. Bus. Rev. 89 (1–2), 62–77.

Rosen, B.N., Sandler, D.M., Shani, D., 1991. Social issues and socially responsible investment behavior: a preliminary empirical investigation. J. Consum. Aff. 25 (2), 221–234.

Russo, M.V., Fouts, P.A., 1997. A resource-based perspective on corporate environmental performance and profitability. Acad. Manage. J. 40 (3), 534–559.

Tirole, J., 1988. The Theory of Industrial Organization. MIT Press, USA.

Waddock, S.A., Graves, S.B., 1997. The corporate social performance—financial performance link. Strategic Manage. J. 18 (4), 303–319.

Wernerfelt, B., 1984. A resource-based view of the Firm. Strategic Manage. J. 5 (2), 171–180.

Wernerfelt, B., 1995. A resource-based view of the firm: ten years after. Strategic Manage. J. 16 (3), 171–174.

Funding and Accounting Systems

CHAPTER 20

The Costs and Benefits of Cost—Benefit Analysis as Applied to Environmental Regulation

Imad A. Moosa
School of Economics, Finance and Marketing, RMIT University, Melbourne, VIC, Australia

Contents

20.1 INTRODUCTION

Cost—benefit analysis (CBA), sometimes called benefit—cost analysis, is an analytical tool that is used to assess the costs and benefits of public projects, including regulatory proposals. It is a systematic process whereby costs and benefits are assessed and measured in order to determine the soundness and viability of a project, decision or policy—hence, it is about feasibility. The technique is also used for comparing alternative projects, policies, or courses of action. The process involves a comparison of the total expected cost of each option against total expected benefit. The costs and benefits are typically expressed in monetary terms and adjusted for the time value of money by discounting the flows of net benefits. In this chapter we examine CBA as applied to environmental regulation.

The primary objective of environmental regulation is to prevent or minimize environmental degradation and the harmful effects it may have on human health and well-being. The accomplishment of this objective (representing the benefits of environmental regulation) does not come at no cost. The costs associated with environmental regulation are borne by the government, the business sector, and consumers: the government incurs the cost of devising, implementing, and monitoring regulatory measures; business firms bear the cost of compliance and any possible adverse effect on the economy; and consumers bear the cost of higher prices of some goods and the removal of others from the

Handbook of Environmental and Sustainable Finance
ISBN 978-0-12-803615-0

market (those deemed not to be environmentally friendly). Several evaluation techniques are used for the purpose of evaluating costs and benefits, including cost-effectiveness analysis, cost-minimization analysis, cost-utility analysis, economic impact analysis, fiscal impact analysis, and social return on investment analysis. However, the most common technique used for this purpose is CBA.

The origin of CBA can be traced back to the work of the French engineer and economist Jules Dupuit (1844, 1853), although Hanley and Spash (1993) claim earlier origins to the U.S. Secretary of the Treasury, Albert Gallatin, in 1808. However, it was Dupuit who defined the way in which benefits and costs are measured and embraced the principle that an investment decision, such as building a road or a bridge, should meet the criterion that benefits exceed costs. More recently, CBA has been the object of presidential executive orders, which (according to Gayer, 2011) "indicate that the executive branch has fully endorsed the use of CBA within regulatory policy-making." President Carter's Executive order 12044 required regulatory agencies to quantify the benefits and costs of regulation. When Ronald Reagan (who was an enthusiastic free marketeer and deregulator) became president in 1981, he was concerned about (or what he perceived to be) excessive regulation. Consequently, he issued Executive Order 12291 to forbid the introduction of regulation unless the potential social benefits exceed the potential social costs. When President Bill Clinton came into office, he replaced Executive Order 12291 with Executive Order 12866 whereby the cost–benefit criterion was amended in such a way as to make sure that "each agency shall assess both the costs and the benefits of the intended regulation." President George Bush (Junior) operated under the framework established by Reagan and Clinton, which made environmental groups increasingly agitated about CBA.

Subsequently, the Obama administration approved rules with net benefits well in excess of $100 billion (Sunstein, 2012).

The use of CBA in environmental regulation is a controversial issue, particularly because the procedure can be used to produce results to support prior views. There are those who totally reject it, those who accept it without questioning, those who think that it is useful but problematical, and those who think that it is so useful and has no alternative that we should strive to improve the technique. What is interesting is that anyone belonging to any of these categories (with respect to the view they hold on CBA) could be an opponent or a proponent of regulation. After all, CBA can be used to demonstrate that an environmental regulation proposal should or should not be adopted. Only environmental fanatics, who think that any cost is worthwhile, reject CBA outright.

The objective of this chapter is to evaluate the costs and benefits (read pros and cons) of CBA. The costs of CBA stem from the fact that CBA can be abused in such a way as to make the wrong decision from a social point of view, just because those decisions are consistent with preconceived, ideologically driven beliefs. The benefits of CBA lie in the fact that CBA makes it possible to adopt a project that looks infeasible in purely commercial terms. This is why CBA is a double-edged sword. These are the points that we will elaborate on in the remainder of this chapter, starting with the identification and measurement of costs and benefits.

20.2 IDENTIFYING AND MEASURING COSTS

The costs of environmental regulation include, among other items, (1) government funding of scientific research; (2) the cost of monitoring; and (3) the amount spent by business to be compliant. The costs of environmental regulation are associated with the price, output, and income effects of regulatory action. Examples of the price effect and output effect are higher prices of environmentally friendly cars and smaller catches of fish. The income effect may reflect the imposition of taxes. Implicit costs are exemplified by the inconvenience of using public transport. In terms of nature, as opposed to source, costs can be classified into explicit accounting costs and implicit economic costs. Accounting costs may be classified into capital (fixed) costs and operating (variable) costs. An example of fixed costs is the construction costs of a sewage treatment plant. Variable costs are directly related to risk containment—for example, the cost of monitoring water quality.

The Environmental Protection Agency (EPA) recognizes several cost categories, according to the ease of estimation. These categories include the following: (1) real resource environmental costs, such as the costs associated with the acquisition, installation, and operation of equipment; (2) government regulatory costs (that is, the costs of monitoring and enforcement of environmental regulation); (3) social welfare losses resulting from the rise in price or decline in output of the goods associated with environmental regulation; (4) transitional costs such as the value of resources displaced due to regulation-induced production; and (5) indirect costs resulting from adverse effects on production quality, innovation, and productivity.

In general, two methods are used to measure the explicit costs of environmental regulation. By following the engineering approach, the cost of containing environmental risk is measured by the least-cost available technology needed to achieve the targeted level of risk. For this purpose, engineers and scientists identify combinations of equipment, labor, and materials needed by polluters to comply with environmental regulation. An alternative option is to use the survey approach whereby the opinions of external experts are utilized. The problem with the survey approach is inherent bias because polluting firms have every incentive to report inflated costs to boost the probability that the proposed regulation will be rejected.

Measuring the costs of environmental regulation is not straightforward, implying that the estimates can be manipulated to produce what is suitable from own perspective. If, for example, a firm installs an efficient piece of equipment that happens to be environmentally friendly, how much of the cost of installing this equipment pertains to regulatory compliance? Likewise, measuring the costs of environmental regulation in terms of its effects on productivity, competitiveness, and growth is problematical (see, for example, Moosa and Ramiah, 2014).

20.3 IDENTIFYING AND MEASURING BENEFITS

In environmental regulation a "benefit" is defined as any gain in human well-being (welfare or utility) which is typically measured by how much an individual is willing to pay to secure that gain or how much an individual is willing to accept in compensation to forgo that gain (Pearce, 1998). The benefits of environmental regulation can be classified into primary (such as improved health) and secondary (such as enhanced productivity). For example, implementing regulation to reduce air pollution brings about primary benefits in terms of reducing mortality, chronic illness, health-care spending, minor illnesses, and the incidence of respiratory problems. The secondary effects take the form of a higher level of welfare. Once the benefits of environmental regulation have been identified, these benefits are measured in monetary terms, which can be problematical.

The society derives utility from environmental quality (clean air, clean water, etc.) through two sources of value: user value and existence value. User value is derived from the usage of or access to an environmental good such as a river or a lake. Existence value is derived from the continuing existence of a high-quality environmental good. User value may be direct or indirect. Direct user value is the commercial or recreational benefit derived from the use of an environmental good (for example, swimming and catching fish in a clean river). Indirect user value is the value derived by someone knowing that someone else is deriving direct value. Existence value is measured by the utility derived by knowing that the river is clean enough for fish to survive. Mitchell and Carson (1989) classify the sources of existence value into (1) vicarious consumption (the notion that individuals value a public good for the benefits accruing to others) and (2) stewardship, which arises from a sense of obligation to preserve the environment for future generations by recognizing the intrinsic value of natural resources.

The monetization of benefits is an exercise that is typically based on the assumption that secondary benefits are insignificant to the extent that they are likely to be offset by secondary costs, which means that emphasis is placed on the estimation of primary benefits. Smith and Krutilla (1982) suggest two approaches to the estimation of monetary values: the physical linkage approach and behavioral linkage approach. The physical linkage approach is based on the technical relation between an environmental resource and the user of the resource. This approach, therefore, is based on a functional relation between environmental damage and the level of risk containment. For example, environmental damage is initially measured in nonmonetary terms, such as the number of premature deaths, then a monetary value is assigned to the nonmonetary benefit of reducing environmental damage.

In the behavioral linkage approach, benefits are quantified by observing actual market behavior or through survey responses about hypothetical markets. This approach comprises the political referendum method (using the actual market of a public good by monitoring voting results from political referenda on proposed changes in environmental policy) and the contingent valuation method, which is based on surveys to inquire about individuals' willingness to pay for environmental improvement. There are also the averting expenditure method, the travel cost method, and the hedonic price method. The averting expenditure method is intended to assess changes in spending on goods and services that are substitutes for environmental quality. According to the travel cost method, valuing a change in environmental quality is based on an assessment of the effect of that change on the demand for a complementary good. The hedonic price method is based on the theory that a good is valued for the attributes it possesses, which can be used to estimate the implicit or hedonic price of an environmental attribute and identify its demand as a means to assign value.

Valuation methods are alternatively classified as the stated preference method, the revealed preference method, and the life satisfaction method. The stated preference method is based on a questionnaire designed to elicit estimates of people's willingness to pay for (or willingness to accept) a particular outcome. In contingent valuation surveys, participants are asked what they are willing to pay for a defined health benefit or for a reduction in risk. For example, Hultkrantze et al. (2006) conducted a study in Sweden to elicit willingness to pay for safety enhancement and found that, for a given outcome, respondents were willing to pay more for a personal traffic safety device than for a public road safety program. An alternative procedure is choice modeling, whereby individuals may be asked if they would be willing to pay a certain amount more for choice A than choice B.

The revealed preference method is based on observing people's behavior in related markets. The life satisfaction method is about reported life satisfaction in surveys. Econometric methods are typically used to estimate the life satisfaction provided by nonmarket goods, and this is subsequently converted into a monetary value by estimating the effect of income on life satisfaction. For example, in wage—risk studies, it is assumed that workers are willing to give up income for improved workplace safety or to require (accept) income for taking on more risk. To disentangle the wage—risk trade-off from other factors that affect wages, it is necessary to employ models that control for differences in labor productivity as well as the quality component of the job. The wage—risk equation is typically specified as a regression of the wage rate on a vector of personal characteristic variables (such as age and the level of education), a vector of job characteristic variables, the probabilities of a fatal or nonfatal injury, and workers compensation for an injury. The coefficients on the risk of a fatality indicate willingness to pay for safety by accepting a lower wage rate or as a willingness to accept a higher level of risk in return for wage compensation. Vassanadumrongdee and Matsuoka (2005) asked survey participants in Bangkok what they would be willing to pay for a reduction of the risk of mortality from traffic accidents and air pollution. They found that, despite various differing perceptions about air pollution and road traffic accidents, willingness to pay to reduce risk was similar for both contexts.

20.4 CALCULATING THE PRESENT VALUE OF NET BENEFITS

The decision whether or not to accept an environmental project is based on the present value of net benefits (that is, net of costs), which is the same metric used to choose between two mutually exclusive projects. This technique is extrapolated from the project appraisal methodology used in corporate finance to evaluate the feasibility of commercial projects on the basis of the net present value of cash flows. The extrapolation of the technique from corporate finance to environmental regulation is criticized on the grounds a technique that is used to evaluate commercial projects (where costs and revenues are easy to measure) cannot be used to evaluate projects that involve human life and well-being.

Since the costs and benefits of environmental regulation do not accrue at the same time, they must be adjusted to reflect the time value of money. This is because people are not indifferent with respect to the timing of costs and benefits—they generally prefer to receive benefits as early as possible and pay to cover costs as late as possible. Another reason for discounting is the opportunity cost of the capital invested in environmental regulation. We will find out that the very operation of discounting provides another reason for criticizing CBA as applied to environmental regulation.

Consider an environmental regulation project that, over a period of n years, gives rise to the costs C_0, C_1, ..., C_n and the corresponding benefits B_0, B_1, ..., B_n. It is assumed that monetary values can be assigned to costs and benefits, which provide yet another justification for criticizing CBA as applied to environmental regulation, particularly the estimation of the monetary value of benefits. Since it is unlikely that benefits arise immediately, it follows that $B_0 = B_1 = \cdots B_m = 0$ for $m < n$. For any time period, t, the flow of (nominal) net benefits is $B_t - C_t$ for $t = 0$, ..., n. The present value of net benefits in nominal terms (PVN) is obtained by discounting the nominal net benefit flows at an appropriate discount rate, which gives

$$PVN = \sum_{t=0}^{n} \frac{B_t - C_t}{(1 + i)^t} \tag{20.1}$$

where costs and benefits are measured in nominal terms (that is, at current rather than constant prices). In Eqn (20.1) i is the discount rate, which is typically an interest rate of some sort. The discount rate should reflect the opportunity cost of the funds used to finance environmental regulation and the rate of return that could be realized through private spending on consumption and investment, assuming the same level of risk. The decision rule would be to accept an environmental regulation proposal if the present value of net benefits is greater than zero or greater than a prespecified benchmark. In the case of choice between two or more mutually exclusive proposals, the proposal with the highest present value is chosen. An example of two mutually exclusive projects in environmental regulation is the choice between carbon tax and tradable permits to reduce the emission of carbon dioxide. Another example from the field of public works involves the choice among three alternatives for the design of road intersection: crossroads with traffic lights, roundabouts, and flyovers.

In Eqn (20.1), the criterion used to judge the feasibility of an environmental project (the present value of net benefits) depends on the estimates of costs and benefits as well as the choice of the discount rate. We have seen that the estimates of costs and benefits (particularly benefits) are not precise, which means that those who oppose regulation tend to overestimate costs and underestimate benefits, and vice versa. Figures 20.1–20.3 show that the present value of net benefits changes with the estimates of costs and benefits and the choice of the discount rate. In Figure 20.1, it is shown that the present value of the net benefits of a project rises as cost estimates decline. In this case, *PVN* changes from negative to positive. In Figure 20.2, the present value of net benefits changes from positive to negative as the estimates of benefits get smaller. Figure 20.3 shows what happens

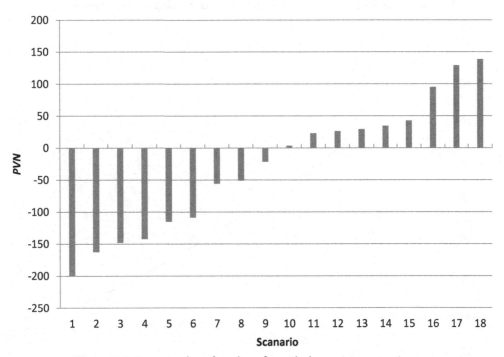

Figure 20.1 Present value of net benefits with decreasing cost estimates.

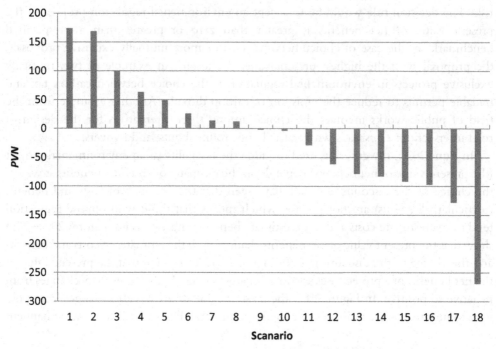

Figure 20.2 Present value of net benefits with decreasing benefit estimates.

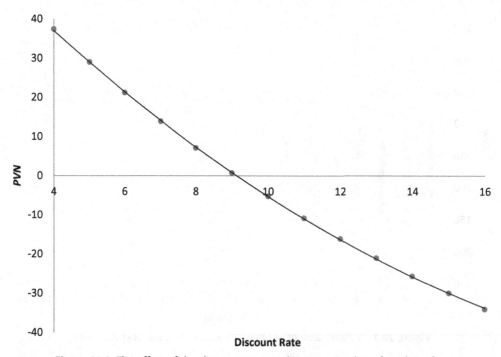

Figure 20.3 The effect of the discount rate on the present value of net benefits.

to the present value of net benefits as the discount rate rises from 4% to 16%. At a discount rate of 9%, the present value of net benefits is close to 0, after which it turns negative.

In Eqn (20.1), net benefits are measured in nominal terms without adjusting for inflation, which would make the present value of net benefits bigger that what it is in real terms. For the very reason that costs and benefits do not necessarily materialize at the same time, adjustment must be made for inflation, which causes an erosion of value over time. Without adjusting for inflation, the costs and benefits that occur later in the life of the project appear higher than they should, which causes bias toward projects with delayed benefits. Furthermore, using Eqn (20.1) would necessarily imply that decision makers are subject to money illusion, considering the nominal (rather than real) values of costs and benefits. If the expected inflation rate is π, it follows that the present value of real net benefit flows (PVR) is given by

$$PVR = \sum_{t=0}^{n} \frac{B_t - C_t}{(1 + \pi)^t (1 + i)^t} \tag{20.2}$$

Equation (20.2) tells us that the present value of real net benefits depends on four parameters: benefits, costs, the discount rate, and the expected inflation rate. The measurement of expected inflation, which can make a lot of difference for the calculation of the present value of real net benefits, can be rather subjective. Actually, it is more difficult to come up with an estimate of the expected inflation rate than the nominal discount rate. Figure 20.4 shows how the present value of nominal net benefits declines as the discount rate rises (downward sloping-curve as in Figure 20.3). However, in Figure 20.4 it is also shown that the real value of net benefits declines as the inflation rate rises, which is indicated by the downward shift in the curve describing the relation between the present value and the discount rate. In Figure 20.5, the discount rate is kept at 2%, but the inflation rate is changed upward from 1% to 12%. In this case, the present value of real net benefits declines, becoming negative at an inflation rate of 7% and higher.

Cooley et al. (1975) argue that since the discount rate is a nominal interest rate, it has the expected inflation rate as a component because, according to the Fisher equation, the nominal rate is equal to the sum of the real rate and the expected inflation rate. The Fisher equation is written as

$$i_t = r_t + \pi_t \tag{20.3}$$

where r is the real discount rate. By substituting Eqn (20.3) into Eqn (20.2), we obtain an alternative measure of net benefits in real terms (PVR^*), which is given by

$$PVR^* = \sum_{t=0}^{n} \frac{B_t - C_t}{(1 + \pi)^t (1 + r + \pi)^t} \tag{20.4}$$

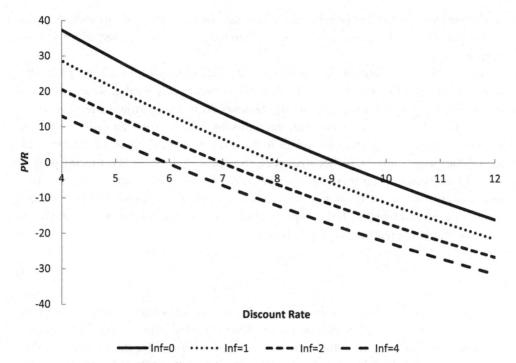

Figure 20.4 The effect of the discount rate with varying inflation rates.

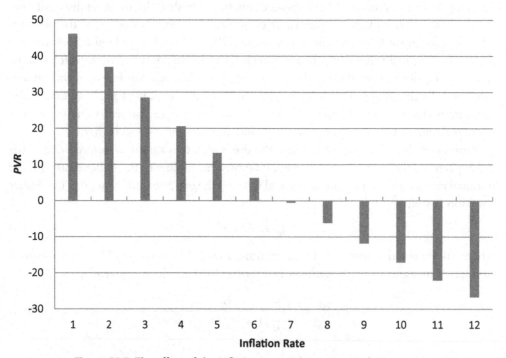

Figure 20.5 The effect of the inflation rate with a constant discount rate.

In Eqn (20.4), the inflation rate appears twice in the denominator when real net benefits are discounted at the nominal rate, i. The alternative procedure would be to discount net benefits at the real discount rate, which is the nominal discount rate minus the inflation rate. In this case, the present value formula becomes

$$PVR^* = \sum_{t=1}^{n} \frac{B_t - C_t}{(1 + \pi)^t (1 + r)^t} = \sum_{t=1}^{n} \frac{B_t - C_t}{(1 + \pi)^t (1 + i - \pi)^t} \tag{20.5}$$

From Eqns (20.1), (20.2) and (20.5) we can compare the values of net benefits realized at time t by using each equation. Consider first Eqns (20.1) and (20.2), which give

$$\frac{PVR_t}{PVN_t} = \frac{1}{(1 + \pi)^t} \tag{20.6}$$

It follows that for $\pi > 0$, $PVR < PVN$ such that

$$\frac{\partial}{\partial \pi}(PVR_t) = \frac{-PVN_t}{(1 + \pi)^{2t}} \tag{20.7}$$

which means that PVR is negatively related to the inflation rate. Consider now Eqns (20.1) and (20.5), which give

$$\frac{PVR_t^*}{PVN_t} = \left[\frac{(1 + i)^t}{(1 + \pi)^t (1 + i - \pi)^t} \right] \tag{20.8}$$

We can readily see that if $i = \pi$, it follows that $PVR^* = PVN$ and that if $i > \pi$, then $PVR^* < PVN$.

It is not straightforward to see how the ratio PVR^*/PVN changes in response to changes in the discount rate and inflation rate. When $t = 1$ Eqn (20.8) gives the following partial derivatives

$$\frac{\partial}{\partial \pi}\left(\frac{PVR^*}{PVN} \right) = \frac{-(1 + i)(i - 2\pi)}{[(1 + \pi)(1 + i - \pi)]^2} \tag{20.9}$$

and

$$\frac{\partial}{\partial i}\left(\frac{PVR^*}{PVN} \right) = \frac{-\pi(1 + \pi)}{[(1 + \pi)(1 + i - \pi)]^2} \tag{20.10}$$

What we are concerned with here is the sensitivity of the ratio PVR^*/PVN with respect to changes in the nominal discount rate and the inflation rate—the ratio would be insensitive if the partial derivatives have values close to 0. A graphical illustration is perhaps easier. In Figure 20.6, we observe the ratios PVR/PVN and PVR^*/PVN corresponding to years 1–10 when the discount rate is 6% and the inflation rate is 4%. As we

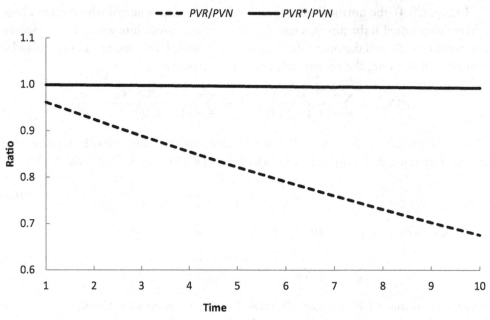

Figure 20.6 *PVR/PVN* and *PVR*[*]*/PVN* ($i = 0.06$, inf $= 0.04$).

can see, PVR/PVN declines while PVR^*/PVN does not change. This result is not specific to the selected values of the discount and inflation rates, but this is not the case. In Figure 20.7, we select extreme values to find out if a big difference between the nominal discount rate and the inflation rate affects the behavior of the ratios. When the discount rate is 19% and the inflation rate is 1%, PVR/PVN falls slowly, but hardly anything happens to PVR^*/PVN. Further evidence on the insensitivity of PVR^*/PVN is provided in Figures 20.8 and 20.9. In Figure 20.8, we observe what happens to the ratio as the nominal discount rate rises from 1 to 10, which is nothing. In Figure 20.9, we observe what happens to the ratio as the inflation rate rises from 1 to 10, which is very little. The lines describing the behavior of PVR^*/PVN are flat with slopes of 0 against the nominal discount rate and inflation rate, which means that the partial derivatives represented by Eqns (20.9) and (20.10) are close to 0.

The conclusion is that PVR^*/PVN is not sensitive to changes in either the inflation rate or the nominal discount rate, which means that $PVR^* \approx PVN$ under any set of conditions. If this is the case then it is not a good idea to discount net benefits at the real discount rate as Cooley et al. (1975) recommend. This means that there is no point in using Eqn (20.5) because it is equivalent to Eqn (20.1). In fact, if the sound procedure is to base CBA on real net benefits, then using Eqn (20.5) produces misleading results, because it amounts to discounting the nominal values of net benefits at the nominal discount rate.

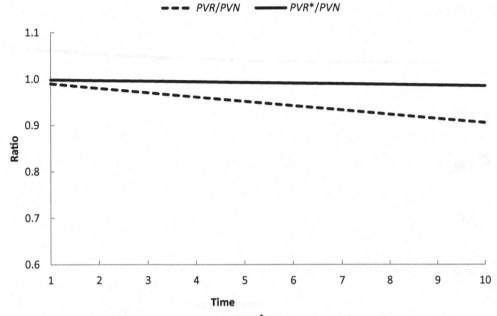

Figure 20.7 *PVR/PVN* and *PVR*/PVN* ($i = 0.19$, inf $= 0.01$).

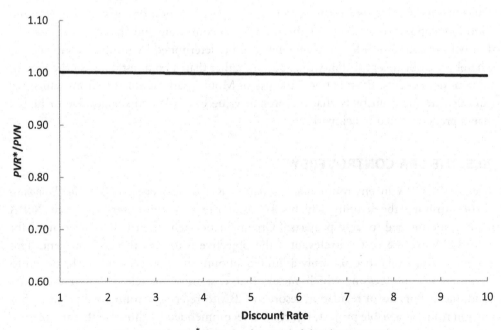

Figure 20.8 *PVR*/PVN* as a function of the discount rate.

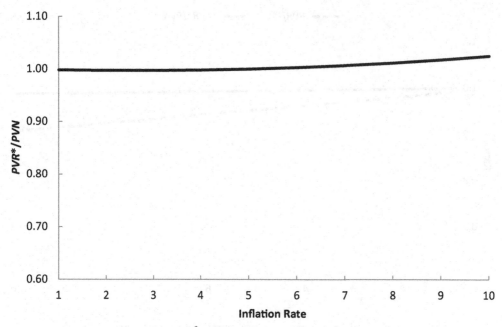

Figure 20.9 PVR^*/PVN as a function of the inflation rate.

Evaluation of the feasibility of an environmental regulation project involves the risk of taking the wrong decision by accepting an infeasible project or rejecting a feasible project. This outcome results from overestimation or underestimation of costs and benefits and choosing inappropriate values for the nominal discount rate and the inflation rate. To deal with this kind of risk, the parameters can be determined by using scenario analysis whereby we obtain a probability distribution, rather than a point estimate, of the present value of net benefits. Even better is the use of Monte Carlo simulation from which we can calculate the probability that the present value of net benefits is positive or higher than a predetermined benchmark.

20.5 THE CBA CONTROVERSY

The use of CBA in environmental regulation is a controversial issue. *The Economist* (2014) attributes the scrutiny which CBA is accorded to the observation that "CBA has become integral to large projects." Green fanatics believe that there is no need for using CBA because cost is irrelevant if the objective is to save the environment. This argument is not valid because only a limited amount of resources can be channeled to environmental protection, which means that cost must be relevant. On the other hand, the opponents of regulation resort to CBA to reject environmental projects. For what in reality is a viable project, rejection is recommended by showing that the present value of net benefits is negative, which can be obtained by manipulating estimated benefits and costs as well as the discount rate and inflation rate. On paper there are more arguments against, rather than for, CBA but the question that the opponents of CBA fail to answer is the following: what is the alternative?

Consider the following arguments for CBA that have been suggested in the literature:

1. The procedure offers a way of achieving superior environmental results at a lower overall cost to the society than other available approaches.
2. It makes the decision-making process more objective and transparent by presenting an explicit set of assumptions and detailed description of the procedures underlying regulatory action.
3. It provides a compromise between the wishes of environmental fanatics and those of the opponents of regulation. No project is accepted without considering costs and no project is rejected just because it is costly.
4. The cost–benefit approach is better equipped than the zero-risk or technology-based approaches at addressing distributional concerns. Unlike other approaches, CBA accounts explicitly for regulatory costs, including the costs that fall on disadvantaged groups. It is best equipped to evaluate the economic efficiency and equity implications of regulatory decisions. We will elaborate on the distributional effects of environmental regulation later on.
5. It is neutral in the sense that it is not biased toward the acceptance or rejection of a project. This is another point that we shall elaborate on.
6. There is no viable alternative. We will consider some potential alternatives later on.

The arguments against the use of CBA are numerous. These include the following, exactly as they appear in the literature:

1. CBA is a deeply flawed method that repeatedly leads to biased and misleading results.
2. The technique offers no clear advantage in making regulatory policy decisions and often produces inferior results (compared to other approaches) in terms of both environmental protection and overall social welfare.
3. Cost–benefit studies are costly in terms of time and money.
4. Superior, time-tested regulatory alternatives are available. We will examine some of these alternatives later on.
5. There are important differences between economic regulation and environmental regulation that must not be overlooked. For example, it is much more difficult to evaluate the benefits of environmental regulation than those of economic regulation.
6. CBA extrapolates the technique of discounted cash flows from finance to environmental decision making. This criticism encompasses two points: the difficulty of estimating the costs and benefits of environmental regulation and the appropriateness of discounting net benefits.

7. CBA is intended to perform for public policy a calculation that markets perform for the private sector.
8. Costs are easier (or less difficult) to measure than benefits, but their quantifiability makes them no more certain or reliable.
9. CBA rests on a series of assumptions and value judgments that cannot be remotely described as objective.
10. Insisting on CBA may represent undemocratic attempts to reorient legislative mandates.
11. Regulatory agencies do not have access to the information pertaining to alternative products and processes and the associated costs—as a result, it is not an easy task to implement the engineering approach to the measurement of compliance cost.
12. It is doubtful if regulatory agencies are capable of estimating the benefits of environmental regulation, which are more difficult to estimate than compliance costs.
13. Monetizing health benefits may not be feasible or even desirable—for example, a human life or limb does not have an established unique market value (although some dollar figures have been put on human organs and life in terms of lost income).
14. It is difficult to assess the effect of a given reduction in a specific pollutant on the health of affected individuals. This task requires some understanding of the causal relation between that pollutant and an array of health outcomes.
15. There is the moral issue of whether or not valuations should change if a rule would mostly benefit future generations. Discounting implies that a human life today is more valuable than a human life in the future. Another moral issue is that of judging mortal danger on the basis of rates of return.
16. The technique is inferior to other methods because the latter do not require the translation of the benefits of regulation into dollar values.
17. The technique ignores the question of who suffers as a result of environmental problems, which means that the conclusions derived from CBA tend to reinforce the existing patterns of economic and social inequality.
18. The use of the efficiency criterion leads to the choice of regulation that produces benefits to the winners that exceed the costs incurred by the losers.
19. The technique is nonneutral, and this is why the opponents of regulation like it.
20. Some would argue that the technique is complex, but this may be a reflection of the complexity of public choice problems.

Some brief responses to these points can be put forward at this stage. On the issue of the time and money required to conduct cost—benefit studies, the question should be not whether or not CBA is costly but whether or not it is cost effective. As far as the difficulty of evaluating benefits is concerned, no one disputes the proposition that evaluating the benefits is hard, but it does not necessarily mean that we should abandon the technique. As for the difficulty of estimating costs, it is not clear how alternative procedures are immune from this criticism.

Between the opponents and proponents of the use of CBA in environmental regulation, there are those who express mixed feelings about the technique—the same feeling as some people have of democracy, that it is the least bad of all available systems of government. Gayer (2011) makes this point clear by concluding as follows:

> Americans rely on environmental regulations to protect the air we breathe and the water we drink. These rules, however, have significant costs to American workers and businesses. Cost-benefit analysis does and should play a central role in designing and choosing among regulatory options. However, the ability of cost-benefit analysis to ensure that Americans get the largest bang for their buck from environmental regulations is hampered by reliance on less-credible empirical research, improper assumptions, and insufficient incorporation of cost-benefit analysis into the decision making process.

It seems that Gayer is in favor of using CBA subject to some qualifications to ensure the soundness of decisions. Ashford (1981) expresses a similar opinion of mixed feelings by arguing that "cost-benefit analysis can be a useful tool, but some regulatory reformers would have us apply it as an indiscriminate decision making rule."

20.6 A DISCUSSION OF CONTROVERSIAL ISSUES

We will now discuss some of the points raised in the cost–benefit controversy. To start with, while discounting makes sense when we talk about the value of money, discounting human life does not sound right. For example, at a discount rate of 5%, one life today is worth almost 40 billion lives 500 years from now—this sounds rather unethical. Heinzerling and Ackerman (2002) criticize the discounting of costs and benefits because discounting "systematically and improperly downgrades the importance of environmental regulation." They make the following points against discounting: (1) while discounting makes sense in comparing alternative financial investments, it cannot be used reasonably to choose between preventing noneconomic harms to present generations and preventing similar harms to future generations and (2) it tends to trivialize long-term environmental risks, minimizing the very real threat that the society faces from potential catastrophes and irreversible environmental damage. Discounting implies that catastrophic events are tolerable if they happen far into the future. It also implies total indifference with respect to future generations and even the selfishness of the present generation.

Heinzerling and Ackerman (2002) also criticize CBA because the procedure involves "pricing the priceless." They write:

> In cost-benefit analysis, therefore, both the cost of, say, putting a scrubber on a power plant to reduce air pollution and the benefits of doing so, including the saving of human lives and the prevention of debilitating and painful diseases, are presented in terms of dollars.

A big portion of the criticism directed at cost–benefit analysis pertains to the difficulty of measuring the benefits of environmental regulation. Heinzerling and Ackerman (2002) are particularly critical of the contingent valuation method, which they describe as "essentially a form of opinion poll" whereby researchers ask a cross-section of the affected population how much they would be willing to pay to preserve or protect something that cannot be bought in a store. They are also critical of valuation based on what people are willing to pay by observing their behavior in other markets, which is used to assign a dollar value to risks to human life (for example, by calculating the "wage premium" paid to workers who accept more risky jobs). For Heinzerling and Ackerman (2002), "human life is the ultimate example of a value that is not a commodity and does not have a price."

Measurement errors can be attributed to several factors, the first of which is overreliance on data from previous projects, which may not be appropriate for the project under consideration. The second factor is that CBA is conducted by a team with a dominant leader who tends to impose his or her subjective views. There is also the problem of confirmation bias among project supporters who look for a reason to proceed. Last, but not least, it is typically the case that heuristics are used inappropriately to assign monetary values to costs and benefits.

A problematical aspect of CBA is the distributional consequences and equity issues as the decision rules do not consider how the benefits and costs are distributed across segments of the society. Distributional considerations give rise to the concepts of environmental justice and racism, which are relevant to the skewness or otherwise of the distribution of costs and benefits. When the economic basis for CBA was almost in place in the 1960s, there was no way to deal with the distribution of costs and benefits among different income groups, ethnic groups, regions, or even between producers and consumers.

The use of the efficiency criterion leads to the choice of regulation that produces benefits to the winners that exceed the costs incurred by the losers. If winners could compensate losers, such that no one is made worse off by the introduction of regulation, the appropriate thing to do would be to select the most efficient regulation according to the results of CBA and then the tax code is used to redistribute net benefits equitably. In practice, however, regulatory policies do not encompass the objective of offsetting transfers through the tax code, which means that a consequence of the efficiency criterion is inequity. Alternatively, "distributional weights" can be used to account for the distributional effects in the calculation of costs and benefits. According to Pearce (1998), "the distributional issue was perhaps never resolved."

Heinzerling and Ackerman (2002) criticize CBA on the grounds that the technique ignores the question of who suffers as a result of environmental problems, which means that the conclusions derived from CBA tend to reinforce the existing patterns of economic and social inequality. They argue that "cost-benefit analysis treats questions about equity as, at best, side issues, contradicting the widely shared view that equity should count in public policy." As a result, CBA would justify the imposition of greater environmental burdens on the poor than on the wealthy, because the poor are likely to express less willingness to pay to avoid environmental harms, simply because they have fewer resources. Consequently, the poor get poorer. The counterargument is that abandoning CBA, or using an alternative project evaluation technique, will not make the poor less poor and the rich less rich. The problem is not the use of CBA but failure to use the tax code and the whole institutional setup that preserves inequality, not only with respect to the effect of regulatory measures. For example, advocates of environmental justice argue that minority communities typically undertake environmentally hazardous activities because they have few alternatives and/or not aware of the underlying risks (Rhodes, 2003). Environmental discrimination refers to a situation where the poor are burdened with more than their fair share of environmental risks while enjoying a small portion of the benefits of environmental regulation.

Gayer (2011) argues that concern about the distributional effects of environmental regulation does not invalidate the desirability of CBA, suggesting that the procedure should account for the benefits and costs of regulatory options on specific demographic groups. The cost–benefit approach, he argues, "is even better equipped than the zero-risk or technology-based approaches at addressing distributional concerns." Unlike other approaches, CBA explicitly accounts for regulatory costs, including those costs that fall on disadvantaged groups. He suggests that CBA is better equipped to account for the effects of regulation on different population groups. For example, a naïve approach might promote an expensive environmental regulation to address pollution in a disadvantaged neighborhood, resulting in an increase in the demand for housing in that neighborhood and pricing the disadvantaged out of the rental market (for example, Banzhaf and Walsh, 2008; Sieg et al., 2004). He concludes that CBA is best equipped to evaluate the economic efficiency and equity implications of regulatory decisions.

Now we turn to the issue of neutrality. As we have seen, CBA may be used to prove that a regulatory proposal is worthwhile or otherwise and that the opponents and proponents of regulation may use CBA to support their claims. In practice, however, there is a tendency for those opposing regulation to insist on the use of CBA. No one opposed regulation more than Ronald Reagan—yet he initiated the need to use the procedure by the US regulatory agencies. Pearce (1998) suggests that this is the case when he argues:

In particular, there was a political backlash against perceived excessive regulation and against clear inconsistencies and irrationalities in public controls. In the view of those concerned about regulation, CBA helps to prevent over-regulation.

Heinzerling et al. (2005) seem to disagree with the view that CBA is neutral, suggesting that the technique was used by the George Bush administration as a "powerful weapon" against the regulatory initiatives proposed by the EPA and other regulatory bodies. They point out that CBA is rooted in the concept of economic efficiency and the doctrines of neoclassical economics, which is why "economics should not be allowed to make the final decision on which regulation will take effect and which will not." For example, Hahn (1996) finds that only 12 out of 40 final regulations and 9 out of 21 proposed regulations passed CBA. He concludes that regulatory agencies have an incentive to exaggerate benefits (but then the regulated firms have an incentive to exaggerate costs).

The opponents of the use of CBA in environmental regulation offer no viable alternative. The points raised by them are equally valid for other techniques, in which case the only alternative is to take decisions on the hunches of a benevolent dictator. So, it may be worthwhile to consider some alternatives. For example, in cost-effectiveness analysis, monetary values are not assigned to outcomes involving improvement in health. But as health improves less is spent on health care and less income is lost because of illness, which means that monetary values can be assigned to improvement in health. The same is implied by insurance against illness. In cost-minimization analysis, costs are calculated with the objective of choosing the least costly course of action; hence the problem of estimating cost is applicable here. And since it is unlikely to find reliable equivalence between two projects, cost-minimization analysis is inappropriate. Cost-utility analysis is similar to cost-effectiveness analysis.

In economic impact analysis, economic impact is typically measured in terms of changes in economic growth (output or value added) and the associated changes in employment and income, which involves the problem of attributing changes in growth to environmental regulation. Economic impact analysis involves more serious measurement problems because it is broader than CBA in the sense that it counts business relocation and the resulting spending multiplier impacts on a given region. The same applies to the social return on investment analysis metric that is used for measuring extra-financial value (that is, environmental and social value not reflected in conventional financial accounts) relative to the resources invested in a project. The length of time and resources it takes to carry out this kind of analysis is significant, depending on the scope of the analysis and the extent to which outcome data are already available.

Stavins (2009) wonders whether CBA "plays a truly useful role in Washington, or is it little more than a distraction of attention from more important perspectives on public policy, or—worst of all—is it counter-productive, even antithetical, to the development, assessment, and implementation of sound policy in the environmental, resource, and energy realms." His overall conclusion is that "cost-benefit analysis can play an important role in legislative and regulatory policy debates on protecting and improving the natural environment, health, and safety." Furthermore, he points out that regulators often face considerable uncertainty, in which case they have to use ranges (perhaps with probabilities) for both costs and benefits, rather than specifying a single number. Arrow et al. (1996) suggest that CBA has a potentially important role to play in helping inform regulatory decision making, but it should not be the sole basis for decision making.

Unlike environmental regulation, homeland security (including the military and spy agencies) receives funding without CBA, presumably on the grounds that the level of risk is intolerable. As for financial regulation, it has been mostly deregulation, put in place without CBA (such as the abolition of the Glass—Steagall Act in the United States). In both cases, CBA is not used, not because benefits are difficult to quantify but because of the power of the military-industrial complex and the financial lobby. The benefits derived from spending on homeland security are not difficult to measure. Excluding the possibility of a foreign invasion, the number of lives saved can be estimated from the frequency and severity of previous attacks. It is doubtful if these benefits are remotely comparable to the number of lives that can be saved by environmental regulation.

While Heinzerling and Ackerman (2002) are justified in their apprehension about the use of CBA, the alternative of eliminating all environmental risks irrespective of cost is infeasible—it may mean reducing risk to zero, which amounts to closing down and putting an end to economic activity. A zero-risk regulatory approach is impossible, which makes the use of CBA the least bad course of action. A middle-ground view on CBA is the following. Despite its shortcomings, we cannot ignore the cost of regulation. Protecting the environment is one of many areas requiring resource allocation, which means that we should consider both costs and benefits—this sounds like using CBA. However, we do not have to use CBA in its strict form whereby we come up with a single figure for the present value of net benefits or the benefit—cost ratio and base a critical decision on that figure alone. There is no doubt that CBA has its limitations—those who conduct CBA must acknowledge these limitations.

20.7 CONCLUSION

CBA, which has been a cornerstone of the regulatory approval process, is an essential tool that is used to weigh benefits against costs. In practice, however, the use of CBA is sometimes inadequate, resulting in the adoption of regulatory measures that impose higher costs than necessary or the rejection of others that should be accepted. The evaluation process contains significant elements of subjectivity and value judgment. This is why CBA can be used to support the arguments of both the opponents and proponents of regulation. And this is why this approach has been condemned by both (opponents and proponents).

The use of CBA to appraise environmental regulation projects is indeed a controversial issue. If it is poorly executed, critics will use poor practice as a basis for criticizing the technique. Although environmental regulation saves life and limb, that does not come at no cost, hence the argument for the use of CBA to weigh costs against benefits. In some cases, however, risk—benefit analysis is more appropriate—that is, if the level of risk cannot be tolerated, risk must be contained at any cost. In other situations, risk can be tolerated without the need for CBA. It is foolish to suggest that CBA should be used to decide whether or not we keep on using cars, given that thousands of lives are lost in car accidents. Then one can only wonder why military expenditure is not subject to CBA when it is justified in terms of saving lives. And while CBA is required for a project to reduce smog in the atmosphere, giving banks (classified as too big to fail) billions of taxpayers' dollars does not require going through the procedure. Is it that the environment is placed at the bottom of our list of priorities just because big businesses complain that environmental regulation is harmful to their bottom lines?

REFERENCES

Arrow, K.J., Cropper, M.L., Eads, G.C., Hahn, R.W., Lave, R.G., Portney, P.R., Russell, M., Schmalensee, R., Smith, V.K., Stavins, R.N., 1996. Is there a role for benefit-cost analysis in environmental, health and safety regulation? Science 272 (5259), 221–222.

Ashford, N.A., April 1981. Alternatives to cost-benefit analysis in regulatory decisions. Annu. N.Y. Acad. Sci. 363, 129–137.

Banzhaf, H.S., Walsh, R.P., 2008. Do people vote with their feet? An empirical test of Tiebout's mechanism. Am. Econ. Rev. 98 (3), 843–863.

Cooley, P.L., Roenfeldt, R.L., Chew, I.-K., 1975. Capital budgeting procedures under inflation. Financ. Manag. 4 (4), 18–27.

Dupuit, A.J., 1844. On the Measurement of the Utility of Public Works (Translated in R. Barback, International Economic Papers, 2, 1952).

Dupuit, A.J., 1853. On utility and its measure—on public utility. J. Econ. 36 (1), 1–27.

Gayer, T., 2011. A Better Approach to Environmental Regulation: Getting the Costs and Benefits Right. The Hamilton Project. Discussion Paper 2011–2016.

Hahn, R.W., 1996. Regulatory reform: what do the Government's numbers tell us? In: Hahn, R.W. (Ed.), Risks, Costs and Lives Saved: Getting Better Results from Regulation. Oxford University Press, Oxford.

Hanley, N., Spash, C., 1993. Cost-benefit Analysis and the Environment. Edward Elgar, Cheltenham.

Heinzerling, L., Ackerman, F., 2002. Pricing the Priceless: Cost-Benefit Analysis of Environmental Protection. Working Paper. Georgetown Environmental Law and Policy Institute.

Heinzerling, L., Ackerman, F., Massey, R., 2005. Applying Cost-benefit Analysis to Past Decisions: Was Environmental Protection Ever a Good Idea? Georgetown Law Faculty Publications. Available at: http://scholarship.law.goergetown.edu.facpub/323/.

Hultkrantze, L., Lindberg, G., Andersson, C., 2006. The value of improved road safety. J. Risk Uncertain. 32 (2), 151–170.

Mitchell, R.C., Carson, R.T., 1989. Using Surveys to Value Public Goods: The Contingent Valuation Method. Resources for the Future, Washington, DC.

Moosa, I.A., Ramiah, V., 2014. The Costs and Benefits of Environmental Regulation. Edward Elgar, Cheltenham.

Pearce, D., 1998. Cost-benefit analysis and environmental policy. Oxf. Rev. Econ. Policy 14 (4), 84–100.

Rhodes, E.L., 2003. Environmental Justice in America. Indiana University Press, Bloomington.

Sieg, H., Smith, V.K., Banzhaf, H.S., Walsh, R.P., 2004. Estimating the general equilibrium benefits of large changes in spatially delineated public goods. Int. Econ. Rev. 45, 1047–1077.

Smith, K., Krutilla, J.V. (Eds.), 1982. Explorations in Natural Resource Economics. John Hopkins University Press, Baltimore.

Stavins, R., 2009. Is Benefit-Cost Analysis Helpful for Environmental Regulation? Available at: http://www.robertstavinsblog.org/2009/07/08/is-benefit-cost-analysis-helpful-for-environmental-regulation/.

Sunstein, C.R., 2012. The Stunning Triumph of Cost-Benefit Analysis. Bloomberg. Available at: http://www.bloomberg.com/news/2012-09-12/the-stunning-triumph-of-cost-benefit-analysis.html.

The Economist, 2014. Why Doing a Cost-Benefit Analysis Is Harder than It Looks. Available at: http://www.economist.com/blogs/economist-explains/2014/04/economist-explains-13.

Vassanadumrongdee, S., Matsuoka, S., 2005. Risk perceptions and value of a statistical life for air pollution and traffic accidents: evidence from Bangkok, Thailand. J. Risk Uncertain. 30 (3), 262–287.

CHAPTER 21

The Crowdfunding of Renewable Energy Projects

Davide Bonzanini, Giancarlo Giudici, Andrea Patrucco
School of Management, Politecnico di Milano, Milan, Italy

Contents

21.1 INTRODUCTION

Funding is a crucial process for renewable energy projects. Uncertainty about future cash flows, due to any inefficiency of the plant, or to biased estimation of the productivity, or simply to low expected profitability, may prevent entrepreneurs and banks from investing in the project.

Crowdfunding, i.e., tapping the crowd of Web surfers in order to collect money, might help to reduce and overcome this gap. Through the Internet, sponsors of "green" projects (mostly nonprofit organizations and companies operating in the energy field) are allowed to get in touch with funders who seek for investment opportunities but also believe in environmental sustainability and in the transition to renewable energy. Thanks to its rapid diffusion around the globe, crowdfunding has been crucial for the success of several eco-friendly projects, such as the installation of plants for the production of renewable energy, the promotion of energy efficiency, sustainable agriculture and tourism, waste reduction.

This chapter shows the results of a study about crowdfunding in the renewable energy field, aimed at discovering which are the main determinants of success for this type of financing. Crowdfunding has the potential to represent the disintermediation of the banking system in the renewable energy business as funders and sponsors get in touch directly. Due to the recent global crisis, financial constraints for entrepreneurial projects have risen (Campello et al., 2010) enhancing the attention of entrepreneurs and policy makers on the crowdfunding phenomenon. Crowdfunding offers several advantages to sponsors of "green" projects. First, it allows to reduce and share the risk among several

Handbook of Environmental and Sustainable Finance
ISBN 978-0-12-803615-0

small investors, avoiding bank debt, and the covenants often requested by credit institutions. Second, crowdfunding may finance very small projects that are typically neglected by professional investors due to the fixed costs of due diligence activity. Then, crowdfunding also helps to overcome the "not in my backyard" phenomenon, through a significant involvement of local communities in the development and financing of the project, through the Web. Hostility may be reduced allowing the local population to become the owner and online controller of a green project. This helps to foster the effective implementation of renewable energy plants.

In detail, we collected data for 84 projects posted from December 2013 to June 2014 on 13 crowdfunding platforms specialized in "green" initiatives. We focus on projects where money is provided in exchange for loans (debt-based) and/or shares (equity-based). All the projects aim at realizing plants for the production of renewable energy (mostly through solar panels). The average requested funding is equal to €452,000. The aggregate power of the plants exceeds 94 MW. The sample projects raised €35 million, with an average success rate of the crowdfunding campaign equal to 95.56%.

We investigate the determinants of the campaign success in order to point out the variables that positively affect the probability that a "green" project will be funded by investors. The study has a relevant impact on the industry of renewable energy generation, because crowdfunding has the potential to represent an interesting source of capital, combining the opportunity of a profit with the desire to contribute to climate action initiatives.

We find that the expected profitability of the project, the presence of a "bonus" for investors, as well as the platform reputation and commitment (in terms of track record and direct participation into the financing) are significantly correlated with the probability of success. The "social capital" accumulated by the promoters of the projects through social network contacts is also important to determine success. Finally we find that clauses providing benefits for local communities in terms of royalties or rewards play a role, but only if the expected profitability of the project is low.

We conclude that investors seem not to take into account the traditional measures of project risks (such as the leverage, or the investment size), probably because their contribution in the project is limited to a low amount of money, if compared to their stock available for investments. Expected profits are found to be the most relevant determinant, while other platform-specific variables, as well as the promoters' activism on the Web, significantly improve the probability of the fundraising success.

The remainder of the chapter is as follows. In Sections 21.2 and 21.3 we review the nascent industry of crowdfunding and the related literature, respectively. Section 21.4 presents statistics about the sample projects. Section 21.5 is devoted to the development of the research hypotheses and to the empirical analysis. Finally Section 21.6 concludes.

21.2 CROWDFUNDING: FACTS AND FIGURES

Expanding on the idea of crowdsourcing (Afuah and Tucci, 2012), crowdfunding implies mobilizing the crowd to finance projects which are posted by their proponents either on personal pages or on dedicated Web sites, known as crowdfunding platforms. Funding is provided by Web users in exchange for some monetary claims on the project revenues, an economic reward, or simply for donation. Suggestions and feedbacks for improving the projects are also welcome by proponents. In other words, crowdfunding involves "an open call, through the Internet, for the provision of financial resources either in form of donation or in exchange for the future product or some form of reward and/ or voting rights" (Belleflamme et al., 2014).

Crowdsourcing and crowdfunding share indeed the same target: the Internet users. However, crowdfunding differs from crowdsourcing, as its main aim is to solicit the contribution of money by the crowd of Internet users, while crowdsourcing focuses mainly on the provision of ideas, feedbacks and solutions to firms' problems, and innovative projects.

The claims of crowdfunders in the project outcome determine different business models of crowdfunding:

- equity-based: crowdfunders become shareholders of the project and have the right to share the profits (sometimes with differential voting power and dividend distribution rights, compared to the sponsor);
- lending-based: funds are paid back and the crowdfunders have the right to receive an interest payment contractually promised;
- donation-based: funds are provided with no other compensation, for philanthropic or sponsorship proposal;
- reward-based: funds are provided in exchange for nonmonetary benefits (for example, a release of the product to be realized, or gifts, vouchers …).

Crowdfunding first gained popularity as a way to fund creative, philanthropic, and social endeavors. This popularity still prevails, but crowdfunding's application for business has gained traction as well. According to Massolution (2013), in 2013 crowdfunding platforms around the world raised a total capital of $5.1 billion, with a projection of $8 billion in 2014. The volume was virtually 0 before 2009.

In the US, reward-based crowdfunding platforms like Kickstarter and Indiegogo raised more than $1 billion altogether and supported more than 100,000 projects in raising funds. The JOBS Act in 2012 also introduced equity crowdfunding; as soon as the law will be fully operating, a further development in the industry is expected.

In Europe the European Commission adopted in March 2014 a communication on crowdfunding entitled "Unleashing the Potential of Crowdfunding in the European Union," detailing an agenda for promoting the industry best practices, raising awareness, monitoring the development of crowdfunding markets and national legal frameworks, and assessing whether any legislative action is necessary. In the communication, crowd-funding is defined as a "one of the newly emerging financing models that increasingly contribute to helping start-ups move up the 'financial escalator' and contribute to building a pluralistic and resilient social market economy."

Focusing on renewable energy projects, the evidence shows that in 2013 funds raised by energy and environmental projects through lending-based and equity-based crowdfunding have been equal to $292 million and $45 million, respectively (Massolution, 2013).

21.3 REVIEW OF THE LITERATURE

Scholars and practitioners agree that crowdfunding holds the potential to change traditional funding practices in general, but—more interestingly—especially for entrepreneurial projects and start-ups by complementing financial support provided by professional investors like venture capitalists and business angels (Agrawal et al., 2013; Giudici et al., 2013; Giudici, 2015). Brabham (2008) and Kleemann et al. (2008) acknowledge that the application of modern information technologies is a prerequisite to the development of crowdfunding. The Web infrastructure is mandatory to reach the networks of consumers and allows to get in contact with masses at virtually no cost. Crowdfunding has become feasible to everyone with the diffusion of the social networks and of crowdfunding platforms where promoters are allowed to present funding opportunities and business plans, and ask for money through conventional payment infrastructure like bank drafts, credit card, or Paypal.

Crowdfunding can help small business and community organizations to overcome the finance gap, mobilizing people and resources through the Web, and taking advantage of the opportunity to reduce transaction costs both for borrowers and for investors, thanks to the collective knowledge shared by participants (Surowiecki, 2004). Lévy (1997) provides a definition of collective intelligence, according to which every individual has a specific know-how that, combined with the one of every group member, allows to achieve the highest level of superior knowledge. Howe (2008) posits that the crowd seems to be more efficient than individuals and small group of people, thanks to its characteristics of organic interactions overcoming the simple sharing and accumulation of knowledge. This assumption can be particularly important in the case of "green" projects, where uncertainty about the technological feasibility and efficiency, the effects on the environment, and the cash flow profitability cannot be easily estimated especially for plants exploiting novel technologies.

Franke and Klausberger (2008) notice that the crowd is a part of a multilateral system and incentives aimed at fostering funders' motivation are needed. A successful showcase of any crowdfunding campaign should be communicative and attractive. Video clips, presentations, banners, and other tools are usually used for that purpose.

Potential investors attribute an important value to proactive communication and to the release of information (e.g., ranking, reports; see Ward and Ramachandran, 2010). Gerber et al. (2011) posit that crowdfunders like to feel part of the community of people supporting a project, to exchange ideas about it, and discuss it on social networks. Proponents' decision to use crowdfunding to finance a project stems from the need for capital, from the possibility to obtain feedbacks and suggestions by crowdfunders, from the fact that posting a project on the Internet and opening it to the scrutiny of a large crowd can be perceived as a signal of quality and trust, and from the possibility of relying on social capital to raise funds. According to Schwienbacher and Larralde (2010) crowdfunders are motivated by both financial goals and intrinsic motivations. Investors aim at participating into innovative projects, obtaining recognition and even fun, joining a community with similar priorities, being among the pioneers of a new business, and expand their personal network.

Burtch et al. (2013) show that the period of the fundraising campaign has an indirect impact on the outcome: the larger is the visibility of the project, the greater are the probabilities of a success. Achieving the critical mass in the early days from the funding start date is an important determinant for successfully closing the campaign. Agrawal et al. (2011) and Mollick (2014) shows that geography matters in determining both the typology of projects submitted to the crowd (that reflects the cultural background of the area) and the funding success. Last but not least, the proponent "social capital" (measured by the number of contacts on social networks) has been identified as an important determinant of success in a crowdfunding campaign (Giudici et al., 2014).

Crowdfunding has its dark side as well. It has been highlighted (Giudici, 2015) that crowdfunders have little bargaining power toward the project promoters. Contrary to professional investors like banks or venture capitalists, they are not able to impose covenants or agreements. The risk of fraud is not negligible and information asymmetries are huge. Crowdfunding campaigns are not associated to offering prospectuses approved by any public authority. Reputation is the only asset on which funders may rely. Equity-based crowdfunding also requires administrative and governance challenges, in order to manage a plethora of shareholders; this will inevitably subtract time and resources to the project. Such criticalities should also be considered, alongside triumphalism for crowdfunding "democracy."

21.4 THE SAMPLE

This section introduces some statistics about projects aimed at producing renewable energy posted on crowdfunding platforms specialized in "green" initiatives. In particular, we focus on projects based on the lending and equity models, i.e., offering a return to investors (either in the form of a loan or in the form of share issue). As at June 2014, around the world we identified 13 crowdfunding platforms specialized in renewable energy projects. The portals are listed in Table 21.1. We also found other five platforms but at the time of the data collection they were still inactive and did not publish any project. The latter Web sites are not listed in Table 21.1.

Geographically, the platforms are predominantly established in Europe (10 out of 13) and in particular in countries where environmental issues are perceived as important such as Germany (5) and the Netherlands (3). The majority of the platforms (69.2%) adopt a lending-based model, two of them adopt the equity-based model while the remaining two are classified as "hybrid" because they allow the publication of projects characterized by different models.

In 10 cases the platforms allow fundraising only on an "all or nothing" basis; this means that money provided by the crowd is available for the promoters only if the initial target amount is reached (or within a certain margin). Other platforms adopt also a "keep it all" model, according to which the promoters are allowed to receive the money for any amount.

Crowdfunders are offered the possibility both to contribute to the reduction of the emission of carbon dioxide, funding the installation of renewable energy plants, and to enjoy a return on the investment. In fact, the energy produced is sold to the national grid at a price that is generally comprehensive of a feed-in tariff (FiT), a mechanism designed by government authorities to encourage investments in renewable energy. The incentive varies depending on the renewable sources used, the technical characteristics of the plant, and may change across time. Revenues will be used to pay back costs and then to reward crowdfunders and any other investors. Under the lending-based system, investors are entitled to receive a contractual interest (fixed or floating) while in an equity-based scheme, investors generally become members of a cooperative and thus co-owners of the plant. The partnerships usually pay annual dividends to their members.

Table 21.1 List of Crowdfunding Platform Specialized in Renewable Energy Projects

Crowdfunding Model	Platform Name	Country
Lending-based	Abundance Generation	The UK
	CollectiveSun	The US
	Duurzaam Investeren	The Netherlands
	Econeers	Germany
	Eollice	Chile
	GreenVesting	Germany
	LeihDeinerUmweltGeld	Germany
	Lumo	France
	Mosaic	The US
Equity-based	crowdEner.gy	Germany
	Windcentrale	The Netherlands
Hybrid	Green Crowd (lending/donation)	The Netherlands
	Green Crowding (equity/reward)	Germany

We collected information about the projects related to the generation of renewable energy posted by the aforementioned crowdfunding platforms. We identified 84 crowd-funding campaigns: 13 adopt equity-based schemes while 71 adopt the lending-based system. Table 21.2 shows the distribution of the projects among the different active platforms, the average funding target, and the aggregate power of the proposed plants. The leading platforms are Mosaic (US, 25 campaigns) and Green Crowd (The Netherlands, 10 campaigns). In terms of total installed power, the largest contribution came from Mosaic again (34 MW) followed by LeihDeinerUmweltGeld (almost 32 MW). Yet we underline that projects posted on the latter portal raise money through both crowd-funding and traditional sources (banks and funds).

The average funding target is very dispersed and varies from €7500 (Lumo) to €2 million (Windcentrale). The mean value for the total sample is €452,000. Also the size of the plant is very dispersed. Solar photovoltaic plants are generally smaller while wind farms are characterized by larger power. The average power size is 1122 kW and the total power for the sample size is larger than 94 MW. We can estimate that such potential may contribute to reduce the emission of more than 2200 tons of carbon dioxide in the atmosphere each year, and to save the equivalent consumption of 29,000 tons of oil each year.

Figure 21.1 splits the sample projects by energy source. Photovoltaic plants account for 76% of the sample. Wind turbines represent 17% of the sample, mainly due to the contribution of the platform Windcentrale. Then biomass plants—sponsored mainly by Econeers and LeihDeinerUmweltGeld—account for 5% of the sample. The latter

Table 21.2 Number of Crowdfunding Campaigns for Renewable Energy Projects, Average Funding Target, and Average Plant Size (Power), by Crowdfunding Platform. Sample: 84 Projects Posted from December 2013 to June 2014

Platform	Published Projects	Average Target (k€)	Total Plant Power (kW)
Abundance Generation	9	950	3973
CollectiveSun	2	34	23.5
crowdEner.gy	4	41	256
Duurzaam Investeren	2	450	14,000
Econeers	4	613	1676
Eollice	4	18	30
Green Crowd	10	58	1133
Green Crowding	1	58	26
GreenVesting	6	136	1477
LeihDeinerUmweltGeld	7	326	31,897
Lumo	2	7.5	55
Mosaic	25	217	34,024
Windcentrale	8	2047	6150
Total Sample	84	452	94,185

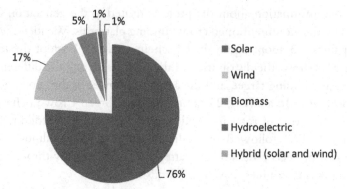

Figure 21.1 Percentage of crowdfunding campaigns by renewable source. Sample: 84 projects posted from December 2013 to June 2014.

platform has also concluded a successful crowdfunding campaign to finance the installation of a wind turbine in a solar park. And last but not least, we have identified one project in the hydroelectric field posted by Green Crowd.

Table 21.3 lists statistics about the total and average capital raised by every single platform, as well as the numbers of funders (when disclosed by the crowdfunding portal). The equity-based Dutch platform Windcentrale is by far leading in both the figures (more than €16 million collected from 15,157 funders). The lending-based platforms

Table 21.3 Crowdfunding Campaigns: Total and Average Capital Raised, Total and Average Number of Funders. Sample: 84 Projects Posted from December 2013 to June 2014

Platform	Raised Capital (k€)	Raised Capital per Project (k€)	Funders	Funders per Project
Abundance Generation	8135	904	3701	411
CollectiveSun	68	34	n.a.	n.a.
crowdEner.gy	165	41	110	28
Duurzaam Investeren	900	450	122	61
Econeers	1231	308	976	244
Eollice	44	11	83	21
Green Crowd	403	40	349	35
Green Crowding	58	58	n.a.	n.a.
GreenVesting	814	136	123	21
LeihDeinerUmweltGeld	2184	312	n.a.	n.a.
Lumo	15	8	79	40
Mosaic	4757	190	7171	287
Windcentrale	16,373	2,047	15,157	1895
Total Sample	35,147	418	>27,872	n.a.

Abundance Generation and Mosaic are next. The total capital raised by the sample campaigns exceeds €35 million. Taking into account also undisclosed numbers, we can estimate that about 30,000 participants invested in the sample projects.

Table 21.4 looks at the success of the crowdfunding campaigns. We define a successful campaign as a campaign that reaches the minimum funding target initially defined by the promoters. Remarkably, the success rate seems very high. Most of the platforms were able to fund all their projects in the renewable energy sector. Only the German platform Econeers seems to exhibit a poor success rate with only two success cases over four proposals. On average, the sample projects reach a funding level equal to 95.56% of the target and 88.09% of the campaigns are successful. We underline that projects presented in some platforms (for example, Mosaic and Green Crowd) will be for sure realized, even if the crowdfunding campaign is not successful, because there is a commitment from other financiers to provide any lacking capital.

With no ambition to propose a statistically significant comparison, we highlight that the average success ratio for generic crowdfunding campaigns under the lending and equity-based scheme are equal to 92% and 44%, respectively (Massolution, 2013).

It is interesting to analyze the expected rate of return for investors. Clearly we have to distinguish between lending-based and equity-based campaigns. Table 21.5 reports the offered interest rate for 71 lending-based campaigns. The values range from 3.4% to 14% (this latter number refers to the Chilean platform Eollice). Numbers are difficult to be compared, because interest rates refer to different countries, with different

Table 21.4 Performance of Crowdfunding Campaigns: Successful versus Unsuccessful Projects. Sample: 84 Projects Posted from December 2013 to June 2014

Platform	Average Completion Rate (%)	Percentage of Successful Projects (%)
Abundance Generation	96	100
CollectiveSun	100	100
crowdEner.gy	100	100
Duurzaam Investeren	100	100
Econeers	56.25	50
Eollice	100	100
Green Crowd	97	87.50
Green Crowding	100	100
GreenVesting	100	100
LeihDeinerUmweltGeld	100	100
Lumo	100	100
Mosaic	95	88
Windcentrale	100	100
Total Sample	95.56	88.09

Table 21.5 Offered Expected Interest Rate in Projects Posted by Lending-Based Platforms. Sample: 71 Projects Posted from December 2013 to June 2014

Offered Interest Rate	Minimum (%)	Median (%)	Maximum (%)	Average (%)
Abundance Generation	6.65	7.35	9	7.38
CollectiveSun	8	8	8	8
Duurzaam Investeren	6	6	6	6
Econeers	4.5	5	5	4.88
Eollice	14	14	14	14
Green Crowd	4	4	8	4.94
GreenVesting	4.8	5.5	6.2	5.5
LeihDeinerUmweltGeld	4	5	6	4.81
Lumo	4.58	4.79	5.01	4.79
Mosaic	3.4	5.4	7	5.08
Total Sample	3.4	5.5	14	5.96

currencies, and debt conditions can be very different (e.g., maturity, seniority, and covenants are not always the same). To this extent, Table 21.6 contains statistics about the debt maturity. We can see that the debt is paid back within 2–24 years. On average the loan period is equal to 10.68 years. Interestingly, a number of platforms (Abundance Generation, CollectiveSun, GreenVesting, and Lumo) only display projects with debt maturity equal to, or longer than 10 years.

Table 21.7 reports the values of the expected return offered by the projects tapping equity capital from the crowd. The values are reported as expectations in the exhibits and memorandum displayed on the Web, given the FiT granted in each different country (generally for a period of 18–20 years).

Table 21.6 Debt Maturity in Projects Posted by Lending-Based Platforms. Sample: 71 Projects Posted from December 2013 to June 2014

Maturity (years)	Minimum	Median	Maximum	Average
Abundance Generation	19	20	24	20.27
CollectiveSun	10	10	10	10
Duurzaam Investeren	n.a.	n.a.	n.a.	n.a.
Econeers	5	7	12	8.88
Eollice	4	4	4	4
Green Crowd	5	6	10	6.96
GreenVesting	10	10	10	10
LeihDeinerUmweltGeld	2	7	15	7
Lumo	15	15	15	15
Mosaic	5	10	12	9.78
Total Sample	2	10	24	10.68

Table 21.7 Expected Return Offered by Projects Posted by Equity-Based Platforms. Sample: 13 Projects Posted from December 2013 to June 2014

Expected Return	Minimum (%)	Median (%)	Maximum (%)	Average (%)
crowdEner.gy	4	6	10.1	5.33
Green Crowding	7	7	7	7
Windcentrale	6.7	6.9	8.4	7.2
Total Sample	4	6.8	10.1	7.0

21.5 THE DETERMINANTS OF THE CAMPAIGN SUCCESS

In this section we aim at analyzing the determinants of the funding success. We believe that this is an important objective, in order to understand how sponsors of "green" projects, and especially of renewable energy production, may speed up the funding process and contribute to climate action ventures. For each sample project we collected information about the technical characteristics, the financial plan, the offering clauses, and the promoter (that it is usually a company operating in the energy or consulting sector).

Our research hypothesis posits that the probability of success in a crowdfunding campaign related to a project in the renewable energy field depends on three different determinants: (1) the project financial characteristics (in term of investment size, financial leverage, expected return, and benefits); (2) the crowdfunding platform reputation and commitment (measured by the number of projects previously displayed on the Web site and by any financial contribution that the platform itself granted to the campaign); and (3) the proponents' social capital (the number of contacts on social networks).

We measure the success of the crowdfunding campaign through the ratio between the amount of money effectively raised at the end of the campaign and the initial target (AVTARGET). Referring to the project financials, we introduce a number of variables. We account for the financial leverage of the project (LEV) assuming that this is may be a signal of risk, warning investors' favor. Then we introduce the target capital collection (SIZE): we hypothesize that projects requiring larger contribution could experience more troubles in reaching the target. We also introduce the size of the minimum contribution requested to crowdfunders (MIN_CONT): the larger this threshold, the larger the probability that investors could renounce to join the campaign. The variable CF controls for the ratio between the crowdfunding collection and the total investment requested to develop the project.

We take into account the expected return offered to investors (EXP_RET) both in terms of interest rate (for lending-based projects) or expected profitability (for equity-based campaigns). Larger yields may more easily attract investors. In some cases, crowdfunders may enjoy a further bonus (for example, in some cases they enjoy a discount if they invest earlier in the project, or if they hold their investment for a long period) or proponents may commit to donate money or rewards to local communities when developing the projects (for example, we register commitments to sell renewable energy produced by the plant at a discount to local public entities, or to pay a royalty, or to involve local enterprises in the manufacturing of the plant). Clearly, this can boost the favor of the crowd toward the project and speed up the fundraising process. We take into account the existence of such clauses with two dummy variables (BONUS and BEN_COM, respectively). Dealing with the platform characteristics, we consider the number of projects previously published in the Web site (N_PROJ) assuming that a larger track record may better attract investors. We also consider any financial contribution that the platform committed in the specific campaign (dummy variable PLAT_-INV). This may serve as a quality signal for investors, when evaluating the project. Finally, we measure the proponents' social capital (SOCIAL) through the cumulated number of contacts and "likes" on social networks (LinkedIn, Facebook, and Twitter). Giudici et al. (2014) highlight that the proponents' social capital is an important variable in determining the probability of crowdfunding success.

The variables included in the study are summarized in Table 21.8, while in Table 21.9 we report basic statistics about the data. Interestingly, on average the sample campaigns

Table 21.8 Description of the Variables Introduced in the Empirical Analysis

Name	Variable	Description
AVTARGET	Funding success	Ratio between capital effectively raised and initial target (%)
BEN_COM	Community benefit	Dummy variable equal to 1 if the project provides a monetary contribution to the benefit of the community where the plant is installed, 0 otherwise
BONUS	Bonus	Dummy variable equal to 1 if the project provides a bonus in addition to the expected return, 0 otherwise
CF	Crowdfunding contribution	Ratio between the target of the crowdfunding campaign and the total capital required to finance the project (%)
EXP_RET	Expected return	Expected return (%) according to energy sales and production, disclosed in the memorandum
LEV	Leverage	Ratio between debt and total capital required to finance the project (%)
MIN_CONT	Minimum contribution	Log of the minimum contribution that an investor has to pay in order to participate in the crowdfunding campaign
N_PROJ	Platform reputation	Log of the cumulated number of projects published by the platform before
PLAT_INV	Platform investment	Dummy variable equal to 1 if the platform invests in the published projects, 0 otherwise
SIZE	Target capital to be collected	Log of the expected crowdfunding campaign proceeds
SOCIAL	Social capital	Log of the total number of Facebook like, LinkedIn followers, and Twitter followers of the sponsor/promoter

Table 21.9 Descriptive Statistics about the Variables Introduced in the Empirical Analysis. Sample: 84 Projects Posted from December 2013 to June 2014

	Average	Minimum	Maximum	Median	Standard Deviation
AVTARGET	95.56	9	158	100	18.018
BEN_COM	0.40	0	1	0	0.494
BONUS	0.15	0	1	0	0.364
CF	78.29	2	100	100	32.891
EXP_RET	6.8502	4	35	5.5	4.97928
LEV	77.97	0	100	100	37.517
MIN_CONT	3.7563	1.79	6.91	2.907	1.46575
N_PROG	2.2433	0.69	3.22	2.197	0.79565
PLAT_INV	0.44	0	1	0	0.499
SIZE	12.03	8.29	14.99	12.14	1.58
SOCIAL	6.0001	0	11.81	6.56	3.60602

are successful, since the mean percentage of the committed capital is 95.56% compared to the initial target (see Table 21.4).

Table 21.10 reports the results of the regression analysis, where we also include variables controlling for the technical characteristics of the project (expected production and renewable source) and expected project duration. Model 1 is an ordinary least squares (OLS) regression introducing all the independent variables above. The results highlight that the financial characteristics of the project (summarized by the leverage, the target crowdfunding size, the fraction of the investment financed through other sources) seem to be not so relevant to determine the success of the campaign. On the contrary the platform track record of previous projects and—to a larger extent—the commitment of the platform managers to participate in the financing of the project are important quality signals for the crowd. The capability of the promoters to build a network through social networks is also significantly and positively correlated. The most relevant determinants seem to be the expected profitability disclosed in the business plan and the existence of bonuses to attract investors. Surprisingly, benefits for local communities seem not to influence the investors' decisions.

Yet, in order to better disentangle the latter finding, we run an alternative OLS regression (Model 2) in which we add a further variable, namely the interaction between the expected return and the community benefit dummy (EXP_RET x BEN_COM). The results highlight that most of the previous findings are still robust. Only the track record of the crowdfunding platform is no more significant at conventional levels. Interestingly, we find that the interaction coefficient is significantly negative, and the coefficient of the BEN_COM variable becomes significant and positively correlated. We conclude that benefits and royalties to local communities and public entities are positively discounted by investors only when the expected profitability is low. There seems to be a "substitution effect": when the expected profitability is sufficiently high, transfers to local communities are irrelevant to convince investors to fund the project. On the contrary,

Table 21.10 The Determinants of Crowdfunding Success: Ordinary Least Squares Regression Results. The Dependent Variable Is AVTARGET, Namely The Ratio Between Capital Effectively Raised And Initial Target. Independent Variables Are Described In Table 21.8. T-Statistics Are Calculated Using White Heteroskedasticity-Consistent Method. Sample: 84 Projects Posted From December 2013 To June 2014

	Model 1		Model 2	
Variable	Coefficient (Standard Error)	T-Stat (p-Value)	Coefficient (Standard Error)	T-Stat (p-Value)
Constant	83.4105	4.51	90.763	4.43
	(18.4973)	(0.000***)	(20.5092)	(0.000***)
BEN_COM	9.6495	1.44	13.5307	1.84
	(6.723)	(0.156)	(7.3536)	(0.065)*
BONUS	28.3752	2.17	29.4122	2.16
	(13.066)	(0.030)**	(13.5856)	(0.034)**
CF	0.1082	0.87	0.1126	0.92
	(0.1250)	(0.390)	(0.1224)	(0.361)
EXP_RET	1.9077	2.37	1.9126	3.02
	(0.8065)	(0.021)**	(0.6333)	(0.002)***
LEV	−1.2646	−1.36	−0.1056	−1.32
	(0.0929)	(0.178)	(0.0798)	(0.190)
MIN_CONT	−2.9336	−0.97	−2.4827	−0.88
	(3.0225)	(0.335)	(2.8224)	(0.382)
N_PROJ	3.5764	1.75*	2.6799	1.52
	(2.0416)	(0.080)	(1.7631)	(0.128)
PLAT_INV	10.9764	1.74	8.4253	1.77
	(6.2955)	(0.086)*	(4.7627)	(0.077)*
SIZE	1.6904	0.73	1.0013	0.44
	(2.3322)	(0.471)	(2.2946)	(0.664)
SOCIAL	1.0337	2.30	1.0806	2.42
	(0.4494)	(0.022)**	(0.4457)	(0.015)**
EXP_RET x BEN_COM	—	—	−1.1780	−1.74
			(0.677)	(0.081)*
Control variables	Yes		Yes	
Sample size	84		84	
R-2 (adj.)	27.84%		28.28%	
F-test	1.91**		1.94**	

***, **, * = Statistically significant at 99%, 95%, 90% level.

for lower levels of profitability, promoters committing to increase the wealth of local institutions through royalties or discounts, or other equivalent means, will better convince the crowd to finance the investment.

We carried out a robustness analysis taking into account alternative variables to measure the size and the financial leverage of the project. We also run a probit regression, where the dependent variable is a dummy variable equal to 1 if the crowdfunding campaign is successful, 0 otherwise. The results are virtually unchanged.

21.6 CONCLUSION

Climate action plans at the local level are crucial for a successful transition to a low carbon emission and sustainable future. To the extent that climate action policy is generating tangible benefits in terms of income, jobs, and better services (such as access to clean energy), the willingness to support renewable energy production will increase. In general, the crowdfunding of green projects may improve access to finance for climate action campaigns both in developed and in growing countries. The credit crunch originated from the global financial crises increased the difficulties to fund projects to develop renewable energy plants. Budget cuts also push governments to reduce FiTs, as many established technologies for producing green energy are close to reach the "grid parity." Therefore, a number of Web platforms specialized in supporting the crowdfunding of renewable energy projects appeared on the stage.

In this study we propose an unprecedented study on such platforms. Analyzing 13 dedicated crowdfunding platforms, we collected data about 84 projects that tapped the crowd to raise finance to build renewable energy plants (mostly photovoltaic modules, but also wind farms and biomass engines). We focus on projects issuing debt and equity capital and we explore the determinants of the campaign success. The results show that expected profitability is a key determinant. Promoters of low-profit projects should consider to include clauses aimed at leaving economic benefits to local communities, through royalties to public institutions and no-profit organizations, or discounts in energy supply, or engaging local workers in building the plant. Co-investments made by platforms are positively considered by crowd investors. Building up a wide network of contacts and "likes" through social networks is also a winning strategy.

We think that the analysis will be particularly useful for policy makers and sponsors of "green" projects, in order to increase the efficiency of crowdfunding campaigns aimed at installing small-scale renewable energy plants. The number of Web platforms specialized in hosting crowdfunding campaigns in the "green" business is growing around the world and this is a unique occasion to boost investments in sustainable initiatives. Yet it will be necessary to develop good practices to protect lenders and shareholders from frauds and opportunistic behavior, and adequate governance rules to manage a plethora of investors once the money has been collected on the Web.

REFERENCES

Afuah, A., Tucci, C.L., 2012. Crowdsourcing as a solution to distant search. Acad. Manage. Rev. 37 (3), 355–375.

Agrawal, A., Catalini, C., Goldfarb, A., 2011. The Geography of Crowdfunding. NBER Working Paper 16820.

Agrawal, A., Catalini, C., Goldfarb, A., 2013. Some Simple Economics of Crowdfunding. NBER Working Paper 19133.

Belleflamme, P., Lambert, T., Schwienbacher, A., 2014. Crowdfunding: tapping the right crowd. J. Bus. Ventur. 29 (5), 585–609.

Brabham, D.C., 2008. Crowdsourcing as a model for problem solving: an introduction and cases. Int. J. Res. New Media Technol. 14 (1), 75–90.

Burtch, G., Ghose, D., Wattal, S., 2013. An Empirical Examination of the Antecedents and Consequences of Contribution Patterns in Crowd-funded Markets. Working Paper, SSRN.

Campello, M., Graham, J.R., Harvey, C.R., 2010. The real effects of financial constraints: evidence from a financial crisis. J. Financ. Econ. 97 (3), 470–487.

Franke, N., Klausberger, K., 2008. Design Communities: Business Model of the Future? Working Paper, SSRN.

Gerber, E.M., Hui, J.S., Kuo, P.Y., 2011. Crowdfunding: Why People Are Motivated to Post and Fund Projects on Crowdfunding Platforms. Northwestern University Creative Action Lab, Sheridan Drive, Evanston.

Giudici, G., 2015. Equity crowdfunding of an entrepreneurial activity. In: Audretsch, D.B., Lehmann, E.E., Meoli, M., Vismara, S. (Eds.), University Evolution, Entrepreneurial Activity and Regional Competitiveness. Springer.

Giudici, G., Guerini, M., Rossi-Lamastra, C., 2013. Crowdfunding in Italy: state of the art and future prospects. Econ. Politica Ind. J. Ind. Bus. Econ. 40 (4), 173–188.

Giudici, G., Guerini, M., Rossi-Lamastra, C., 2014. Why Crowdfunding Projects Can Succeed: The Role of Proponents' Individual and Territorial Social Capital. Working Paper, Politecnico di Milano.

Howe, J., 2008. The rise of crowdsourcing. Wired 14 (6).

Kleemann, F., Voß, G.G., Rieder, K., 2008. Un(der)paid innovators: the commercial utilization of consumer work through crowdsourcing. Sci. Technol. Innovation Stud. 4 (1), 5–26.

Lévy, P., 1997. Collective Intelligence. Plenum Trade, New York and London.

Massolution, 2013. 2013CF – The Crowdfunding Industry Report.

Mollick, E., 2014. The dynamics of crowdfunding: an exploratory study. J. Bus. Ventur. 29 (1), 1–16.

Schwienbacher, A., Larralde, B., 2010. crowdfunding of small entrepreneurial ventures. In: Cumming, D. (Ed.), Handbook of Entrepreneurial Finance. Oxford University Press.

Surowiecki, J., 2004. The Wisdom of Crowds. Anchor Books, New York.

Ward, C., Ramachandran, V., 2010. Crowdfunding the Next Hit: Microfunding Online Experience Goods. Working Paper, Computational Social Science.

CHAPTER 22

Management Accounting and Biodiversity: The Cultural Circuit of Capitalism and the Social Construction of a Perfect Market?

Gillian Vesty[1], Albie Brooks[2], Judy Oliver[3], Sukanta Bakshi[1]

[1]School of Accounting, RMIT University, Melbourne, VIC, Australia; [2]Department of Accounting, University of Melbourne, Melbourne, VIC, Australia; [3]Department of Accounting, Economics and Finance, Swinburne University of Technology Melbourne, VIC, Australia

Contents

22.1 INTRODUCTION

> *So it has come to this. The global biodiversity crisis is so severe that brilliant scientists, political leaders, eco-warriors, and religious gurus can no longer save us from ourselves. The military are powerless. But there may be one last hope for life on earth: accountants.*
>
> **Jonathon Watts, The Guardian, 28 October, 2010 in Gleeson-White (2014, p. ii).**

Can accountants and accounting really be the knight in shining armor? The assumption underlying this opening media claim suggests—as an organizational level influencer—accounting can be influential in the creation of the "perfect market." Of course, perfect markets are never obtainable and there will always be incomplete information (Merton, 1987). But, if we are to bring biodiversity into the corporate accounts, and continue to live in a capitalist society, we need to work a lot harder than just tweak at the edges (Gray, 2010). Accounting's role in the marketplace, unlike economics, is to service the activities of individual companies. As such, the purpose of this chapter is to provide details on the capacity of accounting, and accountants, to connect with the *noise* in the marketplace. Can accounting render biodiversity a homogenous calculative commodity, create low barriers to entry, and provide buyers and sellers with complete knowledge of quantities on offer and prices to match demand? A big ask.

As a theory of capitalism, economics and its quantitative models have been widely adopted by the accounting academy, treated as normative prescriptions to advise management practices (for varying critiques, see Klamer and McCloskey, 1992; Napier, 1996; Gray, 1990). Nevertheless, some of the earlier management accounting developments—residual income, marginal analysis and cost benefit, largely derived from neoclassical economics, and enthusiastically advocated by the academy—were not well articulated at the practice level (Kaplan, 1984; Johnson and Kaplan, 1987). Uncertainty and complex, pragmatic issues motivated a stream of research on the theory-practice gap, as the residual *noise*, largely ignored by economists, was found to influence management decisions and accounting controls at the organizational level (Scapens, 1994, 2006). Questions were asked, by academics, about why practitioners used some concepts and did not use others, particularly those that met academic standards of logical rigor and empirical validity (Seal, 2010, p. 95). As Thrift (2005) explains "I am not at all convinced that managers of capitalist firms—jointly or severally—know what they are doing for quite a lot of the time" (2005, p. 2). Scapens (2006) echoed similar thoughts, wondering what it was about *some* of the theories that managed to bridge the gap between theory and practice. The implementation of economic models (particularly with assumed linear cause-and-effect expectations and rational, profit maximizing outcomes) was frequently confounded by the complex and somewhat messy nature of social organizations (Burchell et al., 1980; Hines, 1988, 1991; Hopwood, 1987). Yet, the normative models continue to be promulgated by the academy, even when their ability to represent reality is questioned.

With accounting called to provide solutions for the biodiversity crisis, there has been renewed interest in the theoretical insights offered by full cost accounting (FCA) (Jones, 2014). FCA is a broad term for an evolving accounting technique that takes into account, in monetary terms, the sustainability impacts of an organization (Bebbington et al., 2001; Bebbington and Gray, 2001; Bebbington et al., 2006; Gray and Milne, 2007; Bebbington, 2007; Unerman et al., 2007; Frame and Cavanagh, 2009; Jones, 2010; Davies, 2014). A plethora of theoretical approaches are available to managers, including the extent to which FCA is adopted (costs can be conventional, hidden, contingent, and/or intangible, and some approaches are considered in terms of being *fuller* cost accounting, rather than *complete*) (Davies, 2014). The technique has evolved overtime, largely generated from a pool of environmental economics measurement techniques, to estimate nonmarket values of externalities, with varying valuation methodologies ("contingent valuations, cost benefit, travel cost method, benefits transfer, dose—response models, hedonic pricing models, and choice modeling") used in high-profile natural resource cases, like Exxon Valdez (Herborn, 2005, p. 520). FCA has, likewise, been formalized in the offset and compensation programs as part of the worldwide construction of biodiversity markets (Madsen et al., 2010, 2011).

To date much of the FCA literature has been around normative developments and strategizing around the potential for FCA in practice. There has been limited research of actual practices considering the "social" organizational influences, and more research is required to determine the way economic theories weave their way into management accounting practice. To study FCA practices within organizations, we draw on Thrift's (2005) cultural circuit of capitalism, in which the producers and consumers of managerial knowledge (i.e., business schools, management consultants, management gurus, and the media) are recognized as influential in the proliferation of the market for biodiversity. To this we add two other institutionalized contributors, professional bodies, and practicing managers to further situate accounting within an organizational decision-making discourse (Seal, 2010).

We use insights from our recently conducted field research and (among the mish-mash of social influences, unresolved issues, and biodiversity concerns) examine the flow of managerial knowledge and accounting concepts toward accomplishing some abstract notion of an economic ideal. Rather than demonstrating the difference between accounting and economic ideals, the focus of this chapter is on the way management accounting contributes to the shaping of concrete markets. We recognize management accounting as grounded in systems thinking, a material linguistic device (or discourse that includes, written, and spoken "texts," including sets of accounts, budgets, financial ratios, symbols, diagrams, photographs, and so forth) designed to serve the purposes of management control within organizations (Seal, 2010). We show the unfolding FCA discourse as it sits alongside economic theory, in the management knowledge creation processes. This activity is viewed in terms of the social construction of perfect market (Garcia-Parpet, 2007). We also address calls for research that investigates the processes by which management accounting knowledge is produced, disseminated, and operationalized (Seal, 2010, p. 96).

In the sections that follow, we begin with details of FCA techniques in management accounting practices. Subsequently, Thrift's (2005) cultural circuit of capitalism is discussed, along with the management accounting extensions proposed by Seal (2010) to help frame this study. We describe the influencers, in more detail—with a particular focus on the way accounting has begun to work on biodiversity as a new market for exchange. We use this background literature to help to explain the management accounting reaction to the biodiversity crisis. The chapter concludes with insights from recent case research, conducted by the authors, to consider the "theoretical enterprise"—a combination of theories *of* capitalism (i.e., economics) and the theories *in* capitalism (i.e., formal and informal management accounting and control mechanisms, such as FCA)—in management discourse. The first is a municipal council operating in a natural world heritage site (WHS), and the second is a multinational listed coal mining company. Both operate in the same state of Australia. The first are the biodiversity market "sellers" and the latter, the biodiversity market "buyers." The overall discourse is framed in terms of the biodiversity market, FCA, and the social creation of buyers and sellers.

22.2 FCA IN MANAGERIAL DECISION MAKING

The concept of FCA has surfaced as an accounting technique, potentially capable of dealing with the biodiversity crisis (Jones, 2010, 2014; Davies, 2014). FCA is not a new concept and was originally developed to overcome the limitations of conventional accounting, by including costs of direct environmental impacts from corporate activities (Gray, 1992; Bebbington et al., 2001; Bebbington and Gray, 2001; Bebbington et al., 2006). If applied as designed, will provide buyers and sellers with more complete (never perfect) pricing knowledge (Bebbington et al., 2006; Gray and Milne, 2007; Unerman et al., 2007; Frame and Cavanagh, 2009). Davies (2014) extends FCA and highlights how costs on biodiversity impacts can be used for both internal and external decision making. For example, PUMA's environmental profit and loss statement, a more recent application of FCA, is designed with the external decision maker in mind (Sabeti, 2011). Biodiversity impacts can be included, either through the costs of avoiding biodiversity impacts and/or the costs of restoring impacts after they happen. Depending on the nature of the organization, the latter option indicates that impacts cannot be completely mitigated (i.e., as in mining activities), and is the least preferred, but sometimes unavoidable option, given they maintain their social license to operate in our current capitalist present.

Much of the environmental sustainability and biodiversity accounting literature is concerned with the external stakeholder, the legitimizing efforts of managers and debates about the extent to which corporate reports actually veil corporate unsustainability (Gray, 1992; Gray and Bebbington, 2001; Bebbington, 1997, 2001; Lehman, 2001; Milne and Patten, 2002; Deegan et al., 2002; O'Dwyer, 2003; Gray et al., 1987, 1991; Gray, 2006; Lamberton, 2005). We contribute with a study of FCA management accounting practices *within* the organization. The view is that FCA for management control purposes is derived from transdisciplinary expectations of achieving a broader set of strategic, future-focused outcomes. When market uncertainty exists, such as the biodiversity crisis, management accounting (underpinned by dynamic budgeting and costing models that operate in continuous time) is increasingly required to effect strategic, proactive responses. Economic information flows are assuming an even greater salience in the management of organizations, with a focus on "being strategic rather than merely having a strategy" (Hopwood, 2009, p. 799).

FCA must be further differentiated from conventional cost accounting, which frequently provides cost information to meet the regulated financial accounting standards. This latter form of accounting information is artificially constrained by regulation, and is backward looking and a slowly evolving process (i.e., note the decades associated with the development and implementation of International Financial Reporting Standards). Conventional financial accounting does not necessarily provide the broader picture required for internal management decisions and, given the debates over the legitimizing "spin" of corporate reports, it is not easy to recognize "good" corporate biodiversity governance from reports that are mere legitimizing tools. Currently, any FCA that is disclosed to external stakeholders, such as PUMA's environmental profit and loss, is voluntary in nature (and more than likely scrutinized by corporate lawyers), as it is outside the regulatory reporting framework.

FCA, when applied in a management accounting context, is more concerned with using the economic data and normative insights and translating them to the strategic management control of the business operations, the supply chain, competitors, as well as individuals within the organization. The role of management accounting is to consume both calculative and qualitative data and integrate them into techniques that influence long-term business decisions, including resource allocation and the management of risk (Simmonds, 1981; Eldenberg et al., 2011).

22.3 THE CULTURAL CIRCUIT OF CAPITALISM: MANAGEMENT ACCOUNTING AND CONTROL

Rather than viewing capitalism as a system of oppression, Thrift (2005) conceptualizes capitalism as dynamic and emergent:

> *For quite a few people, capitalism is not just hard graft. It is also fun. People get stuff from it —*
> *and not just more commodities. Capitalism has a kind of crazy vitality. It doesn't just line its*
> *pockets. It also appeals to gut feelings. It gets involved in all kinds of extravagant symbioses. It*
> *adds to the world as well as subtracts.*
>
>   *Thrift (2005, p. 1)*

This chapter is an exemplar of this crazy vitality. The biodiversity market is a new and vibrant market, being constructed on a proliferation of knowledges, which have largely evaded capitalism until now. The accounting academy is contributing to this emergent market with the construction of academic impacts, publication targets, research income, promotions, and so forth, around biodiversity expertise. The biodiversity market comprises many intangible attributes similar to modern forms of capitalism. There is the emergence of theoretical enterprises, made up of "virtual" notions (organizational networks, the knowledge economy, the new economy, and communities of practice) with "success" measured by the tools of soft capitalism, which somewhat mimics a "touchy-feely replay of Taylorism" (Thrift, 2005, p. 6). Management accounting gurus have contributed with multidimensional performance measurement systems like the balanced scorecard, aimed at the monitoring and controlling the intangible assets including employee skills, knowledge and know-how (Kaplan and Norton, 1996, 2001). Simons (1994, 2013) explains the role of formal accounting systems, such as the balanced scorecard, to encourage organizational learning, stimulate new ideas and strategies. Referring to these as "interactive controls," Simons suggests they are positive and inspirational—these are "the yang: forces representing sun, warmth, and light" (2013,

p. 9). Nevertheless, the tension set by this supposedly "soft" management control practice is somewhat cynical, aimed at improved productivity, profit maximization, and growth (Thrift, 2005). Thus, accordingly soft capitalism would recognize the broader management accounting function as furthering the other versions of Simons' boundary and diagnostic controls, "the yin: forces representing darkness and cold" (2013, p. 9).

Even scarier than soft capitalism, is the biodiversity crisis and the knowledge that the capitalist system must race to construct new markets that deliver sustainable development in a timely manner. As highlighted in the earlier plea from the media, the economic ideals must be converted (by accounting) to concrete realities, so the market contracting can win before we default on our contract with nature (Serres, 1995). This makes the vast repertoire of materials, generated by the academy, all important in the framing of what is consumed and disseminated and adopted in practice. The debates that claim the capitalist ideals are the source of the biodiversity problem (Gray, 2002, 2010) and, that accounting intervention should not be used to address the conflicts inherent in advanced capitalism (Brown and Frame, 2005; Antheaume, 2007), are the *necessary noise*. These academic debates also contribute toward the social construction of perfect markets. The very debates calling for emancipatory and radical change, an ontological shift taking us outside capitalism and away from the innate contradictions between economic growth and the natural resources available to achieve this growth, are also important contributors in the production and consumption of knowledge. As the media become involved and the idle chitchat in the corridors and cafeterias of large corporate buildings voice similar concerns; if the noise is loud enough, practices might be influenced.

Thrift's (2005) cultural circuit of capitalism (as highlighted in Figure 22.1, below) provides a guide to study the dissemination of managerial knowledge. The institutionalized producers and consumers of managerial concepts draw together all sorts of knowledge and opportunities, even from the surprisingly important informal as well as formal interactive control, which "stimulate organizational learning and the emergence of new ideas and strategies" (Simons, 1994, 2013, p. 7). The loop of institutionalized producers and consumers is involved in exchanging ideas, promoting a *particular sort* of knowledge and a *particular sort* of practical theory (Seal, 2010, p. 96). In efforts toward utopian capitalist market ideals, the process and actors can be studied as they address the biodiversity crisis that is impacting decision making over resource allocation and the protection of the resource base of all raw materials, including clean air and fresh running water. The cultural circuit of capital does not offer simple, straightforward, and unproblematic exchange—it is "a feedback loop which is intended to keep capitalism surfing along the edge of its own contradictions" (Thrift, 2005, p. 6).

As highlighted in the opening quote, one way of creating a biodiversity market is the use of *media* to draw attention to the "crisis" and ensure biodiversity impacts are part of the political economy, are well understood, evoke passionate outbursts on a global scale, and expectations of equilibrium in supply and demand, a ubiquitous part of everyday life.

Figure 22.1 Accounting influencers in the cultural circuit of capitalism. *(Adapted from Thrift (2005) and Seal (2010), p. 97.)*

The mature MBA students in *business schools* are ingesting and reproducing academic knowledge on biodiversity. Articles and cases provide practical examples that become part of managerial discourse (for example, the institutionalized Harvard Business School contributions); "sustainability" *management consultants* are packaging their services and selling to their clients. Accounting professional services firms, KPMG, Deloitte, EY, and PwC are included in this category and contribute with vast amounts of technical guidance on accounting for biodiversity. Overlapping is the creation of the *management guru* (management-driven gurus, such as Porter and Eccles and sustainability-driven gurus, such as Gray, Elkington, Epstein, Jones, and so forth)—who is espousing new theories to "hook" their managerial audiences with their solutions to the biodiversity crisis. Seal (2010, p. 7) explains: "As accomplished performers in their own right, management gurus usually combine some technical managerial knowledge with important psychological skills with which they reassure their managerial audiences." As we will be shown later in this chapter, *practicing managers* also prepare technical models and write articles for audiences, both inside and outside the corporate boundaries. Accounting's *professional bodies* have likewise engaged heavily, individually and collectively through groups, such as the Accounting Bodies Network, the Integrated Reporting Council (IIRC), and the Prince's Accounting for Sustainability Project (A4S), to advance accounting for sustainability. Supported by the "big four" accounting firms, the new integrated reporting approaches and new methodologies have been developed to acknowledge a broader repertoire of capital inputs, not only financial in operational activities (IIRC, 2013; Eccles and Serafeim, 2013; Stubbs and Higgins, 2014). They are actively working to ensure sustainability is embedded in business strategies and accounting practices within the firm.

Like a whirlpool, drawing in a plethora of ideas, some theories in capitalism are institutionalized into practice. Together, the theories *of* capitalism (i.e., economics) and the theories *in* capitalism (i.e., formal and informal management accounting and control mechanisms) make up the production and consumption of management knowledge and the "theoretical enterprise" (Thrift, 2005; Seal, 2010).

22.4 THE DEAFENING NOISE: BIODIVERSITY CRISIS SHOCKS FUNCTIONING MARKETS?

There is no doubt, that the biodiversity crisis is the ultimate contradiction to capitalism, but the growing noise in the marketplace can no longer be ignored (Serres, 2014). Potentially, only three choices remain: an attempt to mirror biodiversity's real price in a perfect market, a radical move away from capitalism and its need for growth and consumption altogether, or maintain status quo and more than likely renege on the natural contract (Serres, 1995). If the latter occurs, there will not be much call for accounting or its services, when the species is extinguished (Gray, 1990).

However, as will be found throughout this book, there is a lot of economic, finance, and accounting-related activity, particularly around the first alternative. If this is the route and a "perfect" biodiversity market is to be accomplished, the following must be worked toward:

- Biodiversity must be rendered a homogenous calculative commodity;
- Buyers and sellers have complete knowledge of quantities on offer and prices to match demands; and
- There must be low barriers to entry: the removal of obstacles preventing organizations entering the biodiversity market, thereby increasing competition and decreasing profits.

Each of these comes with the caveat that biodiversity is not necessarily renewable or substitutable and that equilibrium is increasingly becoming unsettled as manmade capital increases to the detriment of natural capital. As highlighted in the State of Biodiversity Markets Report:

> We believe that by providing solid and trustworthy information on prices, regulation, science, and other market-relevant issues, we can help payments for ecosystems services and incentives for reducing pollution become a fundamental part of our economic and environmental systems, helping to make the priceless valuable
>
> **Madsen et al. (2010, p. 4)**

The development of marketlike instruments and payments leads to the question about buyers and sellers and the current state of play *within* organizations. The "theoretical enterprise" and the circuit of producing and consuming managerial knowledge in relation to the social construction of a perfect biodiversity market are relatively unknown. Adding to the vagaries on current practice is that the last few decades have brought about different forms of commodity. Many of these, like software applications, draw on consumer desires for rich experiences and are created as part of impression management, enhanced by marketing and the mass media. This politicization activity further motivates views about the perceptions and value of nature's scarce resources, further drawing on accounting as the means by which biodiversity can become a calculable technique, and included in market transactions (Vesty et al., 2015b). Such as with the creation of the European Union, it is only when biodiversity is homogenized and standardized that it can be rendered calculable, and any (dis)-equilibrium can be recognized and managed (Barry, 2001).

Nevertheless, as concerns about the unsustainability of human activity and awareness of the global biodiversity crisis increases, so has the proliferation of capitalist models by the accounting academy to "better" account for the biodiversity crisis (Freeman and Groom, 2013; Jones, 2014; Khan, 2014). Management accounting has the potential to draw on insights from environmental and ecological economics on the value of global ecosystem services (Gupta and Foster, 1975; Daly, 1985; Turner, 1987; Costanza et al., 1997; McNeely et al., 1990; TEEB, 2008, 2012) and financial accounting offerings, relating to valuing and reporting on corporate externalities (Gray, 1992; Gray and Bebbington, 2001; Tinker and Gray, 2003; Jones, 1996, 2003, 2014; Jones and Solomon, 2013; Houdet, 2008; Houdet et al., 2009; Cuckston, 2013). As highlighted in the introduction, a traditional financial accounting approach is one that is constrained by the need to meet regulated, rule-based disclosures. For example, accounting for agricultural stock (IAS 41) follows accounting principles and methodologies based around the transformation of biological assets (living plants and animals) into agricultural produce. This accounting standard generally requires these assets to be measured at fair value less costs to sell, which might seem a relatively straightforward exercise, when pricing, for example, the sale of timber. However, the challenge for environmentally concerned accounting researchers, with an interest in providing numerical, market-based values on which to build their research is to similarly recognize the use of ecosystem services, such as water, air, biodiversity (Gray, 1991; Jones, 1996, 2014). Similar to cultural and heritage assets, values derived from a capitalist economic system, which maximizes financial returns to private investors, are argued to be arbitrary and nonsensical if they ignore other more important intrinsic values.

Accounting for biodiversity is an important part of the overall "natural capital" discourse (Gray, 1991; Maunders and Burritt, 1991; Jones, 1996, 2010; Jones and Matthews, 2000; Jones, 2014). Thus, environmental economics has contributed to accounting developments with varying methodologies that help make visible the value of nature's products through its provision of "ecosystem goods and services" (Costanza et al., 1997; WBCSD, 2011; Houdet and Germaneau, 2014). Two broad approaches have been followed when accounting for biodiversity (i.e., quantifying and/or financializing flora and fauna, and the habitats they inhabit), a holistic ecosystems approach or a bottom-up, individual valuation methodology (Jones, 2014, p. 2). Jones (1996, 2003) has empirically explored the different approaches and advocates a bottom-up approach, which accounts for the habitats, flora, and fauna in terms of their hierarchical criticality (for other application of Jones, 1996 see Siddiqui, 2013; Khan, 2014). Segregating biodiversity according to its criticality (Jones, 1996) contributes to the construction of the biodiversity market, by segregating the priceless and impossible to value. Jones (2014) uses iconic rhinoceros species to exemplify this point, but highlights the need for greater consensual efforts toward grading and valuation.

The alternative ecosystems approach looks at the ecosystem more holistically and considers the service ecosystems provide in terms of their marketability (WBCSD, 2011). The marketed ecosystem services comprise the inputs used and paid for by businesses while the nonmarketed ecosystem services are less readily quantified and might include intrinsic values associated with aesthetic, spiritual, or recreational enjoyment of ecosystems. Ecosystem services have also been classified as providing provisioning services, cultural services, support services, and regulating services (Jones, 2014; see also Fisher et al., 2009).

Discussed above and explored further in this chapter, FCA is supported by other management accounting techniques (Figge et al., 2002; Epstein, 2008). This includes environmental management accounting, a quantitative method which includes monetary information on environment-related costs, earnings and savings, and physical flows (Burritt et al., 2002; Jasch, 2003; Savage and Jasch, 2005; Jasch, 2009; Bennett et al., 2011). Along with financial performance, the physical information on the use, flows and destinies of energy, water and materials provides insights into the extent to which natural capital, and human waste, has increased or diminished overtime. This activity helps with "pricing" of nonproduct output (waste) by highlighting the purchase costs of materials converted into waste and emissions, as well as address the price of biodiversity and ecosystem services (Gonzalez and Houdet, 2009; Houdet et al., 2009; Vogtländer et al., 2010). These performance measurements are recognized in performance reports, including international standards on the environment (the ISO 14000 series). Likewise, the International Standard on Assurance Engagements (ISAE) 3410, Assurance Engagements on Greenhouse Gas (GHG) Statement extends auditing responsibilities beyond financial reporting and further defines GHG emissions to enhance the quality and consistency of assurance engagements on GHG information.

Overtime, the accounting approaches have expanded beyond their pricing-only focus (the fundamental of bookkeeping and economics) to ensure more complete knowledge and assumptions behind the numbers. Combining quantified events (i.e., physical flows of GHG emissions; waste; percentage of habitat destroyed etc.) with qualitative accounts helps describe corporate value adding or diminishing activities (Jones, 2014; Gray et al., 2014; IIRC, 2013). The qualitative events might also be quantified by ranking order of importance to ensure riskiness of projects is prioritized in decisions (Vesty et al., 2015a).

In concluding this section, financial methods might include fair valuations, cost-based approaches (such as replacement costs or costs avoided as part of FCA) or, sometimes, willingness to pay for ecosystems services is used in typically challenging, nonmarketed ecosystem services. However, it must be noted that much of this activity is underpinned by the need to take a "business case" approach to sustainable development, which recognizes the role the environment plays in the economic success of communities as well as corporations (Bruntland, 1987; Spence, 2010). The normative discourse emerging from these groups aims to unify a broad range of accounting, finance, and environmental economic tools. Strategic management accounting is the linking device to ensure these emerging theories are put into practice. Given the strong calls, and accounting efforts around the "business case" approach, this further confirms the views that the accounting profession has agreed to taking the capitalism path, as outlined at the beginning of this section. The remainder of this chapter is interspersed with illustrations from recent case research conducted by the authors. We draw on two cases, both of which operate in the same state of Australia. The first is a municipal council operating in a natural WHS, and, the second is a multinational listed coal mining company. The first are the biodiversity market "sellers" and the latter, the biodiversity market "buyers."

22.5 THE SELLERS: MUNICIPAL COUNCIL OPERATING IN A WHS

The municipal council operates in a region recognized for its significant representation of Australia's biodiversity, including some rare and threatened species (eucalypts and pine species), large and intact tracts of protected bushland and national parks, complemented with valuable indigenous heritage, wilderness, recreation, aesthetic, and spiritual ecosystem services. Obligations to the World Heritage Convention are managed under national legislation, the *Environment Protection and Biodiversity Conservation Act 1999* and other legislative processes relating to heritage, threatened species, water, environmental protection and planning, and so forth are important in the strategic planning for the future protection of the million plus hectares.

The municipal council is structured in such a way that biodiversity issues must flow seamlessly across the organization. The key departments reporting through to the general manager include:

- Services—key operations from procurement to waste management, from leisure and visitor information to bushfire recovery;
- Community outcomes—asset planning, the environment, and the community are entwined with strategic plans and measured by the business performance team;
- Development and customer service—this department handles building, development, and planning, in terms of compliance and regulation monitoring. It is the customer interface for these services;

- People and systems—this is the strategic planning hub which includes governance and risk, information solutions, human resources, communications, and marketing
- Integrated planning and finance—the chief financial officer (CFO) and team associated with corporate planning and reporting nuances
- The cultural center—dedicated to the indigenous and other important heritage aspects of the national parks.

Each of the departments works closely together. The structure allowed this collaboration, which was evidenced as we began to interview the different managers. Examples such as: "Did you speak with Rebecca, George and Tom?"; "I will walk you over now, so you can talk with Susan"; "Yes. You must go out there and have a look at…"; "When we're ready, I can go and get him and then on the biodiversity side of things, I've got our environmental scientist who is in today…He can come in and speak a little more specifically" and "I looked at your model and went downstairs to ask the others about it…" The interview transcripts ran for pages of uninterrupted enthusiasm about their biodiversity activities. The tape was still recording as one of our authors walked between interviews, continuing to chat about the project with the senior managers as they introduced him to others. At the end of one interview, we commented on the passion that radiated from individuals and groups of individuals:

> *I've got that kind of commitment and passion about looking after the natural environment and I'm very fortunate I've got a team — everyone that tends to work in the environment field, particularly here, does so because they love the natural environment…I could make way more money in the private sector, but I've chosen particularly because my heart's in it to look after the natural environment and for me the waterways, because they tend to be the pulse*
> **Manager Natural Environment**

When examining the theories behind FCA, we found them entering discussions in many different and adhoc ways. They have had multidisciplinary teams of academics on site (for example, an Australian Research Council grant was awarded to build knowledge so they could inform management of the drivers of ecosystem change). This new knowledge facilitated the links between science, policy, and management practice, with the quantitative outcomes of this project, all key contributions to developments in FCA. Management consultants have likewise contributed to the municipal council's database of knowledge on FCA. We were given a copy of an extensive report that mapped the municipal council's biodiversity and valuation studies. Secondary data were accessed directly from the Internet, including links to the local Environmental Protection Agency (EPA) Web site where the FCA model guidelines and spread sheet were available for download.

The accountants explained that they did not use this in their formal accounting, but others did use it within the municipal council for comparative purposes. In comparing with conventionally derived accounting (largely based on reporting standards and market values), they used the FCA variance (those amounts considered as nonmarket values) in decision making about budgeting and resource allocation. Most interviewees, in the finance department, had a broad idea of FCA but it was not part of their accounting repertoire:

> I think it comes down to your definition of FCA and I'm sure you've seen it where we have a sense of financial costs and we call that FCA. But the whole lifecycle is much, much bigger and far more complicated and I think very few people actually apply that level of rigour to their accounting systems. Even people that claim to be doing lifecycle analysis…no one does a scope three lifecycle analysis or very, very, very rarely because it is a massive, massive dataset. Maybe as we get better with input/output analysis and these things become a little bit more standardised we'll be applying more of that sort of thing.
>
> *Manager Finance Department*

Most of the valuation methodologies, outside conventional accounting numbers, were largely found in other divisions, where environmental economics was relied upon to value biodiversity for regulatory monitoring reasons and income generation, including offset calculations. For example, FCA provides a guide for the practicing managers in environmental service provision "…we do all of Part 5 [on the environmental assessments of property developers]…where you look at the different potential environmental impacts and ways to mitigate these." (Manager Asset Planning).

Importantly, they are quite comfortable in their own practicing managers' efforts in developing tools that somewhat reinforce the underlying knowledge provided by academics and consultants. While they spend time using FCA in internal decision making and to develop knowledge so the costs can be reengineered and reallocated in a way that the environmental consumer pays indirectly:

> I think 'x' told you about the State Government has put together a calculator in the last few years, and so when we went to training and had a look at that, we were already capturing all of those things really…and that captured a lot of the full cost accounting, I guess as we define it…[For example] the Local Government Act requires that we charge a reasonable cost for the domestic waste services…And there's some guidelines on what reasonable costs looks like, and what can and can't be included.
>
> *Manager Corporate Planning and Reporting*

And from another manager:

> …we generate our own income through the fees and charges — the gate fees, the user fees essentially. We don't impact on council's budget, we don't take any of council's rate money, and we don't give any back…
>
> *Manager Waste and Cleaning Services*

What we found, driving the biodiversity market, was the realization that they were in a privileged biodiversity position and could recognize the possibility for broader market change. To protect their boundaries, different strategies were required to meet their biodiversity market goals. For example, they were running out of landfill for waste, and given their World Heritage status would not get a license to build any more.

They did not want one anyway, so the solution was to maximize current position and innovate for the future. Examples included:

- A project to capture methane gases from current municipal waste refuse;
- Programs to encourage people to do more on–site composting where neighbors can share compost;
- Make greater use of biodiversity, such as the natural sandstone as buffer zones to help filter and flush pollutants;
- Training the locals (primary school nature awareness programs; high school youth leadership courses on environmental action and residents); enroll volunteers to their bush care, land care, stream watch, and track care (these volunteers look after climbing sites, downhill mountain bike sites, etc.);
- Adopt water sensitive urban design to manage stormwater and urban runoff (i.e., encourage residents to put in rainwater tanks);
- Generate community support and willingness to pay:

> So we were one of the early councils to apply for and get an environment levy. We're in the final year of a 10-year levy but having said that we went back to the community and over 80 percent said they supported some kind of renewed environment levy… Our community were the ones that really pushed for the increased environmental protections that are afforded through a World Heritage listing…
>
> **Manager Natural Environment**

- Motivate State and Federal Governments to issue grants:

> …So the idea was to get our own catchment in order and use it as a demonstration site and apply for State Government and Federal funding to kind of match the resources that we're putting in through our environment levy and so that we're able to get almost like a two to one return for our community who are paying a levy to look after the environment
>
> **Manager Asset Planning**

Finally, and most importantly was to begin to trade in biodiversity. They first developed an inventory of biodiversity:

> We have what's called a 'natural asset plan and part of that is valuing the natural assets…we do the aquatic monitoring so we've got over 10 years of data…we have done a number of vegetation surveys…very recently we redid our vegetation surveys, so that's now all been uploaded to our GIS system. We've also done weed mapping and so our weed teams have these little tablets so when they go out in the field and discover a new weed they can log that on the system or when they've eradicated weeds from an areas they can also record that as well.
>
> **Environmental Scientist**

These values are formalized and used in biodiversity trading:

> I had a quick look at your formula (Appendix 22.1) and so I went down and spoke to our development section and we don't use this method of valuation. Actually we're kind of fortunate in some ways in that because of our large amount of totally protected areas we don't have the kind of possibilities for…those massive developments that will come in re-develop tens of

hectares for either commercial or even residential and then you have to offset that through biodi-versity offset schemes and stuff like that. But other councils do and because we do have some nice protected areas we've been approached by other councils...an example where [another] City Council had a big development being proposed, they couldn't find any suitable offset vege-tation communities similar to the ones that were being impacted...so they funded some reha-bilitation of a vegetation of similar nature up here...

He continued on to say:

So we're aware of them and we have participated but usually we're the recipient of that funding as opposed to us the ones that are trying to find suitable offsets for major developments because we just don't have the suitability. Yes so that's once again — and most of our...ecosystem instead of being valued using that model are through their legislatives. So we use the threatened species listing and the endangered ecological lists. We use the legislative requirements as a no-go zone as opposed to try to value and do the offsets. [As a buyer? (i.e. polluter)] We just basically say no. That's what I asked them downstairs. They said 'No.' We know they're valuable. We just say no.

CFO

So, as for the opening question—Can accountants save the world? In this case we might say "accounting" *is* being used in the biodiversity market creation. But we need to ask, where are the accountants? The problem for accounting is that at this stage, they are largely outside this market activity:

Because you're talking about tens of millions of dollars' worth of yearly operating of an asset and...it's not actually captured by accountants. But that might be actually about embedding finance staff throughout the organisation, rather than having a core finance division. That would be my proposal. I'd probably be shouted down

Senior Accountant

Likewise, the management and accounting gurus (those that have been developing the normative FCA models for accountants) were not raised in our conversation. We wondered if it was because, the gurus themselves came from diverse and fractured back-grounds, with views that have either not reached or not yet convinced practicing ac-counting managers.

22.6 THE BUYERS MULTINATIONAL COAL MINING

The coal mining organization we visited is an Australian publicly listed organization. While our discussions with managers were based on Australian operations, this company has an offshore majority shareholder and is in the global top 10 "pure-play" coal com-panies (based on reserves) in the world. The company operates mines both independently and as joint ventures. In the Australian operations, the coal is mined using either open cut or underground methods. The underground methods comprise either "bord and pillar" operations, or more commonly the "longwall" mining method. Each method requires extensive government approvals with detailed background reports on the proposed infrastructure developments. The coal price is market driven and revenue estimations are also aligned with government mining approvals. The company also participates in some upstream and downstream components of the value chain of the mining industry, such as railway and port developments. Their profitability is determined by pricing, cost management initiatives (such as lean manufacturing initiatives), global supply and demand for coal, the quality of the coal, foreign exchange fluctuations as well as the maturity of their mines and prospect of new mine developments. The latter is sometimes difficult to achieve and stranded assets arise where land purchased (for mining purposes with approvals) has subsequently been rezoned and permits reneged by governments.

We visited the head office on several occasions, having arranged individual meetings and joint discussions with three key managers: The CFO, Strategic Asset Manager, and the Public Relations Manager, who also managed the sustainability strategy for the organization. The head office was relatively small and functional, as the key activity was at the coalface, literally, with several coalmines situated throughout Australia. The following discussion is based on our written notes (rather than recorded interviews, as above). These were largely, informal meetings based on an earlier introduction from one of the Directors.

The first meeting was with the CFO, who explained the accounting function as relatively traditional, with accountants in the field largely involved in the market-based accounting numbers, derived for financial reporting. Any calculations that would be required for investment appraisal would be initial estimates of numbers that would need to be included as initial outlays, terminal values, and cash flows. Thus any biodiversity impacts that were recognized would eventually come to the accountant to be formalized as a financial number (asset, liability, expense, or revenue) to be included in cash flows. The CFO gave us an example of unacceptable *noise*, as the empty coal trucks were being loaded with the initial batch of coal. To prevent the noise disturbing the local communities, including the local fauna, they designed special rubberized buffer mats to go in the bottom of the coal trucks. The cost of the buffer mats (or other similar treatment for a biodiversity-related problem) would enter the accounts in a dynamic reiterative process, beginning first with the recognition of the biodiversity problem, then evaluating the solution and associated mitigation or restoration cost. Thus, all accounting numbers were market-based values. However, underpinning the analysis for all projects is the analysis of risk. Some of this risk assessment might include the use of more qualitative information, on potential biodiversity impacts and actions to mitigate. This type of assessment is conducted in the very early stages of project development and is necessary part of FCA. The risk information influences strategic decisions and thus, the "completeness" of FCA.

On one of our subsequent visits, we met with the CFO and the Strategic Asset Manager and discussed some of the strategies around biodiversity impacts. As indicated, the coal mining company is in a position that requires them to confront biodiversity issues right at the beginning of the project. Even the purchase of land is tenuous, as this does not necessarily mean mining approvals in the future and leaves them with stranded assets to account for. We discussed the accounting techniques taken from the early feasibility stage of the proposition through to inclusion of biodiversity impact costs as cash flows when they were recognized. The Strategic Asset Manager further highlighted the need to be an active participant in the local community and how it helped them to specify the ways in which they would manage their social license to operate. He provided an example of managing the restoration of discontinued mine site. He explained how they actually flooded it to create a large inland lake. The lake was developed into a thriving recreational park, used by the community to host water sporting competitions. This development has helped support the community through tourism, sport, and recreation. There were numerous other examples of projects to support indigenous communities and annual payments to local municipalities as part of their social license to operate.

However, the process was never straightforward, and a politically motivated process. This was particularly noticeable for the coal mining industry as a central contributor to the climate change debate. The mining industry is one of the more highly regulated industries, particularly when it comes to sustainability and biodiversity issues. We witnessed associated angst when the Public Relations Manager arrived late for our joint meeting, having been held up in an earlier meeting with the local government minister. He showed us diagrams of the newly rezoned areas where the majority of previously suitable land for mining had been canceled supposedly by "the whim of the minister," complained the Manager. The diagrams now were covered in red "no go mining zones" after the meeting. The outcome of the regulatory environment, as well as market supply and demand forces, highlights the highly dynamic market that the coal mining organization is operating in.

By way of illustrating how government regulation plays a major role in new mine approvals, an extract is provided in Table 22.1 taken from a government approvals document. This extract specifically relates to the biodiversity impacts when establishing a new mine development. In the coal mining company's jurisdiction, the environmental assessment must detail the impact on air quality, noise, transport, flora and fauna, surface and groundwater management, methods of mining, landscape management, and rehabilitation. The mining companies are also required to undertake extensive public consultation, which results in specific investments in community infrastructure, to meet local preferences.

A key issue highlighted in Table 22.1 is the notion of the biodiversity market through the "biodiversity-offset strategy." An important part of this assessment process requires mining companies to buy offsets relating to biodiversity issues. As indicated in Appendix 22.1, a biobanking scheme exists to facilitate this.

A key aspect of this scheme is a biodiversity assessment methodology which provides values for threatened species, populations, ecological communities, and their habitats.

Table 22.1 Extract from Environmental Report Relating to Biodiversity Impacts at Existing and Proposed Mine Project

Panel A: Existing Mine

Water balance	Water deficit (maximum of 6.8 ML/day) sourced from surface water runoff, groundwater inflows into the mining areas, groundwater extraction from the UG4 borefield and via a water sharing.
Coal transport	Approximately four trains per day on the railway
Biodiversity offsets	1282 ha of native vegetation and 144 ha of endangered ecological communities (EEC). In addition, 153 ha of disturbed lands are to be regenerated with native vegetation and 48 ha of cleared land is to be regenerated with EEC.
Rehabilitation	Rehabilitate 370 ha of land to woodland and 580 ha of land to grassland

Panel B: Proposed Mine Development

Water demands and supply	• Water surplus in initial years (i.e., years 1–5) of 174 ML/annum is predicted under average climatic conditions. Surpluses will be controlled by reducing pump from the northern borefield and by designing the mater management system to contain runoff during high rainfall events; • Water deficits of 599 ML/annum in the remainder of the operating years are predicted under average climatic conditions. Deficits are intended to be met by accessing additional water under a modified Water Sharing Agreement.
Overburden emplacement	Overburden will be emplaced in an out-of-pit emplacement area to the north of the pit.
Coarse reject, tailings management	Generation of 2 Mtpa of coarse reject and tailings, which will be transferred via conveyor for co-disposal with overburden in the pit void.
Hours of operation	24 h a day, 7 days a week.
Biodiversity offset	The project would result in the clearing of 1534 ha of land; of which 632 ha is grassland and 902 ha is native woodland (including 123 ha of EEC). The biodiversity-offset strategy proposed to compensate for this loss includes a total of 4.066 ha of land (including 1168 ha of EEC) within 8 biodiversity-offset areas.
Rehabilitation, final landform, and end land use	The 1534 ha of land that would be cleared will be rehabilitated, including rehabilitating the 632 ha of existing degraded secondary grassland and shrublands to native open woodland and EEC communities. The rehabilitate land will be protected with in perpetuity conservation after mining.

The biodiversity assessment methodology assesses the biodiversity values in terms of the loss of biodiversity or gain in biodiversity values from management actions. Actions might include retention of native vegetation, dead timber, rocks, and natural water flows; replanting or supplementary planting where regeneration is insufficient; management of soil erosion as well as others such as weed control and management of fire, pests, and human disturbance.

In addition to valuing biodiversity losses/gains, the methodology also establishes the circumstances in which biodiversity values can be offset or not by the retirement of biodiversity credits. There are two classes of biodiversity credits calculated: ecosystem credits and species credits. The methodology includes calculations on the number and type of ecosystem credits and species credits that are created when offsetting losses by the improvement of biodiversity values at a designated biobank site.

The valuation model includes the valuation of both direct and indirect biodiversity impacts. Indirect impacts involve the valuing of impacts on water quality and subsequently downstream biodiversity values; increased light or noise that may affect threatened species habitat; or development that may restrict movement of threatened species or populations in surrounding areas. Included in the impact assessment is the demonstration of corporate measures taken to minimize these negative impacts (i.e., controls to prevent erosion; noise and light barriers or structures to allow movement of threatened species or populations). The need for mining companies, such as coal mining company to value, price, and pay for such biodiversity impacts, illustrates the important use of FCA techniques and potential demand for FCA as offsetting grows.

However useful the Appendix 22.1 model might appear, in practice, it was not used by the coal mining company either. While it was required to identify, value, and price its biodiversity effects, the current model for doing this through the biobanking scheme was considered relatively unworkable at the organizational and project level (a point also noted by Madsen et al., 2010, 2011). The Public Relations Manager explained how they had used the model and conducted all the calculations, as suggested. However, he claimed no one had used the model as the price of purchasing and equivalent offset site was three times less costly than the upfront payment required by this scheme. This resulted in coal mining company acquiring offset properties, like those offered by the municipal council, with market-based calculations made outside the model. Although the model was used for comparative purposes, like the municipal council, it was not part of the accountant's repertoire and was largely used by managers to highlight variances for improved resource allocation decisions. Nevertheless, given growing demand for ecosystem credits, this model might end up being the last option for buyers, if there has been a saturation of all equivalent sites for purchase outside the scheme, in perpetuity. Thus, if the model's calculations are accurate, prices have the capacity to increase, at least threefold. Increasing competition and decreasing profits (for buyers and sellers when biodiversity credits are exhausted) is indicative of a developing market for biodiversity.

22.7 CONCLUSION: THE BIODIVERSITY MARKET AND FCA

The information age and the rise of new intangible wealth have resulted in the words "crises" playing a significant role in motivating accounting change. This chapter opened with calls for "accountants" to help save the planet from the "biodiversity crisis." It was argued that accounting could help put a price on nature's capital, currently used in the value creation activities of corporations. Nevertheless, there are crises even within accounting. Firstly, there is a view that the statutory financial reports only conveys 20–30% of a firm's value, instead of the 90% witnessed 40 years ago (Gleeson-White, 2014). Secondly, there are greater concerns that even the voluntary disclosures are mere legitimizing accounts, offering no further insights of organizational activities (Gray, 1992; Deegan et al., 2002). In response to these concerns, we contribute with insights from *management accounting*, offering nonregulated accounting methodologies to inform managerial practices. For example, management accounting is recognized for developments in FCA and others including activity-based costing, lean accounting, just-in-time, return on investment, value-based management, balanced scorecards, and so forth. Our contention is that if management accounting discourse, associated with FCA, is part of the cultural circuit of capitalism only then will it be embraced in practice. FCA use within the organization can contribute to the biodiversity crisis, in a timely manner, certainly faster than if we wait for it to be regulated in corporate accounts.

In this chapter, we focused on FCA innovations by managers operating as *buyers* and *sellers* in the biodiversity market. In activities bringing nonmarket biodiversity values to FCA accounting practices, we have also indirectly shown the value associated with other intangible capital within the organization, including the expert knowledge skills and capabilities that comprise the cultural circuit of capitalism. Just like biodiversity impacts, these skills are also difficult to value but are important contributory mechanisms to the biodiversity market. The management decision making associated with potential biodiversity impacts are not obviously transparent in the balance sheet or financial reports or salaries of senior managers. However, the managerial knowledge creation activities can certainly be revealed using Thrift's (2005) cultural circuit of capitalism.

As we worked through the cultural circuit of capitalism, we highlighted the ways in which managers within organizations embraced FCA. In-house frameworks, to assist in biodiversity valuations, were developed by the practicing managers, with the support of management consultants. They helped the environmental scientists with lists and quantitative figures (hectares; numbers) relating to critical, or endangered natural capital to be used in market trades, like-for-like. For example, our selling organization would receive funds from a polluter to maintain "x" hectares of natural capital that matched the land being destroyed in development somewhere else. In this way, it was the activity costs, rather than the value of nature that was considered in the trade. Our buying organization, also purchased equivalent offset land, again accounted for in quantitative (not financial) values. The price recognized in the books was the purchase and maintenance costs associated with the equivalent offset property.

Nevertheless, the Appendix 22.1 formula, developed by the local government department, was not formally used in the biodiversity market trading process. It was criticized for its imperfections, instead used as a conversation piece. It was applied within both of our case organizations, but used to generate a variance that could then inform market negotiations. For example, the buyers would generate the price from the model but purchase cheaper offsets elsewhere. Similarly, EPAs version of FCA was used by the selling organization, but only internally for comparative purposes. Similar to the biodiversity-offset model, EPAs FCA model helped to reveal the variance between direct calculated costs and modeled full costs which enabled the municipal council to negotiate other ways to purchase or manage revenues. They found other ways to manage income through grant activity, environmental levies, and their capabilities as an offset provider. This was necessary as consumer prices were capped, which inhibited the inclusion of full costs in direct rate charges.

Other influential sources of knowledge did not necessarily come from the accountant, the gurus, business schools, or even the accounting profession. Instead, the accountant's role at these two case sites was viewed more traditionally. The "accountant" continues to play a transaction-focused bookkeeping role while much of the full cost "accounting" is being performed by others, such as scientists and environmental engineers, heavily influenced by environmental economics and their individual valuation tools. Even if the economically derived valuation methodologies were not considered perfect representations, they played a role in managerial experimentation and elaborations of the internal accounts used for decision making. In this way, the practicing managers (nonaccountants) were highly visible in the development of innovative methodologies and approaches to the biodiversity market. Unfortunately, for the accountant—there is still a long way to go before their gurus and activities of the professional bodies are part of the institutionalized practice. There is still a lot more work to be done to increase the influence of FCA in elite business schools and hence thinking of senior managers, so it becomes a part of enacting new managerial discourse. At the moment, the tension between conventional market-based strategies and FCA approaches is observed. The activities come from the willingness of passionate employees, like those in the municipal council, to take a reduction in salaries so they can contribute to the wider social good. At this stage, it is still left to the regulator and associated government authorities to manage the biodiversity market. Perhaps, they should be included in the cultural circle of capitalism.

In conclusion, it appears as if much of the FCA discourse is outside the accounting domain, and part of the cultural circuit of capitalism, only through the activities of environmental practicing managers that have a strong influence on biodiversity market practice. The looplike activity must return some of the emphasis back to the accounting academy and its gurus so they can better perform in the biodiversity market, by managing the theory-practice gaps and unanticipated crises. This ultimately forces invention.

ACKNOWLEDGMENTS

The authors would like to thank the anonymous managers who willingly contributed to support this case research. We would also like to acknowledge CPA Australia for their funding support. We are extremely grateful to Rob Gray and Lee Parker for their valuable contributions to earlier drafts of this chapter.

REFERENCES

Antheaume, N., 2007. Full Cost Accounting: Adam Smith Meets Rachel Carson. Sustainability Accounting and Accountability. Routledge, London, 211−225.

Barry, A., 2001. Political Machines: Governing a Technological Society. A&C Black, New York.

Bebbington, J., 1997. Engagement, education and sustainability: a review essay on environmental accounting. Account. Auditing Account. J. 10 (3), 365−381.

Bebbington, J., 2001. An overview of accounting for externalities. In: Freedman, M., Jaggi, B. (Eds.), Advances in Environmental Accounting. Association of Chartered Certified Accountants, London, pp. 19−27.

Bebbington, J., 2007. Accounting for Sustainable Development Performance. Elsevier, London.

Bebbington, J., Gray, R., 2001. An account of sustainability: failure, success and a reconceptualisation. Crit. Perspect. Account. 12 (5), 557−605.

Bebbington, J., Brown, J., Frame, B., 2006. Accounting technologies and sustainability assessment models. Ecol. Econ. 61 (2−3), 224−236.

Bebbington, J., Gray, R., Hibbitt, C., Kirk, E., 2001. Full Cost Accounting: An Agenda for Action − ACCA Research Report No. 73. Association of Certified Chartered Accountants, London.

Bennett, M., Schaltegger, S., Zvezdov, D., 2011. Environmental management accounting. Rev. Manag. Account. Res. S:53−84.

Brown, J., Frame, B., 2005. Democratizing Accounting Technologies: The Potential of the Sustainability Assessment Model (SAM). Victoria University of Wellington, New Zealand. Centre for Accounting, Governance and Taxation Research, Working Paper Series No. 15.

Brundtland, G.H., 1987. Report of World Commission on Environment and Development: Our Common Future. UN World Commission on Environment and Development, Oxford, 1−300.

Burchell, S., Clubb, C., Hopwood, A., Hughes, J., Nahapiet, J., 1980. The roles of accounting in organizations and society. Account. Organ. Soc. 5 (1), 5−27.

Burritt, R., Hahn, T., Schaltegger, S., 2002. Towards a comprehensive framework for environmental management accounting-links between business actors and EMA tools. AAR 12 (2), 39−50.

Costanza, R., d'Arge, R., de Groot, R., Farber, S., Grasso, M., Hannon, B., Limburg, K., Naeem, S., O'Neill, R., Paruelo, J., Raskin, R., Sutton, P., van den Belt, M., 1997. The value of the world's ecosystem services and natural capital. Nature 387 (15), 253−260.

Cuckston, T.J., 2013. Bringing tropical Forest biodiversity conservation into financial accounting calculation. Account. Auditing Account. J. 26 (5), 688−714.

Daly, H.E., 1985. Ultimate confusion: the economics of Julian Simon. Futures 446−450.

Davies, J., 2014. Full cost accounting − integrating biodiversity. In: Jones, M. (Ed.), Accounting for Biodiversity. Routledge, New York and London.

Deegan, C., Rankin, M., Tobin, J., 2002. An examination of the corporate social and environmental disclosures of BHP from 1983–1997: a test of legitimacy theory. Account. Auditing Account. J. 15 (3), 312–343.

Eccles, R.G., Serafeim, G., 2013. The performance frontier: innovating for a sustainable strategy. Harv. Bus. Rev. 91 (5), 50–60.

Eldenburg, L.G., Brooks, A., Oliver, J., Vesty, G., Wolcott, S., 2011. Management Accounting, second ed. John Wiley & Sons, Milton.

Epstein, M.J., 2008. Making Sustainability Work. Best Practices in Managing and Measuring Corporate Social, Environmental, and Economic Impacts. Greenleaf.

Figge, F., Hahn, T., Schaltegger, S., Wagner, M., 2002. The sustainability balanced scorecard–linking sustainability management to business strategy. Bus. Strategy Environ. 11 (5), 269–284.

Fisher, B., Turner, R.K., Morling, P., 2009. Defining and classifying ecosystem services for decision making. Ecol. Econ. 68 (3), 643–653.

Frame, B., Cavanagh, J., 2009. Experiences of sustainability assessment: an awkward adolescence. Account. Forum 33 (3), 195–208.

Freeman, M.C., Groom, B., 2013. Biodiversity valuation and the discount rate problem. Account. Auditing Account. J. 26 (5), 715–745.

Garcia-Parpet, M.F., 2007. In: MacKenzie, D., Muniesa, F., Siu, L. (Eds.), Do Economists Make Markets? On the Performativity of Economics, pp. 20–53.

Gleeson-White, J., 2014. Six Capitals: The Revolution Capitalism Has to Have – Or Can Accountants Save the Planet? Allen & Unwin, Australia.

Gonzalez, G., Houdet, H., 2009. Accounting for Biodiversity and Ecosystem Services from a Management Accounting Perspective. Integrating Biodiversity into Business Strategies at a Wastewater Treatment Plant in Berlin. Veolia Environnement–Orée, 18 p. Accessed in November.

Gray, R., 1990. Accounting and economics: the psychopathic siblings: a review essay. Br. Account. Rev. 22 (4), 373–388.

Gray, R., 1992. An exploration of the challenge of gently accounting for accountability, transparency and sustainability. Account. Organ. Soc. 17 (5), 571–580.

Gray, R., 2002. The social accounting project and accounting organizations and society privileging engagement, imaginings, new accountings and pragmatism over critique? Account. Organ. Soc. 27 (7), 687–708.

Gray, R., 2006. Social, environmental and sustainability reporting and organisational value creation? Whose value? Whose creation? Account. Auditing Account. J. 19 (6), 793–819.

Gray, R., 2010. Is accounting for sustainability actually accounting for sustainability... and how would we know. Account. Organ. Soc. 35 (1), 47–62.

Gray, R., May 1991. The Accounting Profession and the Environmental Crisis (or Can Accountancy Save the World?). University of Dundee. Working Paper Series, p. 57.

Gray, R., Bebbington, J., 2001. Accounting for the Environment. SAGE Publications, London.

Gray, R., Milne, M., 2007. Future prospects for sustainability reporting. In: Unerman, J., Bebbington, J., O'Dwyer, B. (Eds.), Sustainability Accounting and Accountability. Routledge, London, pp. 185–207. New York and London.

Gray, R., Owen, D., Maunders, K., 1987. Corporate Social Reporting: Accounting and Accountability. Prentice-Hall International.

Gray, R.H., Owen, D.L., Maunders, K.T., 1991. Accountability, corporate social reporting and the external social audits. Adv. Public Interest Account. 4, 1–21.

Gray, R., Adams, C.A., Owen, D., 2014. Accountability, Social Responsibility and Sustainability: Accounting for Society and the Environment. Pearson Education Limited.

Gupta, T.R., Foster, J.H., 1975. Economic criteria for freshwater wetland policy in Massachusetts. Am. J. Agric. Econ. 57 (1), 40–45.

Herbohn, K., 2005. A full cost environmental accounting experiment. Account. Organ. Soc. 30 (6), 519–536.

Hines, R.D., 1988. Financial accounting: in communicating reality, we construct reality. Account. Organ. Soc. 13 (3), 251–261.

Hines, R.D., 1991. The FASB's conceptual framework, financial accounting and the maintenance of the social world. Account. Organ. Soc. 16 (4), 313–331.

Houdet, J., 2008. Integrating Biodiversity into Business Strategies, the Biodiversity Accountability Framework. FRB – Orée, Paris.

Houdet, J., Germaneau, C., 2014. Accounting for biodiversity and ecosystem services from an EMA perspective. In: Jones, M. (Ed.), Accounting for Biodiversity. Routledge, New York and London.

Houdet, J., Pavageau, C., Trommetter, M., Weber, J., 2009. Accounting for Changes in Biodiversity and Ecosystem Services from a Business Perspective. Preliminary Guidelines towards a Biodiversity Accountability Framework. Ecole Polytechnique, Department of Economics, 63 p.

Hopwood, A.G., 1987. The archeology of accounting systems. Account. Organ. Soc. 12 (3), 207–234.

Hopwood, A.G., 2009. The economic crisis and accounting: implications for the research community. Account. Organ. Soc. 34 (6–7), 797–802.

IIRC (Integrated Reporting Council), 2013. Consultation Draft of the International IR Framework. Available at: http://www.theiirc.org/wp-content/uploads/Consultation-Draft/Consultation-Draft-of-the-InternationalIRFramework.pdfhttp://www.theiirc.org/wp-content/uploads/Consultation-Draft/Consultation-Draft-of-the-InternationalIRFramework.pdf.

Jasch, C., 2003. The use of environmental management accounting (EMA) for identifying environmental costs. J. Clean. Prod. 11 (6), 667–676.

Jasch, C., 2009. Material Flow Cost Accounting. Springer, Dordrecht.

Johnson, H.T., Kaplan, R.S., 1987. Relevance Lost: The Rise and Fall of Management Accounting. Harvard Business School Press, Boston.

Jones, M.J., 1996. Accounting for biodiversity. Br. Account. Rev. 28, 281–303.

Jones, M.J., 2003. Accounting for biodiversity: operationalising environmental accounting. Account. Auditing Account. J. 16 (5), 762–789.

Jones, M.J., 2010. Accounting for the environment: towards a theoretical perspective for environmental accounting and reporting. Account. Forum 34 (2), 123–138.

Jones, M. (Ed.), 2014. Accounting for Biodiversity. Routledge, London and New York.

Jones, M.J., Matthews, J., 2000. Accounting for Biodiversity: A Natural Inventory of the Elan Valley Nature Reserve. Chartered Association of Certified Accountants.

Jones, M.J., Solomon, J.,F., 2013. Problematising accounting for biodiversity. Account. Auditing Account. J, 26 (5), 668–687.

Kaplan, R.S., 1984. Yesterday's accounting undermines production. Harv. Bus. Rev. 62 (4), 95–101.

Kaplan, R.S., Norton, D.P., 1996. Using the balanced scorecard as a strategic management system. Harv. Bus. Rev. 74 (1), 75–85.

Kaplan, R.S., Norton, D.P., 2001. Transforming the balanced scorecard from performance measurement to strategic management: Part I. Account. Horizons 15 (1), 87–104.

Khan, T., 2014. Kalimantan's biodiversity: developing accounting models to prevent its economic destruction. Account. Auditing Account. J. 27 (1), 150–182.

Klamer, A., McCloskey, D., 1992. Accounting as the master metaphor of economics. Eur. Account. Rev. 1 (1), 145–160.

Lamberton, G., 2005. Sustainability accounting—a brief history and conceptual framework. Account. Forum 29 (1), 7–26.

Lehman, G., 2001. Reclaiming the public sphere: problems and prospects for corporate social and environmental accounting. Crit. Perspect. Account. 12, 713–733.

Madsen, B., Carrol, N., Moore Brands, K., 2010. State of Biodiversity Markets Report: Offset and Compensation Programs Worldwide. Available at: http://www.ecosystemmarketplace.com/documents/acrobat/sbdmr.pdf (accessed 20.11.14.).

Madsen, B., Carroll, N., Kandy, D., Bennett, G., 2011. 2011 Update: State of Biodiversity Markets. Forest Trends, Washington, DC. Available at: http://www.ecosystemmarketplace.com/reports/2011_update_sbdm.

Maunders, K.T., Burritt, R.L., 1991. Accounting and ecological crisis. Account. Auditing Account. J. 4 (3), 9–26.

Merton, R.C., 1987. A simple model of capital market equilibrium with incomplete information. J. Finance 42 (3), 483–510.

Milne, M.J., Patten, D.M., 2002. Securing organizational legitimacy: an experimental decision case examining the impact of environmental disclosures. Account. Auditing Account. J. 15 (3), 372—405.

McNeely, J.A., Miller, K.R., Reid, W.V., Mittermeier, R.A., Werner, T.B., 1990. Conserving the World's Biological Diversity. World Bank, Washington, DC.

Napier, C., 1996. Accounting and the absence of a business economics tradition in the United Kingdom. Eur. Account. Rev. 5 (3), 449—481.

O'Dwyer, B., 2003. Conceptions of corporate social responsibility: the nature of managerial capture. Account. Auditing Account. J. 16 (4), 523—557.

Sabetti, H., 2011. The for-benefit enterprise. Harv. Bus. Rev. 89 (11), 98—104.

Savage, D., Jasch, C., 2005. International Guidance Document on Environmental Management Accounting. International Federation of Accountant (IFAC), New York.

Scapens, R.W., 1994. Never mind the gap: towards an institutional perspective on management accounting practice. Manag. Account. Res. 5 (3), 301—321.

Scapens, R.W., 2006. Understanding management accounting practices: a personal journey. Br. Account. Rev. 38 (1), 1—30.

Seal, W., 2010. Managerial discourse and the link between theory and practice: from ROI to value-based management. Manag. Account. Res. 21 (2), 95—109.

Serres, M., 1995. Conversations on Science, Culture, and Time. University of Michigan Press.

Serres, M., 2014. Time of crisis: what the financial crisis revealed and how to reinvent our lives and future. Bloomsbury Academic, New York.

Siddiqui, J., 2013. Mainstreaming biodiversity accounting: potential implications for a developing economy. Account. Auditing Account. J. 26 (5), 779—805.

Simmonds, K., 1981. Strategic management accounting. Manag. Account. 59 (4), 26—30.

Simons, R., 1994. How new top managers use control systems as levers of strategic renewal. Strategic Manag. J. 15 (3), 169—189.

Simons, R., 2013. Levers of Control: How Managers Use Innovative Control Systems to Drive Strategic Renewal. Harvard Business Press, Boston.

Spence, A., 2010. Public Perceptions of Climate Change and Energy Futures in Britain: Summary Findings of a Survey Conducted from January to March 2010. Cardiff University.

State of NSW and Office of Environment and Heritage, 2014. Biobanking Assessment Methodology 2014. Available at: http://www.environment.nsw.gov.au/resources/biobanking/140661BBAM.pdf.

Stubbs, W., Higgins, C., 2014. Integrated reporting and internal mechanisms of change. Account. Auditing Account. J. 27 (7), 1068—1089.

TEEB, 2008. An Interim Report, European Communities. Available at: http://www.teebweb.org/publication/the-economics-of-ecosystems-and-biodiversity-an-interim-report/.

TEEB, 2012. In: Bishop, J. (Ed.), The Economics of Ecosystems and Biodiversity in Business and Enterprise. Earthscan, London and New York.

Thrift, N., 2005. Knowing Capitalism. Sage, London.

Tinker, T., Gray, R., 2003. Beyond a critique of pure reason: from policy to politics to praxis in environmental and social research. Account. Auditing Account. J. 16 (5), 727—761.

Turner, M.G., 1987. Spatial simulation of landscape changes in Georgia: a comparison of 3 transition models. Landsc. Ecol. 1 (1), 29—36.

Unerman, J., Bebbington, J., O'Dwyer, B., 2007. In: Sustainability Accounting and Accountability. Routledge, Abingdon.

Vogtländer, J., Van der Lugt, P., Brezet, H., 2010. The sustainability of bamboo products for local and Western European applications. LCAs and land-use. J. Clean. Prod. 18 (13), 1260—1269.

Vesty, G., Brooks, A., Oliver, J., 2015a. Contemporary Capital Investment Appraisal from a Management Accounting and Integrated Thinking Perspective: Case Study Evidence. CPA, Australia.

Vesty, G., Telgenkamp, A., Roscoe, P.J., 2015b. Creating numbers: carbon and capital investment. Account. Auditing Account. J. 28 (3), 302—324.

WBCSD (World Business Council for Sustainable Development), 2011. Guide to Corporate Ecosystem Valuation: A Framework for Improving Corporate Decision-Making. Switzerland, Geneva.

Appendix 22.1:[1] Ecosystem Credits—Determining the Current Site Value Score for a Vegetation Zone at the Development or Biobank Site (Source: Office of Environment and Heritage, NSW, 2014, p. 69)

$$SV_c = \frac{\left(\sum_{v=o}^{j}(a_v w_v) + 5((a_d a_g) + (a_b a_i) + (a_h a_j) + (a_c a_k)) \right) \times 100}{c}$$

Where,

SV_c is the current site value score of the vegetation zone

a_v is the attribute score for the vth *site attribute* $(a-j^*)$

a_k is equal to $(a_d + a_e + a_f)/3$, the average score for attributes d, e, and f

w_v is the weighting for the vth *site attribute* $(a-j)$

c is the maximum score that can be obtained given the attributes $a-j$ that occur in the vegetation type when in benchmark condition (the maximum score varies depending on which attributes occur in the vegetation type under assessment)

*$(a-j)$ represents different site attributes, e.g., native plant species richness, native ground cover (grasses) to which a weighting is applied based on the site value score.

---

[1] "The Biobanking program was formally implemented in New South Wales by the Office of Environment and Heritage (NSW OEH) in the fall of 2009. While transactions have occurred, activity within Biobanking has fallen short of expectations... The total value of credits sold by the program cumulatively is AUD 2.8 million (or USD 2.5 million)... The value of credits sold only in 2010 was AUD 1.6 million (or USD 1.5 million)... Yet, despite these promising figures, demand is outstripping supply. There is a reported shortage of 22,000 ecosystem credits and 5000 endangered species credits" (Masden et al., 2011).

INDEX

Note: Page numbers followed by "b", "f" and "t" indicate boxes, figures and tables respectively.

Printed in the United States
by Baker & Taylor Publisher Services

Printed in the United States
By Bookmasters